NANO PLASMONICS

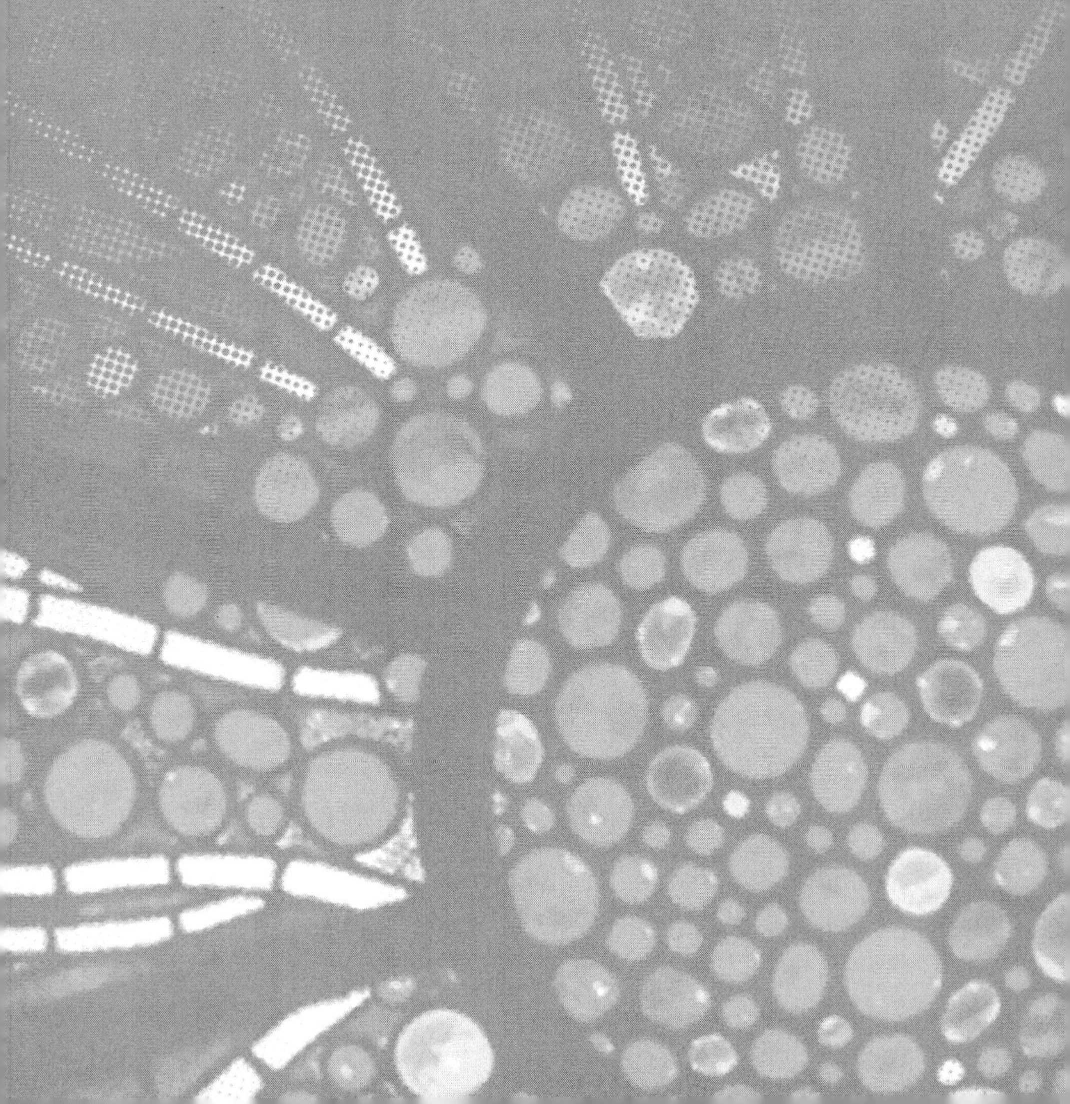

NANO PLASMONICS

Vasily Klimov

Translated from the Russian by
Anna Sharonova

PAN STANFORD PUBLISHING

Published by

Pan Stanford Publishing Pte. Ltd.
Penthouse Level, Suntec Tower 3
8 Temasek Boulevard
Singapore 038988

Email: editorial@panstanford.com
Web: www.panstanford.com

British Library Cataloguing-in-Publication Data
A catalogue record for this book is available from the British Library.

Nanoplasmonics

ISBN 978-981-4267-16-8 (Hardcover)
ISBN 978-981-4267-42-7 (eBook)

Printed in the USA

Contents

Preface

Nanoplasmonics is a component of nano-optics, which is optics at the nanometer scale. The subject of nanoplasmonics is optical properties of metal nanoparticles and nanostructures determined by electron oscillations around a crystal lattice. Primarily, the importance of nanoplasmonics is based on two factors. First, localization and enhancement of optical fields can occur owing to the presence of a spatial nanoscale substance (the lightning rod effect). Second, frequencies of metal nanoparticle eigenoscillations are located in optical spectrum in ranges from ultraviolet to infrared. Coexistence of these two properties of metal nanoparticles causes an interesting and complex physics, which forms the basis for numerous applications.

More specifically, the properties of metal nanoparticles and related physics are determined, first of all, by their shape: any variation from a trivial spherical form leads to the origin of new properties and peculiarities. Almost full control over shapes of nanoparticles allows one to talk about full control over their optical properties. The adjustability of plasmonic nanoparticles allows for their effective interaction between themselves and with light, atoms, and molecules. In other words, this book deals with the investigation of the influence of the geometry and shapes of nanoparticles on their properties. This aspect is extremely important because until now for simplicity many researchers have considered wave propagations in infinite systems which have no peculiarities related to real nanoparticles of finite volume. Moreover, sometimes such infinite systems have pathological properties.

Despite the rapid growth of nanoplasmonics, its main achievements are scattered over many journal publications, and one of the main goals of this project is to make a first-time systematic presentation of results obtained over the world. Though many results in this area were obtained by me personally, this book is based, of course, not just on my results. I have tried to make the book as self-sufficient as possible from the theoretical point of view, and I expect that many formulas can be used immediately by both theorists and experimentalists in nanoplasmonics and related areas. Theoretical methods and approaches presented in the book can be also applied in further original investigations. As to the experimental aspects of nanoplasmonics, this work contains, first of all, an overview of fundamental experiments and applications having reputable substantiation and interpretation.

Although I have tried to make the book comprehensible to a wide audience, some parts of it are rather complicated and need special grounding. That is why for better perception the book is supplied with a lot of photographs, pictures, plots, and diagrams. These illustrations will be especially useful for those readers who have just entered into this domain of science. From this point of view, the electronic edition will be most useful because the majority of illustrations in it are in full color.

Although currently nanoplasmonics is of great importance, I decided to issue this book in English only after I felt assured that its tone is correct and the book is demanded by the audience. More than 1500 copies of two Russian editions have already been sold, and this circulation an be regarded as a success for a scientific book. I believe that the English version of the book will be even in greater demand because it is extended and updated to include further development of nanoplasmonics.

Vasily Klimov
July 2014

Acknowledgments

First of all, I want to thank the staff of Pan Stanford Publishing who patiently worked on my manuscript for several years. I wish to express my sincere gratitude to my wife, Mrs. Irina Klimova, who has been a constant source of inspiration, providing support and encouragement for this project in spite of her own very busy professional schedule.

I also would like to extend my thanks to Mrs. Anna Sharonova, who has allowed me to make clear a lot of intricate places in the manuscript from the linguistic point of view during translation from the Russian language.

I would like to acknowledge my PhD students Ilya Zabkov, Ilya Treshin, and Andrey Pavlov for proofreading the text and for preparing some fascinating illustrations.

Finally, I am deeply grateful to the famous physicist Vladilen Stepanovich Letokhov (1939–2009), who, many years ago, drew my attention to new areas of science that did not exist at that time—nano-optics and nanoplasmonics.

Vasily Klimov
July 2014

Chapter 1

Introduction

Ubi materia, ibi geometria
Where there is matter, there is geometry
—Johannes Kepler

Nanoplasmonics is a component of nano-optics, which is optics at the nanometer scale. The subject of nanoplasmonics is optical properties of metal nanoparticles and nanostructures determined by conduction electron oscillations around a crystal lattice. Primarily, importance of nanoplasmonics is based on two factors. First, localization and enhancement of optical fields can occur due to the presence of a spatial nanoscale substance (the lightning rod effect). Second, frequencies of metal nanoparticle eigenoscillations are located in the optical spectrum in ranges from ultraviolet to infrared. Coexistence of these two properties of metal nanoparticles causes an interesting and complex physics, which forms the basis for numerous applications.

The first scientific research studies of plasmon oscillations in nanoparticles and nanostructures were carried out in the early 20th century. In 1902, Robert Wood discovered unexpected properties in the spectra of light reflected from metal gratings (Wood's anomalies, (Wood, 1902)). Almost at the same time, Maxwell

Nanoplasmonics
Vasily Klimov

ISBN 978-981-4267-16-8 (Hardcover), 978-981-4267-42-7 (eBook)
www.panstanford.com

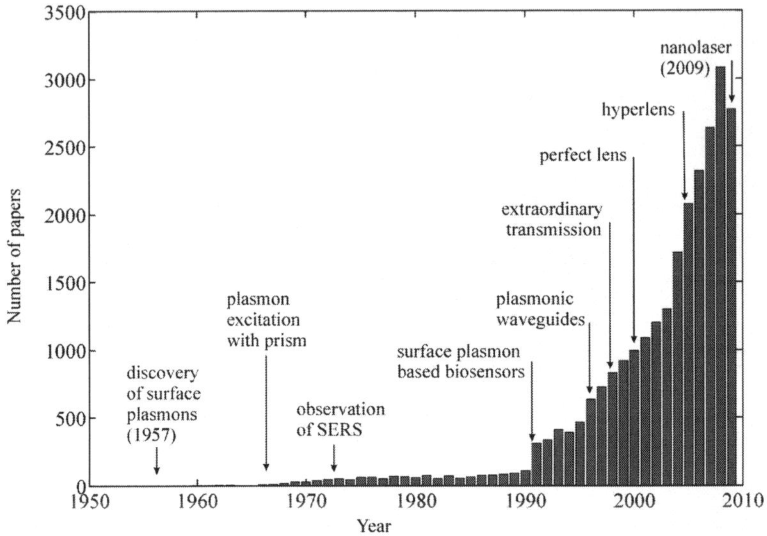

Figure 1.1 The number of annual publications containing the word combination "surface plasmon" in the title or abstract and the years of their publication (based on information from www.sciencedirect.com).

Garnet (1904) described the bright colors observed in glasses with metal nanoparticles with the help of the Drude theory of metals and the description of nanosphere properties proposed by Rayleygh (1871). Formally, optical properties of nanoparticles are determined completely by solutions of Maxwell's equations with corresponding boundary conditions on a particle surface. The solutions of Maxwell's equations found by Mie (1908); Sommerfeld (1909), and other authors in the early 20th century are exactly applicable for a description of plasmonic phenomena in spherical nanoparticles or on a metal–dielectric interface. Thus, at first sight one cannot understand why a profound interest in optics of metal nanoparticles, that is, nanoplasmonics, arose specifically at the close of the 20th and in the early 21st century. Increase of interest in surface and localized plasmons is demonstrated in Fig. 1.1, where a growing number of annual publications containing the word combination "surface plasmon" in the title or abstract is shown.

Primarily, the growth of this interest is related to the fact that modern nanotechnologies allow synthesizing and producing of nanoparticles of almost arbitrary shape and composition.

Secondly, modern diagnostic technologies (scanning microscopes of different types) allow characterizing of properties of single nanoparticles and nanostructures in detail.

Thirdly, evolution of computers and computational methods allows predicting and optimizing of properties of nanoparticles and nanoparticle-based nanodevices.

All these factors provoke the origin of a new physics. In its turn, the new physics paves the way for development of new devices on the nanometer scale.

More specifically, the properties of nanoparticles and related physics are determined, first of all, by their shape: any variation from a trivial spherical form leads to the origin of new properties and peculiarities. Almost full control over shapes of nanoparticles allows one to talk about the full control over their optical properties and, above all, over their spectra. The adjustability of plasmonic nanoparticles provides their effective interaction between themselves and with light, atoms, and molecules.

On the basis of these interactions, one can develop a wide range of nanodevices (and some of them are already being developed)—first of all, sensors to be applied in biology and medicine. There are a number of ideas describing construction of true integrated optoelectronic nanodevices, in which nanometric scales of separate elements are combined with optical frequencies of their operation on the basis of plasmonic nanoparticles and nanostructures. It is the optical frequencies that offer a hope to speed up substantially computers having processor elements of the nanometer scale but functioning at frequencies of five orders less than optical frequencies.

In all of these applications, both surface plasmons (i.e., plasmons propagating on a metal–dielectric interface) and localized plasmons related to oscillations of conduction electrons in nanoparticles are significant.

Surface plasmon physics and applications are developed considerably and described in the literature (Agranovich and Mills, 1982; Raether, 1988). Yet, physics of plasmon oscillations in nanoparticles is rather more complex and developed far less, and results obtained in this field are dispersed in a variety of articles.

The purpose of this book is a systematic description of the properties and applications of plasmon oscillations in nanoparticles. Besides, an analogy between nanoplasmonics and atom physics will be considered, that is, the analogy between spectra and wave functions of usual atoms and molecules, on the one hand, and between plasmon oscillations in isolated nanoparticles and clusters of nanoparticles, on the other hand. The analogy allows one to consider the plasmonic atoms and plasmonic molecules paradigm for description and prediction of plasmon oscillation properties in nanoparticles of very different shapes. Also, particular attention will be given to a description of ordinary atoms and molecules interaction with plasmonic nanoparticles, which is of crucial importance for many applications.

Nanoplasmonics is a new synthetic area of scientific research, rooted in physics, optics, chemistry, biology, and other studies; therefore the material cannot be presented without involving of concepts of these branches of knowledge. Nevertheless, I have tried to make the book comprehensible for all specialists and senior students looking for a prompt acquaintance with this actively developing area of nanophotonics and nanoplasmonics and for a description of the latest results in this field.

Chapter 2 is an introduction to the world of plasmonic nanoparticles, describing the role of nanoparticles in human history and, briefly, main modern methods of nanoparticles synthesis. An important component of the chapter is a range of photographs and illustrations graphically demonstrating the abundant variety of the world of nanoparticles.

Chapter 3 summarizes briefly main equations describing electrodynamics of continuous media, optical properties of metals, and bulk plasmon properties. The chapter also contains a brief introduction to the physics of surface and localized plasmons.

In Chapter 4, the properties of surface plasmons, that is, of waves propagating along a metal–dielectric interface and in more complex layered structures, are considered. The methods of excitation and observation of surface plasmons are also discussed in the chapter.

Generally, Chapters 2–4 are introductory, and those who are eager to hit upon the forefront of nanoplasmonics can begin reading from Chapter 5, where the main concepts of optical oscillations in nanoparticles are explicated. A mathematical apparatus forming a clear notion of localized plasmons and plasmonic atoms is also described in the chapter.

Chapters 6–9 deal with localized plasmons in nanoparticles of very different shapes from spherical in Chapter 6 up to more complex Platonic solids in Chapter 9. In all these chapters, single nanoparticles with optical properties resembling properties of usual atoms are considered. It may be said that the chapters deal with plasmonic atoms. In all these chapters, analytical expressions for the Green's function of electromagnetic field and plasmonic spectra are stated.

Chapter 10 concerns plasmon oscillations in clusters of nanoparticles and, first of all, in a system of two spherical nanoparticles. A cluster of nanoparticles is characterized by strongly localized plasmons in a gap between them as a result of their interaction. Such plasmons are absent in single nanoparticles, and their properties remind us of the properties of ordinary molecules consisting of atoms. Strongly localized plasmon oscillations in clusters of nanoparticles are naturally referred to as "plasmonic molecules." In this chapter, plasmon oscillations in a cluster of two nanoparticles of an irregular shape are also considered. Here, significant attention is also paid to plasmon wave propagation along a chain of nanoparticles and to focusing of plasmons by self-similar clusters, that is, clusters with components similar to the whole.

Chapter 11 is devoted to description of properties of nanoparticles made of "unusual" materials, that is, materials that do not occur in nature. Recently, materials of this type (metamaterials) have become widely distributed due to nanotechnology advances.

In Chapter 12, basic experimental facts and the plasmon theory of extraordinary light transmission through arrays of nanoapertures in a metal film are presented. This effect has been discovered recently and now is regarded as a basis for various promising nanodevices. Properties of light transmission through a single nanohole in a flat or nanostructured metal film are also considered in the chapter.

Chapter 13 concerns existing as well as emerging applications of plasmonic nanoparticles and nanostructures in nanodevices and nanoinstruments of different purposes.

In the first section of the chapter, results related to tumor therapy and visualization performed with the help of nanoparticles are presented. Nanoparticles exhibit a wonderful property to accumulate in tumors, allowing one to destroy cancer cells through heating layered gold nanoparticles by infrared radiation. In the second section, biosensors based on surface plasmons are considered. Strong localization of surface plasmons makes them extremely sensitive to a change of dielectric properties near the interface. Many commercial devices enabling operative detection of one or another biocomponents are based on this property.

In the third section of Chapter 13, biosensors based on localized plasmons, which are also very sensitive to the environment, are examined. According to this principle, two types of biosensors are developed. In one type, the analyzed substance causes plasmonic nanoparticles agglomeration, resulting in their absorption spectrum change. In another type, the analyzed substance accumulates on the surface of nanoparticles, provoking a change of local permittivity and that of the solution absorption spectrum. The biosensors considered in the third section are effective for many purposes. Inhomogeneous broadening of plasmon resonance is an important disadvantage of these sensors due to difficulties in providing equal sizes and forms for millions of nanoparticles. The biosensors using a spectrum of a single particle are free from this disadvantage (Section 13.4).

The element base for plasmon-integrated circuits (Section 13.5) is the most important direction of nanoplasmonic application development. The elements of the named integrated circuits have nanoscale sizes and optical frequencies of operation. That gives a hope of the construction of plasmon processors and other devices of data processing and transmission. In this section, both passive elements (waveguides, splitters, mirrors) and active ones (gates and switches) are considered.

In Section 13.6, applications based on nanoparticles' influence on atoms and molecules emission are considered. The section is likely to be the most interesting when development of nanodevices with an atom or a molecule functioning as the basic element is in the center of attention. Here, nanoantennas allowing adjusting of purely optical components with plasmonic ones are examined also.

In Section 13.7, the results of the construction of superlenses and hyperlenses that are capable of focusing light with nanometric resolution are presented. The basic element of the lenses is metal nanofilms with permittivity equal to -1 or anisotropic metamaterials. In this section, the applications of negative refraction and chiral media for drug enantiomer resolution and for effective excitation of atoms and molecules are also discussed.

In Section 13.8, recent results in development of invisible cloaks are presented. The basic element of such systems is strongly anisotropic plasmonic metamaterials and metamaterials with negative refractive index with properties designed within the so-called transformation optics theory.

Appendix A1 contains main results of the theory of radiation of atoms and molecules located near nanoparticles. The theory forms the basis for many applications of nanoparticles and is referred to in a range of the chapters. In Appendix A2, a brief summary of modern computational methods applied in nanoplasmonics is stated. Appendix A3 presents a list of some terms and abbreviations used commonly in nano-optics and nanoplasmonics.

References

Agranovich, V.M., Mills, D.L. (1982) *Surface Polaritons. Electromagnetic Waves at Surfaces and Interfaces* (North-Holland, Amsterdam, New York).

Maxwell Garnett, J.C. (1904) Colours in metal glasses and in metallic films, *Philos. Trans. R. Soc. London*, **203**, 385–420.

Mie, G. (1908) Beiträge zur Optik trüber Medien, speziell kolloidaler Metallösungen, *Ann. Phys. (Leipzig)*, **25**, 377–445.

Raether, H. (1988) *Surface Plasmons on Smooth and Rough Surfaces and on Gratings (Springer Tracts in Modern Physics)* (Springer-Verlag, Berlin).

Rayleigh, Lord (1871) On the light from the sky, its polarization and colour, *Philos. Mag.*, **41**, 107–120, 274–279 (reprinted in *Scientific Papers* by Lord Rayleigh, Vol. I: 1869–1881, No. 8, Dover, New York, 1964).

Sommerfeld, A. (1909) Über die Ausbreitung der Wellen in der drahtlosen Telegraphie, *Annalen der Physik (4th series)*, **333**, 665–736.

Wood, R.W. (1902) On a remarkable case of uneven distribution of light in a diffraction grating spectrum, *Proc. Phys. Soc. London,* **18**, 269–275.

Chapter 2

The World of Nanoparticles

2.1 Role of Micro- and Nanoparticles in the History of Our Civilization

At first sight, the outside world consists of big objects: buildings, trees, mountains, etc. Yet upon a closer view, micro- and nanoparticles play a significant part in our environment. The particles surround us everywhere. We tread the earth that feeds us—it consists of separate particles. The earth's atmosphere and different basins, soil and cosmic dust, clouds and granite, wood and clothes, and brick and concrete are but separate examples of the prevalence and importance of particles and nanoparticles. Food products—bread, milk, meat, butter, and fish—can be regarded as micro- and nanoparticle systems. A human being itself consists of particles forming blood, bones, and tissues. Generally, any field of human activity deals with numerous types of micro- and nanoparticles, and therefore studying micro- and nanoparticles is the most important mission of different sciences: physics, chemistry, biology, mathematics, mechanics, and others.

A significant place among all micro- and nanoparticles is occupied by metal nanoparticles that are, first of all, distinctively

Nanoplasmonics
Vasily Klimov

ISBN 978-981-4267-16-8 (Hardcover), 978-981-4267-42-7 (eBook)
www.panstanford.com

Figure 2.1 The Roman Licurgus Cup. Light absorbed by gold-silver alloy nanoparticles provides the red color of transmitting light (a) and green color of scattered light (b) (http://www.thebritishmuseum.ac.uk/explore/online_tours/general/the_art_of_glass/the_lycurgus_cup.aspx).

characterized by optical properties and the ability of effective light scattering and absorption.

Precious and semiprecious stones always had a great aesthetic, mystic, and economic value in human life, and their coloring is defined by negligible amounts of admixtures of heavy metals and their oxides. Often, the admixtures can be considered as nanoparticles.

In ancient times, people learned to use metal nanoparticles for practical purposes. Archeological excavations demonstrate that colored stained-glass windows were manufactured in ancient Egypt in the 2nd millennium BC and in ancient Rome from the 1st century AD. The Licurgus Cup, shown in Fig. 2.1, is a wonderful example of this civilization advance. This cup was made by craftsmen of ancient Rome in the 4th century AD and now is kept in the British Museum in London.

The cup changes its color from red to greenish when illuminated by white light from different sides. For a long time, the nature of these colors was unclear, but now it is known that the mentioned effects of light are caused by light scattering by the gold and silver nanoparticles that are contained in glass (Liz-Marzan, 2004).

The influence of the nanoparticles' sizes and forms on light scattering is graphically demonstrated in Fig. 2.2, where light scattering by silver nanoparticles of different shapes and sizes is

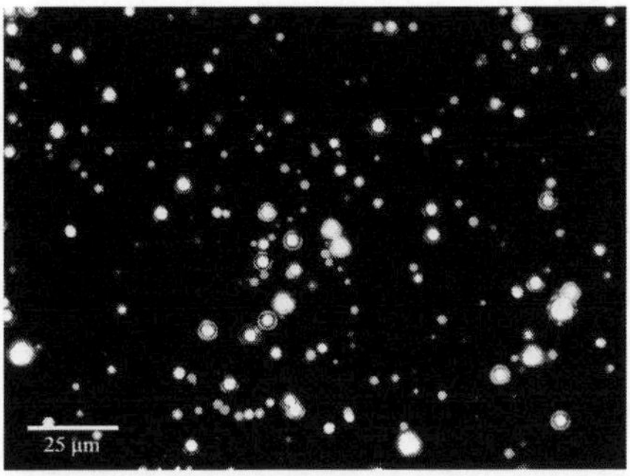

Figure 2.2 Image of silver nanoparticles of different form and size obtained by the dark-field microscopy method (McFarland and Van Duyne, 2003).

visualized by means of the dark-field microscopy method. The figure depicts that the smallest particles scatter blue light effectively, and as sizes of the particles are increasing, the maximum of light scattering is shifting to the red region of the spectrum.

From the quantitative viewpoint, the shift is clearly visible in Fig. 2.3, where the dependence of the light-scattering spectrum on a particle's size and form is shown. Analysis of the figure leads to a conclusion that nanotriangles effectively scatter red light, pentagons effectively scatter green light, and spherical nanoparticles effectively scatter blue light.

In the Middle Ages, nanoparticles were used extensively for manufacturing of stained-glass windows for Gothic cathedrals. In Fig. 2.4, bright colors of the stained glass are shown. Today, they have not lost their original view, owing to metal nanoparticles, too. In the process of the production of colored stained glass, metal oxides were grinded fine, mixed with low-melting glass and water, wine, or vegetable oil, and then deposited in a thin layer upon the glass, which was annealed in a furnace after each layer. Iron provided green color; cobalt, blue; gold, ruby; and selenium, all kinds of tints from soft pink to saturated red.

The color of natural emeralds and rubies is determined by admixtures of chromium compounds. In the 17th century, German alchemist Johann von Löwenstern-Kunckel (1630–1703) developed a technology of "gold ruby glass" (Cranberry glass) that looked as good as genuine rubies. This know-how was considered to die

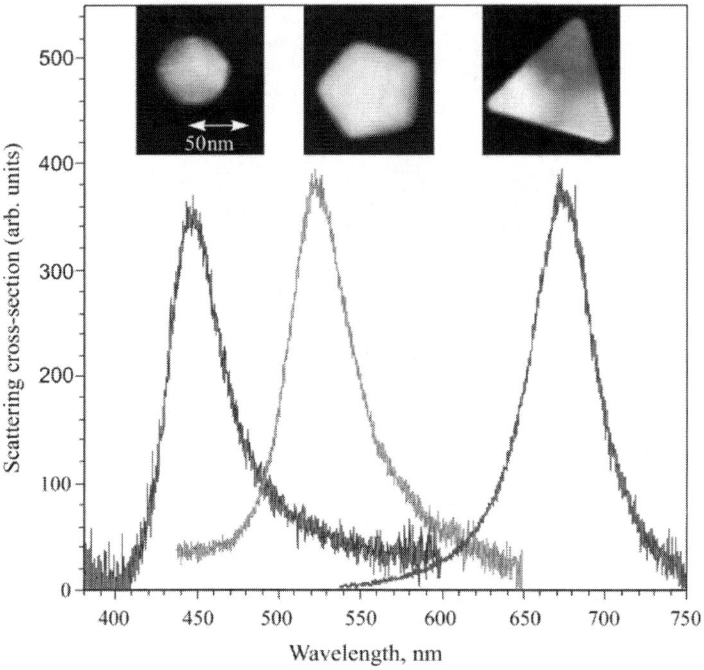

Figure 2.3 Spectra of individual nanoparticles and their TEM images (Mock et al., 2002). *Abbreviation*: TEM, transmission electron microscopy.

with Kunckel, since the data he left was confusing and did not contain the description of the most important technical conditions allowing one to obtain this valuable glass of unusually intensive color. The secrets of the "gold ruby glass" were rediscovered by famous Russian naturalist Mikhail Lomonosov. After more than 70 experiments with making use of gold compounds, a careful study of methods describing injection of gold in combination with different components (minium, magnesia, stannic lime, horn silver, etc.) into glass, and taking into consideration fusion temperature, Lomonosov developed independently the procedure of obtaining "gold ruby glass," which was beyond the power of his contemporaries in Western Europe. According to the technology, "gold ruby glass" can be produced by adding a negligible quantity (0.001%–0.1%) of splintered gold particles of a 4–30 μm diameter to regular glass.

Figure 2.4 Stained-glass windows of the Gothic cathedral in Chartres (France).

In May 1937, the decision was reached at the top level of Russian authorities to install ruby stars that can be seen from afar and point the development direction for the whole planet, on the five Kremlin towers, toward the 20th anniversary of the Great October Socialist Revolution. In Russia, ruby glass had been never before produced in industrial quantities, and that was not the only difficulty. According to technical requirements, the glass was supposed to have a definite density, transmit red rays of a certain wavelength, be resistant to sudden temperature drops, etc. A special ruby glass meeting the stated conditions was invented by Nikanor Kurochkin, who proposed to melt glass with the addition of selenium and gold nanoparticles, and the Kremlin stars are shining unceasingly from November, 7 of 1937 until now (Fig. 2.5).

Nanoparticles made of titania have a great significance, too. These particles absorb ultraviolet radiation, owing to a wide forbidden band. Moreover, TiO_2 nanoparticles acquire catalytic properties that cause microbe destruction upon contact with nanoparticles. Thus, use of the nanoparticles in the production of textiles provides a

Figure 2.5 Photograph of the ruby glass star of the Vodovzvodnaya Tower of the Moscow Kremlin.

disinfectant effect, boosting sales of "nanotechnological" underwear, in its turn. Besides, due to effective absorption of ultraviolet radiation TiO_2 nanoparticles are widely used in many sunscreen creams and lotions and also as a pigment for excellent white paints. One more application of TiO_2 nanoparticles is the production of self-cleaning surfaces. Water ceases to wet surfaces covered by these nanoparticles (lotus effect). Like a lotus flower, the nanoparticles reduce wettability by forming nanoroughnesses in conjunction with their own hydrophobic properties. On a surface like this, water forms drops and moves without wetting it. Moreover, when moving, the drops capture dirt particles and thus clean surfaces. This property is of prime importance in the production of skyscraper windows that can hardly be cleaned by usual means.

Many other examples of nanoparticle application in the late 20th century can be given, but true prosperity of nanoparticle application, especially metal, begins just now, when nanotechnology development allows synthesizing of nanoparticles of practically arbitrary shapes, as well as constructing different nanostructures on their basis, in which nanoparticles interact and cooperative effects appear. Relying on that sort of nanoparticles, smart nanosensors, nanodetectors, data processing nanodevices, and other nanodevices can be developed. Modern and prospective applications of metal nanoparticles and nanostructures will be presented in more detail in Chapter 13.

2.2 Modern Methods of Nanoparticle Synthesis

Now, there is a large variety of methods of dielectric, semiconductor, and metal nanoparticle and nanostructure synthesis (Gusev, 2005; Suzdalev, 2005; Daniel and Astruc, 2004; Murray et al., 2000; Dutta and Hofmann, 2003). Some methods were known for quite a long time, some appeared recently. Right now, a number of methods allow one to synthesize nanoparticles of a strictly defined shape and size, and it is very important for nanotechnological applications. Here, as well as in other nanotechnological approaches, two crucially important ways can be used. The first one is the "bottom-up" method based on synthesis of nanoparticles and nanostructures by adding of single atoms, one by one. Another is the "top-down" method based on cutting someway or other of a certain nanoparticle or a nanostructure from a macroscopic piece of a substance. Further sections describe synthesis methods both within "bottom-up" (Sections 2.2.1–2.2.4) and "top-down" (Section 2.2.5) technologies.

2.2.1 *Methods Based on Chemical Reactions in a Solid Body*

Apparently, these methods are the most ancient since they were used in ancient Egypt in the 2nd millennium BC and in ancient Rome from the 1st century AD.

From the modern viewpoint, these technologies used reactions of thermal decomposition of salts or metal complexes that are accompanied by formation of a metal nucleating center or metal oxides, with their further growth provoked by agglomeration. The size of the nanoparticles generated in such a way can vary in a wide range from 1 to 100 nm, depending on agglomeration conditions. Studies by means of the electron microscope reveal that the glass of the Licurgus Cup (Fig. 2.1) contains gold nanoparticles of 70 nm diameter (Liz-Marzan, 2004).

Solid-state nanoparticles can be also obtained as a result of photochemical reactions, for example, involving silver chalkogenides. In this type of reactions, nucleating centers emerge and then they grow, forming metal nanoparticles of the size of 10–100 nm.

2.2.2 *Nanochemistry Methods*

Nanochemistry is a rather new scientific discipline regarding chemical reactions at the nanometer scale as an object of study. The purpose of nanochemistry is the production of chemical products of nanometric sizes from 1 to 100 nm (Murray et al., 2000).

The problem is to find chemical reactions and approaches that could allow full control of the content, size, and shape of synthesized nanoparticles. Furthermore, nanochemistry is supposed to provide the possibility of fabrication of nanoparticles and nano-objects with complex structures that enable combining various useful functions at the nanoscale.

Nanochemistry proposes the following possibilities:

- fabrication of nanoparticles of a wide assortment of metals, semiconductors, glass, and polymers;
- fabrication of multilayered nanoparticles (core-shell type);
- nanostructuring of surfaces, providing them with functional properties, and self-organizing of nanoparticles on the surfaces arranged in such a way;
- organization of periodic and aperiodic functional structures of nanoparticles;
- in situ fabrication of nanosensors, nanoprobes, and other nanodevices.

Below, two main directions of modern nanochemistry that is based on colloidal chemistry for the most part will be presented.

2.2.2.1 Precipitation from colloid solutions

Apparently, precipitating from colloid solutions was the first effective method of obtaining dispersed nanoparticles. In the Faraday Museum of Royal Institution of Great Britain (London), established in 1973 by Elisabeth II, two phials of gold colloid solution (Fig. 2.6) prepared by Michael Faraday in the first half of the 19th century, are kept. These solutions preserve their stability for nearly 200 years. The obtaining procedure and properties of gold colloid solutions were described by Faraday (1857).

Figure 2.6 Colloidal solutions of gold obtained by Faraday (the Faraday Museum at the Royal Institution of Great Britain, www.rigb.org/rimain/heritage/faradaypage.jsp).

Regularly, nanoparticles are obtained from colloidal solutions by a chemical reaction between solution components and its interruption at a certain moment of time (Rosetti et al., 1984; Herron et al., 1989; Herron et al., 1993; Kuczynski and Thomas, 1983). Hereby, colloidal clusters of the size of 1–100 nm emerge. They can exist in a liquid phase for a long time without precipitation, due to a weak intercluster interaction, charge repulsion, and surface passivation. Colloidal clusters in the liquid phase can be divided into two groups, lyophilic (hydrophilic) and lyophobic (hydrophobic).

Lyophilic clusters can sorb on their surface environment molecules forming durable solvate complexes with them. Clusters of this type are surrounded by an envelope that remains preserved even when separate clusters coagulate or transform to gel nanosystems. The most typical representatives of hydrophilic clusters are oxides of silicium, ferrum, and other metals.

Lyophobic clusters do not adsorb dissolvent molecules on their surface. Though their surface can be modified by dissolvent ions, herewith it gets a positive or negative charge.

As an example of colloidal gold preparation, the following reaction can be mentioned:

$$2HAuCl_4 + 3H_2O_2 \rightarrow 2Au + 8HCl + 3O_2$$

Cadmium sulphide (CdS) nanocrystals, which are important in quantum dots physics, can be obtained by precipitation from a mix of cadmium perchlorate $Cd(ClO_4)_2$ and sodium sulphide Na_2S:

$$Cd(ClO_4)_2 + Na_2S \rightarrow CdS \downarrow +2NaClO_4$$

CdS nanoparticle growth is interrupted by a spasmodic pH increase.

For synthesis of III–IV semiconductor nanocrystals (InP, InAs, GaP, GaAs), effective synthesis methods are developed as well (Alivisatos et al., 2001; Battaglia and Peng, 2002; Lucey et al., 2005).

Colloidal particles of metal oxides are obtained by salt hydrolysis (Bleier and Cannon, 1986; Franz and Schwier, 1990; Chertov et al., 1993). For instance, TiO_2 nanoparticles are prepared easily by titanium tetrachloride hydrolysis:

$$TiCl_4 + 2H_2O \rightarrow TiO_2 \downarrow + 4HCl$$

Forming of metal and semiconductor clusters of very small size dispersion (and also monodisperse) is possible inside molecular sieve (zeolite) pores. Cluster isolation in pores is preserved even by high-temperature heating. For example, semiconductor clusters $(CdS)_4$ were synthesized inside zeolite cavities (Wang and Herron, 1987, 1988). Properties of the clusters obtained in ultrathin channels and zeolites pores, in particular, are analyzed in the review (Bogomolov, 1978). Larger semiconductor nanoparticles are synthesized by addition of supplementary molecules to the initial small cluster previously stabilized by organic ligands in a colloid solution. This synthesis of large nanoparticles can be considered inorganic compound polymerization.

Nanoparticles can be obtained also by supersonic procession of colloidal solutions containing large particles.

Precipitation from colloidal solutions allows one to synthesize nanoparticles of compound composition, that is, nanocrystal heterostructures. In this case, a core and a shell of a compound nanoparticle are formed of semiconductive substances with different electronic structures. Preparation of heterostructures, for example, CdSe/ZnS, ZnS/CdSe, HgS/CdS, ZnS/ZnO, and TiO_2/SnO_2, is a result of controlled precipitation of one type of semiconductor molecules on preliminary synthesized nanoparticles made from another semiconductor (Kortan et al., 1990; Haesselbarth et al., 1993; Kamat and Patrick, 1992; Bedja and Kamat, 1995). These heteronanoparticles can be covered by a layer of yet another semiconductor, and so on. The possibility of layered quantum nanostructure preparation is of great importance for the creation of biologically secure nanocrystal quantum dots with high-quantum fluorescence efficiency.

Among all methods of isolated nanoparticles obtaining, the method of precipitation from colloid solutions is the most selective and allows obtaining of stabilized nanoclusters with very small size distribution, thus having a substantial significance for use of nanoparticles as catalysts or in micro- and nanoelectronic devices.

Chemical synthesis of large metal clusters with the use of colloidal solutions was considered in detail by Schmid (1995) and Roldugin (2000).

There are different chemical techniques of nanoparticles obtaining in colloidal solutions, but anyway particles are to be protected from coalescence. Stabilization of colloid particles and clusters is achieved by ligand molecules. Thiols, triphenylphosphine and its derivatives, and phenanthroline are commonly used as ligands. In ligand-stabilized clusters, a metal core is distinguished as well as an external shell of metal atoms partially connected with ligand molecules. Cluster protection by means of an external shell is shown in Fig. 2.7: the surface of a colloidal 12 nm Au nanoparticle (the dark region) is covered by a light shell of ligand molecules $P(m\text{-}C_6H_4SO_3Na)_3$.

Apparently, metal clusters consisting of 55 atoms placed in two shells are the smallest particles retaining some metal properties, but scanning tunnel spectroscopy gives evidence of electronic band splitting in that kind of particles at room temperature already.

2.2.2.2 The reverse micelle synthesis

Recently, water/oil microemulsions (reverse micelles) have been used for metal nanoparticles obtaining by reduction from salts. Generally, a system of reverse micelles consists of two immiscible

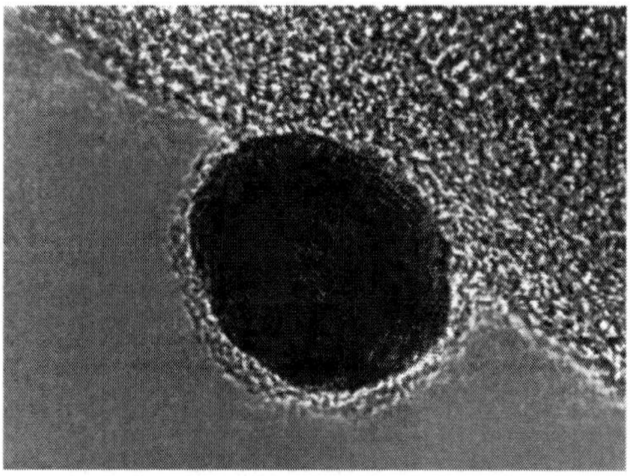

Figure 2.7 A colloidal particle of gold of approximately 11 × 13 nm covered by a $P(m\text{-}C_6H_4SO_3Na)_3$ ligand shell (Schmid, 1995).

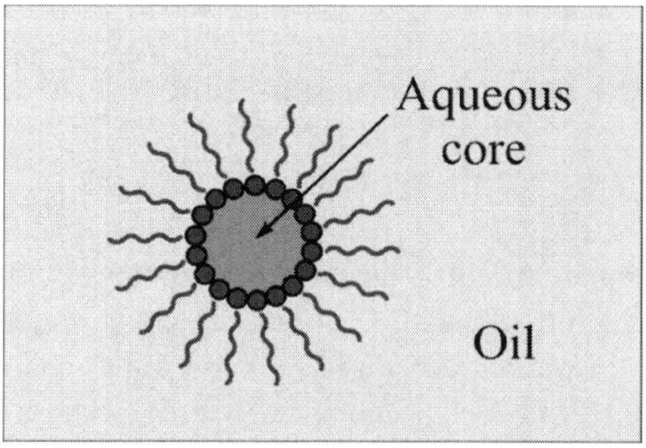

Figure 2.8 Schematic plot of a chemical reactor in a reverse micelle.

liquids, water and oil, where the water phase is dispersed in nanodrops surrounded by a monolayer membrane of a surface-active agent. Usually, the continuous phase consists of isooctane or hexane, while sodium bis(2-ethylhexyl)sulfosuccinate (AOT) acts as a surface-active agent. Different salt solutions (and even cadmium acetate or sodium sulphate) can be included in a reverse micelle besides pure water (Petit et al., 1993). In this case, nanodrops of salt solutions act like microreactors in nanoparticles synthesis, while the droplet size limits naturally the size of resulting nanoparticles. In Fig. 2.8, a water core of a reverse micelle nanoreactor dispersed in oil is shown.

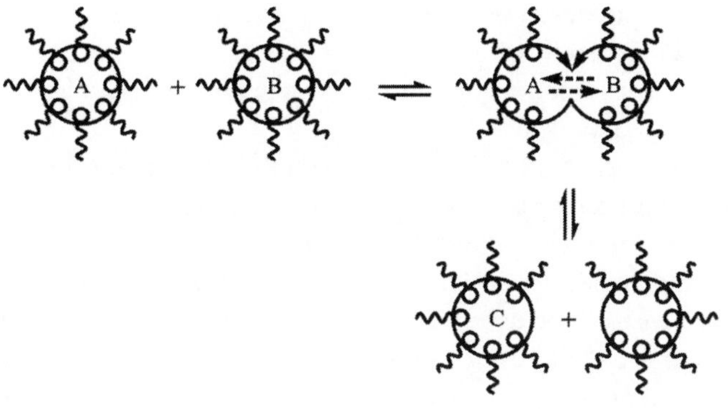

Figure 2.9 Scheme of a reaction proceeding in a reverse micelle.

Essential is the fact that micelles can take a spherical or nanocylinder form of a finite length, depending on the surface-active agents concentration, pH, and other factors.

In a microemulsion system, reverse micelles collide constantly and can coalesce and collapse, exchanging their contents permanently. Reverse micelles are also used for obtaining solid nanoclusters. For this purpose, one mixes two identical emulsion systems with reverse-micelle-containing substances A and B (Fig. 2.9). As a result of the exchange, a new compound, C, emerges. The size and the form of the cluster are defined by the reverse micelle radius and form.

For the moment, the reverse micelle method is an important resource for synthesis of monodisperse clusters of silver, gold, platinum, cobalt, iron, and their compounds. Metal nanoclusters can be obtained from microemulsions through injection of a reducing agent, for example, sodium borohydrate and hydrazine, or through fumigation of H_2S and H_2. Accordingly, metal clusters, Pd, Pt, Rh, and Ir (3–5 nm), and bimetal nanoclusters were obtained. Precipitation reactions in microemulsions result in obtaining carbonates, oxides, sulphides, etc. To obtain Ag_2S clusters, one uses mixtures of two types of reverse micelles, which contain Ag combined with surface-active agents and Na_2S in the water phase. Ag_2S clusters of 3.0–5.8 nm emerge as a result of coalescence accompanied by exchange of aqueous medium content.

In particular, substance A might be cadmium acetate and substance B, sodium sulphide. As a result of reverse micelle interaction, cadmium sulphide emerges, which is a nucleus of a nanocrystal quantum dot.

Vargaftik et al. (1985) have obtained an icosahedral palladium cluster containing 561 atoms of Pd in five filled shells through reduction of $Pd(OAc)_2$ by hydrogen in the presence of phenanthroline. Ligand molecules are situated at icosahedron vertexes and edges.

To release nanoclusters from a reverse micelle, one destroys it by thioles and then dissolves, filters, and dries obtained compounds. In this way, nanoclusters of up to 10 nm, passivated by thioles, can be obtained. In turn, two- and three-dimensional nanostructures can be obtained from nanoclusters of that kind.

The reverse micelle method can also be applied to a multistage synthesis of multilayer nanoparticles (Lal et al., 2000).

2.2.3 *Gas-Phase Synthesis of Nanoparticles (Gusev, 2005)*

Usually, isolated nanoparticles are obtained by metal, alloy, or semiconductor evaporation at a controlled temperature in an atmosphere of an inert low-pressure gas with consequent steam condensation on a cold surface. This is the simplest method of nanocrystal powder obtaining. In contrast to evaporation in vacuum, atoms of a substance evaporated in an inert atmosphere lose their kinetic energy faster due to collisions with gas atoms and form clusters.

The first research studies in this field were carried out by Kohlschutter and Ehlers (1912), Kohlschutter and Noll (1912). Studying Zn, Cd, Se, and As evaporation in vacuum as well as in an atmosphere of hydrogen, nitrogen, and carbonic acid has shown that the size of nanoparticles obtained depends on the gas pressure and its atomic weight. Harris et al. (1948) have evaporated gold from a heated tungsten wire under a nitrogen pressure of 40 Pa and obtained spherical nanoparticles of 1.5–10 nm diameter (the average size is about 4 nm) in a condensate. They have found that the nanoparticles' size depends on the gas pressure and less on the evaporation rate. At a different gas pressure, the condensation of aluminum vapors in H_2, He, and Ar atmospheres allowed one to obtain nanoparticles of 20–100 nm and to determine the average aerosol particle size dependence on gas pressure (Gen et al., 1959). In particular, it was shown that increase of gas pressure from 10–130 to 360–400 Pa leads to particles size growth. Under a higher pressure, the nanoparticles' size stays almost unchangeable.

Obtained by evaporation and condensation, nanocrystal particles of 20 nm radius and less have a spherical form, while larger particles can be facetted. There are no dislocations in isolated nanocrystals, but disclinations that are more energy effective than dislocations in small crystals can emerge (Vladimirov and Romanov, 1986).

Evaporation-condensation setups differ by the way of evaporated metal input, the way of evaporation energy supply, the working atmosphere, the condensation procedure, and the procedure of powder collection obtaining.

A metal can be evaporated from a cup. On the other hand, a metal can be delivered in a heating and evaporation zone by a fluid jet, as a wire, or as a metal powder. Energy supply can be performed directly by heating, current passing through the wire, an electroarc discharge in plasma, inductive heating through high-frequency and super-high-frequency current, laser radiation, electron beam heating, etc. Evaporation and condensation can occur in vacuum, a motionless buffer gas, and a gas stream, including a plasma stream.

Condensation of a vapor-gas mixture of 5.000–10.000 K occurs after entering a chamber (of a large section and volume) filled by a cold buffer gas. Cooling takes place due to both expansion and contact with cold atmosphere.

Collection of the nanocrystal powder obtained by condensation is an independent task because due to their small size, particles are constantly in Brownian motion, staying suspended in gas without precipitating under the action of gravity. To collect obtained powders, one uses specific filters and centrifugal precipitation. In some cases, liquid film capturing is applied.

Depending on metal evaporation conditions (gas pressure, substrate position, and temperature), condensation can proceed both in the volume and on the reaction chamber surface. Spherical particles forms are typical for volume condensates, whereas surface condensate particles are facetted. At the same evaporation and condensation conditions, metals of higher melting temperature form particles of smaller size.

Adjusting the composition of the gas phase containing two or more elements besides the buffer gas, one can grow monocrystal nanoparticles of different shapes and compositions. Bulakh (1974), Bulakh and Sheftall (1974) demonstrated by an example of cadmium sulphide that deviation of Cd:S concentrations from one in a gas phase, that is, from stoichiometry, allows obtaining of CdS particles in forms of prismatic slabs, plates, ribbons, needles, whiskers, tubes, and others. These results have shown that temperature and the proportion of initial components of the gas phase are the main factors defining the shape of small particles obtained.

Gas-phase synthesis allows one to obtain particles of the size from two to several hundreds of nanometers. Smaller particles of controlled size can be obtained by separation of clusters, depending on their mass in a time-of-flight mass spectrometer. Thus, antimony, bismuth, and lead clusters containing 650, 270, and 400 atoms, respectively, were obtained (Muhlbach et al., 1981).

Highly dispersed silver and cuprum precipitations were obtained on glass by metal evaporation in inactive atmosphere under 0.01–1.13 Pa (Petrov, 1969). Using the same method, Li_n clusters containing from 15 to 2 lithium atoms were obtained (Kimoto and Nishida, 1977).

The method of vapor condensation in a buffer gas is frequently used for science purposes—for obtaining of small amounts of nanopowders. Nanopowders synthesized by this method agglomerate weakly and conglomerate at comparatively low temperatures.

2.2.4 *Nanolithography Methods of Metal Nanoparticle and Nanostructure Synthesis*

The word "lithography" is derived from the Greek words "lithos" (stone) and "grapho" (to write). Earlier, this word denoted printing with the help of relief stones. Now, this term has a wider meaning. Nevertheless, the basic element of modern lithographic methods is a relief mask, which can be material (e.g., crystal) or immaterial (e.g., standing light wave). There are a lot of methods allowing one to

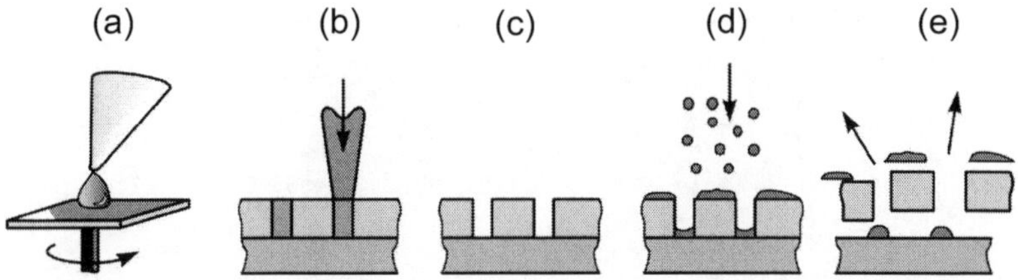

Figure 2.10 Scheme of the process of electron beam lithography. (a) Application of a resist layer on a substrate, (b) irradiation of the predetermined resist regions by an electron beam, (c) removal of the irradiated regions, (d) sputtering of metal, and (e) removal of remains of the resist and of the metal located on the resist.

manufacture masks of that kind and to synthesize nanoparticles and nanostructures with their help, but we will consider only the most significant ones.

2.2.4.1 Electron beam and ion beam lithography

Within this approach to nanostructure production, an electron beam of a scanning electronic microscope is used for irradiation of particular regions of a positive resist located on a substrate (Fig. 2.10a,b). Irradiated regions define the nanostructures' form. In the next stage, the irradiated regions of the resist are removed by chemical treatments (Fig. 2.10c). Then, a metal layer of necessary thickness is sputtered on the obtained mask (Fig. 2.10d). In the final process stage, the mask covered with metal is removed, and as a result metal nanoparticles or nanostructures remain on the substrate.

Generally, ion beam lithography is analogous to the electron beam one. The main difference is that due to a larger ion mass an ion beam focuses better in the matter and the resist bonds are destroyed more rapidly in this case. Both methods combine the top-down technology, used during mask production, and the bottom-up technology, used during metal sputtering.

Besides, the ion beam can be used for immediate deposition of atoms on a substrate, thus forming nanoparticles and nanostructures by beam scanning only without any mask application. This case

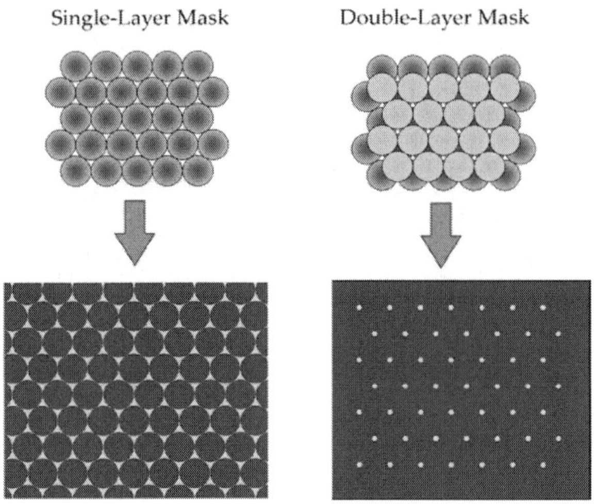

Figure 2.11 Schematic plot of single- and double-layered nanosphere masks and nanoparticles arrays obtained with the help of nanosphere lithography nanoparticles.

is an example of the bottom-up technology. Typical accuracy of ion beam lithography is about 5 nm.

2.2.4.2 Nanosphere lithography

Use of nanosphere lithography appears to be prospective for obtaining periodically located nanoparticles of several predefined shapes (Hulteen and Van Duyne, 1995) (Fig. 2.11). This approach is based on the presence of channels in one or even two close-packed layers of nanospheres. By sputtering metal atoms on a structure of this kind, only atoms passed through the channels in the nanosphere lattice reach the substrate. After nanospheres and the metal on their surface have been removed, necessary metal nanostructures remain at the substrate. In the case of a one-layer covering of the substrate, these structures consist of *nanotriangles* at the nodes of the hexagonal lattice (Fig. 2.11). In the case of two close-packed nanosphere layers, *nanohexagons* (practically, disks) remain on the substrate, located at the nodes of a hexagonal lattice of a greater period, too.

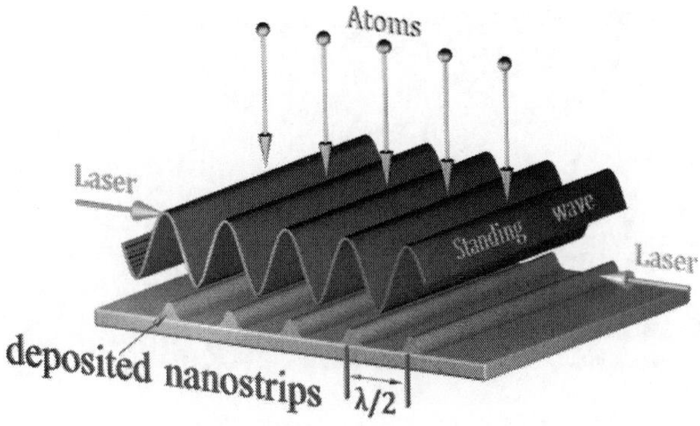

Figure 2.12 Scheme of atom beam focusing by standing light wave nodes.

2.2.4.3 Atom nanolithography with the help of optical fields

In the previous sections, the synthesis of nanostructures was in fact conditioned by the control of separate atom motions through different material masks. However, recently nano-optics development has enabled the control of atoms' motion by means of optical fields (Balykin et al., 2006; McClelland et al., 1993; Bradley et al., 1999).

Depending on detuning of the laser field frequency from the resonance frequency of electron oscillations in a neutral atom, an atom would be pulled in or pushed out of the weaker optical field toward the stronger one and vice versa. For the first time, the possibility of atom beam focusing into nanostripes was demonstrated with the help of a laser standing wave. The geometry of this experiment is shown in Fig. 2.12, while in Fig. 2.13 the atomic force microscopy (AFM) image of Cr nanostripes obtained is shown. By means of a standing wave, emerging as a result of a superposition of several laser beams, even more complex periodic structures can be derived.

A similar procedure was applied for fabrication of Al (McGowan et al., 1995) and Na (Timp et al., 1992) nanostructures.

2.2.4.3.1 The atomic "camera obscura" method

An original method of atom nanolithography using the well-known principle of camera obscura was proposed by Balykin et al., (2006).

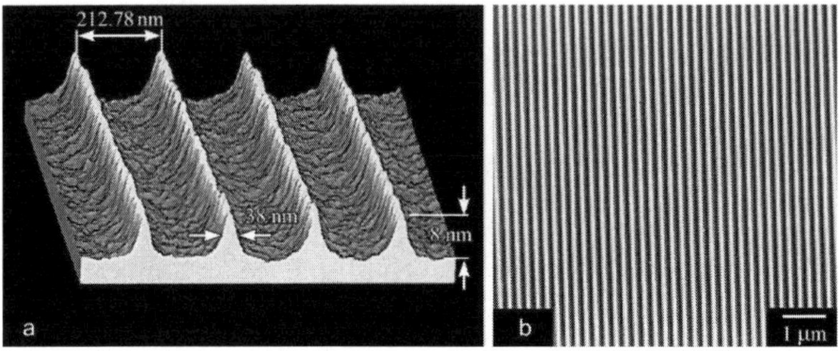

Figure 2.13 AFM image of chromium nanostripes obtained as a result of atom beam focusing by standing light wave nodes. (a) Cr nanostripes of 38 nm width on a silicon substrate. The plotting area size is 1 × 1 μm; (b) Cr nanostripes on a sapphire substrate. The plotting area size is 8 × 8 μm. Reprinted from (Bradley et al., 1999), copyright (1999), with permission from Elsevier.

Implemented by the authors, an atomic "camera obscura" (atomic pinhole camera) allowed formation of an identical nanostructure array, where a typical size of a separate nanostructure was less than 50 nm, on the surface of a solid body.

A schematic experimental setup with an atomic "camera obscura" is shown in Fig. 2.14.

A Cr atoms beam is transmitted through a macroscopic metal mask, thus forming, analogously to optics, a "radiant object" of predefined geometry. Atoms passed through mask apertures propagate in straight lines like light rays and then reach a thin membrane (thickness $h = 5$ μm) located at $L = 90$ mm from the mask with a great number ($N \approx 3 \times 10^7$ cm^{-2}) of conical apertures (of $d = 50$ nm input hole diameter and $D \approx 350$ nm exit diameter). Each membrane aperture is the atoms' "camera obscura," forming an individual image of an "object" on the surface of the substrate located at $l = 5$ μm behind the membrane. In this geometry, a set of "object's" images diminished by approximately $n = L/l =$ 18,000 times are formed by Cr atoms sputtered on the surface. A remarkable peculiarity of this approach is the possibility to form millions of *identical* nanoparticles and nanostructures of complex forms.

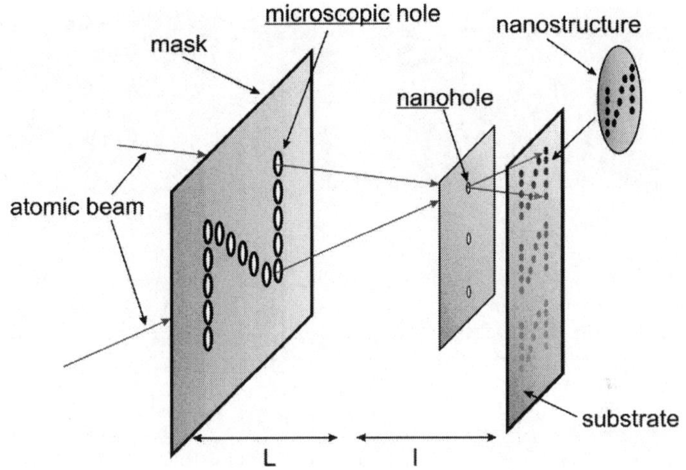

Figure 2.14 Scheme illustrating the operation of atomic projection lithography.

2.2.5 *Forming of Arbitrary Three-Dimensional Metal Nanostructures with the Help of Focused Ion Beams*

An outstanding feature of focused ion beams (FIBs) is their ability to modify substances (destroy chemical and crystal bonds) with a high spatial resolution (up to 5 nm) and at a great depth (up to 10 μm) (see, e.g., (Munnik et al., 2003)). Substance bonds can be modified so intensively that many substances become positive resists per se. Thus, one can form a small amount of nanoparticles and nanostructures of any complexity consequently by directing an FIB on desired regions and subsequently removing modified areas by one or another method.

It should be mentioned that a three-dimensional metal or other material processing with the help of FIBs is a typical example of top-down nanotechnology.

2.3 Nanoparticles and Nanostructures Gallery

In the previous section, main modern technologies of nanoparticles and nanostructures synthesis were described. In the present section,

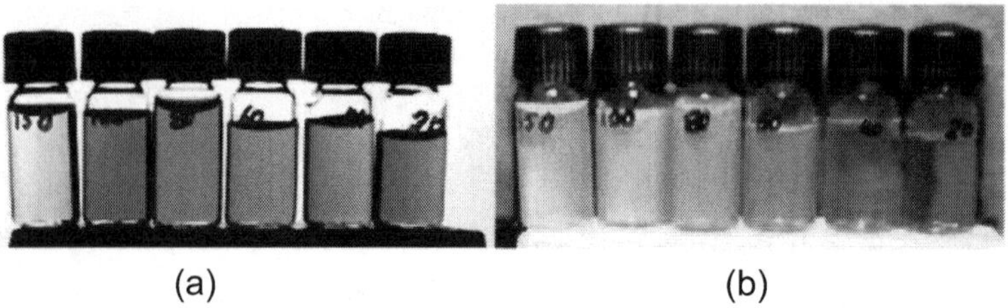

Figure 2.15 Colloid solution of commercially available spherical gold nanoparticles of 150, 100, 80, 60, 40, and 20 nm (from left to right). (a) Irradiation from behind and (b) irradiation from the front. The bright colors in (a) are caused by absorption in nanoparticles, while the colors in (b) are caused by back diffuse scattering (Sönnichsen, 2001).

a set of images of nanoparticles and nanostructures of most different shapes is presented as a demonstration of these technologies effectiveness. The great variety of nanoparticles' forms determines a greater variety of their optical properties and a possibility of development of different optical, optoelectronic, and mechanical devices on their basis.

Spherical nanoparticles appear to be the easiest to produce, and now spherical nanoparticles of small size dispersion are available commercially (see, e.g., http://www.nanopartz.com). In Fig. 2.15, colloidal solutions of gold nanoparticles of 20–150 nm are shown, irradiated from behind (Fig. 2.15a) and from the front (Fig. 2.15b). The different solution colors on Fig. 2.15a are conditioned by extinction, while the colors in Fig. 2.15b are conditioned by back diffuse scattering. In spite of minimum gold nanoparticle concentration ($<10^{-2}$ wt%), colloid solutions have bright colors, especially when irradiated from behind (Fig. 2.15a). A noticeable difference of colors for light transmission and back-scattering is caused by different dependencies of scattering and absorption cross sections on the light wavelength. The theory of optical properties of spherical nanoparticles is considered in detail in Chapter 6.

Clusters of spherical nanoparticles can be produced as well as isolated spherical nanoparticles. Thus, in Fig. 2.16 different clusters of gold nanoparticles are shown, where gold nanoparticles of 5 nm diameter are bound to each other by phenylacetylene nanobridges of different structures (Feldheim, 2001).

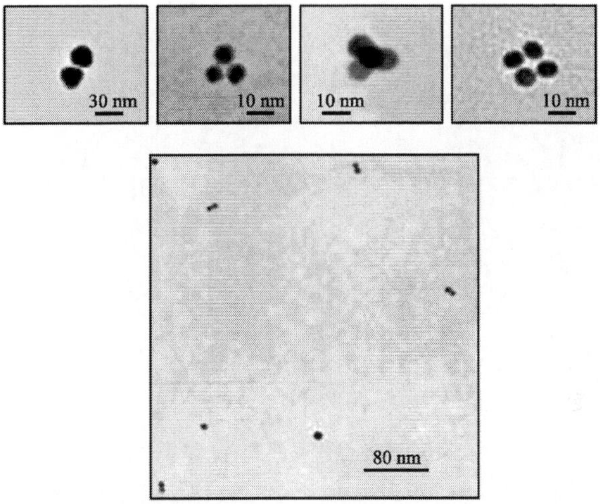

Figure 2.16 TEM images of gold clusters bound by different organic molecules. On the lower image several dimers of gold nanoparticles are shown. Reproduced from (Feldheim, 2001), by permission of the Electrochemical Society.

More diverse, close-packed clusters of polystyrene nanospheres of 844 nm diameter were obtained by Manoharan et al. (2003). In this case, nanospheres binding is provided by van der Waals forces.

Besides simple spherical nanoparticles and their clusters, spherically layered nanoparticles (nanoshells or nanomatryoshkas; "matryoshka" is a Russian nested doll) can be synthesized (Prodan et al., 2003). Asymmetrical layered nanostructures (nanoeggs and nanorice) can be manufactured by the same method (Wang et al., 2006). With the help of colloidal chemistry methods described in Section 2.2, one can also synthesize nanoparticles of a more complex form. Gold nanorods (Mohamed et al., 2000) are shown in Fig. 2.17, while silver nanocubes are shown in Fig. 2.18a. In turn, silver nanocubes can be used to obtain gold nanoboxes, that is, hollow nanocubes (Fig. 2.18b) (Sun and Xia, 2002).

Chemical methods of synthesis allow one to obtain nanoparticles of even more exotic forms. Thus, ZnO tetrapod ("four-legged") crystals are shown in Fig. 2.19. They have a remarkable feature:

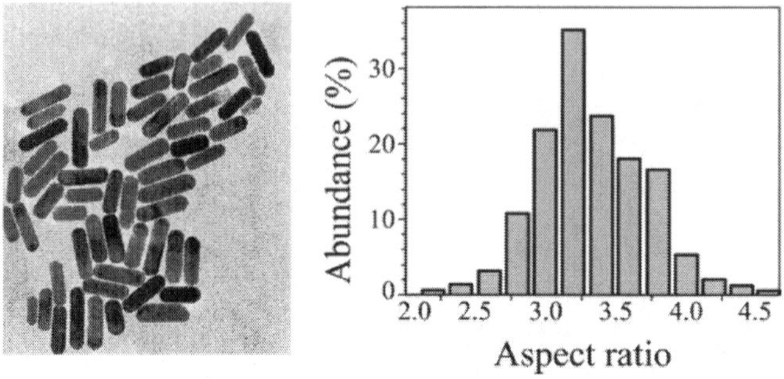

Figure 2.17 TEM image of gold nanorods obtained and bar graph of particle distribution over aspect ratio (Mohamed et al., 2000).

Figure 2.18 (a) SEM images of silver nanocubes with edges of 175 nm. (b) SEM images of hollow gold cubes (nanoboxes). On the inset, there is an enlarged image of a nanocube, with triangle face lying on a silicon substrate. Courtesy of Younan Xia.

one of the "legs" is always normal to any surface wherever the nanoparticles are placed. Earlier, this property was used widely (antitank hedgehogs). Modern nanoscience pins its hopes of creation of standardized nanoneedles for apertureless scanning microscopes of high resolution on the particles of this shape.

There is an opinion that the one of the most important human achievements—the wheel—has no analogues in nature. However, that is not quite true. In Fig. 2.20, a crystal nanoring of zinc oxide is shown. Rings of that kind are formed from nanostripes spontaneously by long-range electrostatic forces, while short-range

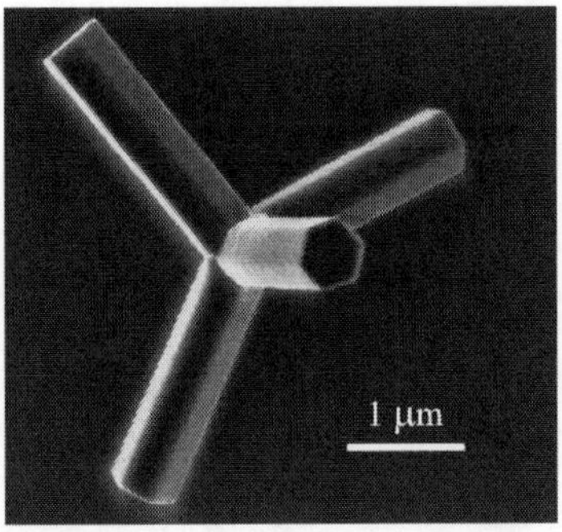

Figure 2.19 SEM image of ZnO crystals in a tetrapod form. Reprinted from (Newton and Warburton, 2007), copyright (2007), with permission from Elsevier.

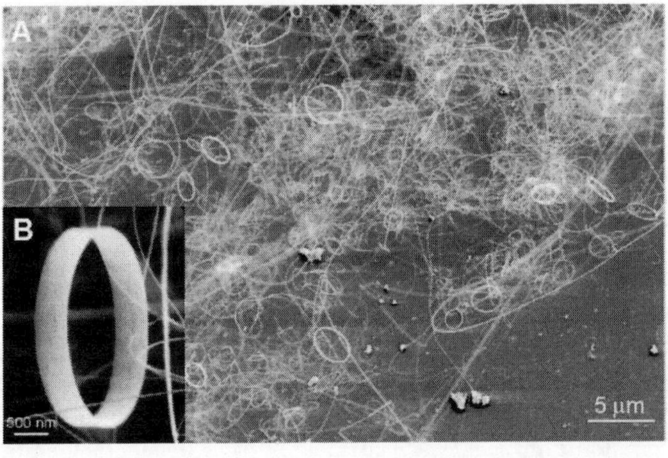

Figure 2.20 Zinc oxide nanorings. From (Kong et al., 2004), reprinted with permission from AAAS.

chemical bonds lead to ring closure. These nanorings can be applied in different nanodevices, in particular, nanoelectromechanical ones.

(Zhang et al., 2003) discovered nanospirals and nanosprings of usual silica (Fig. 2.21). It turns out that these particles are quite stable and their behavior is identical to that of usual macroscopic springs. This property allows one to hope that nanosprings can be used in nanoscale mechanical devices.

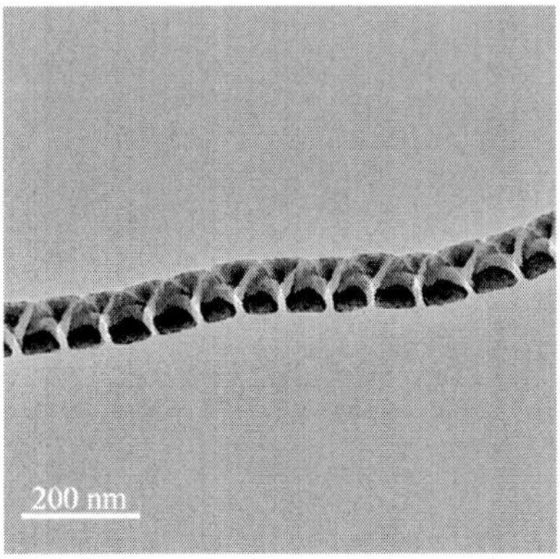

Figure 2.21 Silica nanospring (Kalaugher, 2003).

Until now, we have given examples of rather complex nanoparticles of different shapes, which were synthesized with the help of various chemical methods (the bottom-up technology). Even more complex nanoparticles and nanostructures can be synthesized by means of physical methods.

Thus, nanosphere lithography and atomic nanolithography allow one to obtain many regularly structured nanoparticles of the same form simultaneously. In Fig. 2.22, the image of a regular lattice of silver nanotriangles, obtained by means of nanosphere lithography, is shown.

Analogous arrays of Cr nanoparticles can be obtained with the help of atomic lithography through focusing an atom beam by nodes of a standing optical wave, formed by two perpendicular beams (Fig. 2.23).

Atom nanolithography also allows obtaining of arrays of millions of nanoparticles of a more complex shape than circular or triangular. In Figs. 2.24a and 2.25b, a letter "X" of approximately 100 nm and a letter "λ" of 50 nm, obtained by atomic projection lithography without use of focusing optical beams (camera obscura), are shown.

With the help of electron and ion beam lithography, creation of nanoparticles and nanostructures of almost arbitrary shape is possible. In Fig. 2.25, images of golden disks with different distances between their centers are shown. These clusters are obtained by

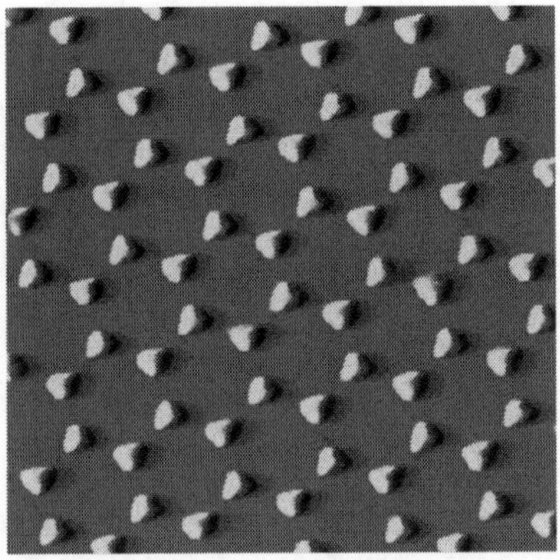

Figure 2.22 AFM image of regularly structured silver nanotriangles of 50 nm height and approximately 150 nm side length formed by means of nanosphere lithography (nanosphere diameter is 400 nm) (Haynes et al., 2000).

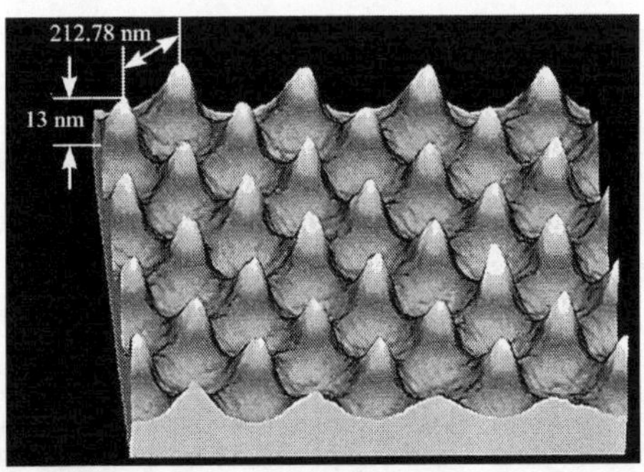

Figure 2.23 Two-dimensional array of chromium nanodots obtained through atom beam focusing by two orthogonal standing waves. Reprinted from (Bradley et al., 1999), copyright (1999), with permission from Elsevier.

methods of electron lithography (Rechberger et al., 2003). The possibility of creation of nanodiscs pairs with an arbitrary distance between them allows one to investigate their interaction, and it has a crucial significance for manufacturing of optical nanodevices.

With the help of focused ion beam (FIB) lithography, which is often denoted as ion milling, it is also possible to manufacture

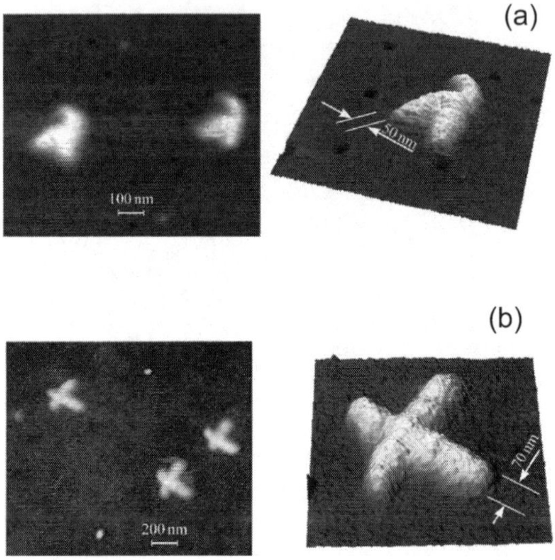

Figure 2.24 Image of metal nanoparticles obtained by atomic projection lithography. (a) A particle in the form of "λ" of 50 nm; (b) a particle in the form of "X" of approximately 100 nm (Balykin et al., 2006).

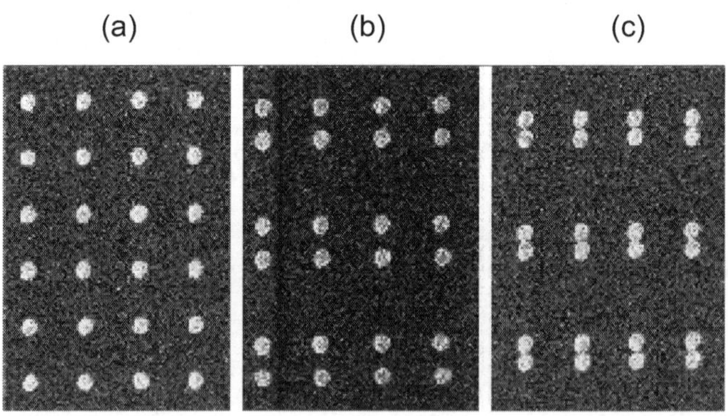

Figure 2.25 Lattice formed by nanodiscs pairs with different distances between their centers: (a) 450 nm, (b) 300 nm, and (c) 150 nm. The nanodiscs' diameter is 150 nm, and the height is 17 nm. The image is obtained by a scanning electronic microscope. Reprinted from (Rechberger et al., 2003), copyright (2003), with permission from Elsevier.

Figure 2.26 SEM image of a nanohole array in a gold film on a quartz substrate. Apertures of 200 nm diameter are drilled by FIBs (Krishnan et al., 2001).

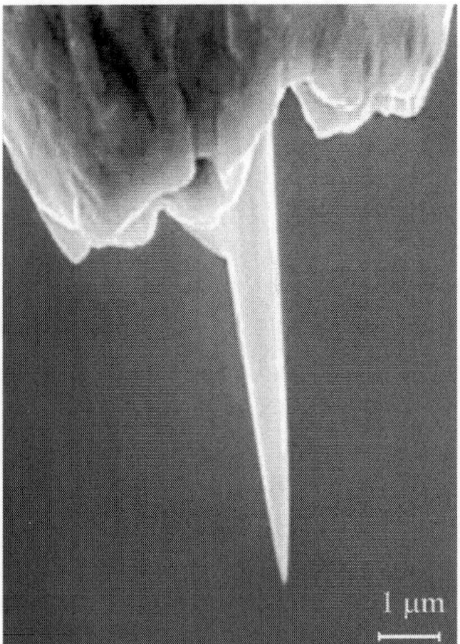

Figure 2.27 SEM image of a gold needle fabricated by an FIB. Reproduced from (Novotny and Stranick, 2006), with permission of Annual Reviews.

nanohole arrays (Fig. 2.26). Lattices of that kind have extraordinary light and atom transmission and focusing and can be used in numerous applications.

And, finally, in Fig. 2.27 a needle of apertureless scanning microscopes is shown. Such tips were obtained with the help of three-dimensional lithography by means of FIBs. Such needles are the main elements of scanning microscopes and allow one to obtain both optical and topographical images of nanoparticles and nanostructures with nanometer resolution.

2.4 Conclusion

Thus, now there is a variety of methods and technologies allowing one to obtain any quantities of plasmon nanoparticles and nanostructures of almost arbitrary shape and composition. This circumstance allows one to set the task of creation of brand-new, smart optical, optoelectronic, mechanical, and other nanodevices and nanoinstruments. To develop such devices, numerous investigations are necessary in very different directions. Theoretical and experimental investigations of nanoparticles' and nanostructures' optical properties are of special significance because they form a basis for a majority of smart nanodevices and help optimize their operation. Above that, comprehension of nanoparticles' and nanostructures' optical properties allows one to use effectively different scanning microscopes, more precisely nanoscopes, and see what is happening in the process of nanodevice synthesis and operation. The rest of the chapters of the book are devoted to these problems.

References

Alivisatos, A.P., Peng, X., Manna, L. (2001) Process for forming shaped group III-V semiconductor nanocrystals, and product formed by process, US Patent 6, 303, 736.

Balykin, V.I., Borisov, P.A., Letokhov, V.S., Melentiev, P.N., Rudnev, S.N., Cherkun, A.P., Akimenko, A.P., Apel, P.Y., Skuratov, V.A. (2006) Atom "pinhole camera" with nanometer resolution, *JETP Lett.*, **84**, 466–469.

Balykin, V.I., Klimov, V.V., Letokhov, V.S. (2006) Atom nanooptics, in *Handbook of Theoretical and Computational Nanotechnology*, eds. Michael Rieth, Wolfram Schommers, **7**, 1–78.

Battaglia, D., Peng, X. (2002) Formation of high quality InP and Inas nanocrystals in a noncoordinating solvent, *Nano. Lett.*, **2**, 1027–1030.

Bedja, I., Kamat, P.V. (1995) Capped semiconductor colloids. Synthesis and photoelectrochemical behavior of TiO_2 capped SnO_2 nanocrystallites, *J. Phys. Chem.*, **99**, 9182–9188.

Bleier, A., Cannon, R. (1986) Nucleation and growth of uniform monoclinic zirconium dioxide. Better ceramics through chemistry, *MRS Symp. Proc.*, **73**, eds. C.I. Brinker, D.E. Clark, D.R. Ulrich (Pittsburg MRS 1986 P), 71–78.

Bogomolov, V.I. (1978) Liquids in ultrathin channels (Filament and cluster crystals), *Soviet Phys. Uspekhi*, **21**, 77–83.

Bradley, C.C., Anderson, W.R., McClelland, J.J., Celotta, R.J. (1999) Nanofabrication via atom optics, *Appl. Surf. Sci.*, **141**, 210–218.

Bulakh, B.M. (1974) Cadmium sulphide crystallization from gas phase, *Cryst. Growth*, **10** (Nauka, Moscow), 98–114.

Bulakh, B.M., Sheftall, N.N. (1974) Mechanism of cadmium sulphide crystal growth from gas phase, *Collect. Art.: Cryst. Growth*, **10** (Nauka, Moscow) 115–134.

Chertov, V.M., Litvin, V.I., Mitroniuk, I.F., Tsyrina, V.V. (1993) Synthesis and texture of xerogels on the basis of ultrafine powders of aluminium oxide and monohydroxide, *Neorg. Mater.*, **29**, 1019–1020.

Daniel, M.-C., Astruc, D. (2004) Gold nanoparticles: assembly, supramolecular chemistry, quantum-size-related properties, and applications toward biology, catalysis, and nanotechnology, *Chem. Rev.*, **104**, 293–346.

Dutta, J., Hofmann, H. (2003) Self-Organization of Colloidal Nanoparticles, Encyclopedia of Nanoscience and Nanotechnology, ed. H.S. Nalwa, **X**, 1–23.

Faraday, M. (1857) Experimental relations of gold (and other metals) to light, *Philosoph. Trans. R. Soc.*, **147**, 145–181.

Feldheim, D. (2001) Assembly of metal nanoparticle arrays using molecular bridges, *Electrochem. Soc. Interface*, **Fall 2001**, 22–25.

Franz, G., Schwier, G. (1990) Starting materials for advanced ceramics—production and properties, in *Raw Materials for New Technologies*, ed. M. Kursten (Nagele and Obermuller, Stuttgart), 139–158.

Gen, M.Y., Ziskin, M.S., Petrov, Y.I. (1959) Study of Size Distribution of Al Aerosol in Dependence of Operational Conditions, *Doklady AN SSSR*, **127**, 366–368 [in Russian].

Gusev, A.I. (2005) Nanomaterials, Nanostructures, Nanotechnologies (Fismatlit, Moscow).

Haesselbarth, A., Eychmuller, A., Eichberger, R., Giersig, M., Mews, A., Weller, H. (1993) Chemistry and photophysics of mixed cadmium sulfide/mercury sulfide colloids, *J. Phys. Chem.*, **97**, 5333–5340.

Harris, L., Jeffries, D., Siegel, B.M. (1948) An electron microscopy study of gold smoke deposits, *Appl. Phys.*, **19**, 791–794.

Haynes, C.L., Haes, A.J., Van Duyne, R.P. (2000) Nanosphere lithography: synthesis and application of nanoparticles with inherently anisotropic structures and surface chemistry, *Mater. Res. Soc. Symp. Proc.*, **635**, C6.3/1–C6.3/6.

Herron, N., Calabrese, J.C., Forneth, W.E., Wang, Y. (1993) Crystal structure and optical properties of $Cd_{32}S_{14}(SC_6H_5)_{36}$. DMF4, a cluster with a 15 angstrom cadmium sulfide core, *Science*, **259**, 1426–1428.

Herron, N., Wang, Y., Eddy, M.M., Stucky, G.D., Cox, D.E., Moller, K., Bein, T. (1989) Structure and optical properties of cadmium sulfide superclusters in zeolite hosts, *J. Am. Chem. Soc.*, **111**, 530–540.

Hulteen, J.C., Van Duyne, R.P. (1995) Nanosphere lithography: a materials general fabrication process for periodic particle array surfaces, *J. Vac. Sci. Technol. A*, **13**(3), 1553–1558.

Kalaugher, L. (2003) Nanosprings jump into place, http://nanotechweb.org/cws/article/tech/17297.

Kamat, P.V., Patrick, B. (1992) Photophysics and photochemistry of quantized zinc oxide colloids, *J. Phys. Chem.*, **96**, 6829–6834.

Kimoto, K., Nishida, I. (1977) A study of lithium clusters by means of a mass analyzer, *J. Phys. Soc. Japan*, **42**, 2071–2072.

Kohlschutter, V., Ehlers C. (1912) Versuche uber Kondensation von Metalldampfen, *Z. Electrochem.*, **18**, 373–380.

Kohlschutter, V., Noll, N. (1912) Uber feine Metallzerteilungen, *Z. Electrochem.*, **18**, 419–428.

Kong, X.Y., Ding, Y., Yang, R., Wang, Z.L. (2004) Single-crystal nanorings formed by epitaxial self-coiling of polar nanobelts, *Science*, **303**, 1348–1351.

Kortan, A.R., Hull, R., Opila, R.L., Bawendi, M.G., Steigerwald, M.L., Carrol, P.J., Brus, L.E. (1990) Nucleation and growth of cadmium selenide on zinc sulfide quantum crystallite seeds, and vice versa, in inverse micelle media, *J. Am. Chem. Soc.*, **112**, 1327–1332.

Krishnan, A., Thio, T., Kim, T.J., Lezec, H.J., Ebbesen, T.W., Wolff, P.A., Pendry, J., Martin-Moreno, L., Garcia-Vidal, F.J. (2001) Evanescently coupled resonance in surface plasmon enhanced transmission, *Opt. Commun.*, **200**, 1–7.

Kuczynski, J., Thomas, J.K. (1983) Surface effects in the photochemistry of colloidal calcium sulfide, *J. Phys. Chem.*, **87**, 5498–5503.

Lal, M., Levi, L., Kim, K.S., He, G.S., Wang, X., Min, Y.H., Pakatchi, S., Prasad, P.N. (2000) Silica nanobubbles containing an organic dye in a multilayered organic/inorganic heterostructure with enhanced luminiscence, *Chem. Mater.*, **12**, 2632–2639.

Liz-Marzan, L.M. (2004) Nanometals: formation and color, *Mater. Today*, Feb, 26–31.

Lucey, D.W., MacRae, D.J., Furis, M., Sahoo, Y., Cartwright. A.N., Prasad, P.N. (2005) Monodispersed InP quantum dots prepared by colloidal chemistry in a noncoordinating solvent, *Chem. Mater.*, **17**, 3754–3762.

Manoharan, V.N., Elsesser, M.T., Pine, D.J. (2003) Dense packing and symmetry in small clusters of microspheres, *Science*, **301**, 483–487.

McClelland, J.J., Scholten, R.E., Palm, E.C., Celotta, R.J. (1993) Laser focused atomic deposition, *Science*, **262**, 877–880.

McFarland, A.D., Van Duyne, R.P. (2003) Single silver nanoparticles as real-time optical sensors with zeptomole sensitivity, *Nano Lett.*, **3**, 1057–1062.

McGowan, R.W., Giltner, D.M., Lee, S.A. (1995) Light force cooling, focusing, and nanometer-scale deposition of aluminum atoms, *Opt. Lett.*, **20**, 2535–2537.

Mock, J.J., Barbic, M., Smith, D.R., Schultz, D.A., Schultz, S. (2002) Shape effects in plasmon resonance of individual colloidal silver nanoparticles, *J. Chem. Phys.*, **116**, 6755–6759.

Mohamed, M.B., Volkov, V., Link, S., El-Sayed, M.A. (2000) The "lightning" gold nanorods: fluorescence enhancement of over a million compared to the gold metal, *Chem. Phys. Lett.*, **317**, 517–523.

Muhlbach, I., Recknagel, E., Sattler, K. (1981) Inert gas condensation of Sb, Bi and Pb clusters, *Surf. Sci.*, **106**, 188–194.

Munnik, F., Benninger, F., Mikhailov, S., Bertsch, A., Renaud, P., Lorenz, H., Gmür, M. (2003) High aspect ratio, 3D structuring of photoresist materials by ion beam, *LIGA Microelectron. Eng.*, **67–68**, 96–103.

Murray, C.B., Kagan, C.R., Bawendy, M.G. (2000) Synthesis and characterization of monodisperse nanocrystals and close packed nanocrystal assemblies, *Ann. Rev. Mater. Sci.*, **30**, 545–610.

Newton, M.C., Warburton, P.A. (2007) ZnO tetrapod nanocrystals, *Mater. Today*, **10**, 50–54.

Novotny, L., Stranick, S.J. (2006) Near-field optical microscopy and spectroscopy with pointed probes, *Annu. Rev. Phys. Chem.*, **57**, 303–331.

Petit, C., Lixon, P., Pileni, M.P. (1993) In-situ synthesis of silver nanocluster in AOT reverse micelles, *J. Phys. Chem.*, **97**, 12974–12983.

Petrov, Y.I. (1969) Light absorption by Ag, Cu, Al and Se small particles, *Opt. Spectrosc.*, **27**, 665–673.

Prodan, E., Radloff, C., Halas, N.J., Nordlander, P. (2003) A hybridization model for the plasmon response of complex nanostructures, *Science*, **302**, 419–422.

Rechberger, W., Hohenau, A., Leitner, A., Krenn, J.R., Lamprecht, B., Aussenegg, F.R. (2003) Optical properties of two interacting gold nanoparticles, *Opt. Commun.*, **220**, 137–141.

Roldugin, V.I. (2000) Quantum-sized metal colloid systems, *Uspekhi Khimii*, **69**, 899–923.

Rossetti, R., Ellison, J.L, Gibson, J.M., Brus, L.E. (1984) Size effects in the excited electronic states of small colloidal CdS crystallites, *J. Chem. Phys.*, **80**, 4464–4469.

Schmid, G. (1995) Chemical synthesis of large metal clusters and their properties, *Nanostruct. Mater.*, **6**, 15–24.

Sönnichsen, von C. (2001) *Plasmons in Metal Nanostructures*, PhD dissertation (Ludwig-Maximilians University, Munich).

Sun, Y., Xia, Y. (2002) Shape-controlled synthesis of gold and silver nanoparticles, *Science*, **298**, 2176–2179.

Suzdalev, I.P. (2006) Nanotechnology: Physical Chemistry of Nanoclusters, Nanostructures, and Nanomaterials (Komkniga, Moscow).

Timp, G., Behringer, R.E., Tennant, D.M., Cunningham, J.E., Prentiss, M., Berggren, K.K. (1992) Using light as a lens for submicron, neutral-atom lithography, *Phys. Rev. Lett.*, **69**, 1636–1639.

Vargaftik, M.N., Chovilin, A.L., Likholobov, V.A., Stolyarov, I.P., Zagorodnikov, V.P. (1986) Structure of a palladium cluster used as a catalyst for the oxidative acetoxylation of olefins, *Dokl. Phys. Chem. (Engl. Transl.)*, **284**, 896–899.

Vladimirov, V.I., Romanov, A.E. (1986) *Disclination in Crystals* (Nauka, Leningrad), 223 p.

Wang, H., Wu, Y., Lassiter, B., Nehl, C.L., Hafner, J.H., Nordlander, P., Halas, N.J. (2006) Symmetry breaking in individual plasmonic nanoparticles, *PNAS*, **103**, 10856–10860.

Wang, Y., Herron, N. (1987) Optical properties of cadmium sulfide and lead (II) sulfide clusters encapsulated in zeolites, *J. Phys. Chem.*, **91**, 257–260.

Wang, Y., Herron, N. (1988) Photoluminescence and relaxation dynamics of cadmium sulfide superclusters in zeolites., *J. Phys. Chem.*, **92**, 4988–4994.

Zhang, H.F., Wang, C.-M., Buck, E.C., Wang, L.-S. (2003) Synthesis, characterization, and manipulation of helical SiO_2 nanosprings, *Nano Lett.*, **3**, 577–580.

Chapter 3

Introduction to Electrodynamics of Metals

There is a variety of excellent books on electrodynamics of continuous media, including metals (Landau and Lifshitz, 1960; Jackson, 1975). Nevertheless, in this chapter we present fundamental facts of this area for the readers' convenience. We will begin with Maxwell's equations for continuous media, and then we will refer to main facts of the Drude–Sommerfeld theory of metals (Drude, 1900; Sommerfeld and Bethe, 1933; Ashcroft and Mermin, 1976). In the conclusion of the chapter, we will demonstrate that propagation of electromagnetic waves—bulk plasmons—is possible in metals, and then we will discuss their properties.

3.1 Maxwell's Equations and the Propagation of Electromagnetic Waves

Practice has shown that the interaction between metals and light can be reliably described in terms of classical Maxwell's equations. These equations are also completely applicable for nanoparticles and nanostructures, since at room temperature the spacing between electron levels stays tight as compared with thermal excitation

Nanoplasmonics
Vasily Klimov

ISBN 978-981-4267-16-8 (Hardcover), 978-981-4267-42-7 (eBook)
www.panstanford.com

energy $k_B T$ due to high electron density. Thus, the optics of metal nanoparticles considered in this book will be described with the help of the classical Maxwell's equations. Yet, this does not exclude the existence of complex and even unexpected effects owing to electric permittivity dependence on frequency. In fact, even very small particles with an electron-free path becoming comparable with a nanoparticle size can be described by Maxwell's equations. However, in this case one should use modified versus bulk electric permittivity values (see Section 3.4).

It is well known in ordinary life that metals reflect light effectively and thus do not allow light propagation inside it. In microwave and infrared bands, metals are used frequently for fabrication of waveguides and cavities. At these frequencies, Joule losses inside metals are small and approximation of a perfect conductor (with infinite conductivity) is extremely effective. As the frequency increases, the absorption increases, too, and in the optical region the approximation is not quite adequate. In the ultraviolet band, some metals (Na) turn out to be transparent, while others (Ag, Au) are heavily absorbing in this region due to interband transitions.

This complex behavior of metals can be described with the help of electric permittivity $\varepsilon(\omega)$, which is the fundamental notion of the whole book. The physical reason for the strong dependence of electric permittivity on frequency is a change in phase of the currents induced in metal relative to the incident light phase in the frequency region close to the so-called plasmon frequency, which will be defined further (see Eq. (3.32) et seq.).

Before discussing metal optical properties, let us remember Maxwell's equations for continuous media. It is these macroscopic equations that define precisely all the effects and phenomena interesting for us. The main advantage of Maxwell's macroscopic equations is that they do not concern microscopic charges and currents inside a substance, because they vary on the scales that are substantially smaller than nanoparticles' and other bodies' sizes. Peculiarities of transition from microscopic to macroscopic Maxwell's equations can be found in many textbooks (Jackson, 1975; Landau and Lifshitz, 1960).

Thus, let us take macroscopic Maxwell's equations in the form:

$$\operatorname{rot}\mathbf{E} = -\frac{1}{c}\frac{\partial \mathbf{B}}{\partial t}; \quad \operatorname{div}\mathbf{B} = 0$$
$$\operatorname{rot}\mathbf{H} = \frac{1}{c}\frac{\partial \mathbf{D}}{\partial t} + \frac{4\pi}{c}\mathbf{j}; \quad \operatorname{div}\mathbf{D} = 4\pi\rho \tag{3.1}$$

These equations relate macroscopic fields **E** (electric field strength), **D** (electric displacement), **B** (magnetic induction), and **H** (magnetic field strength) to densities of free currents **j** and charges ρ. c is the speed of light in vacuum.

These macroscopic fields are related to each other through electric **P** and magnetic **M** polarization:

$$\mathbf{D} = \mathbf{E} + 4\pi\mathbf{P}; \quad \mathbf{H} = \mathbf{B} - 4\pi\mathbf{M} \tag{3.2}$$

Physically, **P** and **M** polarizations are electric and magnetic dipole moments per unit volume, and in the absence of magnetic polarization internal (microscopic) currents and charges can be found from the equations:

$$\rho_{\text{micro}} = -\operatorname{div}\mathbf{P}; \quad \mathbf{j}_{\text{micro}} = \frac{\partial \mathbf{P}}{\partial t} \tag{3.3}$$

A great advantage of Maxwell's macroscopic equations is that macroscopic fields include all polarization effects.

In the case of linear, uniform, and isotropic media, the equations relating inductions and strengths of fields can be presented in a general form:

$$\mathbf{D}(\mathbf{r}, t) = \int d\mathbf{r}' dt' \varepsilon\left(\mathbf{r} - \mathbf{r}', t - t'\right)\mathbf{E}\left(\mathbf{r}', t'\right)$$
$$\mathbf{B}(\mathbf{r}, t) = \int d\mathbf{r}' dt' \mu\left(\mathbf{r} - \mathbf{r}', t - t'\right)\mathbf{H}\left(\mathbf{r}', t'\right) \tag{3.4}$$

In these equations, $\varepsilon(\mathbf{r}, t)$ and $\mu(\mathbf{r}, t)$ (in general, tensors) are referred to as electric permittivity and magnetic permeability. Dependencies of $\varepsilon(\mathbf{r}, t)$ and $\mu(\mathbf{r}, t)$ on spatial and temporal coordinates are denoted as spatial and temporal dispersions. For the Fourier components, that is, for integrals of the type:

$$\mathbf{D}(\mathbf{k}, \omega) = \int d\mathbf{r} dt e^{-i(\mathbf{kr} - \omega t)}\mathbf{D}(\mathbf{r}, t) \tag{3.5}$$

constitutive equations take a simpler form:

$$\mathbf{D}(\mathbf{k}, \omega) = \varepsilon(\mathbf{k}, \omega)\,\mathbf{E}(\mathbf{k}, \omega)$$
$$\mathbf{B}(\mathbf{k}, \omega) = \mu(\mathbf{k}, \omega)\,\mathbf{H}(\mathbf{k}, \omega) \tag{3.6}$$

Instead of ε and μ, electric and magnetic susceptibilities are frequently used, which relate electric and magnetic polarizations to field strengths:

$$\mathbf{P}(\mathbf{k}, \omega) = \kappa(\mathbf{k}, \omega)\,\mathbf{E}(\mathbf{k}, \omega)$$
$$\mathbf{M}(\mathbf{k}, \omega) = \chi(\mathbf{k}, \omega)\,\mathbf{H}(\mathbf{k}, \omega) \tag{3.7}$$

and thus electric and magnetic susceptibilities are related to electric permittivity and magnetic permeability by the equations:

$$\varepsilon = 1 + 4\pi\kappa$$
$$\mu = 1 + 4\pi\chi \tag{3.8}$$

The last important constitutive equation relates bound (polarization) current to electric field density:

$$\mathbf{j}(\mathbf{k}, \omega) = \sigma(\mathbf{k}, \omega)\,\mathbf{E}(\mathbf{k}, \omega) \tag{3.9}$$

where σ is metal conductivity.

The relation between electric permittivity and conductivity is quite conventional, and apparently the following equation is the most convenient:

$$\varepsilon = \varepsilon' + i\varepsilon'' = \varepsilon' + \frac{4\pi i\sigma}{\omega} \tag{3.10}$$

where ε' and, ε'' are the real and imaginary parts of electric permittivity. With this definition, Eq. (3.9) follows directly from Eq. (3.3).

Above, we have considered regular or "classical" constitutive equations relating polarizations to field strengths. However, for the moment the technologies of manufacture of artificial nanostructured materials, which have been already named "metamaterials," are developing rapidly (Lindell et al., 1994; Vinogradov 2001; Smith et al., 2004; Cai and Shalaev, 2010). In these materials, constitutive equations can have a more complex form than Eq. (3.7) or Eq. (3.6). For example, inductions can be defined both by electric and magnetic field strengths:

$$\mathbf{D}(\mathbf{k}, \omega) = \varepsilon(\mathbf{k}, \omega)\,\mathbf{E}(\mathbf{k}, \omega) + [\xi(\mathbf{k}, \omega)\,\mathbf{H}(\mathbf{k}, \omega)]$$
$$\mathbf{B}(\mathbf{k}, \omega) = \mu(\mathbf{k}, \omega)\,\mathbf{E}(\mathbf{k}, \omega) + [\zeta(\mathbf{k}, \omega)\,\mathbf{H}(\mathbf{k}, \omega)] \tag{3.11}$$

In media of this type, a richer assortment of effects arises, and they are investigated intensively at the present time (see Chapter 11 and Chapter 13, Sections 7 and 8).

Usually (but not always), one may neglect spatial dispersion and write constitutive equations as:

$$\mathbf{D}(\mathbf{k},\omega) = \varepsilon(\omega)\,\mathbf{E}(\mathbf{k},\omega)$$
$$\mathbf{B}(\mathbf{k},\omega) = \mu(\omega)\,\mathbf{H}(\mathbf{k},\omega) \qquad (3.12)$$

Neglecting the spatial dispersion means that substance response to external fields is spatially localized, that is, does not depend on the fields of neighboring points.

In the optical frequencies region, complex electric permittivity $\varepsilon = \varepsilon' + i\varepsilon''$ of a nonmagnetic medium can be determined, for example, through examination of properties of the light reflected by the surface of a metal or other substance and through measurement of the complex refractive index $\tilde{n}$:

$$\tilde{n} = \sqrt{\varepsilon} = n + i\kappa \qquad (3.13)$$

In Eq. (3.13), n is a usual refractive index and κ is an extinction coefficient that determines optical absorption.

From Eq. (3.13), one can easily find the obvious relations:

$$\varepsilon' = n^2 - \kappa^2$$
$$\varepsilon'' = 2n\kappa \qquad (3.14)$$

and vice versa:

$$n^2 = \frac{\varepsilon'}{2} + \frac{1}{2}\sqrt{\varepsilon'^2 + \varepsilon''^2}$$
$$\kappa = \frac{\varepsilon''}{2n} \qquad (3.15)$$

The extinction coefficient κ is related to the absorption coefficient in the Bouguer–Lambert–Beer law describing exponential decrease of light intensity in media:

$$\mathrm{I}(x) = \mathrm{I}_0 \mathrm{e}^{-\alpha x}$$

by the relation:

$$\alpha(\omega) = \frac{2\kappa(\omega)\,\omega}{c} \qquad (3.16)$$

Thus, the imaginary part of electric permittivity defines absorption in a substance. The time-averaged energy absorption (dissipation) rate is defined by the field work over induced currents (Landau and Lifshitz, 1960):

$$\frac{dQ}{dVdt} = \frac{1}{2}\mathrm{Re}\left(\mathbf{j}\left(\mathbf{r},\omega\right)\mathbf{E}^*\left(\mathbf{r},\omega\right)\right)$$
$$= \frac{1}{2}\sigma\left(\omega\right)\left|\mathbf{E}\left(\mathbf{r},\omega\right)\right|^2 = \frac{\omega}{8\pi}\varepsilon''\left|\mathbf{E}\left(\mathbf{r},\omega\right)\right|^2 \quad (3.17)$$

In the case of low losses $\left|\varepsilon'\right| \gg \left|\varepsilon''\right|$, the real part of the complex refractive index characterizes the change of phase velocity of wave propagation in a medium, c_{m}, in comparison with velocity of light in vacuum, c:

$$c_{\mathrm{m}} = c/n \quad (3.18)$$

After the inductions and field strengths have been related in a practically significant case of nonmagnetic media without sources, one can combine the Maxwell's equations and obtain the following equation:

$$\mathrm{rot\,rot}\,\mathbf{E}\left(\mathbf{r},t\right) = -\frac{\partial^2\mathbf{D}\left(\mathbf{r},t\right)}{c^2\partial t^2} \quad (3.19)$$

which after the space-time Fourier transform

$$\mathbf{E}\left(\mathbf{k},\omega\right) = \int d\mathbf{r}dt\exp\left(i\mathbf{kr} + i\omega t\right)\mathbf{E}\left(\mathbf{r},t\right) \quad (3.20)$$

can be written as:

$$-\mathbf{k}\left(\mathbf{kE}\left(\mathbf{k},\omega\right)\right) + k^2\mathbf{E}\left(\mathbf{k},\omega\right) = \varepsilon\left(\omega,\mathbf{k}\right)\frac{\omega^2}{c^2}\mathbf{E}\left(\mathbf{k},\omega\right) \quad (3.21)$$

The solutions of Eq. (3.21) are divided into two classes, depending on electric field polarization. For transverse waves, that is, waves with wavevector perpendicular to the electric field strength, $\mathbf{kE}\left(\mathbf{k},\omega\right) = 0$, Eq. (3.21) takes the form:

$$k^2\mathbf{E}\left(\mathbf{k},\omega\right) = \varepsilon\left(\omega,\mathbf{k}\right)\frac{\omega^2}{c^2}\mathbf{E}\left(\mathbf{k},\omega\right) \quad (3.22)$$

and wave propagation is possible only when the dispersion relation

$$k^2 = \varepsilon\left(\omega,\mathbf{k}\right)\left(\frac{\omega}{c}\right)^2 \quad (3.23)$$

is satisfied.

For longitudinal waves, the wavevector and field strength are parallel, $\mathbf{E} \parallel \mathbf{k}$, and Eq. (3.21) reduces to

$$\varepsilon(\omega, \mathbf{k}) \frac{\omega^2}{c^2} \mathbf{E}(\mathbf{k}, \omega) = 0 \tag{3.24}$$

where a nontrivial solution is possible only when the following condition is met:

$$\varepsilon(\omega, \mathbf{k}) = 0 \tag{3.25}$$

which implies that longitudinal oscillations are possible only at the frequencies for which electric permittivity is equal to zero.

In the presence of spatial dispersion, generally speaking, electric permittivity is a tensor ε_{ik}, even in an isotropic medium, and preferential direction is formed by the wavevector $\mathbf{k}$. When a medium is not only isotropic but also has an inversion center, the tensor ε_{ik} can be composed of the components of the $\mathbf{k}$ vector and the Kronecker tensor δ_{ik}. The general form of this tensor can be expressed as:

$$\varepsilon_{ik}(\omega, \mathbf{k}) = \varepsilon_t(\omega, k)\left(\delta_{ik} - \frac{k_i k_k}{k^2}\right) + \varepsilon_l(\omega, k)\frac{k_i k_k}{k^2} \tag{3.26}$$

where $\varepsilon_l(\omega, k)$ and $\varepsilon_t(\omega, k)$ depend only on the absolute value of the wavevector (and on frequency ω). If the electric field strength $\mathbf{E}$ is directed along the wavevector, the induction is $\mathbf{D} = \varepsilon_l \mathbf{E}$, and if $\mathbf{E} \perp \mathbf{k}$, then $\mathbf{D} = \varepsilon_t \mathbf{E}$. Accordingly, the values $\varepsilon_l(\omega, k)$ and $\varepsilon_t(\omega, k)$ are denoted as longitudinal and transverse electric permittivities. Thus, the presence of spatial dispersion enables propagation of both transverse and longitudinal modes in a medium, governed by the dispersion laws:

$$\begin{aligned} k_t^2 &= \left(\frac{\omega}{c}\right)^2 \varepsilon_t(\omega, k_t) \\ \varepsilon_l(\omega, k_l) &= 0 \end{aligned} \tag{3.27}$$

correspondingly. The specific expressions of $\varepsilon_l(\omega, k)$ and $\varepsilon_t(\omega, k)$ are stated below in the end of the next section. Effects of the spatial dispersion are to be taken into account when considering nanoparticles of less than 10 nm size (see Chapter 6, Section 6.5).

3.2 The Drude–Sommerfeld Theory of Optical Properties of Metals

To explain optical properties of metals, one is to use the solid-state (condensed matter) theory. In particular, in a rather wide-frequency region optical properties of metals can be explained with the help of the plasma model or the free-electron model (Drude, 1900; Sommerfeld and Bethe, 1933; Ashcroft and Mermin, 1976), where the gas of free electrons moves relative to a positively charged crystal lattice. For alkaline metals, the domain of applicability of these models extends to the ultraviolet region, while for noble metals interband transitions take place in a visible region and restrict the domain of applicability of the simple plasma model. In the model of free electrons, inhomogeneity of the lattice potential and electron–electron interaction are ignored, and the whole influence of band theory is reduced to input of the effective electron mass m^* instead of the free-electron mass (the quasi-free-electrons model). In external electromagnetic field $\mathbf{E}$, the equation of the electron motion is of the form:

$$m^* \left(\ddot{\mathbf{x}} + \gamma \dot{\mathbf{x}}\right) = -e\mathbf{E}\left(t\right) \tag{3.28}$$

where $\gamma = 1/\tau$ is the frequency of electrons collisions. The value $\tau = 1/\gamma$ is called relaxation time of free-electron gas and equals 10^{-14} s by the order of magnitude.

In the case of a monochromatic external field $\mathbf{E}\left(t\right) = \mathbf{E}_0 e^{-i\omega t}$, the solution of the Eq. (3.28) has the form:

$$\mathbf{x}\left(t\right) = \frac{e}{m^* \left(\omega^2 + i\omega\gamma\right)} \mathbf{E}\left(t\right) \tag{3.29}$$

Electron displacement $\mathbf{x}\left(t\right)$ leads to the origin of a dipole moment of macroscopic polarization:

$$\mathbf{P} = -n_0 e\mathbf{x} = -\frac{n_0 e^2}{m^* \left(\omega^2 + i\omega\gamma\right)} \mathbf{E} \tag{3.30}$$

where n_0 is the concentration of conduction electrons.

It is apparent from Eq. (3.30) that the susceptibility κ of metal is defined by the relation:

$$\kappa = -\frac{n_0 e^2}{m^* \left(\omega^2 + i\omega\gamma\right)} \tag{3.31}$$

Table 3.1 Drude parameters for noble metals Au, Ag, and Cu. (Sönnichsen, 2001)

	Au	Ag	Cu
ω_{pl}, eV	9.1	9.1	8.8
τ, fs	29	40	40

Correspondingly, the electric permittivity $\varepsilon = 1 + 4\pi\kappa$ takes the form:

$$\varepsilon(\omega) = 1 - \frac{\omega_{\text{pl}}^2}{\omega^2 + i\omega\gamma} \tag{3.32}$$

where $\omega_{\text{pl}}^2 = 4\pi n_0 e^2/m^*$ is plasmon frequency of the free electron gas. The real and imaginary components of electric permittivity are of the form:

$$\varepsilon' = 1 - \frac{\omega_{\text{pl}}^2\tau^2}{1 + \omega^2\tau^2}$$

$$\varepsilon'' = \frac{\omega_{\text{pl}}^2\tau}{\omega\left(1 + \omega^2\tau^2\right)} \tag{3.33}$$

The relaxation time τ can be found from measurement of conductivity at low frequencies (direct current) by using the $\tau = \sigma m^*/\left(n_0 e^2\right)$ relation. All the parameters sufficient for Au, Ag, and Cu description within the framework of the Eq. (3.32) model (Drude parameters) are stated in Table 3.1 (Sönnichsen, 2001).

Knowing the electric permittivity behavior at different frequencies can be useful. In a high-frequency region, that is, when $\omega\tau \gg 1$, the decrement is rather small and electric permittivity turns out to be a real value:

$$\varepsilon = 1 - \frac{\omega_{\text{pl}}^2}{\omega^2} \tag{3.34}$$

Let us remark that the imaginary part of electric permittivity cannot be ignored for noble metals due to interband transitions (see below). In the case of low frequencies $\omega \ll \gamma$, $\varepsilon'' \gg \varepsilon'$, both real and imaginary parts of the complex refractive index turn out to be of the same order:

$$n \approx \kappa = \sqrt{\frac{\varepsilon''}{2}} = \sqrt{\frac{\omega_{\text{pl}}^2\tau}{2\omega}} \tag{3.35}$$

In this region, metals are generally absorbing with the absorption index of:

$$\alpha = \sqrt{\frac{2\omega_{\mathrm{pl}}^2 \tau \omega}{c^2}} \tag{3.36}$$

Taking the conductivity as $\sigma_0 = ne^2\tau/m^* = \omega_{pl}^2\tau/4\pi$, one can write the equation in the form:

$$\alpha = \frac{2}{c}\sqrt{2\pi\sigma_0\omega} \tag{3.37}$$

Application of the Burger–Lambert–Beer absorption law leads to the conclusion that field density is decreasing as follows:

$$\mathbf{E} \sim e^{-z/\delta} \tag{3.38}$$

where the thickness δ of so-called skin depth is:

$$\delta = \frac{c}{\sqrt{2\pi\sigma_0\omega}} \tag{3.39}$$

A more rigorous consideration of a low-frequency region in terms of the Boltzmann equation shows that this simple description is actually valid as long as the electron-free path length is small relative to the skin layer thickness:

$$l = v_{\mathrm{F}}\tau \ll \delta \tag{3.40}$$

where v_{F} is the Fermi velocity. In Table 3.2, values of the Fermi velocity for typical metals are stated.

At room temperature, $l \approx 10$ nm, $\delta \sim 100$ nm, and Eq. (3.40) is fulfilled, thus confirming correctness of the simple model of free electrons. However, at low temperatures the mean free path of electrons can increase by several orders of magnitude, leading to changes of the depth of field penetration in metal. This phenomenon is called the anomalous skin effect. For nanoparticles, the Drude model can be applied with precaution because the particles' size can be smaller than the free path of electron (see Section 3.4).

Table 3.2 Fermi velocities for typical metals at room temperatures (Ashcroft and Mermin, 1976)

Metal	Li	Na	K	Rb	Cs	Cu	Ag	Au
v_{F}, 10^8 cm/s	1.29	1.07	0.86	0.81	0.75	1.57	1.39	1.40

Due to simple analytical dependency on frequency, the Drude–Sommerfeld theory can be effectively used both in analytical and in numerical investigations of monochromatic plasmon oscillations in nanostructures and nanoparticles. Moreover, this theory can be also applied for investigations of plasmon oscillations in the time domain using Eq. (3.28) The last circumstance is especially important for solution of Maxwell's equations by the finite-difference time domain (FDTD) scheme (Kashiva and Fukai, 1990, see also Appendix A2).

The Drude–Sommerfeld model requires a more precise description when consideration of effects of spatial dispersion is necessary. In this case, it is usually enough to modify the expressions for longitudinal electric permittivity (see Eqs. (3.26) and (3.32)):

$$\varepsilon_l\left(\mathbf{k},\omega\right)=1-\frac{\omega_{\mathrm{pl}}^2}{\omega\left(\omega+i\gamma\right)}\Rightarrow 1-\frac{\omega_{\mathrm{pl}}^2}{\omega\left(\omega+i\gamma\right)-\beta^2k^2} \qquad (3.41)$$

where $\beta^2 = 3/5v_{\mathrm{F}}^2$, v_{F} is the Fermi velocity of electrons in metal (see Table 3.2). The complete theory of electric permittivity taking into account the spatial dispersion can be found, for instance, in (Lindhard, 1954).

3.3 Optical Properties of Real Metals

In the model of free electrons, Eq. (3.32), electric permittivity tends to unity at high frequencies $\omega \gg \omega_{\mathrm{pl}}$. However, for noble metals (Au, Ag, Cu) a contribution of an ion lattice turns out to be substantial, and the more suitable expression for the electric permittivity of these significant for plasmonic metals has the form:

$$\varepsilon=\varepsilon_\infty-\frac{\omega_{\mathrm{pl}}^2}{\omega^2+i\omega\gamma} \qquad (3.42)$$

where the contribution of bound electrons and lattice is described by ε_∞. This expression is useful for not very large frequencies. In the asymptotic case, $\omega/\omega_{\mathrm{pl}} \to \infty$ the model of free electrons should be used again.

The parameters of the generalized Drude model, Eq. (3.42), found by comparison of experimental data for optical frequencies, are shown in Table 3.3 (Sönnichsen, 2001).

Table 3.3 Drude parameters for Au and Ag in the optical frequencies region (Sönnichsen, 2001)

	Au	Ag
ω_{pl}, *eV*	9.0	9.1
τ, *fs*	10	36
ε_∞	9.84	3.7

It is important to note that values of the parameters in Table 3.3 differ from their values based on conductivity in the low-frequency region (see Table 3.1).

Applicability limits of the generalized expression Eq. (3.42) can be seen from Figs. 3.1 and 3.2, where experimental values of electric permittivity for gold and silver (Johnson and Christy, 1972) and their approximations by means of Eq. (3.42) are shown. The figures display that in the region of rather high frequencies ($\hbar\omega > 2$ eV for gold and $\hbar\omega > 4$ eV for silver), even the generalized model of free electrons describes the properties of real metals poorly.

To take interband transitions into account phenomenologically, one may add the additional equation for bound electron displacements **y** to Eq. (3.28):

$$m_0\left(\ddot{\mathbf{y}} + \gamma_0\dot{\mathbf{y}} + \omega_0^2\mathbf{y}\right) = -e\mathbf{E} \tag{3.43}$$

where ω_0 is the frequency of interband transitions $\omega_0 = \Delta E/\hbar$.

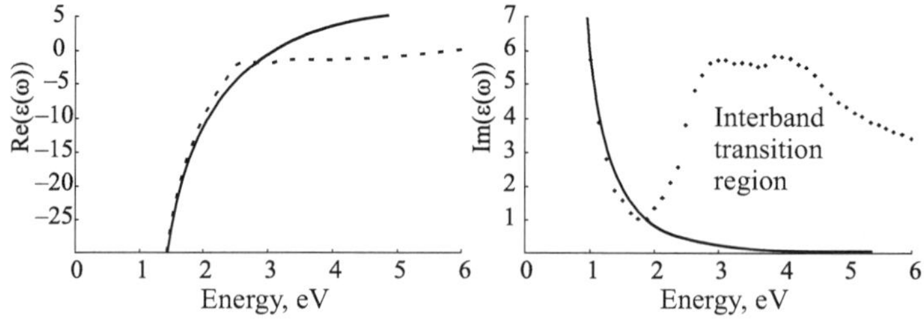

Figure 3.1 Electric permittivity of gold as a function of light frequency: the dotted line shows experimental data from Johnson and Christy (1972), while the solid line represents the generalized model of free electrons, Eq. (3.42). Interband transitions for $\hbar\omega > 2$ eV result in incorrectness of the model in this region.

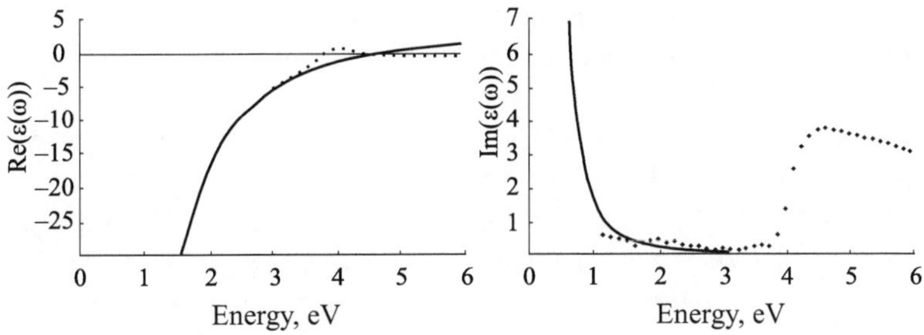

Figure 3.2 Electric permittivity of silver as a function of light frequency: the dotted line shows experimental data from Johnson and Christy (1972), while the solid line represents the generalized model of free electrons, Eq. (3.42). Interband transitions for $\hbar\omega > 4$ eV result in incorrectness of the model in this region.

As a result, effective electric permittivity of metals can be defined by an expression of the type:

$$\varepsilon(\omega) = \varepsilon_{\infty} - \frac{\omega_{pl}^2}{\omega^2 + i\omega\gamma} - \frac{A_0}{\omega^2 + i\omega\gamma_0 - \omega_0^2} - \ldots \tag{3.44}$$

where more resonance terms can be added, if needed. Thus, this generalization is based on the assumption that both free and bound electrons contribute to polarization of metal.

3.4 Electric Permittivity of Small Particles

The presented above Drude parameters and experimental values of electric permittivity were defined for samples of macroscopic scale (so-called bulk values). In the case of nanoparticles, the length of the electron free path is comparable to a particle's size, and conduction electrons scattering on the surface will provoke increase of collision frequency and decrease of relaxation time. This effect is especially significant for nanoparticles characterized by a high ratio between the surface area and volume. The increase is proportional to the Fermi velocity v_F and can be described by the empirical formula (Kawabata and Kubo, 1966; Genzel and Kreibig, 1980):

$$\tau_{\text{effective}}^{-1} = \tau^{-1} + C\frac{v_F}{r} \tag{3.45}$$

where r is the effective radius of a nanoparticle and C ~ 1 is the dimensionless phenomenological parameter. In any case, the reduction of the particles' size results in an increase of the imaginary part of the electric permittivity and, correspondingly, in greater losses in the nanoparticle. Within the framework of the correction Eq. (3.45), the macroscopic description remains valid down to 2 nm size order. When nanoparticles' sizes are being reduced further, effects of size quantization become significant, that is, the de Broglie wavelength of electrons becomes comparable to the size of the nanoparticle and the band structure becomes qualitatively different. In the present book, properties of such particles are not considered.

Regarding small nanoparticles (smaller than 10 nm), one should also consider the effects of spatial dispersion, Eq. (3.41). An example of this consideration in the case of spherical nanoparticles will be given in Section 6.5.

3.5 Dispersion in Free-Electron Gas and Bulk Plasmons

After the optical properties of metals have been determined, one can describe propagation of waves in this substance with the help of Maxwell's macroscopic equations stated in Section 3.1.

In the case of a metal infinite in all dimensions, the equation of monochromatic transverse ($\operatorname{div}\mathbf{E} = 0$) electromagnetic oscillations has the form:

$$\omega^2 \varepsilon(\omega) = c^2 k^2 \tag{3.46}$$

In the case of the Drude dispersion law, the expression has the form:

$$\varepsilon(\omega) = 1 - \frac{\omega_{\text{pl}}^2}{\omega^2} \tag{3.47}$$

and the solution of the dispersion Eq. (3.46) has the form:

$$\omega^2 = \omega_{\text{pl}}^2 + c^2 k^2 \tag{3.48}$$

or

$$\omega = \sqrt{\omega_{\text{pl}}^2 + c^2 k^2} \tag{3.49}$$

where the wavevector k has an arbitrary value. The solution, Eq. (3.49), is shown in Fig. 3.3.

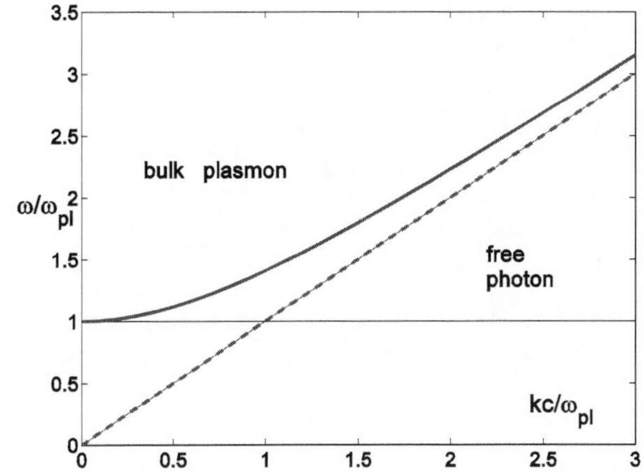

Figure 3.3 The dispersion law for bulk transverse plasmons, Eq. (3.49). The red dashed line corresponds to the dispersion of free photons, $\omega = ck$.

The figure displays that when the wavevectors are small, the frequency of bulk plasmon oscillations does not become zero, like electromagnetic oscillations do in free space, but tends to a certain constant, which is called bulk plasmon frequency (see Eq. (3.32)). The other peculiarity of the dispersion law Eq. (3.48) is that electric permittivity takes a positive value (as it is shown in Fig. 3.4) when the law Eq. (3.49) is fulfilled, and thus in this frequency band metal is transparent and similar to dielectric.

Plasmon oscillations characterized by the dispersion law Eq. (3.48) are called transverse bulk plasmons, and quasiparticles of the same name with energy $\hbar\omega(\mathbf{k})$ may be assigned to them according to laws of quantum mechanics.

Longitudinal plasmons can also exist in metals. The dispersion equation for these plasmons is crucially related to spatial dispersion in metal, that is, to dependence of electric permittivity on the wavevector:

$$\varepsilon(\omega, \mathbf{k}) = 0 \tag{3.50}$$

The theory of longitudinal plasmons is represented, for example, in (Kittel, 2005; Lindhard, 1954). It implies that the explicit form of the dispersion law for longitudinal plasmons is:

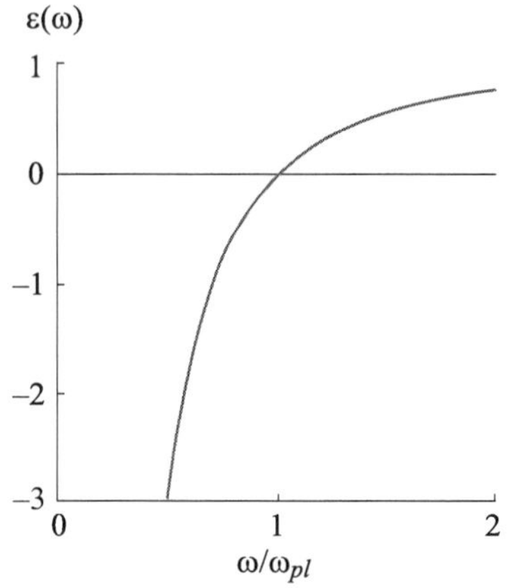

Figure 3.4 Electric permittivity dependence in the Drude model on frequency, Eq. (3.34). Bulk plasmons exist in the transparency region $\omega/\omega_{\text{pl}} > 1$.

$$\omega^2 = \omega_{\text{pl}}^2 + \frac{3v_{\text{F}}^2 k^2}{5} \tag{3.51}$$

where v_{F} is the Fermi velocity.

Plasmon oscillations characterized by the dispersion laws Eqs. (3.50) and (3.51) are called longitudinal bulk plasmons, and quasiparticles of the same name may be assigned to them according to laws of quantum mechanics.

In the region of small wavevectors, $\mathbf{k} \to 0$, the difference between transverse and longitudinal bulk plasmons disappears and both types of plasmons come out to be related to homogeneous (independent of coordinates) electron oscillations around the lattice at ω_{pl} frequency. These oscillations can be easily explained physically.

Indeed, let us consider a metal layer of an arbitrary thickness L and an area S in which all electrons as a whole are displaced by a distance u (Fig. 3.5). In this case, on the lower and top surfaces of the metal surface charges will appear:

$$\sigma_S = \pm e n_0 u \tag{3.52}$$

These charges induce the electric field directed along the displacement of electrons:

$$E = 4\pi\sigma_S = 4\pi e n_0 u \tag{3.53}$$

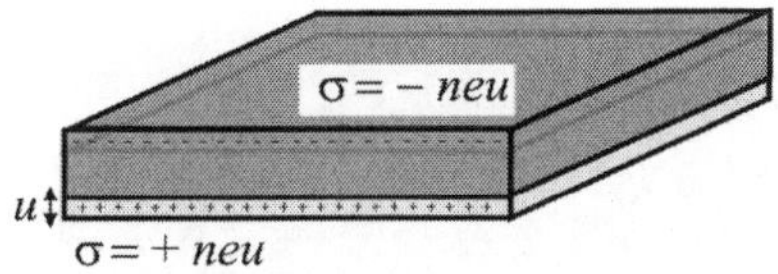

Figure 3.5 Model of homogeneous plasmon oscillations.

Thus, the equation of free oscillations of all the electrons regarded as a whole takes the form:

$$LSn_0m\ddot{u} + en_0LSE = LSn_0m\left(\ddot{u} + \frac{4\pi e^2 n_0}{m}u\right) = 0 \qquad (3.54)$$

Eq. (3.54) displays that the electron gas will oscillate with frequency $\omega_{\rm pl}^2 = 4\pi e^2 n_0/m$, owing to electrostatic forces.

This implies that the $\omega_{\rm pl}$ frequency is the eigenfrequency of electrons oscillations in a metal layer of an arbitrary thickness. In more complex geometries, the oscillations frequencies can differ but their physical meaning remains the same, and it is they that determine the variety of plasmon properties of metal nanoparticles.

Bulk plasmons can be excited by passing charged particles through a foil. In this case, the absorption spectrum will have peaks at frequencies proportional to $\hbar\omega_{\rm pl}$. For metals, $\hbar\omega_{\rm pl} \sim 10$ eV, while for semiconductors $\hbar\omega_{\rm pl} < 0.5$ eV, depending on the doping level. It is important to remember that bulk plasmons were discovered in the experiments of this kind (see, e.g., (Pines, 1963)).

References

Ashcroft, N., Mermin, N. (1976) *Solid State Physics* (Holt, Rinehart and Winston, New York).

Cai, M., Shalaev, V. (2010) *Optical Metamaterials/Fundamentals and Applications* (Springer, New York).

Drude, P., (1900) Zur Elektronentheorie der Metalle, *Ann. Phys.*, **306**, 566–613.

Genzel, L., Kreibig, U. (1980) Dielectric function and infrared absorption of small metal particles, *Z. Phys. B*, **37**, 93–101.

Jackson, J.D. (1975) *Classical Electrodynamics* (Wiley, New York).

Johnson, P., Christy, R. (1972) Optical constants of the noble metals, *Phys. Rev. B*, **6**, 4370–4379.

Kashiwa, T., Fukai, I. (1990) A treatment by the FD-TD method of the dispersive characteristics associated with electronic polarization, *Microwave Opt. Technol. Lett.*, **3**(6), 203–205.

Kawabata, A., Kubo, R. (1966) Electronic properties of fine metallic particles. II. Plasma resonance absorption, *J. Phys. Soc. Japan*, **21**, 1765–1772.

Kittel, C. (2005) *Introduction to a Solid State Physics* (Wiley, New York).

Kopitzki, K. (1993) *Einführung in die Festkörperphysik* (Teubner, Stuttgart).

Landau, L.D., Lifshitz, E.M. (1960) *Electrodynamics of Continuous Media* (Addison-Wesley).

Lindell, I.V., Sihvola, A.H., Tretyakov, S.A., Viitanen, A.J. (1994) *Electromagnetic Waves in Chiral and Bi-Anisotropic Media* (Artech House, Boston, London).

Lindhard, J. (1954) On the properties of a gas of charged particles, Kgl. Danske Videnskab. Selskab, *Mat.-Fys. Medd.*, **28**, 1–57.

Pines, D. (1963) Elementary Excitations in Solids: lectures on phonons, electrons, and plasmons (W. A. Benjamin, New York), p. 299.

Smith, D.R., Pendry, J.B., Wiltshire, M.C.K. (2004) Metamaterials and negative refractive index, *Science*, **305**, 788–792.

Sommerfeld, A., Bethe, H. (1933) Elektronentheorie der Metalle, in *Handbuch der Physik*, Vol. 24-2 (Springer-Verlag, Heidelberg), pp. 333–622.

Sönnichsen, von C. (2001) *Plasmons in Metal Nanostructures*, PhD thesis (Ludwig-Maximilians University, Munich).

Vinogradov, A.P. (2001) *Electrodynamics of Composite Materials* (URSS Editorial, Moscow) (in Russian).

Chapter 4

Surface Plasmons

Bulk plasmons, considered in the previous chapter, can exist only in the transparency region of metals and cannot be localized strongly, since at a given frequency their wavelengths are greater than light wavelengths in free space. In this respect, plasmon oscillations of this kind are irrelevant to nanoplasmonics. More interesting plasmon oscillations arise on the metal–dielectric interface. As it will be shown further, plasmons of this kind can exist in the opacity region ($\varepsilon' < 0$) of metal and have a strong spatial localization at optical frequencies. Such oscillations can occur in two dimensions (e.g., planar metal–dielectric interface) and even in one dimension (metal nanowire). Plasmons of this type are called surface plasmons, and it is their properties that will be considered in this chapter.

4.1 Two-Dimensional Surface Plasmons

Surface waves, now designated as "surface plasmons" (Stern and Ferrell, 1960), were investigated in the beginning of the 20th century (Zenneck, 1907; Sommerfeld, 1909) with respect to wireless telegraphy and therefore are called Zenneck or Sommerfeld waves sometimes. These investigations were carried out in the region of

Nanoplasmonics
Vasily Klimov

ISBN 978-981-4267-16-8 (Hardcover), 978-981-4267-42-7 (eBook)
www.panstanford.com

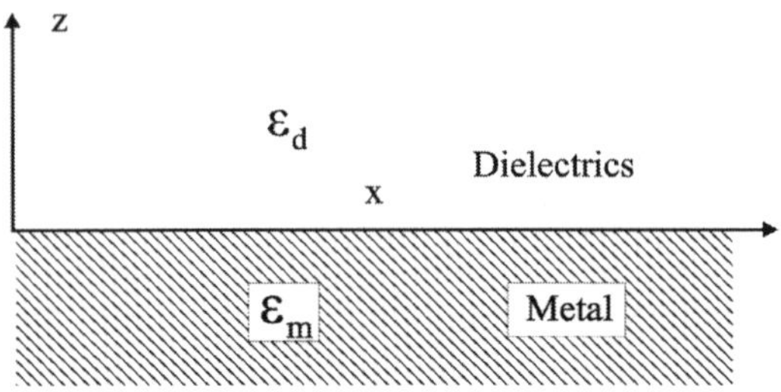

Figure 4.1 The geometry of surface plasmons on a metal–dielectric interface. The waves propagate along the x-axis of the Cartesian coordinate system.

radiofrequencies that are substantially smaller than optical ones. In this frequency band, metal properties are described well by the Drude free-electron model, while surface waves hardly differ from waves in free space. But we are considering the optical frequency region, and many actual results achieved at that time are inapplicable in this case. On the other hand, many mathematical results stay definitely valid for any frequencies and form the basis of any analysis of surface waves.

4.1.1 *Surface Plasmons on a Plane Metal–Dielectric Interface*

The simplest case is one of the metal–dielectric planar interface (Fig. 4.1). In this case, a homogeneous system of Maxwell's equations has the solution:

$$\mathbf{E}_{\mathrm{m}} = E_0 \left\{1,\, 0,\, k_{\mathrm{sp}}/k_{\mathrm{zm}}\right\} \exp\left\{i\left(k_{\mathrm{sp}}x - k_{\mathrm{zm}}z - \omega t\right)\right\}$$
$$\mathbf{H}_{\mathrm{m}} = -E_0 \frac{\varepsilon_{\mathrm{m}}}{k_{\mathrm{zm}}}\left(\frac{\omega}{c}\right)\{0,\, 1,\, 0\} \exp\left\{i\left(k_{\mathrm{sp}}x - k_{\mathrm{zm}}z - \omega t\right)\right\} \quad (4.1)$$

in the metal-filled half-space, and

$$\mathbf{E}_{\mathrm{d}} = E_0 \left\{1,\, 0,\, -k_{\mathrm{sp}}/k_{\mathrm{zd}}\right\} \exp\left\{i\left(k_{\mathrm{sp}}x + k_{\mathrm{zd}}z - \omega t\right)\right\}$$
$$\mathbf{H}_{\mathrm{d}} = E_0 \frac{\varepsilon_{\mathrm{d}}}{k_{\mathrm{zd}}}\left(\frac{\omega}{c}\right)\{0,\, 1,\, 0\} \exp\left\{i\left(k_{\mathrm{sp}}x + k_{\mathrm{zd}}z - \omega t\right)\right\} \quad (4.2)$$

in the dielectric-filled half-space. In Eqs. (4.1) and (4.2),

$$k_{\mathrm{zm}} = \sqrt{\varepsilon_{\mathrm{m}}(\omega)\left(\frac{\omega}{c}\right)^2 - k_{\mathrm{sp}}^2}, \quad k_{\mathrm{zd}} = \sqrt{\varepsilon_{\mathrm{d}}(\omega)\left(\frac{\omega}{c}\right)^2 - k_{\mathrm{sp}}^2} \quad (4.3)$$

are the wavevector components in metals (k_{zm}) or in dielectrics (k_{zd}) transverse to the interface. These components are purely imaginary and provide localization of the surface plasmons near the interface.

This solution is usually referred to as surface plasmon polaritons (SPPs), or simply surface plasmons. It is essential that in this system the surface plasmons have only one polarization (transverse magnetic [TM]).

From the condition of continuity of tangential components of electric and magnetic field strengths, one can easily find the dispersion equation:

$$\frac{\varepsilon_{\mathrm{m}}}{k_{z\mathrm{m}}} + \frac{\varepsilon_{\mathrm{d}}}{k_{z\mathrm{d}}} = 0 \tag{4.4}$$

allowing existence and propagation of sourceless waves.

Combining Eqs. (4.4) and (4.3), one can represent the dispersion equation in the explicit form:

$$k_{\mathrm{sp}}(\omega) = \frac{\omega}{c}\sqrt{\frac{\varepsilon_{\mathrm{m}}(\omega)\,\varepsilon_{\mathrm{d}}(\omega)}{\varepsilon_{\mathrm{m}}(\omega) + \varepsilon_{\mathrm{d}}(\omega)}} \tag{4.5}$$

where the longitudinal wavevector of surface plasmon, k_{sp}, is an explicit function of frequency. When the dispersion equation, Eq. (4.5), is satisfied, the wavevector components directed perpendicularly to interface may be written as:

$$k_{z,j}^2 = \frac{\varepsilon_j^2}{\varepsilon_{\mathrm{m}} + \varepsilon_{\mathrm{d}}}\left(\frac{\omega}{c}\right)^2, \; j = \mathrm{m,d} \tag{4.6}$$

One should mention that often there is no need to solve Maxwell's equations for every new geometry to find corresponding dispersion equations for SPPs. If the reflection or refraction coefficients are known for some geometry, the dispersion laws can be obtained automatically by equating the denominators of the coefficients to zero. In particular, the reflection coefficient of p-polarized waves from two media interfaces is specified by the well-known Fresnel formula:

$$r_{12}^{p} = \frac{\varepsilon_2 k_{z1} - \varepsilon_1 k_{z2}}{\varepsilon_2 k_{z1} + \varepsilon_1 k_{z2}} \tag{4.7}$$

Equating the denominator of Eq. (4.7) to zero, we obtain Eq. (4.4) immediately.

After obtaining Eqs. (4.5) and (4.6), one can consider the conditions defining the existence of surface polaritons. For simplicity, let us suppose that we can ignore imaginary parts of dielectric functions, that is, losses. We are looking for propagating waves. That means, that the component of the wavevector lying along the interface, k_{sp}, is to be real, and the radicand in Eq. (4.5) is to be positive.

On the other hand, we are to find the waves localized near the interface. Therefore, the wavevector component, normal to the surface, is to be purely imaginary. In this case, the exponential decay of fields takes place while moving from the interface. This can be achieved only in the case of the negative denominator of Eq. (4.6). Combining these conditions, we come to the conclusion that localized surface plasmons can exist only when the conditions

$$\varepsilon_{\mathrm{m}}(\omega) \cdot \varepsilon_{\mathrm{d}}(\omega) < 0, \qquad \varepsilon_{\mathrm{m}}(\omega) + \varepsilon_{\mathrm{d}}(\omega) < 0, \tag{4.8}$$

are fulfilled.

Eq. (4.8) implies that the existence of surface plasmons is possible when one of the dielectric functions is negative and exceeds another in absolute value. As it was shown in the previous chapter, metals, especially noble metals like gold and silver, have a greater negative electric permittivity and a rather small imaginary part. Thus, on the interface of metal and air or glass, surface plasmons can exist.

The dispersion law (Eq. (4.5)) on the metal–air interface is illustrated in Fig. 4.2. The figure shows that when a metal, characterized by the Drude dispersion law

$$\varepsilon_{m}(\omega) = 1 - \frac{\omega_{\mathrm{pl}}^{2}}{\omega^{2}} \tag{4.9}$$

fills the $z < 0$ half-space and a dielectric fills the $z > 0$ half-space, two types of eigenoscillations can exist in the system. One of them with $\omega > \omega_{\mathrm{pl}} \quad (\varepsilon_{\mathrm{m}}(\omega) > 0)$ is a modification of bulk plasmons. The only distinction is that in the short wavelength limit $k_{\mathrm{sp}} \to \infty$, the asymptote of the dispersion equation is:

$$\omega = ck_{\mathrm{sp}}\sqrt{\frac{\varepsilon_{\mathrm{d}} + 1}{\varepsilon_{\mathrm{d}}}} \tag{4.10}$$

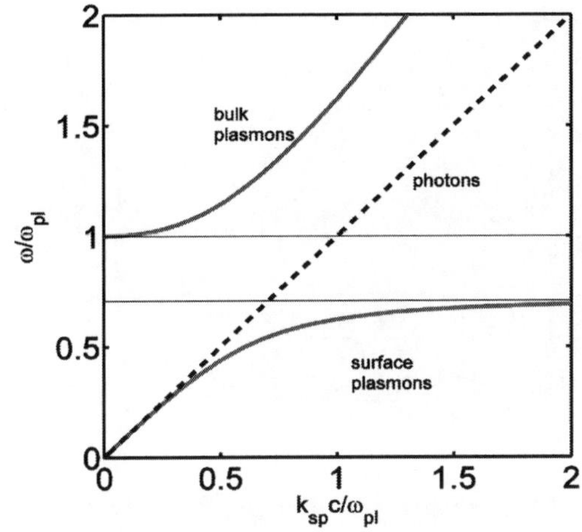

Figure 4.2 The dispersion law of plasmon oscillations in a metal-dielectric system, Eq. (4.5), in the case of the Drude dispersion, Eq. (4.9). The lower curve corresponds to surface plasmons, while the upper curve corresponds to bulk plasmons; the dashed line corresponds to dispersion of photons in free space.

rather than

$$\omega = ck_{\text{sp}}$$

for true bulk plasmons (see Chapter 3).

The branch with

$$\omega < \omega_{\text{sp}} = \frac{\omega_{\text{pl}}}{\sqrt{\varepsilon_{\text{d}} + 1}}$$

is more interesting. This is the branch that corresponds to surface plasmons, which have a range of important properties.

First, for a given frequency, the wavelength of the surface plasmon is smaller than that of a photon in free space and thus the surface plasmon excitation by usual photons is impossible due to conservation of momentum along the interface.

Second, the surface plasmon can exist only in the region of the frequencies with negative electric permittivity (Fig. 3.4, Chapter 3), that is, on the metal–dielectric interface.

And finally, the most important property of surface plasmons is their strong confinement near the interface (Fig. 4.3). The

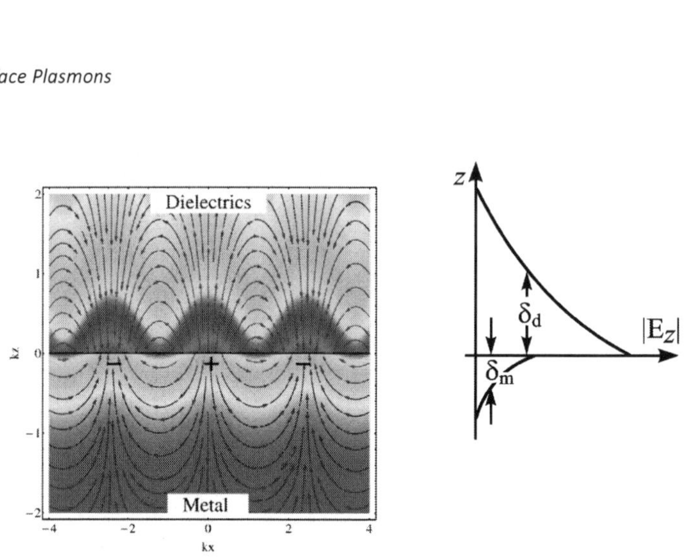

Figure 4.3 The illustration of the surface plasmon electric field distribution on a metal–dielectric interface. (Left) Streamlines of the electric fields. (Right) Exponential decay of the plasmon electric field in dielectrics and metals.

confinement occurs, since for surface plasmons

$$k_{\mathrm{sp}} > \frac{\omega}{c}\sqrt{\varepsilon_{\mathrm{d}}}$$

and the z-components of the wavevector Eq. (4.3) have purely imaginary values, and as a result the fields decay exponentially away from the interface (see Fig. 4.3 for an illustration). It is this localization that is the basis of many applications of surface plasmons.

By now, we have considered the case of a metal without losses. However, practically metals have noticeable losses, $\varepsilon_{\mathrm{m}} = \varepsilon'_{\mathrm{m}} + i\varepsilon''_{\mathrm{m}}$, and surface plasmons decay along the axis of propagation, correspondingly.

In the case of not very large k_{sp}, that is, in the case $\left(\varepsilon'_{\mathrm{m}} + \varepsilon_{\mathrm{d}}\right)^2 \gg \varepsilon''^2_{\mathrm{m}}$, the decay length of surface plasmons can be found from the imaginary part of the surface plasmon wavevector k_{sp}:

$$\delta_{\mathrm{sp}} = \frac{1}{2k''_{\mathrm{sp}}} \approx \frac{c}{\omega}\left(\frac{\varepsilon'_{\mathrm{m}} + \varepsilon_{\mathrm{d}}}{\varepsilon'_{\mathrm{m}}\varepsilon_{\mathrm{d}}}\right)^{3/2} \frac{{\varepsilon'_{\mathrm{m}}}^2}{\varepsilon''_{\mathrm{m}}}. \tag{4.11}$$

In conclusion of this section, let us represent some estimations of the propagation length (along the surface) of plasmons and degrees of their confinement (along the z-axis) in a dielectric, which are characterized by δ_{sp} and $\delta_d = |k_{z,d}|^{-1}$, correspondingly. As it is seen from the dispersion relations, both characteristics depend on frequency to a great extent. The surface plasmons of the frequency close to the plasma frequency ω_{pl} have high localization along the z-axis and a small propagation length due to losses, which are always present in real metals. Using the expressions stated above, one can see that surface plasmons have decay length $\delta_{sp} \approx 20$ μm and $\delta_d = 200$ nm on a plane silver–air interface for a frequency corresponding to wavelength in vacuum $\lambda_{vac} = 450$ nm. At the telecommunications frequency corresponding to the wavelength in vacuum $\lambda_{vac} = 1.5$ μm, we have $\delta_{sp} \approx 1$ mm and $\delta_d \approx 3$ μm, accordingly. There is a general regularity: the smaller the propagation length, the better the localization of surface plasmons. In a metal itself, the depth of penetration and the localization degree remain equal to the skin depth (about 15 nm for silver) in a wide frequency region.

In the case of real metals, losses and deviations from the Drude model caused by interband transitions result in essential changes in dispersion curves compared with the case of a Drude metal

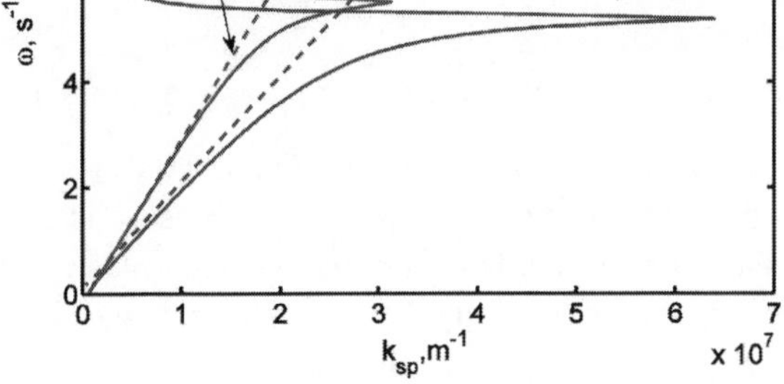

Figure 4.4 The dispersion law for surface plasmons on a silver–air interface (the red line) and on a silver–silica interface (the blue line). Due to losses in metal, the curves for bulk and surface plasmons now are parts of the same curve. The dashed lines correspond to photons in air and in silica, correspondingly.

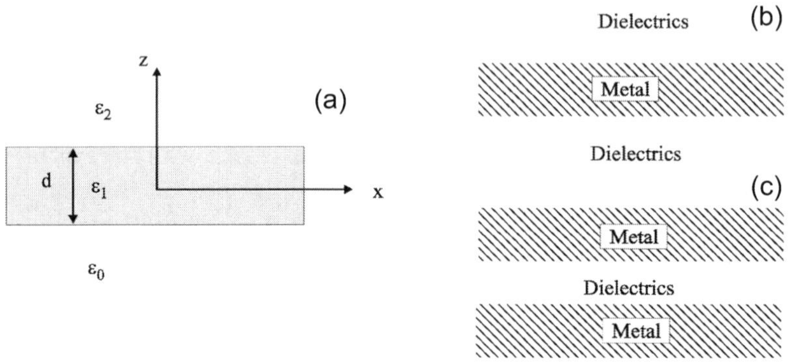

Figure 4.5 Geometry of surface plasmons in an arbitrary three-layered medium (a), in a metal film between dielectric half-spaces (b), and in a dielectric film between metal half-spaces (c).

(Fig. 4.2). In Fig. 4.4, the dependence of the real part of the surface plasmons' wavevector on frequency for a silver–air interface is shown. The most significant distinction is that the surface and bulk plasmons become parts of the same curve, though their behaviors taken separately coincide more or less with those of the dispersion curves for surface and bulk plasmons in the case of the Drude metal. The portion of the curve between ω_{pl} and $\omega_{\text{pl}}/\sqrt{1+\varepsilon_{\text{d}}}$ has no analogue in the case of the Drude metal and corresponds to quasi-localized leaking waves.

4.1.2 *Surface Plasmons in Planar Layered Media*

Existence of two-dimensional surface plasmons is not specific to the metal–dielectric interface only. Surface plasmons can exist also in other layered systems. For example, surface plasmons can exist in a dielectric film surrounded by metal half-spaces or in a metal film surrounded by dielectric half-spaces (Fig. 4.5). More complex geometry of systems results in more complex laws of surface plasmon dispersion (Fuchs and Kliever, 1965, 1966; Kliever and Fuchs, 1966a, 1966b; Vinogradov et al., 2010).

In the case of the three-layered medium, shown in Fig. 4.5a, nontrivial solutions of Maxwell's equations can exist only when the conditions are satisfied:

$$1 + r_{21}^{p} r_{10}^{p} \exp\left(2ik_{z1}d\right) = 0 \tag{4.12}$$

where

$$r_{ij}^{p} = \frac{\varepsilon_j k_{zi} - \varepsilon_i k_{zj}}{\varepsilon_j k_{zi} + \varepsilon_i k_{zj}},\ i,\ j = 0,\ 1,\ 2 \tag{4.13}$$

are the reflection coefficients of p-polarized waves from the i–j boundary and

$$k_{zi} = \sqrt{\varepsilon_i\left(\omega\right)\left(\frac{\omega}{c}\right)^2 - k_{\mathrm{sp}}^2}. \tag{4.14}$$

The dispersion equation Eq. (4.12) can be found both by solving Maxwell's equations directly and by equating the denominators of reflection (or transmission) coefficients to zero. The reflection coefficients for a medium of this type can be found in a great many textbooks also.

Practically, two special cases of a three-layered system with $\varepsilon_2 = \varepsilon_0$ have particular importance. The case of $\varepsilon_2 = \varepsilon_0 > 0,\ \varepsilon_1 < 0$ corresponds to the dielectric-metal-dielectric heterostructure, while the case $\varepsilon_2 = \varepsilon_0 < 0,\ \varepsilon_1 > 0$ corresponds to the metal-dielectric-metal heterostructure. In this symmetrical case, the dispersion equation Eq. (4.12) splits into two simpler ones:

$$\frac{\varepsilon_1 k_{z2} - \varepsilon_2 k_{z1}}{\varepsilon_1 k_{z2} + \varepsilon_2 k_{z1}} e^{ik_{z1}d} = \pm 1 \tag{4.15}$$

which correspond to symmetric or antisymmetric behavior of the x-component of the electric field relative to the $z \rightarrow -z$ transformation.

In the antisymmetric case, the dispersion equation Eq. (4.15) with the "$-$" sign can be written as

$$\frac{\varepsilon_1}{\varepsilon_2} = \frac{k_{z1}}{k_{z2}} \tanh\left(ik_{z1}\frac{d}{2}\right) \tag{4.16}$$

while the symmetric modes ("$+$" sign in Eq. (4.15)) have the dispersion equation:

$$\frac{\varepsilon_1}{\varepsilon_2} = \frac{k_{z1}}{k_{z2}} \coth\left(ik_{z1}\frac{d}{2}\right) \tag{4.17}$$

For a large enough thickness of the middle layer, that is, for $|k_{z1}d| \gg 1$, $\tanh\left(ik_{z1}\frac{d}{2}\right) \rightarrow -1$ and $\coth\left(ik_{z1}\frac{d}{2}\right) \rightarrow -1$, the

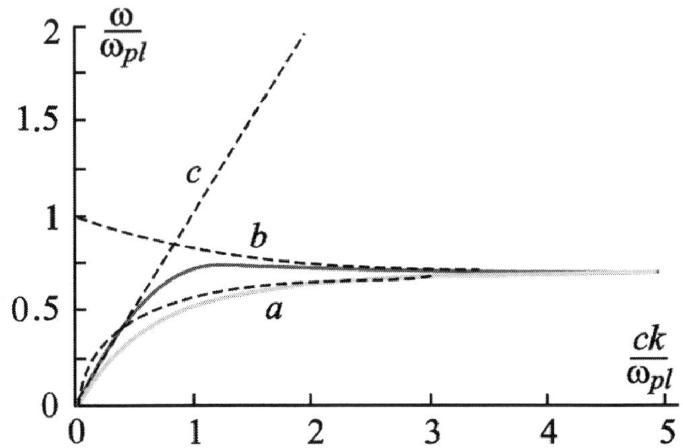

Figure 4.6 The dispersion law of plasmon oscillations, Eqs. (4.16) and (4.17), in a metal film in the case of the Drude dispersion law for metal and for $d\omega_{\rm pl}/(2c) = 0.5$. (a) Symmetric surface plasmons ω_{-}, (b) antisymmetric surface plasmons ω_{+}, and (c) photons in free space. Quasi-static approximation is shown by dashed lines.

dispersion laws in Eqs. (4.16) and (4.17) transform to the law of dispersion of surface plasmons on a metal–dielectric interface, Eq. (4.4).

In the case of a metal film between dielectric half-spaces, essential dispersion takes place only for $\varepsilon_1 = \varepsilon_1(\omega) < 0$, while ε_2 may be regarded as a positive value independent of frequency. In the case when the metal is described by the Drude model, the dispersion laws described by Eqs. (4.16) and (4.17) are shown in Fig. 4.6.

Quasistatic approximations for both branches are depicted by dashed lines in this figure. This approximation is described by the expressions:

$$\omega_{+} = \frac{\omega_{\rm pl}}{\sqrt{1 + \varepsilon_2 \tanh\left(k_{\rm sp} d/2\right)}} \text{ (odd)}$$
$$\omega_{-} = \frac{\omega_{\rm pl}}{\sqrt{1 + \varepsilon_2 \coth\left(k_{\rm sp} d/2\right)}} \text{ (even)} \qquad (4.18)$$

which can be obtained by disregarding retardation ($c \to \infty$ or $k_z = ik_{\rm sp}$) in Eqs. (4.16) and (4.17). As one can see from this figure, it is the high-frequency (odd) branch that suffers from the retardation substantially.

In the region of small wavenumbers k_{sp}, Eqs. (4.16) and (4.17) have the asymptotics:

$$\omega_{+} = \frac{ck_{\mathrm{sp}}}{\sqrt{\varepsilon_2}}\left(1 - \frac{\varepsilon_2}{2}\left(\frac{ck_{\mathrm{sp}}}{\omega_{\mathrm{pl}}\sqrt{\varepsilon_2}}\right)^2 \tanh^2\left(\frac{\omega_{\mathrm{pl}}d}{2c}\right) + \ldots\right)$$

$$\omega_{-} = \frac{ck_{\mathrm{sp}}}{\sqrt{\varepsilon_2}}\left(1 - \frac{\varepsilon_2}{2}\left(\frac{ck_{\mathrm{sp}}}{\omega_{\mathrm{pl}}\sqrt{\varepsilon_2}}\right)^2 \coth^2\left(\frac{\omega_{\mathrm{pl}}d}{2c}\right) + \ldots\right) \quad (4.19)$$

which show that in this regime the dispersion of surface plasmons is close to the dispersion of photons in a bulk dielectric with the electric permittivity $\varepsilon_2 > 0$, and the influence of metal film becomes weak.

For large values of wavenumber k_{sp}, the frequencies of both branches tend to the frequency of the surface plasmon $\omega_{\mathrm{sp}} = \omega_{\mathrm{pl}}/\sqrt{\varepsilon_2 + 1}$ on the metal–dielectric interface:

$$\omega_{\pm} = \omega_{\mathrm{sp}}\sqrt{1 \pm \frac{2\varepsilon_2}{1+\varepsilon_2}e^{-k_{\mathrm{sp}}d}} \quad (4.20)$$

The electric field inside the film for the high-frequency (antisymmetric) branch Eq. (4.16) depends on coordinates as follows:

$$E_x \sim \sin\left(k_{1z}z\right)\exp\left(ik_{\mathrm{sp}}x\right)$$

$$E_z \sim \frac{ik_{\mathrm{sp}}}{k_{1z}}\cos\left(k_{1z}z\right)\exp\left(ik_{\mathrm{sp}}x\right) \quad (4.21)$$

In the case of the low-frequency (symmetric) branch, (see (4.17)), we have

$$E_x \sim \cos\left(k_{1z}z\right)\exp\left(ik_{\mathrm{sp}}x\right)$$

$$E_z \sim -i\frac{k_{\mathrm{sp}}}{k_{1z}}\sin\left(k_{1z}z\right)\exp\left(ik_{\mathrm{sp}}x\right), \quad (4.22)$$

accordingly.

Outside the film, the fields decay exponentially:

$$E_x \sim \exp\left(ik_{\mathrm{sp}}x + ik_{2z}|z|\right)$$

$$E_z \sim \frac{k_{\mathrm{sp}}}{k_{2z}}\exp\left(ik_{\mathrm{sp}}x + ik_{2z}|z|\right). \quad (4.23)$$

Besides, according to Eq. (4.21), the oscillations in the high-frequency branch of the spectrum are polarized mainly along the z-axis (when $k_{1z}d \ll 1$, $E_x \approx 0$, $E_z = \mathrm{const.}$). According to

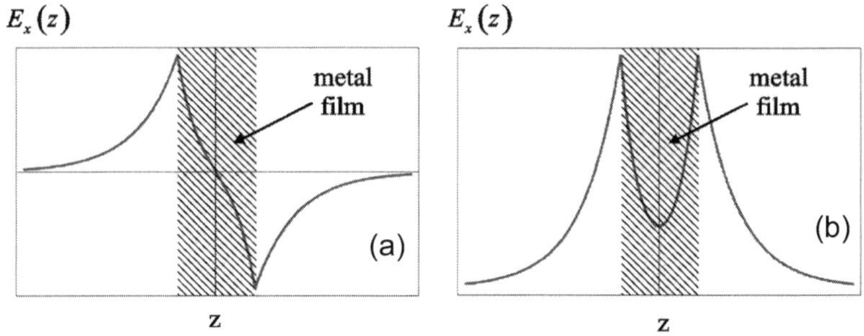

Figure 4.7 The behavior of the x-component of the electric field, $E_x(z)$, as a function of z for antisymmetric (high-frequency) (a) and symmetric (low-frequency) modes (b).

Eq. (4.22), the low-frequency oscillations are polarized mainly along the x-axis (when $k_{1z}d \ll 1$, $E_z \approx 0$, $E_x = \text{const.}$). The behavior of the x-component of the electric field, $E_x(z)$, as a function of z for antisymmetric and symmetric modes is shown in Fig. 4.7.

An interesting property of antisymmetric modes is that a decrease of metal thickness leads to delocalization of plasmons, and as a result the modes transform to a plane electromagnetic wave propagating in the dielectric. In the case of real absorbing metals, this results in a radical increase of the propagation length of surface plasmons and can have a lot of practical applications (see Chapter 13). Vice versa, when the thickness of the metal film decreases, the symmetric modes become more localized and their propagation length decreases in the case of real metals as a result.

In the case of a dielectric film between metal half-spaces, the dispersion equations Eqs. (4.16) and (4.17) remain the same but the dispersion law changes essentially (see Fig. 4.8). This figure displays that when the thickness of the dielectric gap is small enough, there are two surface plasmons in the system, with frequency tending to $\omega_{\text{sp}} = \omega_{\text{pl}}/\sqrt{\varepsilon_1 + 1}$, and one bulk plasmon. In the case of a dielectric film of a small thickness, the quasistatic approximation, Eq. (4.24), describes these plasmons well in contrast to the surface plasmons

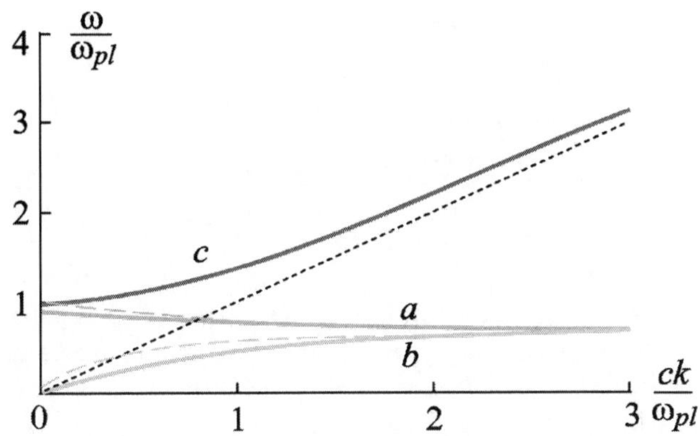

Figure 4.8 The dispersion law of plasmon oscillations in a dielectric film ($\varepsilon_1 = 1$) between metal half-spaces, Eqs. (4.16) and (4.17), in the case of the Drude dispersion law for metal and for $d\omega_{\rm pl}/(2c) = 0.5$. (a) Symmetric surface plasmons, (b) antisymmetric surface plasmons, and (c) bulk plasmons. Quasistatic approximation and photons in a dielectric are shown by the dashed and dotted lines, respectively.

in a metal film.

$$\omega_+ = \frac{\omega_{\rm pl}}{\sqrt{1+\varepsilon_1 \tanh(k_{\rm sp}d/2)}} \text{(even)}$$

$$\omega_- = \frac{\omega_{\rm pl}}{\sqrt{1+\varepsilon_1 \coth(k_{\rm sp}d/2)}} \text{(odd)} \qquad (4.24)$$

When the gap between the metals increases, both surface plasmons transform to a surface plasmon on the metal–dielectric interface gradually. It is important that additional resonance modes appear when the thickness of the dielectric layer increases, since the system obtained becomes a Fabri–Pero resonator.

4.2 One-Dimensional Surface Plasmons

Surface plasmons in systems unlimited only in one direction, that is, wires and fibers, have even more interesting properties. In particular, it was proposed to use plasmonic nanowires for nanoguiding of a light beam (Ashley and Emerson, 1974; Takahara et al., 1997; Weeber et al., 1999; Dickson and Lyon, 2000; Krenn et al., 2002). Owing to an additional dimensional parameter (the

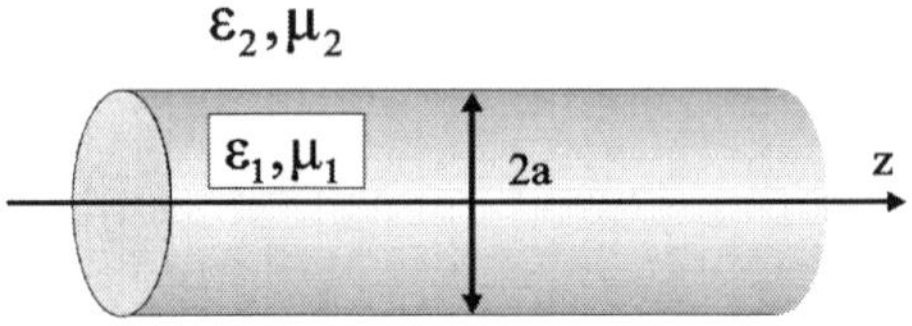

Figure 4.9 The geometry of a metal wire.

wire diameter), the spectrum becomes partially discrete. In this context, the surface plasmons in one-dimensional systems have properties of both two-dimensional systems (continuous frequency spectrum for longitudinal modes) and zero-dimensional systems, which are the properties of nanoparticles of a finite volume (discrete spectrum for transversal oscillations).

4.2.1 *Plasmons in Metal Wires of a Circular Cross Section*

The formal theory of surface plasmon oscillations in circular wires (Fig. 4.9) was developed as far back as in the works of Sommerfeld (1909) for the purposes of radio engineering (see also (Novotny and Hafner, 1994) for optical applications).

The dispersion equation describing all possible surface plasmons and other modes in a cylinder with radius a with electric permittivity and magnetic permeability ε_1 and μ_1 placed in a host medium with parameters ε_2 and μ_2 is given by Stratton (1941):

$$\left[\frac{\mu_1}{u}\frac{J_n'(u)}{J_n(u)}-\frac{\mu_2}{v}\frac{H_n^{(1)'}(v)}{H_n^{(1)}(v)}\right]\left[\frac{\varepsilon_1}{u}\frac{J_n'(u)}{J_n(u)}-\frac{\varepsilon_2}{v}\frac{H_n^{(1)'}(v)}{H_n^{(1)}(v)}\right]$$
$$=n^2h^2\left(\frac{1}{u^2}-\frac{1}{v^2}\right)^2 \qquad (4.25)$$

where

$$n=0,\,1,\,2\ldots;u=\sqrt{k_1^2-h^2}a;v=\sqrt{k_2^2-h^2}a$$

$$k_1^2=\varepsilon_1(\omega)\,\mu_1(\omega)\left(\frac{\omega}{c}\right)^2;k_2^2=\varepsilon_2(\omega)\,\mu_2(\omega)\left(\frac{\omega}{c}\right)^2$$

and it is assumed that components of all fields are proportional to the factor

$$e^{i(hz+n\theta-\omega t)} \tag{4.26}$$

(h is the longitudinal wavenumber).

In a general case, one cannot separate TM waves and transverse electric (TE) waves. However, in the most interesting case of axisymmetric waves ($n = 0$) the dispersion law of TM waves is defined by the equation:

$$\frac{\varepsilon_1}{u}\frac{J_0'(u)}{J_0(u)} = \frac{\varepsilon_2}{v}\frac{H_0^{(1)'}(v)}{H_0^{(1)}(v)} \tag{4.27}$$

while the dispersion law of TE waves is defined by the equation:

$$\frac{\mu_1}{u}\frac{J_0'(u)}{J_0(u)} = \frac{\mu_2}{v}\frac{H_0^{(1)'}(v)}{H_0^{(1)}(v)} \tag{4.28}$$

In Fig. 4.10, the dispersion laws defined by Eqs. (4.27), (4.28), and (4.25) are shown on the assumption that $\varepsilon_1(\omega) = 1 - \omega_{\text{pl}}^2/\omega^2$, $\mu_1 = \mu_2 = \varepsilon_2 = 1$.

From these figures, one can see that plasmon modes can exist only for TM waves as in the case of planar layered media (bottom curve in Fig. 4.10a). Herewith, for any radius of the metal cylinder there is only one plasmon branch having the behavior defined by the law:

$$\omega \sim \frac{hc}{\sqrt{\varepsilon_2}} \tag{4.29}$$

when $h \to 0$, and by the law:

$$\omega \to \omega_{\text{sp}} = \omega_{\text{pl}}/\sqrt{1+\varepsilon_2} \tag{4.30}$$

when $h \to \infty$.

In the case of cylinders of a small radius $a\omega_{\text{pl}}/c \ll 1$, that is, nanocylinders, the law of plasmon modes dispersion, Eq. (4.25), is simplified and takes the form:

$$\omega_n = \omega_{\text{pl}}\sqrt{\frac{I_n'(ha)\,K_0(ha)}{I_n'(ha)\,K_n(ha) - \varepsilon_2 I_n(ha)\,K_n'(ha)}} \tag{4.31}$$

which is shown in Fig. 4.10a by the dotted lines. Generally, the quasi-static approximation describes the dispersion law for nanowires well for $h > \omega/c$.

The increase of the cylinder radius does not change the situation in general, but additional weakly localized modes arise in the region

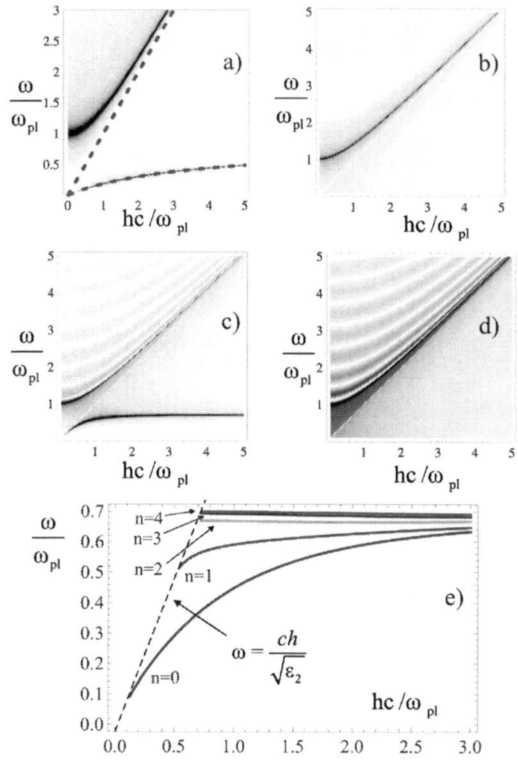

Figure 4.10 The dispersion law (Eq. (4.25)) of surface plasmons in a nonmagnetic metal wire in air in the case of the Drude dispersion. (a) TM_0 mode $a\omega_{\text{pl}}/c = 0.2$ (quasi-static approximation and the dispersion law of photons in free space are shown by dashed lines); (b) TE_0 mode for the wire with $a\omega_{\text{pl}}/c = 1$; (c) TM_0 mode for the wire with $a\omega_{\text{pl}}/c = 5$; (d) TE_0 mode with $a\omega_{\text{pl}}/c = 5$; and (e) plasmon modes of higher orders ($n = 0 \ldots 4$) for the wire with $a\omega_{\text{pl}}/c = 1$.

$\omega > \omega_{\text{pl}}$.These modes correspond to usual waveguide modes (Fig. 4.10c,d). The hybrid plasmonic modes of higher orders for the wire with $a\omega_{\text{pl}}/c = 1$ are shown in Fig. 4.10e. From this figure, one can see that only modes with $n = 0{,}1$ have a substantial dispersion. One should also have in mind that in the case of a lossless material the propagation length of surface plasmons in nanowires is infinite, that is, there are no radiation losses at all for them.

4.2.2 *Plasmons in Nanowires of Other Cross Sections*

Surface plasmons can also exist in wires of other cross sections; however, one failed to obtain analytical results here, and for the determination of plasmon properties different asymptotical and numerical approaches should be used.

In particular, in the case of an elliptic cylinder with $\varepsilon = \varepsilon(\omega)$ and a- and b-semiaxes and situated in a medium of electric permittivity ε_2, there are two plasmon branches instead of one, whose frequencies are defined by the relations (Guzatov and Klimov, 2007) when $h \to \infty$:

$$\omega_{\text{sp}} = \frac{\omega_{\text{pl}}}{\sqrt{1+\frac{a\varepsilon_2}{b}}}$$

$$\omega_{\text{sp}} = \frac{\omega_{\text{pl}}}{\sqrt{1+\frac{b\varepsilon_2}{a}}} \tag{4.32}$$

Naturally, for $a = b$ we obtain from Eq. (4.32) the frequency of surface plasmons in a circular wire, Eq. (4.30).

In the other cases, plasmon oscillations in nanowires should be calculated numerically. Thus, in Fig. 4.11 the distribution of the electric field in a silver nanowire of a triangular cross section with the longitudinal wavenumber $h = 0$ is shown (Kottmann et al., 2001). The figures display clearly that the structure of the fields near the nanowire is changing as the wavelength of an excitation field changes. Figure 4.11a represents the strongest resonance at the wavelength 412 nm. Upon excitation of this resonance, the incident field amplitude increases by 200 times near two vertices adjoining the side parallel to polarization of the incident field, while near the third vertex the field is close to zero. This structure has a simple physical explanation. Along the triangle side parallel to the polarization of the incident field, strong currents are induced, resulting in enhancement of the field intensity in the vertices adjoining this side.

As the propagation direction of the incident wave changes, while its wavelength remains the same, (Fig. 4.11b), the spatial structure of the field stays unaffected but the maximal field amplitude changes. In Fig. 4.11c, the distribution of a field in the same geometry as in Fig. 4.11a is shown but at the wavelength of 600 mn. In this case, the mode structure changes, and maximal amplitude

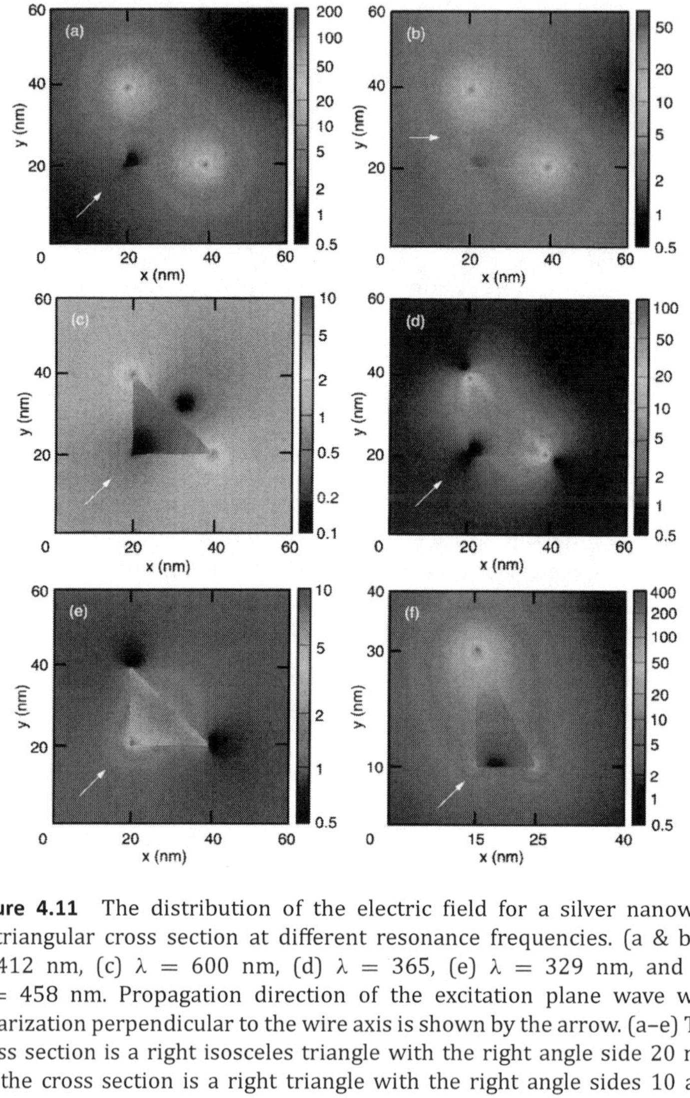

Figure 4.11 The distribution of the electric field for a silver nanowire of triangular cross section at different resonance frequencies. (a & b) $\lambda = 412$ nm, (c) $\lambda = 600$ nm, (d) $\lambda = 365$, (e) $\lambda = 329$ nm, and (f) $\lambda = 458$ nm. Propagation direction of the excitation plane wave with polarization perpendicular to the wire axis is shown by the arrow. (a–e) The cross section is a right isosceles triangle with the right angle side 20 nm; (f) the cross section is a right triangle with the right angle sides 10 and 20 nm (Kottmann et al., 2001).

exceeds only tenfold the amplitude of the incident field. Apparently, in this case the field enhancement is not of the resonance nature but is determined by the lightning-rod effect.

The field distribution near a triangle vertex also depends on a type of excited resonance. As the field amplitude decays gradually when moving away from the vertex in Fig. 4.11a, there is the minimum near the maximum in Fig. 4.11d ($\lambda = 365$ nm). In the case of this resonance in the vertices, oscillating electric dipoles appear leading to the field minimum near the maximum.

An even more interesting situation occurs with the resonance at the wavelength of 329 nm (Fig. 4.11e). The field enhancement is not large ($\sim$10) here, but this enhancement emerges near the triangle vertex oriented along the wave propagation and perpendicular to the polarization. The decrease of the angle between triangle sides (Fig. 4.11f) leads, naturally, to an even greater enhancement of the incident field.

Generally, these figures display that electric fields of resonance plasmon modes of nanowires have very complex structures and are characterized by a great enhancement of incident fields. Both features are especially important for the development of different nanodevices. As to the physical sense of these resonances, the situation is rather complicated. In particular, it is not clear which plasmon types correspond to mentioned resonances. Kottmann et al. (2001) supposed that the resonance in Fig. 4.11e corresponds to excitation of bulk plasmons, but yet the final answer to the question of the nature of the determined resonances can be found in investigations of the dispersion law in such systems.

4.3 Excitation of Surface Plasmons

Surface plasmons on a lossless metal–dielectric interface are nonradiative eigenwaves, and therefore they cannot be excited directly by light waves propagating freely in a transparent medium. Indeed, in Fig. 4.2 one can see that at the same frequency ω (or energy $\hbar\omega$) the wavevector (or momentum) of a photon at the interface cannot be equal to the wavevector of a surface plasmon,

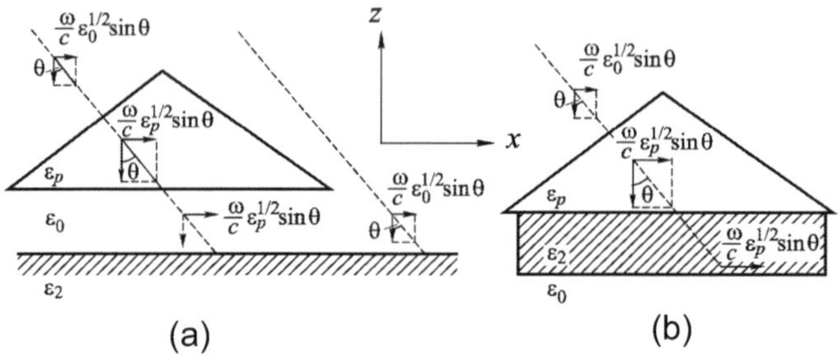

Figure 4.12 The excitation of surface plasmons with the help of the ATIR method. (a) The Otto geometry and (b) the Kretschmann geometry.

and thus the law of energy momentum conservation does not allow the photon to transform to a plasmon and vice versa.

However, when the photon momentum component parallel to the surface is increased by some means, the excitation of a surface plasmon becomes possible.

There are many methods of surface plasmon excitation. Here, we will consider only the most important of them, such as attenuated total internal reflection (ATIR or ATR) (Otto, 1968; Kretschmann, 1971; see Fig. 4.12), the surface diffraction grating (Fig. 4.13), and the method of nanolocalized light sources (Fig. 4.14).

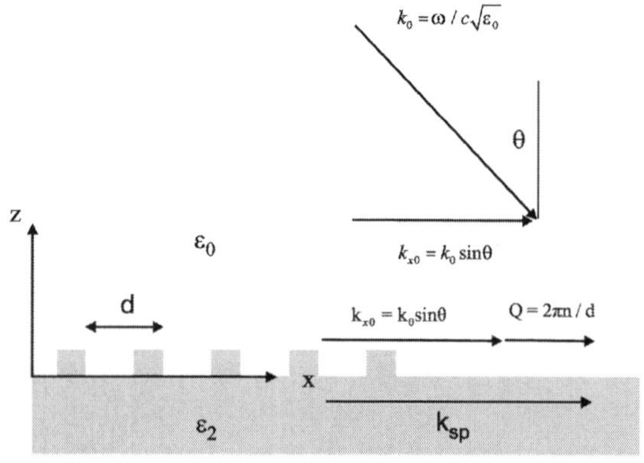

Figure 4.13 The excitation of surface plasmons with the help of gratings on a metal surface.

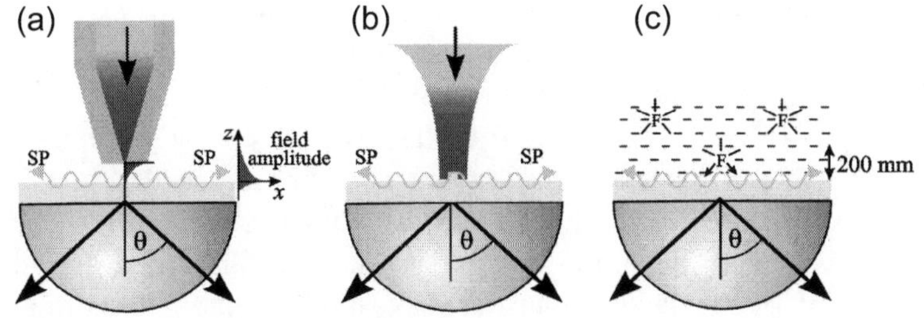

Figure 4.14 The excitation of surface plasmons with the help of nanolocalized light sources. (a) Light from an aperture of a scanning microscope (Hecht et al., 1996), (b) light scattered by a nanoparticle (Novotny et al., 1997), and (b) light of single fluorescent molecules (Lakowicz, 2004).

4.3.1 *Attenuated Total Internal Reflection*

To excite surface plasmons on a flat surface, light waves of a given frequency in a medium of ε_0 permittivity must have the same wavevector as that of surface plasmons and, consequently, a phase velocity less than $c_0 = c/(\varepsilon_0)^{1/2}$. Otto (1968), by suggestion of the method of "slow photon" generation, opened the door to essential development of optical investigations of surface plasmons. The main idea of the method, named ATIR or ATR, is shown schematically in Fig. 4.12a. Before entering the medium with ε_0, an incident light wave first comes to a prism of electric permittivity $\varepsilon_p > \varepsilon_0$. According to the Snellius law, the tangential component of the wavevector is continuous upon the interface crossing, and when the angles of incidence θ are larger than the total internal reflection angle, $\theta > \arcsin\left(\sqrt{\varepsilon_0}/\sqrt{\varepsilon_p}\right)$, we obtain $k_x = (\omega/c)\sqrt{\varepsilon_\mathrm{p}}\sin\theta > (\omega/c)\sqrt{\varepsilon_0}$. But under these conditions, the normal wavevector component k_{z0} becomes purely imaginary, and the wave quickly decays in the direction z in the medium with ε_0 in the distance of about the light wavelength. Thus, the prism is to be necessarily placed at the distance d of the order of or even less than the wavelength λ from the plasmonic medium with permittivity ε_2. As a result, the phase velocity of the incoming light at the interface between air (ε_0) and metal (ε_2) will be slower than the phase velocity in free space and surface plasmons can be effectively excited. This approach to the use of the ATIR method for excitation of surface plasmons is referred to as the Otto method.

Since the Snellius law and Eq. (4.5) are symmetric about ε_0 and ε_2, one can change the positions of these media relatively the prism. In Fig. 4.12b, it is shown how surface plasmons can be excited with the help of the ATIR method on a metal–air interface ($\varepsilon_2/\varepsilon_0$) when the active (metal) medium is represented by a thin film layered on the prism. Due to absorption in an active medium, the film thickness must be rather small so that a sufficient quantity of energy could reach the interface $\varepsilon_2/\varepsilon_0$. This method is known as the Kretschmann method (Kretschmann, 1971) and used widely at the present time.

4.3.2 *The Surface Diffraction Grating Method*

It is well known that as a result of the interaction of light with a diffraction grating, the incident plane wave with the wavevector parallel to the surface $k_{x0} = \omega/c\sqrt{\varepsilon_0}\sin\theta$ (θ is an angle of incidence) is transformed into an infinite set of plane waves with the wavevectors

$$\mathbf{k}_x = \mathbf{k}_{x0} + \mathbf{Q} \tag{4.33}$$

where $\mathbf{Q} = 2\pi n/d$, $n = 0, \pm1, \pm2$, etc., is the vector in the reciprocal space and d is the period of the diffraction grating. Then the value of n can be found to satisfy the condition $k_x = k_{sp} > k_{x0}$, and surface plasmon excitation becomes possible. This explains, for example, well-known Wood's anomalies in diffraction gratings found experimentally in the beginning of the 20th century.

Thus, surface plasmons can be excited with the help of a periodic relief by changing, for example, an angle of incidence to satisfy Eq. (4.5). In this case, the minimum must be observed in the reflection coefficient considered as a function of the angle. Making use of this effect, the dispersion laws of surface plasmons for Al (Teng and Stern, 1967) and for Au and Al (Ritchie et al., 1968) were found.

Rough surfaces can be always considered a set of periodic gratings with different periods, and consequently, surface plasmons can be excited with their help. By measuring the changes of the light reflection coefficient at normal incidence on rough Ag films,

Jasperson and Schnatterly (1969) proved that surface plasmons actually can be excited on rough surfaces.

4.3.3 *The Method of Nanolocalized Light Sources*

Scanning near-field microscopy and other modern instruments pave new ways of surface plasmon excitation (Hecht et al., 1996; Lakowicz, 2004; Ditlbacher et al., 2002). New approaches to surface plasmon excitation are based on the fact that the wave component parallel to the surface, necessary for surface plasmon excitation, is present near any strongly localized light source, such as an aperture of a scanning microscope, nanoparticles, and even fluorescent molecules. If such a light nanosource is brought close to a metal surface, local excitation of surface plasmons is possible.

In Fig. 4.14 (Novotny and Hecht, 2006), the principle scheme for these methods is shown. A metal film is located on a semispherical glass prism to observe plasmon emission that occurs, for example, due to their decay. Surface plasmon excitation occurs with the help of a scanning microscope aperture (Fig. 4.14a), light scattering by nanoparticles (Fig. 4.14b), and fluorescent molecules, situated in a solution over the surface.

4.4 Observation of Surface Plasmons

Thus, we have discussed how surface plasmons can be excited with the help of light. The observation can be performed in different ways.

The most direct method is the detection of reflected light in an ATIR experiment of any configuration stated above. Within this approach, the dependence of the p-polarized light reflection coefficient R_p on the incidence angle or wavelength is measured.

Ideally, $R_p = 0$, and all energy of light wave transforms to surface plasmons. In the absence of radiative decay (e.g., caused by roughness), light transmittance tends to zero and optical absorption has the Lorentzian shape.

It is interesting to trace the physical nature of the reflection coefficient vanishing upon the surface plasmon excitation. Let us mention primarily that the surface plasmons excited by any ATIR method are leaking modes in principle, that is, they lose energy not only due to losses in the metal but also due to energy leakage back to the prism. Due to destructive interference, the minimum of the reflected light intensity occurs between this leaking emission and the part of the exciting light beam that would be present in the absence of a metal film. When the thickness of a metal film is optimal, the destructive interference can be total, the reflected beam vanishes, and the leaking field cannot be detected.

Analyzing this geometry on the basis of Fresnel equations, (Kretschmann, 1971; Raether, 1988) have shown that this optimal case takes place when the radiative linewidth of the surface plasmon $\Gamma_{\rm LR}$, related to the radiation leakage back to the prism, is equal to the radiationless linewidth, $\Gamma_{\rm abs} = {\rm Im}k_{0\rm sp}$, where $k_{0\rm sp}$ is the wavevector of surface plasmon, Eq. (4.5). For a metal film of thickness d_1 and with the permittivity $|\varepsilon_1'| \gg 1$, $|\varepsilon_1'| \gg \varepsilon_1''$, the reflection coefficient can be approximated by the expression (Raether, 1988):

$$R = \left|r_{01}^{p}\left(k_{0\rm sp}\right)\right|^2 \left(1 - \frac{4\Gamma_{\rm LR}\Gamma_{\rm abs}}{\left(k_{\rm sp} - \left(k_{0\rm sp} + {\rm Re}(\Delta k_{\rm sp})\right)\right)^2 + (\Gamma_{\rm LR} + \Gamma_{\rm abs})^2}\right)$$

$$\Delta k_{\rm sp} = \left[\frac{\omega}{c}\frac{2}{1+|\varepsilon_1'|}\left(\frac{|\varepsilon_1'|}{|\varepsilon_1'|-1}\right)^{3/2} \exp\left(-2k_{0\rm sp}d_1\right)\right] r_{01}^{p}\left(k_{0\rm sp}\right) \qquad (4.34)$$

where r_{01}^{p} is the reflection coefficient, Eq. (4.13). Evidently, in the geometry of ATIR, the resonance wavevector of the surface plasmon $k_{\rm sp} = k_{0\rm sp} + \Delta k_{\rm sp}$ differs from the resonance wavevector of the surface plasmon $k_{0\rm sp}$ for a semi-infinite metal by some value $\Delta k_{\rm sp}$, the imaginary part of which describes the radiation leakage of the surface plasmon, $\Gamma_{\rm LR} = Im\Delta k_{\rm sp}$. From Eq. (4.34), one can see that for $\Gamma_{\rm LR} = \Gamma_{\rm abs}$ and for $k_{\rm sp} = k_{0\rm sp} + {\rm Re}\left(\Delta k_{\rm sp}\right)$ the reflection coefficient is equal to zero, indeed.

A more sophisticated method can be used for detection of surface plasmons, that is, one can measure not only reflected wave amplitudes, but also the changes of its phase δ_p. This can be implemented with the help of ellipsometry, which defines two parameters $\psi = arctg\left(R_p/R_s\right)^{1/2}$ and $\Delta = \delta_p - \delta_s$, the index s is related to s-polarized light. Since R_s and δ_s are practically constant, all the ψ and Δ information on surface plasmons is related to R_p and δ_p.

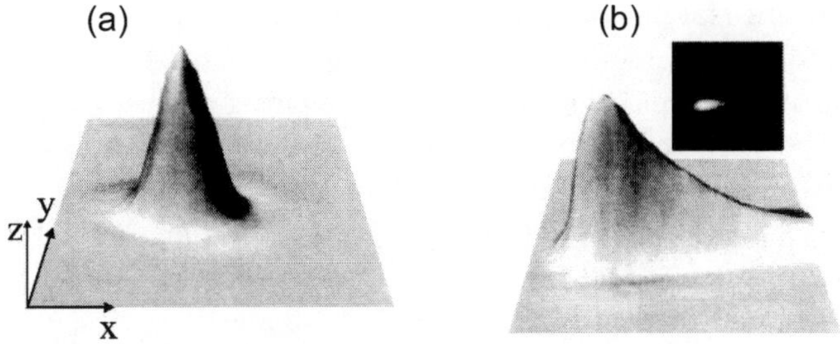

Figure 4.15 The scan image of the near field of a He-Ne laser beam ($\lambda = 633$ nm) incident on the inner side of the prism without a metal film (a) and with a silver film of 53 nm thickness (b). The angle of incidence is larger than the angle of total internal reflection. The exponentially vanishing "tail" in (b) is determined by surface plasmons propagating away from the excitation region. The image is obtained with the help of a near-field optical scanning microscope (Dawson et al., 1994; Dawson et al., 2001).

The methods described above say nothing about the surface plasmon spatial structure. In recent years, owing to the development of nano-optics and scanning microscopes, the possibility of detailed microscopic investigation of the surface plasmons' spatial structure and properties has emerged. In experiments of this kind, surface plasmons are excited through a prism from below (Kretschmann geometry), and the surface distribution of electric fields is measured with the help of a scanning microscope with a needle having an aperture of 50 nm diameter and being able to scan over a metal film surface with high precision. By measuring the intensity of light at an output of the scanning microscope, one can estimate the intensity of the surface plasmon electric field at a given point. In Fig. 4.15, the spatial image of a surface plasmon obtained by this method is shown.

References

Ashley, J.C., Emerson, L.C. (1974) Dispersion relations for non-radiative surface plasmons on cylinders, *Surf. Sci.*, **41**, 615–618.

Dawson, P., Puygranier, B.A.F., Goudonnet, J.-P. (2001) Surface plasmon polariton propagation length: a direct comparison using photon scanning tunneling microscopy and attenuated total reflection, *Phys. Rev. B*, **63**, 205410.

Dawson, P., de Fornel, F., Goudonnet, J.P. (1994) Imaging of surface plasmon propagation and edge interaction using a photon scanning tunneling microscope, *Phys. Rev. Lett.*, **72**, 2927–2930.

Ditlbacher, H., Krenn, J.R., Felidj, N., Lamprecht, B., Schider, G., Salerno, M., Leitner, A., Aussenegg, F.R. (2002) Fluorescence imaging of surface plasmon fields, *Appl. Phys. Lett.*, **80**, 404.

Dickson, R.M., Lyon, L.A. (2000) Unidirectional plasmon propagation in metallic nanowires, *J. Phys. Chem. B*, **104**, 6095–6098.

Fuchs, R., Kliewer, K.L., Pardee, W.J. (1966) Optical properties of an ionic crystal slab, *Phys. Rev.*, **150**, 589–596.

Fuchs, R., Kliewer, K.L. (1965) Optical modes of vibration in an ionic crystal slab, *Phys. Rev. A*, **140**, 2076–2088.

Guzatov, D.V., Klimov, V.V. (2007) Spontaneous emission of an atom placed near a nanobelt of elliptical cross-section, *Phys. Rev. A*, **75**, 052901.

Hecht, B., Bielefeldt, H., Novotny, L., Inouye, Y., Pohl, D.W. (1996) Local excitation, scattering, and interference of surface plasmons, *Phys. Rev. Lett.*, **77**, 1889–1892.

Jasperson, S.N., Schnatterly, S. (1969) Photon-surface-plasmon coupling in thick Ag foils, *Phys. Rev.*, **188**, 759.

Kliewer, K.L., Fuchs, R. (1966a) Optical modes of vibration in an ionic crystal slab including retardation. I. Nonradiative region, *Phys. Rev.*, **144**, 495–503.

Kliewer, K.L., Fuchs, R. (1966b) Optical modes of vibration in an ionic crystal slab including retardation. II. Radiative region, *Phys. Rev.*, **150**, 573–588.

Kottmann, J.P., Martin, O.J.F., Smith, D.R., Schultz, S. (2001) Dramatic localized electromagnetic enhancement in plasmon resonant nanowires, *Chem. Phys. Lett.*, **341**, 1–6.

Krenn, J.R., Lamprecht, B., Ditlbacher, H., Schider, G., Salerno, M., Leitner, A., Aussenegg, F.R. (2002) Non diffraction limited light transport by gold nanowires, *Europhys. Lett.*, **60**, 663–669.

Kretschmann, E. (1971) Die bestimmung optischer Konstanten von Metallen durch Anregung von Oberflächenplasmaschwingungen, *Z. Phys.*, **241**, 313.

Lakowicz, J.R. (2004) Radiative decay engineering 3. Surface plasmon-coupled directional emission, *Anal. Biochem.*, **324**, 153–169.

Novotny, L., Hafner, C. (1994) Light propagation in a cylindrical waveguide with a complex, metallic, dielectric function, *Phys. Rev. E*, **50**, 4094–4106.

Novotny, L., Hecht, B., Pohl, D.W. (1997) Interference of locally excited surface plasmons, *J. Appl. Phys.*, **81**(4), 1798–1806.

Novotny, L., Hecht, B. (2006) *Principles of Nano-Optics* (Cambridge University Press, Cambridge).

Otto, A. (1968) Excitation of nonradiative surface plasma waves in silver by the method of frustrated total reflection, *Z. Phys.*, **216**, 398–410.

Raether, H. (1988) *Surface Plasmons* (Springer-Verlag, Berlin).

Ritchie, R.H., Arakawa, E.T., Cowan, J.J., Hamm, R.N. (1968) Surface-plasmon resonance effect in grating diffraction, *Phys. Rev. Lett.*, **21**, 1530.

Sommerfeld A. (1909) Über die Ausbreitung der Wellen in der drahtlosen Telegraphie, *Ann. Phys. (4th series)*, **28**, 44.

Stern, E.A., Ferrell, R.A. (1960) Surface Plasma Oscillations of a Degenerate Electron Gas, *Phys. Rev.*, **120**, 130.

Stratton, J.A. (1941) *Electromagnetic Theory* (McGraw-Hill, New York).

Takahara, J., Yamagishi, S., Taki, H., Morimoto, A., Kobayashi, T. (1997) Guiding of a one-dimensional optical beam with nanometer diameter, *Opt. Lett.*, **22**, 475–477.

Teng, Y.Y., Stern, E.A. (1967) Plasma radiation from metal grating surfaces, *Phys. Rev. Lett.*, **19**, 511.

Vinogradov, A.P., Dorofeenko, A.V., Nechepurenko, I.A. (2010) Propagating in the film along the interface. Analysis of plasmonic Bloch waves and band structures of 1D plasmonic photonic crystals, *Metamaterials*, **4**, 181.

Weeber, J.C., Dereux, A., Girard, C., Krenn, J.R., Goudonnet, J.P. (1999) Plasmon polaritons of metallic nanowires for controlling submicron propagation of light, *Phys. Rev. B*, **60**, 9061–9068.

Zenneck, J. (1907) Über die Fortpflanzung ebener elektromagnetischer Wellen längs einer ebenen Leiterfläche und ihre Beziehung zur drahtlosen Telegraphie, *Ann. Phys. (4th series)*, **23**, 846.

Chapter 5

The Theory of Plasmon Oscillations in Nanoparticles

Unlike surface or bulk plasmons, oscillations of conduction electrons in metal nanoparticles are called localized plasmons or, simply, plasmons. In a nanoparticle of a specified form, plasmon oscillations have only a discreet set of frequencies, which are called plasma resonance frequencies or, simply, plasmon frequencies. This set of frequencies has a deep analogy with a discrete spectrum of radiation of atoms and molecules.

There are two main approaches to the theoretical calculation of plasmon frequencies and corresponding spatial distributions of electromagnetic fields inside and outside a nanoparticle.

The first approach is to solve the system of Maxwell's equations with specified excitation sources or their integral analogs with the help of one of existing analytical and numerical methods for different frequencies and then to calculate the dependencies of metal particle characteristics on frequency. For example, it can be a scattering or absorption cross section. Then, peak values of cross sections can be attributed to plasmon resonances, and the corresponding distribution of the electromagnetic field describes the structure of localized plasmons.

Nanoplasmonics
Vasily Klimov

ISBN 978-981-4267-16-8 (Hardcover), 978-981-4267-42-7 (eBook)
www.panstanford.com

Now, a wide range of effective numerical methods is elaborated to solve Maxwell's equations in the presence of nanoparticles and a complex nanoenvironment, allowing one to calculate electromagnetic fields near nanobodies and metal nanoparticles (see Appendix A2).

Nevertheless, the direct solution of Maxwell's equations is not optimal for investigation of resonances in general and plasmon resonances in particular. For example, the numerical solution of Maxwell's equation for an electromagnetic fields scattering by a nanoparticle of one material cannot be used directly for defining plasmon properties of the particle of the same shape made of other material. On the other hand, in the context of this solution of Maxwell's equations it is difficult to find precise values of plasmon frequencies and other plasmon characteristics since the resonance behavior of the scattering cross section, as an example, is influenced not only by the plasmon frequency but also by radiative plasmon decay (see below) and by the Joule losses in the particle's material.

A more convenient and powerful approach to describe the plasmon properties of nanoparticles is the approach based on determination of the resonant values of electric permittivity of the nanoparticles' material. This method is referred to as a generalized method of eigenoscillations, or the "ε-method" (Agranovich et al., 1999). This approach, based on fundamental properties of plasmons, allows not only describing of particles of the same shape, but of different materials, but also understanding of the concept of localized plasmons itself. The focus of this approach is influence of the nanoparticles' shape on their electromagnetic and optical properties.

5.1 The "ε-Method" of Maxwell's Equations Solutions for Particles of an Arbitrary Size

As in the case of bulk and surface plasmons, a system of Maxwell's equations is to be solved to describe their properties and applications. Generally, one seeks a solution of the problem of diffraction of a monochromatic field formed by given currents **j** on the body of electric permittivity ε and magnetic permeability equal to 1. The sought fields **E** and **H** satisfy Maxwell's equations ($k = \omega/c$):

$$\mathrm{rot}\mathbf{H} + ik\varepsilon(\omega)\mathbf{E} = \frac{4\pi}{c}\mathbf{j}, \quad \mathrm{rot}\mathbf{E} - ik\mathbf{H} = 0, \tag{5.1}$$

inside the nanoparticle and:

$$\mathrm{rot}\mathbf{H} + ik\mathbf{E} = \frac{4\pi}{c}\mathbf{j}, \quad \mathrm{rot}\mathbf{E} - ik\mathbf{H} = 0, \tag{5.2}$$

outside it (in vacuum). At the boundary of the nanoparticle, the continuity of tangential components of **E** and **H** should be also provided. If the particle is not in a closed cavity, then the fields **E** and **H** should also satisfy Sommerfeld radiation condition:

$$\begin{aligned} E_\theta = H_\varphi \underset{kR\to\infty}{\longrightarrow} f(\theta,\varphi)\frac{\exp(ikR)}{kR}, \\ H_\theta = -E_\varphi \underset{kR\to\infty}{\longrightarrow} g(\theta,\varphi)\frac{\exp(ikR)}{kR}, \end{aligned} \tag{5.3}$$

where R, θ, and φ are the spherical coordinates with a center situated in the center of the particle and f and g are the scattering amplitudes. Besides, if **j** sources are situated at infinity, then the scattered fields $\mathbf{E}_{sc} = \mathbf{E} - \mathbf{E}^0$ and $\mathbf{H}_{sc} = \mathbf{H} - \mathbf{H}^0$ should satisfy the radiation condition, and the incoming fields $\mathbf{E}^0$ and $\mathbf{H}^0$ are defined below—Eq. (5.4). All the fields introduced below satisfy these conditions, and hereafter we will not specify it again.

Let us introduce the fields $\mathbf{E}^0$ and $\mathbf{H}^0$ excited by the same sources **j** in the absence of a scattering particle and satisfying Maxwell's equations in the whole space:

$$\begin{aligned} \mathrm{rot}\mathbf{H}^0 + ik\mathbf{E}^0 &= \frac{4\pi}{c}\mathbf{j}, \\ \mathrm{rot}\mathbf{E}^0 - ik\mathbf{H}^0 &= 0. \end{aligned} \tag{5.4}$$

Certainly, $\mathbf{E}^0$ and $\mathbf{H}^0$ are continuous on the particle surface.

The generalized method of eigenoscillations, or the "ε-method," is based on eigenfunctions $\mathbf{e}_n(\mathbf{r})$, $\mathbf{h}_n(\mathbf{r})$, which satisfy the homogeneous equations:

$$\begin{aligned} \mathrm{rot}\mathbf{h}_n + ik\varepsilon_n\mathbf{e}_n &= 0, \\ \mathrm{rot}\mathbf{e}_n - ik\mathbf{h}_n &= 0, \end{aligned} \tag{5.5}$$

inside the particle and

$$\begin{aligned} \mathrm{rot}\mathbf{h}_n + ik\mathbf{e}_n &= 0, \\ \mathrm{rot}\mathbf{e}_n - ik\mathbf{h}_n &= 0, \end{aligned} \tag{5.6}$$

outside it. The continuity of tangential components on the particle surface S and the Sommerfeld radiation conditions, Eq. (5.3), should be also provided. The eigenfunctions $\mathbf{e}_n$, $\mathbf{h}_n$ describe eigenoscillations of a body of the volume V^+ and electric permittivity ε_n, occurring on the given frequency $\omega = ck$, where c is the speed of light. Eigenvalues of the problem Eqs. (5.5) and (5.6) are the ε_n numbers.

Eigenfunctions corresponding to different eigenvalues, ε_n, are orthogonal in the sense:

$$\int_{V^+} \mathbf{e}_n \mathbf{e}_m dV = 0 \quad (n \neq m), \tag{5.7}$$

where the integration is over the particle's volume.

To prove Eq. (5.7), first one should multiply the first equation in Eq. (5.5) by $\mathbf{e}_m$ and the second one by $\mathbf{h}_m$, then add the first result and the second one, permutate n and m subscripts in the line obtained, and subtract the result from the initial result of addition. The equality obtained may be written as:

$$ik\left(\varepsilon_n - \varepsilon_m\right) \mathbf{e}_n \mathbf{e}_m = \operatorname{div}\left([\mathbf{e}_n \mathbf{h}_m] - [\mathbf{e}_m \mathbf{h}_n]\right). \tag{5.8}$$

Integrating this expression over the nanoparticle's volume V^+, we obtain the integral Eq. (5.7) on the left and the integral over the surface of the particle S on the right:

$$\int_S \left([\mathbf{e}_n \mathbf{h}_m] - [\mathbf{e}_m \mathbf{h}_n]\right) \mathbf{n} dS \tag{5.9}$$

In Eq. (5.9), the fields are taken over the inner side of the particle surface, and $\mathbf{n}$ is the external normal to the nanoparticle surface. Let us apply the same operation to Eq. (5.6). Since it does not contain ε_n, outside the nanoparticle (in V^-), the fields $\{\mathbf{e}_n, \mathbf{h}_n\}$ and $\{\mathbf{e}_m, \mathbf{h}_m\}$ are related by the correlation simpler than Eq. (5.8), and in V^- the left part of the equality in Eq. (5.8) is equal to zero. Integrating this expression over V^-, we obtain two surface integrals, Eq. (5.9): the first one is over the particle external surface and the second over the infinitely distant sphere. The integrals taken over an infinitely distant sphere vanish due to the radiation condition. Consequently, the integral Eq. (5.9) is equal to zero being taken over the outer side of the particle surface S, too. It contains the field components tangential to the surface only, which are continuous according to the boundary conditions. It means that the integral taken over the inner side of the particle surface S (the right side of Eq. (5.8)) is equal to zero, too, that is, the orthogonality, Eq. (5.7), takes place, actually.

It is very important that for systems in thermodynamic equilibrium, the imaginary part of ε_n is *negative* for the chosen time dependence of fields, $e^{-i\omega t}$. This fact should not be confused with the *positive* imaginary part of the permittivity of real materials. To verify this, one has to multiply both equations in Eq. (5.5) and its complex conjugated variant by $\mathbf{e}_n^*$, $-\mathbf{h}_n^*$ and $\mathbf{e}_n$, $\mathbf{h}_n$, correspondingly, and add. As a result we obtain an equality analogous to Eq. (5.8):

$$ik\left(\varepsilon_n - \varepsilon_n^*\right)|\mathbf{e}_n|^2 = \operatorname{div}\left(\left[\mathbf{e}_n^*\mathbf{h}_n\right] + \left[\mathbf{e}_n\mathbf{h}_n^*\right]\right). \qquad (5.10)$$

Analogously, outside the nanoparticle we have:

$$0 = \operatorname{div}\left(\left[\mathbf{e}_n^*\mathbf{h}_n\right] + \left[\mathbf{e}_n\mathbf{h}_n^*\right]\right). \qquad (5.11)$$

Now, integrating Eq. (5.10) over the nanoparticle volume V^+ and Eq. (5.11) over the surrounding volume V^- and making use of the continuity of tangential components, one can obtain:

$$\mathrm{Im}\varepsilon_n k \int\limits_{V^+} |\mathbf{e}_n|^2 \, dV = -\mathrm{Re} \int\limits_{S_\infty} \left[\mathbf{e}_n^*\mathbf{h}_n\right] \mathbf{n} dS, \qquad (5.12)$$

where the integration in the right part is over an infinitely distant surface.

The integral in the right part is positive and equal to the radiated power. This implies that the imaginary part ε_n is negative or equal to zero. The eigenvalues of ε_n have a zero imaginary part only for bodies placed in closed cavities without losses in walls.

As to the real part of ε_n, the situation is much more complex. The electric permittivity spectrum for particles of finite volume always remains discrete, but its values can be both positive and negative. In Fig. 5.1, the dependencies of real parts ε_n on the particle parameter size $ka = 2\pi a/\lambda$ (where a and λ are a typical size of a particle and wavelength correspondingly) are shown. From this figure, one can see that in the case of nanoparticles ($2\pi a/\lambda \to 0$), finite eigenvalues of electric permittivity are possible only in the region of negative values, and these values correspond to localized

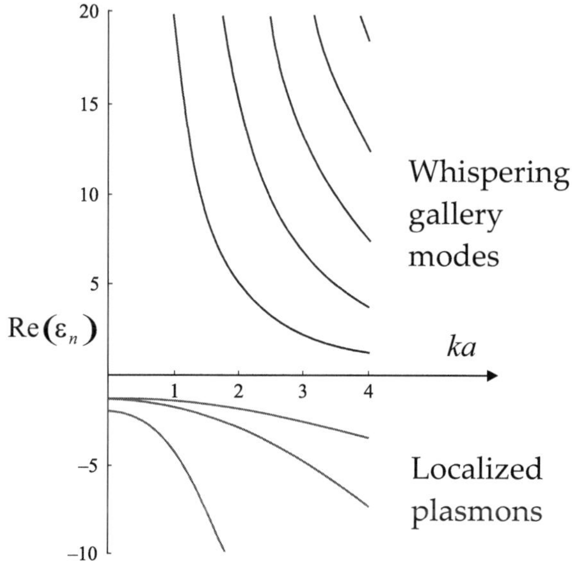

Figure 5.1 Schematic dependency of real parts of ε_n on the nanoparticles' size parameter $ka = 2\pi a/\lambda$.

plasmon oscillations excited in metal nanoparticles, which can have negative values of electric permittivity. Positive eigenvalues of electric permittivity for nanoparticles have large values tending to infinity as $2\pi a/\lambda \to 0$. These modes correspond to "whispering gallery" modes (Braginsky et al., 1989; Schiller and Byer, 1991; Collot et al., 1993; Gorodetsky et al., 1996; Hill and Benner, 1987).

As the particle's size increases, plasmon curves bend toward larger negative values and then disappear, since the increase of the particle size defines the increase of imaginary parts of ε_n due to more effective radiation of light by a nanoparticle. On the contrary, the curves corresponding to "whispering gallery" modes become more distinct since as the size increases, the effects of total internal reflection, lying in the basis of "whispering gallery" modes, become more effective.

After finding the solution of the eigenvalue problem for electric permittivity, ε_n, Eqs. (5.5) and (5.6), the solution of the original diffraction problem Eq. (5.1), Eq. (5.2) can be found as the expansion over eigenfunctions $\mathbf{e}_n(\mathbf{r}), \mathbf{h}_n(\mathbf{r})$:

$$\mathbf{E} - \mathbf{E}^0 = \sum_n A_n \mathbf{e}_n,$$
$$\mathbf{H} - \mathbf{H}^0 = \sum_n A_n \mathbf{h}_n. \tag{5.13}$$

Note that the coefficients A_n in both series, Eq. (5.13), are equal. Indeed, if the first line in Eq. (5.13) is substituted into Eq. (5.1), then making use of Eqs. (5.4) and (5.5), we obtain the second line in Eq. (5.13).

For any coefficients A_n, these series satisfy the second equation in Eq. (5.1) in the region V^+ (inside the particle), both equations in Eq. (5.2) in the region V^- (outside a particle), and the boundary conditions on the particle surface. The expansion coefficients A_n can be found by the requirement that Eq. (5.13) satisfies the Eq. (5.1) in the region V^+. Substituting Eq. (5.13) in the first equation in Eq. (5.1), in the region V^+ we obtain:

$$\sum_n A_n \left(\varepsilon_n - \varepsilon\left(\omega\right)\right) \mathbf{e}_n = \left(\varepsilon\left(\omega\right) - 1\right) \mathbf{E}^0. \tag{5.14}$$

Making use of the orthogonality condition Eq. (5.7), we obtain the sought explicit expression for coefficients A_n:

$$A_n = \frac{\varepsilon\left(\omega\right) - 1}{\varepsilon_n - \varepsilon\left(\omega\right)} \frac{\int_{V+} \mathbf{e}_n \mathbf{E}^0 dV}{\int_{V+} \mathbf{e}_n^2 dV}. \tag{5.15}$$

Substituting this expression in Eq. (5.13), we obtain the final solution of Maxwell's equations within the context of the "ε-method":

$$\mathbf{E}\left(\mathbf{r}\right) = \mathbf{E}^0\left(\mathbf{r}\right) + \sum_n \mathbf{e}_n\left(\mathbf{r}\right) \frac{\varepsilon\left(\omega\right) - 1}{\varepsilon_n - \varepsilon\left(\omega\right)} \frac{\int_{V+} \mathbf{e}_n \mathbf{E}^0 dV}{\int_{V+} \mathbf{e}_n^2 dV}, \tag{5.16}$$

where $\varepsilon(\omega)$ describes the dependence of electric permittivity on a specific material of the particle.

The form of this solution implies that $\varepsilon(\omega)$ enters Eq. (5.16) in a rather simple way, allowing quick calculations for nanoparticles of the same form but of different materials. This is very important for optimization of nanodevices.

The most important feature of Eq. (5.16) is the presence of the resonance factor $\varepsilon_n - \varepsilon(\omega)$ in the denominator. For frequencies ω_n such that $\varepsilon_n - \varepsilon(\omega_n) = 0$, plasmon resonance occurs in the system, and only one term dominates in the solution:

$$\mathbf{E}(\mathbf{r}) \approx \mathbf{e}_n(\mathbf{r}) \frac{(\varepsilon(\omega) - 1)}{(\varepsilon_n - \varepsilon(\omega))} \frac{\int\limits_{V+} \mathbf{e}_n \mathbf{E}^0 dV}{\int\limits_{V+} \mathbf{e}_n^2 dV}. \tag{5.17}$$

In this case, it can be said that on the frequency $\omega \approx \omega_n$ the excitation of a localized plasmon occurs. The spatial structure of this plasmon is described by the eigenfunction $\mathbf{e}_n(\mathbf{r})$, which is independent of a specific material.

Due to the presence of a small resonant denominator, the excitation electric field is strongly enhanced near the plasmonic nanoparticle. This *resonant* enhancement of the local field is usually characterized by the factor G:

$$G = \max\left|\frac{\mathbf{E}^2}{\mathbf{E}_0^2}\right|, \tag{5.18}$$

and is very important for many applications, including single-molecule detection.

Another reason for a huge enhancement of the excitation field is the so-called lightning rod effect. This effect occurs at the points where the radius of the curvature of the particle surface is much smaller than the light wavelength. The small curvature radius of a sharp corner or an edge results in a quasi-electrostatic "crowding" of many electric field lines, which, in turn, leads to a tremendous field enhancement. The lightning rod effect is a *nonresonant* one, that makes it applicable over a wide spectroscopic range. The illustration of the "lightning rod" effect is shown in Fig. 5.2.

The plasmon resonance width as a function of frequency depends essentially on imaginary parts both of ε_n and of $\varepsilon(\omega)$. It is very important that imaginary parts of ε_n and $\varepsilon(\omega)$ must be always of different signs, and thus the resonance width of Eq. (5.17) is defined by the sum of their absolute values. As it will be shown later, in the case of nanoparticles the imaginary part $\varepsilon_n'' \sim (ka)^3$ becomes minute. In this case, the plasmon resonance width is generally defined by losses inside particles, that is, by the imaginary part of $\varepsilon(\omega)$. In the cases when the imaginary part $\varepsilon(\omega)$ is small,

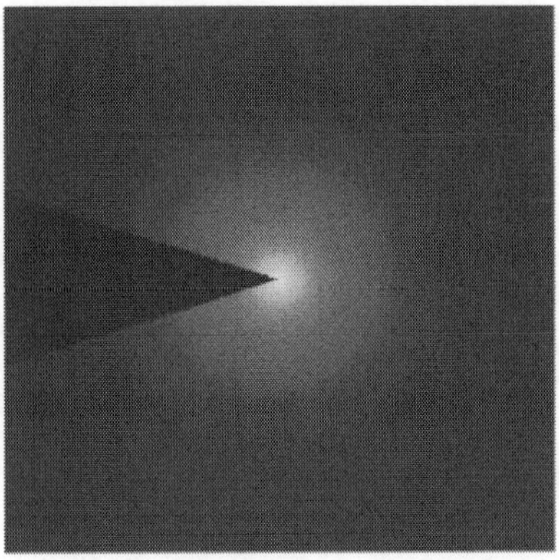

Figure 5.2 Enhancement of the electric field near a sharp corner (the lightning rod effect).

consideration of the imaginary part ε_n is necessary, and a variety of interesting effects appear when quadrupole, octupole, etc., radiation becomes essential instead of a dipole one (Klimov, 2002; Tribelsky and Luk'yanchuk, 2006).

Thus, in the context of the "ε-method" plasmon oscillations appear naturally as a result of the solution of the spectral problem where electric permittivity is an eigenvalue. Another important feature of the described method is that plasmon properties depend in fact only on the particle's shape rather than on its specific material. Both the factors make the "ε-method" extremely useful in most diverse applications and problems. Moreover, the "ε-method" can be regarded as a constructive definition of localized plasmons.

5.2 Application of the "ε-Method" to a Solution of Maxwell's Equations for Nanoparticles

The "ε-method" is especially effective in the case of nanoparticles, that is, in the case when the typical size of a nanoparticle a is smaller than the wavelength λ of the exciting light:

$$ka = \frac{2\pi a}{\lambda} \ll 1. \tag{5.19}$$

In this case, to solve Maxwell's equations Eqs. (5.5) and (5.6), the quasi-static perturbation theory can be used (Rayleigh, 1871;

Stevenson, 1953a, 1953b). In the context of this approach, all the fields are sought as a power series over the size parameter ka, where a is the typical size of a nanoparticle.

$$\mathbf{e} = \mathbf{e}^{(0)} + k\mathbf{e}^{(1)} + k^2\mathbf{e}^{(2)} + \dots ,$$
$$\mathbf{h} = \mathbf{h}^{(0)} + k\mathbf{h}^{(1)} + k^2\mathbf{h}^{(2)} + \dots . \tag{5.20}$$

Electric permittivity $\varepsilon(\omega)$ should not be expanded in powers $k = \omega/c$, that is, frequency.

Substituting these expansions in Maxwell's equations and equating the components with the same powers ka to zero, the system of Maxwell's equations can be reduced to a set of potential theory problems. In particular, eigenvalues ε_n and eigenfunctions $\mathbf{e}_n$ of the "ε-method" can be found by solving the equations

$$\operatorname{div}(\varepsilon(\mathbf{r})\,\mathbf{e}(\mathbf{r})) = 0,$$
$$\operatorname{rot}\mathbf{e}(\mathbf{r}) = 0, \tag{5.21}$$

which can be reduced to the solution of Laplace equations by substituting of $\mathbf{e}_n = -\nabla\varphi_n$:

$$\Delta\varphi_n^{in} = 0, \qquad \Delta\varphi_n^{out} = 0,$$
$$\varphi_n^{in}\big|_S = \varphi_n^{out}\big|_S,$$
$$\varepsilon_n \frac{\partial\varphi_n^{in}}{\partial\mathbf{n}}\bigg|_S = \bar{\varepsilon}\,\frac{\partial\varphi_n^{out}}{\partial\mathbf{n}}\bigg|_S . \tag{5.22}$$

In Eq. (5.22), $\bar{\varepsilon}$ is electric permittivity of the space surrounding the nanoparticle, φ_n^{in} and φ_n^{out} are potentials of eigenfunctions inside and outside the nanoparticle correspondingly, and $\partial\varphi_n/\partial\mathbf{n}|_S$ denotes the normal derivative at the nanoparticles' surface. The last equation in Eq. (5.22) provides continuity of the electrical induction normal components or tangential components of the magnetic field. At the same time, the form of the solution of Maxwell's equation with given exciting currents Eq. (5.16) remains the same. Note that in the nanoparticle case there is no need to find magnetic fields at all for a description of electric plasmon oscillations.

In nanoparticles, the quasistatic description of plasmon resonances, Eq. (5.22), is essentially simpler than the total system of Maxwell's equation (Eq. (5.1)) since the Laplace and Poisson equations are to be solved instead of the Helmholtz ones. And the solution of the Laplace and Poisson equations can be found for nanoparticles of most diverse forms.

The most important feature of the quasi-static description (Eq. (5.22)) is that it allows us to deal with plasmon oscillations only. The other modes of particles ("whispering gallery" modes, see Fig. 5.1) do not occur in this description and do not complicate obtaining and interpreting of results.

Mathematically, the quasi-static "ε-method" differs from the full "ε-method" in that its resonance values of electric permittivity are *negative real* numbers. The absence of the imaginary part in the resonance values of electric permittivity ε_n is due to the fact that there is no radiation in quasi-static approximation and the imaginary parts ε_n appear only due to radiation losses.

Therefore, we come to a conclusion that it is the plasmon part of the spectrum that is described well by the quasi-static "ε-method," which thus can be used effectively for determination of localized plasmon properties.

Naturally, using ε_n and $\mathbf{e}_n = -\nabla\phi_n$ found in quasi-static approximation allows a reasonable description of Eq. (5.17) only in the near-field zone. To find the fields in the Fraunhofer zone, the procedures of near- and far-field linkage are to be used with the help of the total system of Maxwell's equations. This procedure can be executed by a range of methods.

For example, within a quasi-static solution one can find the dipole momentum $\mathbf{d}$ or dipole polarizability α of the nanoparticle, in which one or another localized plasmon is excited. After that with the help of the full system of Maxwell's equations one can find different characteristics of the nanoparticle radiation in the far-field zone, where the key role belongs to retardation effects not taken into account in the quasi-static approximation.

Integrating general solution of Maxwell's equation Eq. (5.16) over the nanoparticle's volume, a general expression for the polarizability tensor $\alpha_{\mu\nu}$ can be obtained through eigenfunctions and eigenvalues of the "ε-method":

$$\alpha_{\mu\nu} = \frac{\varepsilon - \bar{\varepsilon}}{4\pi} \sum_m C^m_{\mu\nu} \frac{\varepsilon_m/\bar{\varepsilon} - 1}{\varepsilon_m - \varepsilon}, \tag{5.23}$$

where $\bar{\varepsilon}$ is electric permittivity of the space surrounding the nanoparticle. The strength of the dipole plasmon resonances $C^m_{\mu\nu}$ can be expressed through the plasmons' eigenfunctions in a very simple way:

$$C^m_{\mu\nu} = \frac{\int\limits_{V+} \mathbf{e}_{m,\mu} dV \int\limits_{V+} \mathbf{e}_{m,\nu} dV}{\int\limits_{V+} \mathbf{e}^2_m dV}. \tag{5.24}$$

Note that in some symmetric particles, only a finite number of the sum terms in Eq. (5.23) is different from zero since only few modes have a nonzero dipole momentum. For example, in ellipsoids only three modes have a nonzero dipole moment.

The general expression for the plasmons' "strengths" allows finding of the general sum rules (Fuchs, 1975; Fuchs and Liu, 1976):

$$\begin{aligned} &\sum_m C^m_{\mu\nu} = V\delta_{\mu\nu}, \\ &\sum_m \frac{\bar{\varepsilon}}{\bar{\varepsilon} - \varepsilon_m} \sum_\mu C^m_{\mu\mu} = V, \end{aligned} \tag{5.25}$$

where V is the nanoparticle volume. Other sum rules are discussed in (Pitarke et al., 2007).

Knowing the nanoparticle polarizability, the fields in the Fraunhofer zone, scattering cross sections, and a range of the other important characteristics can be found. For example, scattering and absorption cross sections are expressed through the polarizability α (Landau and Lifshitz, 1960):

$$\begin{aligned} \sigma^{\text{abs}} &= 4\pi \bar{k} \text{Im} \mathbf{e}^*_\mu \alpha_{\mu\nu} \mathbf{e}_\mu, \\ \sigma^{\text{sca}} &= \frac{8\pi}{3} \bar{k}^4 \mathbf{e}^*_\nu \alpha^*_{\mu\nu} \alpha_{\mu\nu'} \mathbf{e}_{\nu'}, \end{aligned} \tag{5.26}$$

where $\mathbf{e}$ is the vector of the incident radiation polarization and $\bar{k} = \sqrt{\bar{\varepsilon}}\omega/c$ is the wavenumber of the space surrounding the nanoparticle.

Moreover, knowing the dipole momentum and polarizability of the nanoparticle plasmon mode in quasi-static approximation, the imaginary part $\text{Im}(\varepsilon_n)$ can be found, which cannot be defined in this approximation immediately (Meier and Wokaun, 1983; Wokaun et al., 1982). Below, the generalized derivation of the analytical expression for the imaginary part $\text{Im}(\varepsilon_n)$ is represented, which is applicable to arbitrary anisotropic nanoparticles.

Indeed, oscillating dipole plasmon momentum provokes the radiation and therefore radiation damping, which is not related to the Joule losses inside a nanoparticle substance. The radiation damping is related to a radiation reaction force acting on conduction electrons in a nanoparticle. As in the case of radiative damping of a single oscillating charged particle (Jackson, 1975), the radiation influence on a localized plasmon can be described with the help of the self-field strength:

$$\mathbf{E}_{\text{rad}} = \frac{2}{3} i k^3 \mathbf{d} \sqrt{\bar{\varepsilon}}, \tag{5.27}$$

where **d** is the dipole moment of the plasmon:

$$\mathbf{d}_{\mu} = \bar{\varepsilon} \alpha^{m}_{\mu\nu} \left(\mathbf{E} + \mathbf{E}_{\text{rad}}\right)_{\nu}, \tag{5.28}$$

where

$$\alpha^{m}_{\mu\nu} = \frac{\varepsilon - \bar{\varepsilon}}{4\pi} \frac{\varepsilon_m/\bar{\varepsilon} - 1}{\varepsilon_m - \varepsilon} C^{m}_{\mu\nu} \tag{5.29}$$

is the dipole polarizability of the mth plasmon mode in the quasi-static approximation. By solving Eq. (5.28) for **d**, the effective plasmon polarizability can be found with regard to radiation damping:

$$\alpha^{m,\text{ eff}}_{\mu\nu} \approx \alpha^{m}_{\mu\nu} + \frac{2}{3} i \bar{k}^3 \alpha^{m}_{\mu\rho} \alpha^{m}_{\rho\nu}. \tag{5.30}$$

On the other hand, Eq. (5.29) is the expression defining effective polarizability, $\alpha^{m,\,eff}_{\mu\nu}$, which, however, is based on exact solutions of Maxwell's equations Eqs. (5.5) and (5.6):

$$\alpha^{m}_{\mu\nu} = \frac{\varepsilon - \bar{\varepsilon}}{4\pi} \frac{\varepsilon^{\text{exact}}_{m}/\bar{\varepsilon} - 1}{\varepsilon^{\text{exact}}_{m} - \varepsilon} C^{m}_{\mu\nu} \approx \frac{\varepsilon - \bar{\varepsilon}}{4\pi} \frac{\varepsilon_m/\bar{\varepsilon} - 1}{\varepsilon_m - \varepsilon} C^{m}_{\mu\nu} \left(1 - \frac{\Delta\varepsilon_m}{\varepsilon_m - \varepsilon}\right), \tag{5.31}$$

where $\Delta\varepsilon_m = \varepsilon^{\text{exact}}_{m} - \varepsilon_m$ is a correction to the electric permittivity resonance value, determined by plasmon dipole radiation, and a resonance condition $\varepsilon_m \approx \varepsilon$ is assumed. By comparing this expression with Eq. (5.30), the correction $\Delta\varepsilon_m$ can be found:

$$\frac{\Delta\varepsilon_m}{\bar{\varepsilon}} = -\frac{2}{3} i \bar{k}^3 \frac{\left(\varepsilon_m/\bar{\varepsilon} - 1\right)^2}{4\pi} \sum_{\rho} C^{m}_{\rho\rho}. \tag{5.32}$$

In particular, in the case of a sphere of radius a there are only three modes with a nonzero dipole moment, for each of these modes $\sum_{\rho} C^{m}_{\rho\rho} = (4\pi/3)a^3$, and instead of Eq. (5.32) we have:

$$\Delta\varepsilon^{\text{sphere}}_{m} = -2i \left(\bar{k}a\right)^3 \bar{\varepsilon}. \tag{5.33}$$

Let us emphasize that since the radiation reaction Eq. (5.27) is related to the dipole moment, the corrections in Eq. (5.32) are true only for the modes with a dipole momentum, that is, in the case of $C^m_{\mu\nu} \neq 0$.

Generally speaking, for any plasmon mode the corrections to the real parts of the order $\Delta\varepsilon_m \approx (\bar{k}a)^2\bar{\varepsilon}$ also should be taken into account, but there is no simple method of their calculation. Besides, these corrections are purely real and cause only a resonance frequency shift, while Eq. (5.32) describes resonance broadening, which is of crucial importance.

5.3 Implementations of the "ε-Method" of the Maxwell's Equations' Solution

5.3.1 *Analytical Solutions*

In the case of a finite-volume particle of an arbitrary size, analytical solutions exist only for spherical and spherically layered systems (Mie, 1908; Debay, 1909). This solution will be considered in detail in Chapter 6. Even in this case, however, extraction of plasmon resonances from Mie's solution or reduction of this solution to Eq. (5.16) is a complex problem.

More sophisticated analytical solutions were developed for prolate or oblate spheroids (Asano Yamamoto, 1975; Farafonov, 1983; Voshchinnikov Farafonov, 1993). In this case, the plasmon analysis is also very complicated.

Thus, the numerical methods are the main instrument of the "ε-method" equations' solution or particles of arbitrary large sizes.

The "ε-method" equations are essentially simpler in the case of nanoparticles, and corresponding solutions take the form of Eq. (5.16) immediately. In the quasi-static case, that is, in the case of nanoparticles, the "ε-method" equations in Eq. (5.22) can be solved by the method of separation of variables analytically for spheres, spheroids, ellipsoids, clusters of two spherical nanoparticles, and some other geometries. All these geometries are enormously important for applications, and plasmon oscillations in nanoparticles of these shapes will be considered in the following chapters.

5.3.2 Integral Form of the "ε-Method" and Its Numerical Solutions

Commonly, integral forms of equations in Eq. (5.22) are more convenient for finding electric permittivity eigenvalues and the corresponding eigenpotential distribution. For brevity, we assume that a particle is situated in vacuum. The generalization for an arbitrary dielectric medium with permittivity $\bar{\varepsilon}$ is performed by substituting $\varepsilon(\omega)$ and ε_m by $\varepsilon(\omega)/\bar{\varepsilon}$ and $\varepsilon_m/\bar{\varepsilon}$, correspondingly.

To obtain the integral form of the "ε-method," let us present Eq. (5.21) in the form of:

$$\operatorname{div}\mathbf{e}(\mathbf{r}) = -\operatorname{div}\left((\varepsilon(\mathbf{r}) - 1)\,\mathbf{e}(\mathbf{r})\right), \tag{5.34}$$

or, taking into account $\mathbf{e} = -\nabla\varphi$, in the form of:

$$\Delta\varphi(\mathbf{r}) = \operatorname{div}\left((\varepsilon(\mathbf{r}) - 1)\,\mathbf{e}(\mathbf{r})\right). \tag{5.35}$$

An arbitrary solution of the Poisson equation Eq. (5.35) can be presented as:

$$\varphi(\mathbf{r}) = \varphi_0 - \frac{1}{4\pi}\int \frac{d\mathbf{r}'}{|\mathbf{r}-\mathbf{r}'|}\operatorname{div}\left\{\left[\varepsilon(\mathbf{r}') - 1\right]\mathbf{e}(\mathbf{r}')\right\}, \tag{5.36}$$

where φ_0 is the potential of external field and the integration is over the whole space.

After integrating by parts and using the fact that $\varepsilon(\mathbf{r}) = 1$ outside the particle and $\varepsilon(\mathbf{r}) = \varepsilon$ inside it, we obtain the final integral formulation of the "ε-method" for nanoparticles:

$$\varphi(\mathbf{r}) = -\frac{\varepsilon - 1}{4\pi}\int\limits_{V^+} d\mathbf{r}'\nabla'\varphi(\mathbf{r}')\,\nabla'\frac{1}{|\mathbf{r}-\mathbf{r}'|}, \qquad \mathbf{r}, \mathbf{r}' \in V^+. \tag{5.37}$$

The integral equation obtained is a weakly singular homogeneous Fredholm equation of the second kind, which can be solved both numerically and analytically. The analytical solution of Eq. (5.37) relating to a nanoparticle of an arbitrary ellipsoid form will be considered in Chapter 8.

To solve Eq. (5.37) numerically, the integral is approximated with the help of the finite sum, resulting in a usual problem on eigenvalues.

In the case of a cubic mesh, the discrete version of the integral equation Eq. (5.37) takes the form:

$$\varphi_i = \frac{\varepsilon - 1}{4\pi}\sum_{j\neq i}(\nabla\varphi)_j\frac{\mathbf{r}_j - \mathbf{r}_i}{\left|\mathbf{r}_j - \mathbf{r}_i\right|^3}\Delta V_j. \tag{5.38}$$

For cells of the other form, the contribution of the term with $i = j$ will not be equal to zero, and the following expression should be used instead of Eq. (5.38):

$$\varphi_i = \frac{\varepsilon - 1}{4\pi} \left\{ \sum_{j \neq i} (\nabla\varphi)_j \frac{\mathbf{r}_j - \mathbf{r}_i}{\left|\mathbf{r}_j - \mathbf{r}_i\right|^3} \Delta V_j \right.$$
$$\left. +\varphi_i \left(\int_{\Delta S_i} ds' \mathbf{n}\left(\mathbf{s}'\right) \frac{\left(\mathbf{s}' - \mathbf{r}_i\right)}{\left|\mathbf{s}' - \mathbf{r}_i\right|^3} - 4\pi \right) \right\}, \qquad (5.39)$$

where ΔS_i is the surface of ΔV_i and $\mathbf{n}(\mathbf{s}')$ is a normal vector to ΔS_i at point $\mathbf{s}'$.

The integral equation relates potential values in all inner points of the nanoparticle. Eq. (5.22) can also be reduced to an integral equation relating only the values of the surface charge σ on the surface of the nanoparticle. This formulation of the "ε-method" was proposed by (Fuchs, 1975; Ouyang and Isaacson, 1989a, 1989b).

To obtain the "ε-method" equations in the form of surface integral equations, let us note that the surface charge is defined by the expression (Jackson, 1975):

$$\sigma = \mathbf{nP} = \frac{\varepsilon - 1}{4\pi} \mathbf{nE}, \qquad (5.40)$$

where $\mathbf{n}$ is the external normal to the particle surface and $\mathbf{P}$ is polarization. For such surface charges, the electric potential and the electric field strength can be presented by the Poisson integral in any point except the surface S of the particle:

$$\varphi(\mathbf{r}) = \int_S ds' \frac{\sigma(\mathbf{s}')}{|\mathbf{r} - \mathbf{s}'|}, \quad \mathbf{r} \notin S,$$
$$\mathbf{e}(\mathbf{r}) = \int_S ds' \sigma\left(\mathbf{s}'\right) \frac{\mathbf{r} - \mathbf{s}'}{|\mathbf{r} - \mathbf{s}'|^3}, \quad \mathbf{r} \notin S, \qquad (5.41)$$

where integration is performed over the surface S of the particle.

It is well known from potential theory that the normal component of the electric field experiences discontinuity on the particle surface:

$$\varepsilon \mathbf{n}\mathbf{e}^- = \mathbf{n}\mathbf{e}^+, \qquad (5.42)$$

where $\mathbf{ne}^{\pm}$ is the normal component of the electric field at the internal or external boundary. On the other hand, from the theory of double-layer potential (Kellogg, 1929; Mikhlin, 1970) it is known that the values of electric field normal components on the internal and external sides of the surface are related by the formulas:

$$\mathbf{n}\left(\mathbf{r}\right)\mathbf{e}^{\pm}\left(\mathbf{r}\right)=\int\limits_{S}d\mathbf{s}'\sigma\left(\mathbf{s}'\right)\frac{\mathbf{n}\left(\mathbf{r}\right)\left(\mathbf{r}-\mathbf{s}'\right)}{\left|\mathbf{r}-\mathbf{s}'\right|^{3}}\pm 2\pi\sigma\left(\mathbf{r}\right),\quad \mathbf{r}\in S. \quad (5.43)$$

Substituting Eq. (5.43) in Eq. (5.42), we obtain the following integral equation for the surface charge:

$$\sigma\left(\mathbf{r}\right)=\frac{1}{2\pi}\frac{\varepsilon-1}{\varepsilon+1}\int\limits_{S}d\mathbf{s}'\sigma\left(\mathbf{s}'\right)\frac{\mathbf{n}\left(\mathbf{r}\right)\left(\mathbf{r}-\mathbf{s}'\right)}{\left|\mathbf{r}-\mathbf{s}'\right|^{3}},\quad \mathbf{r},\mathbf{r}'\in S. \quad (5.44)$$

The equation is an integral Fredholm equation of the second kind with a weakly singular kernel, which is also asymmetric in a general case. Nevertheless, (Ouyang and Isaacson, 1989a, 1989b) have shown that its eigenvalues:

$$\lambda=\frac{\varepsilon-1}{\varepsilon+1} \quad (5.45)$$

are real numbers.

Moreover, the eigenfunctions in Eq. (5.44) form a complete orthonormal system:

$$\int\limits_{S}d\mathbf{s}\int d\mathbf{s}'\frac{\sigma_{n}\left(\mathbf{s}\right)\sigma_{m}\left(\mathbf{s}'\right)}{\left|\mathbf{s}-\mathbf{s}'\right|}=\delta_{nm}. \quad (5.46)$$

(Mayergoyz et al., 2005) have shown that for three-dimensional nanoparticles with a smooth convex surface S, the eigenvalues of the integral equation Eq. (5.44) can be estimated with the help of the following inequality:

$$\left|\frac{\varepsilon-1}{\varepsilon+1}\right|>Q=\frac{1}{1-A/(4\pi Rd)}, \quad (5.47)$$

where A is the nanoparticle area, R is the maximal radius of curvature of its surface, and d is the maximal radius of the nanoparticle.

Using this relation, the estimation of resonance values of electric permittivity can be found:

$$\frac{1+Q}{1-Q}<\varepsilon<\frac{1-Q}{1+Q}. \quad (5.48)$$

In the case when the nanoparticle material follows the Drude law, the estimation of plasmon frequencies of convex isolated nanoparticles will have the following form:

$$\frac{Q-1}{2Q} < \frac{\omega^2}{\omega_{pl}^2} < \frac{Q+1}{2Q}, \tag{5.49}$$

which implies that the plasmon frequencies' bandwidth in nanoparticles is less than $\omega_{\max} - \omega_{\min} = \omega_{pl}/\sqrt{1 - A/4\pi Rd}$.

For example, in the case of a nanosphere of radius a the nanoparticle area is $A = 4\pi a^2$, the maximal radius of curvature of this surface is $R = a$, the maximal diameter of the nanoparticle is $d = 2a$, and the parameter $Q = 2$ and according to Eq. (5.48):

$$-3 < \varepsilon < -1/3. \tag{5.50}$$

This estimation is in good agreement with the exact analytical solution for the nanosphere (see Chapter 6):

$$\varepsilon_n = -\frac{n+1}{n}, \quad n = 1, 2, 3\ldots, \tag{5.51}$$

which lies in the range $-2 \leq \varepsilon < -1$.

If the surface integral, Eq. (5.44), is replaced by a finite sum of N small fragments of the surface ΔS_i, $i = 1\ldots N$, the system of homogeneous linear equations with an eigenvalue Eq. (5.45) will be obtained:

$$\sigma_i = \frac{\lambda}{2\pi}\sum_j R_{ij}\sigma_j, \tag{5.52}$$

where $\sigma_i = \sigma(\mathbf{s}_i)$ is the value of the surface charge in the center of the ith cell and

$$R_{ij} = \begin{cases} \frac{\mathbf{n}_i(\mathbf{s}_i - \mathbf{s}_j)\Delta S_j}{|\mathbf{s}_i - \mathbf{s}_j|^3}, & i \neq j, \\ 0, & i = j. \end{cases} \tag{5.53}$$

The presented in Eq. (5.44) formulation of the "ε-method" is suitable for finding the resonance values of electric permittivity without finding the potential inside and outside the nanoparticle. If these potentials are to be found, the eigenvalues obtained can be substituted into Eq. (5.41) and the potentials can be found by direct integration.

This simplest method of the "ε-method" equations' solution was implemented by Fuchs (1975) for the description of plasmon oscillations of cubic nanoparticles.

The application of the surface equations method to nanoparticles with sharp edges and vertices has peculiarities that result in slow convergence of a numerical solution. To find the spectrum in this case, the equation adjoint to the integral equation Eq. (5.44) should be used:

$$\tau(\mathbf{r}) = \frac{1}{2\pi}\frac{\varepsilon - 1}{\varepsilon + 1}\int_S d^2\mathbf{r}'\tau(\mathbf{r}')\frac{\mathbf{n}(\mathbf{r}')(\mathbf{r}' - \mathbf{r})}{|\mathbf{r} - \mathbf{r}'|^3}, \quad \mathbf{r}, \mathbf{r}' \in S. \tag{5.54}$$

Eq. (5.54) has the same spectrum as Eq. (5.44), but its numerical solution convergence is much better, especially in the case of nonsmooth surfaces. Physically, Eq. (5.54) describes surface density of the dipoles, which results in the same potentials as the surface charges in Eq. (5.44) do. The effective method of the solution of Eq. (5.54) was presented in (Mayergoyz et al., 2005; Mayergoyz and Zhang, 2005). In this case, the discrete analogue of Eq. (5.54) has the form:

$$\tau_i = \frac{\lambda}{2\pi}\sum_{j=1}^{N} U_{ij}\tau_j, \tag{5.55}$$

where

$$U_{ij} = \begin{cases} \frac{\mathbf{n}_j(\mathbf{s}_j - \mathbf{s}_i)\Delta S_j}{|\mathbf{s}_i - \mathbf{s}_j|^3}, & i \neq j, \\ 2\pi - \sum\limits_{l=1, l\neq i}^{N} U_{il}, & i = j. \end{cases} \tag{5.56}$$

For further accuracy enhancement, Eq. (5.55) can be rewritten in a form that enables the full charge equality to zero independently from calculation accuracy (Mayergoyz and Zhang, 2005):

$$\tau_i = \frac{\lambda}{2\pi}\sum_{j=1}^{N}\left(U_{ij} - \frac{1}{N}\sum_{i=1}^{N} U_{ij}\right)\tau_j. \tag{5.57}$$

5.4 The Analogy between Localized Plasmons and Atoms and Molecules

Above, it was shown that localized plasmons can be characterized by discrete values of electric permittivity (or plasmon frequency) and by corresponding localized potential distributions. These properties of localized plasmons resemble greatly the properties of quantum dynamics of electrons in atoms and molecules described by the Schrodinger equation:

$$\left(-\frac{\hbar^2}{2m}\Delta + U(\mathbf{r})\right)\psi_n(\mathbf{r}) = E_n\psi_n(\mathbf{r}). \tag{5.58}$$

Discrete bound states of electrons are characterized by the negative values of energy eigenvalues E_n (an analog to ε_n) and by localized (integrable) wavefunctions $\psi_n(\mathbf{r})$ (an analog to $\varphi_n(\mathbf{r})$).

The analogy between plasmon oscillations in spherically symmetric particles and in hydrogen-like atoms is especially noticeable.

Indeed, as it will be shown further (Chapter 6), in the case of a spherical nanoparticle of radius a the solution of the "ε-method" equations has the form:

$$\begin{aligned} &\varphi_n(\mathbf{r}) \sim Z_n(r)\, Y_{nm}(\theta, \varphi), \\ &\varepsilon_n + 1 = -\frac{1}{n}, \quad n = 1, 2, \ldots . \end{aligned} \tag{5.59}$$

In Eq. (5.59), $Y_{nm}(\theta, \varphi)$ are usual spherical harmonics and Z_n is the radial part of the eigenfunction:

$$Z_n(r) = \begin{cases} r^n, & r < a, \\ \dfrac{a^{2n+1}}{r^{n+1}}, & r > a. \end{cases} \tag{5.60}$$

In the case of a hydrogen atom, the solution of Eq. (5.58) takes the form analogous to Eq. (5.59):

$$\begin{aligned} &\psi_n(\mathbf{r}) \sim Z_{nl}(r)\, Y_{lm}(\theta, \phi), \\ &E_n = -\frac{m_e e^4}{\hbar^2}\frac{1}{2n^2}, \quad n = 1, 2, \ldots , \end{aligned} \tag{5.61}$$

where the radial part of the wavefunction is described by the expression:

$$Z_{nl}(r) \sim (r/a_0)^l\, F\left(-n + l + 1, 2l + 2, \frac{2r}{na_0}\right) e^{-\frac{r}{na_0}}, \tag{5.62}$$

and $a_0 = \hbar^2/(m_e e^2)$ is the Bohr radius and F is the confluent hypergeometric function (Davydov, 1965). The analogy between Eqs. (5.61) and (5.59) is evident, and angular dependencies are exactly the same. The dependencies of radial functions of Eq. (5.60)

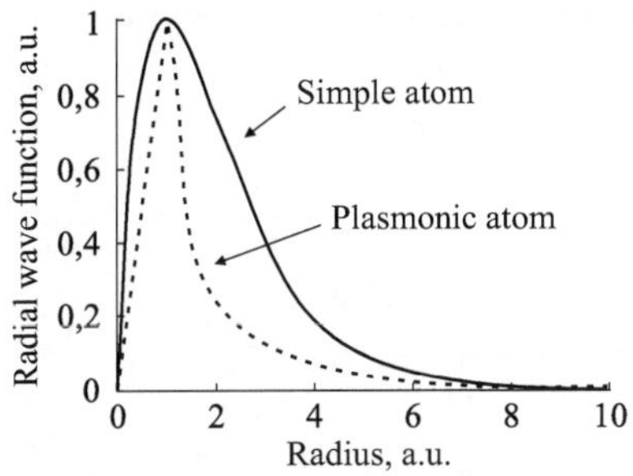

Figure 5.3 The illustration of the analogy between the radial wavefunction of a spherical plasmon and the $2p$ state of a hydrogen-like atom.

for $n = 1$ (dipole oscillations) and Eq. (5.62) for $l = 1$, $n = 2$ ($2p$ state) are shown in Fig. 5.3. The figure displays evidently that the radial functions have very similar behavior.

Therefore, we may talk about a certain analogy between localized plasmons and atoms.

An analogy of this kind can be drawn between plasmon oscillations in clusters of two nanoparticles and electron dynamics in two-atom molecules. Continuing the analogy, plasmon oscillations in nanoparticles may be naturally referred to as plasmonic atoms, and bound states of plasmons atoms that can coexist in nanoparticle clusters may be denoted as plasmonic molecules.

This analogy permits the productive use of many atom physics methods and approaches for a description of plasmon oscillations in nanoparticles.

It should be mentioned that the specified analogy is not perfect, and plasmon physics is more abundant, since unlike spherically symmetric atoms, nanoparticles of the most diverse forms exist or can be produced, enabling control of their spectra and potential distribution, that is, control of the wavefunction of plasmon atoms. On the other hand, the possibility of production of nanoparticles with predefined properties allows arranging of their effective interaction with atomic and molecular systems and thus creating effective nanodevices on this basis. Mathematically, the physics of localized plasmons is more complex and diverse since Eqs. (5.44) and (5.54), describing plasmon dynamics, are not self-adjoint in distinction from the Schrodinger self-adjoint Eq. (5.58) describing atoms, and many results of usual quantum mechanics cannot be applied in the case of plasmons.

References

Agranovich, M.S., Katsenelenbaum, B.Z., Sivov, A.N., Voitovich, N.N. (1999) *Generalized Method of Eigenoscillation in Diffraction Theory* (Wiley, Berlin).

Apell, S.P., Echenique, P.M., Ritchie, R.H. (1996) Sum rules for surface plasmon frequencies, *Ultramicroscopy*, **65**, 53–60.

Asano, S., Yamamoto, G. (1975) Light scattering by a spheroidal particle, *Appl. Opt.*, **14**, 29–49.

Braginsky, V.B., Gorodetsky, M.L., Ilchenko, V.S. (1989) Quality-factor and nonlinear properties of optical whispering-gallery modes, *Phys. Lett. A*, **137**, 393–396.

Collot, L., Lefevre, V., Brune, M., Raimond, J.-M., Haroche, S. (1993) Very high-Q whispering gallery modes resonances observed on fused silica microspheres, *Europhys. Lett.*, **23**, 327–333.

Davydov, A.S. (1965) *Quantum Mechanics* (Pergamon).

Debay, P. (1909) Der Lichtdruck auf Kugeln von beliebigem Material, *Ann. Phys.*, **30**, 57–135.

Farafonov, V.G. (1983) Diffraction of a plain electromagnetic wave on a dielectric spheroid, *Differ. Equations*, **19**, 1765.

Fuchs, R. (1975) Theory of the optical properties of ionic crystal cubes, *Phys. Rev. B*, **11**, 1732–1740.

Fuchs, R., Liu, S.H. (1976) Sum rule for the polarizability of small particles, *Phys. Rev. B*, **14**, 5521–5522.

Gorodetsky, M.L., Savchenkov, A.A., Ilchenko, V.S. (1996) On the ultimate Q of optical microsphere resonators, *Opt. Lett.*, **21**, 453–455.

Hill, S.C., Benner, R.E. (1987) Morphology-dependent resonances, in *Optical Effects Associated with Small Particles*, ed. P.W. Barber, R.K. Chang (World Scientific, Singapore).

Jackson, J.D. (1975) *Classical Electrodynamics*, 2nd ed. (John Wiley & Sons, New York).

Kellogg, O.D. (1929) *Foundation of Potential Theory* (Berlin).

Klimov, V.V. (2002) Spontaneous emission of an excited atom placed near a "left-handed" sphere, *Opt. Commun.*, **211**, 183–196.

Landau, L.D., Lifshitz, E.M. (1960) *Electrodynamics of Continuous Media* (Pergamon, Oxford).

Mayergoyz, I.D., Fredkin, D.R., Zhang, Z. (2005) Electrostatic (plasmon) resonances in nanoparticles, *Phys. Rev. B*, **72**, 155412.

Mayergoyz, I.D., Zhang, Z. (2005) Modeling of the electrostatic (plasmon) resonances in metallic and semiconductor nanoparticles, *J. Comput. Electr.*, **4**, 139–143.

Meier, M., Wokaun, A. (1983) Enhanced fields on large metal particles: dynamic depolarization, *Opt. Lett.*, **8**, 581–583.

Mie, G. (1908) Beiträge zur optik truber medien, speziell kolloida ler metallösungen, *Ann. Phys.*, **25**, 377–445.

Mikhlin, S.G. (1970) *Mathematical Physics, an Advanced Course* (North-Holland, Amsterdam).

Ouyang, F., Isaacson, M. (1989a) Surface plasmon excitation of objects with arbitrary shape and dielectric constant, *Philos. Mag. B*, **60**, 481–492.

Ouyang, F., Isaacson, M. (1989b) Accurate modeling of particle-substrate coupling of surface plasmon excitation in EELS, *Ultramicroscopy*, **31**, 345–349.

Pitarke, J.M., Silkin, V.M., Chulkov, E.V., Echenique, P.M. (2007) Theory of surface plasmons and surface-plasmon polaritons, *Rep. Prog. Phys.*, **70**, 1–87.

Rayleigh Lord (1871) On the light from the sky, its polarization and colour, *Philos. Mag.*, **41**, 107–120, 274–279 (reprinted in *Scientific Papers* by Lord Rayleigh, Vol.1: 1869–1881#8, Dover, New York, 1964).

Schiller, S., Byer, R.L. (1991) High-resolution spectroscopy of whispering gallery modes in large dielectric spheres, *Opt. Lett.*, **16**, 1138–1140.

Stevenson, A.F. (1953a) Electromagnetic scattering by an ellipsoid in the third approximation, *J. Appl. Phys.*, **24**, 1143–1151.

Stevenson, A.F. (1953b) Solution of electromagnetic scattering problems as power series in the ratio (dimension of scatterer)/wavelength, *J. Appl. Phys.*, **24**, 1134–1142.

Tribelsky, M.I., Luk'yanchuk, B.S. (2006) Anomalous light scattering by small particles, *Phys. Rev. Lett.*, **97**, 263902.

Voshchinnikov, N.V., Farafonov, V.G. (1993) Optical properties of spheroidal particles, *Astrophys. Space Sci.*, **204**, 19–86.

Wokaun, A., Gordon, J.P., Liao, P.F. (1982) Radiation damping in surface-enhanced Raman scattering, *Phys. Rev. Lett.*, **48**, 1574–1574.

Chapter 6

Optical Properties of Spherical Particles

Spherical geometry appears to be the most important geometry, since in a first approximation many particles may be regarded as spherical ones. On the other hand, any electrodynamics problems concerning spherical particles can be solved analytically by analogy with the well-known Mie solution for plane wave diffraction on a sphere (Mie, 1908). All that gives a special importance and generality to investigations of optical properties of spherical particles. In this respect, in the present chapter we will not restrict ourselves to plasmon properties of nanoparticles but will also introduce main analytical results concerning arbitrary spherical particles and analyze them with reference to different applications. A special attention will be paid to the case of excitation of spherical particles by point sources of light (molecules and quantum dots) since this case is of particular significance for prospective high-tech applications of nano-optics and nanophotonics.

Nanoplasmonics
Vasily Klimov

ISBN 978-981-4267-16-8 (Hardcover), 978-981-4267-42-7 (eBook)
www.panstanford.com

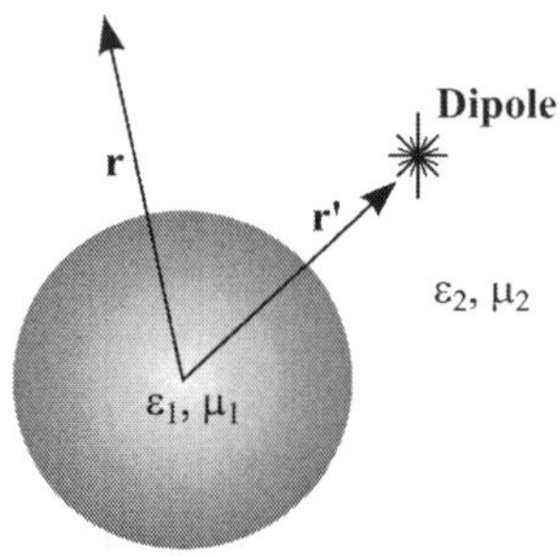

Figure 6.1 The geometry of the problem of dipole radiation near a spherical particle.

6.1 Excitation of a Spherical Particle by a Dipole Source of Light

To be definite, let us consider the case of a dipole source of light situated outside a sphere of radius a with electric permittivity ε_1 and magnetic permeability μ_1 (Fig. 6.1). The case with the source situated inside a particle can be considered similarly. If the source fields are denoted by $\mathbf{E}^{\mathrm{dip}}$, $\mathbf{B}^{\mathrm{dip}}$, then the fields inside and outside the sphere are denoted by $\mathbf{E}^{(1)}$, $\mathbf{B}^{(1)}$ and $\mathbf{E}^{\mathrm{dip}} + \mathbf{E}^{(2)}$, $\mathbf{B}^{\mathrm{dip}} + \mathbf{B}^{(2)}$, correspondingly.

Since the problem has a spherical symmetry, the expression for the fields can be written as an expansion over vector spherical harmonics (Jackson, 1975). Herewith, the main difficulty is the expansion of the source field in vector spherical harmonics with the center related to the center of the spherical particle. To find this expansion, let us note that the expressions for the charge density and for the current of monochromatic dipole source having the dipole moment $\mathbf{d}_0$ and situated at the point $\mathbf{r}'$ have the form:

$$\begin{aligned} \rho &= -(\mathbf{d}_0\nabla)\delta^{(3)}(\mathbf{r}-\mathbf{r}')e^{-i\omega t}, \\ \mathbf{j} &= -i\omega\mathbf{d}_0\delta^{(3)}(\mathbf{r}-\mathbf{r}')e^{-i\omega t}, \end{aligned} \tag{6.1}$$

which is in agreement with the charge conservation law. Later, we will omit the factor of monochromatic dependence on time.

If the current distribution of this kind is placed in an infinite dielectric medium with dielectric and magnetic permittivities ε_2, μ_2, then the expression for the fields takes the form (see, e.g., (Vainshtein, 1988)):

$$\begin{aligned} \mathbf{E}^{\text{dip}} &= k_2^2\mathbf{\Pi}^0 + \nabla(\nabla\mathbf{\Pi}^0), \\ \mathbf{B}^{\text{dip}} &= -ik_0\varepsilon_2\mu_2\left[\nabla\mathbf{\Pi}^0\right], \end{aligned} \quad (6.2)$$

where the electric Hertz vector $\mathbf{\Pi}^0$ is described by the expression

$$\mathbf{\Pi}^0 = \frac{\mathbf{d}_0}{\varepsilon_2}\frac{e^{ik_2|\mathbf{r}-\mathbf{r}'|}}{|\mathbf{r}-\mathbf{r}'|}, \quad (6.3)$$

and ∇ is the "nabla" Hamilton operator affecting all the functions $\mathbf{r}$ after it and $k_0 = \omega/c$; $k_2 = \sqrt{\mu_2\varepsilon_2}\omega/c$.

To find the expansion of dipole electric and magnetic fields in vector spherical harmonics, let us first consider the radial component of the magnetic field. Using the identity:

$$\left(\mathbf{r}\mathbf{B}^{\text{dip}}\right) = -ik_0\mu_2\left(\mathbf{r}\left[\nabla\mathbf{d}_0\right]\right)\frac{e^{ik_2|\mathbf{r}-\mathbf{r}'|}}{|\mathbf{r}-\mathbf{r}'|} = -ik_0\mu_2\left(\mathbf{d}_0\left[\nabla'\mathbf{r}'\right]\right)\frac{e^{ik_2|\mathbf{r}-\mathbf{r}'|}}{|\mathbf{r}-\mathbf{r}'|}, \quad (6.4)$$

(∇' is the "nabla" Hamilton operator affecting all the functions of $\mathbf{r}'$ after it) and the standard expansion (Jackson, 1975):

$$\frac{e^{ik_2|\mathbf{r}-\mathbf{r}'|}}{|\mathbf{r}-\mathbf{r}'|} = 4\pi k_2 i\sum_{n=0}^{\infty}\sum_{m=-n}^{n} j_n(k_2r)h_n^{(1)}(k_2r')Y_{nm}^*(\vartheta',\varphi')Y_{nm}(\vartheta,\varphi), \quad (6.5)$$

which is true for $r < r'$, the expression for the radial component of the magnetic field can be expanded in scalar spherical harmonics $Y_{nm}(\vartheta,\varphi)$:

$$\left(\mathbf{r}\mathbf{B}^{\text{dip}}\right) = \sum_{n=1}^{\infty}\sum_{m=-n}^{n} a_{nm}j_n(k_2r)Y_{nm}(\vartheta,\varphi), \quad (6.6)$$

where

$$a_{nm} = -i4\pi k_0^2\sqrt{\varepsilon_2\mu_2}\mu_2 h_n^{(1)}(k_2r')\left(\mathbf{d}_0\hat{\mathbf{L}}'\right)Y_{nm}^*(\vartheta',\varphi'), \quad (6.7)$$

and $\hat{\mathbf{L}}$ is the angular moment operator $\hat{\mathbf{L}} = -i\left[\mathbf{r}\nabla\right]$.

Here, usual expressions for spherical harmonics $Y_{nm}(\vartheta,\varphi)$ and the spherical Bessel functions $j_n(z)$, $h_n^{(1)}(z)$ are used (Abramowitz and Stegun, 1972).

Now multiplying the Maxwell's equation

$$\left[\nabla\mathbf{E}^{\text{dip}}\right] = ik_0\mathbf{B}^{\text{dip}} \tag{6.8}$$

by $\mathbf{r}$, we obtain the equation that allows finding of the transverse electric (TE) component of the dipole field:

$$\left(\hat{\mathbf{L}}\mathbf{E}_{\text{TE}}^{\text{dip}}\right) = k_0\left(\mathbf{B}^{\text{dip}}\mathbf{r}\right). \tag{6.9}$$

Using the properties of the angular moment operator, $\hat{\mathbf{L}}^2 Y_{nm}(\vartheta, \varphi) = n(n+1)\, Y_{nm}(\vartheta, \varphi)$, it is easy to verify that the TE part of the field will be described by the expression

$$\mathbf{E}_{\text{TE}}^{\text{dip}} = k_0 \sum_{n=1}^{\infty}\sum_{m=-n}^{n} \frac{a_{nm}}{n(n+1)} j_n(k_2 r)\hat{\mathbf{L}}Y_{nm}(\vartheta, \varphi). \tag{6.10}$$

The corresponding component of the magnetic field can be found by substituting this expression into the Maxwell's expression in Eq. (6.8):

$$\mathbf{B}_{\text{TE}}^{\text{dip}} = -i\sum_{n=1}^{\infty}\sum_{m=-n}^{n} \frac{a_{nm}}{n(n+1)}\left[\nabla\left(j_n(k_2 r)\hat{\mathbf{L}}Y_{nm}(\vartheta, \varphi)\right)\right]. \tag{6.11}$$

Now let us consider the transverse magnetic (TM) components of the dipole field. For this purpose, first we are to find the radial component of the electric field. It is not difficult to verify that in this case we have the identity:

$$\left(\mathbf{r}\mathbf{E}^{\text{dip}}\right) = \frac{1}{\varepsilon_2}\left(\mathbf{r}\left[\nabla\left[\nabla\mathbf{d}_0\right]\right]\right)\frac{e^{ik_2|\mathbf{r}-\mathbf{r}'|}}{|\mathbf{r}-\mathbf{r}'|} = \frac{1}{\varepsilon_2}\left(\mathbf{d}_0\left[\nabla'\left[\nabla'\mathbf{r}'\right]\right]\right)\frac{e^{ik_2|\mathbf{r}-\mathbf{r}'|}}{|\mathbf{r}-\mathbf{r}'|}. \tag{6.12}$$

Substituting the expansion of Eq. (6.5) in Eq. (6.12), we obtain the expression for the radial component of the electric field in terms of scalar spherical harmonics:

$$\left(\mathbf{r}\mathbf{E}^{\text{dip}}\right) = \sum_{n=1}^{\infty}\sum_{m=-n}^{n} b_{nm} j_n(k_2 r) Y_{nm}(\vartheta, \varphi), \tag{6.13}$$

where

$$b_{nm} = 4\pi\frac{k_0\sqrt{\mu_2}}{\sqrt{\varepsilon_2}}\left(\mathbf{d}_0\left[\nabla'\hat{\mathbf{L}}'\right]\right) h_n^{(1)}(k_2 r') Y_{nm}^*(\vartheta', \varphi'). \tag{6.14}$$

Knowing the radial component of the electric field, let us find TM components of the field by the perfect analogy with the TE case:

$$\mathbf{B}_{\text{TM}}^{\text{dip}} = -k_0\varepsilon_2\mu_2\sum_{n=1}^{\infty}\sum_{m=-n}^{n} \frac{b_{nm}}{n(n+1)} j_n(k_2 r)\hat{\mathbf{L}}Y_{nm}(\vartheta, \varphi), \tag{6.15}$$

$$\mathbf{E}_{\text{TM}}^{\text{dip}} = -i \sum_{n=1}^{\infty} \sum_{m=-n}^{n} \frac{b_{nm}}{n(n+1)} \left[\nabla \left(j_n(k_2 r) \hat{\mathbf{L}} Y_{nm}(\vartheta, \varphi) \right) \right]. \quad (6.16)$$

Assembling the expressions Eqs. (6.10), (6.11), (6.15), and (6.16), we obtain the resultant vector spherical harmonics expansion of the electric dipole field in an infinite medium with permittivity ε_2 and permeability μ_2 (Klimov et al., 1996a, 1996b; Chew et al., 1976; Chew, 1987):

$$\mathbf{B}^{\text{dip}}(r < r', \omega) = -k_0 \varepsilon_2 \mu_2 \sum_{n=1}^{\infty} \sum_{m=-n}^{n} \frac{b_{nm}}{n(n+1)} j_n(k_2 r) \hat{\mathbf{L}} Y_{nm}(\vartheta, \varphi)$$
$$- i \sum_{n=1}^{\infty} \sum_{m=-n}^{n} \frac{a_{nm}}{n(n+1)} \left[\nabla \left(j_n(k_2 r) \hat{\mathbf{L}} Y_{nm}(\vartheta, \varphi) \right) \right], \quad (6.17)$$

$$\mathbf{E}^{\text{dip}} \left(r < r', \omega \right) = -i \sum_{n=1}^{\infty} \sum_{m=-n}^{n} \frac{b_{nm}}{n(n+1)} \left[\nabla \left(j_n \left(k_2 r \right) \hat{\mathbf{L}} Y_{nm} \left(\vartheta, \varphi \right) \right) \right]$$
$$+ k_0 \sum_{n=1}^{\infty} \sum_{m=-n}^{n} \frac{a_{nm}}{n(n+1)} j_n \left(k_2 r \right) \hat{\mathbf{L}} Y_{nm} \left(\vartheta, \varphi \right). \quad (6.18)$$

Let us note that this approach to expansion in spherical harmonics is also applicable for describing of interaction of multipolar sources of any order both with a simple and a layered sphere.

Full fields **E**, **H** outside the sphere arise from the dipole source solution already found and the solution of homogeneous problem outside the sphere $\mathbf{E} = \mathbf{E}^{\text{dip}} + \mathbf{E}^{(2)}$, $\mathbf{B} = \mathbf{B}^{\text{dip}} + \mathbf{B}^{(2)}$. Inside the sphere, we have $\mathbf{E} = \mathbf{E}^{(1)}$, $\mathbf{B} = \mathbf{B}^{(1)}$ where $\mathbf{E}^{(1)}$, $\mathbf{B}^{(1)}$ are solutions of homogeneous Maxwell's equations inside the sphere.

The solution of the homogeneous problem both inside and outside a microsphere can be developed in a regular manner (Jackson, 1975). In the case of the external region, the scattered fields can be written as:

$$\mathbf{B}^{(2)} = -k_0 \varepsilon_2 \mu_2 \sum_{n=1}^{\infty} \sum_{m=-n}^{n} \frac{\beta_{nm}^{(2)}}{n(n+1)} h_n^{(1)}(k_2 r) \hat{\mathbf{L}} Y_{nm}(\vartheta, \varphi)$$
$$- i \sum_{n=1}^{\infty} \sum_{m=-n}^{n} \frac{\alpha_{nm}^{(2)}}{n(n+1)} \left[\nabla \left(h_n^{(1)}(k_2 r) \hat{\mathbf{L}} Y_{nm}(\vartheta, \varphi) \right) \right], \quad (6.19)$$

$$\mathbf{E}^{(2)} = -i\sum_{n=1}^{\infty}\sum_{m=-n}^{n}\frac{\beta_{nm}^{(2)}}{n(n+1)}\left[\nabla\left(h_n^{(1)}(k_2r)\hat{\mathbf{L}}Y_{nm}(\vartheta,\varphi)\right)\right]$$
$$+k_0\sum_{n=1}^{\infty}\sum_{m=-n}^{n}\frac{\alpha_{nm}^{(2)}}{n(n+1)}h_n^{(1)}(k_2r)\hat{\mathbf{L}}Y_{nm}(\vartheta,\varphi), \qquad (6.20)$$

where the expansion coefficients $\alpha_{nm}^{(2)}$, $\beta_{nm}^{(2)}$ must be defined from the continuity condition for tangential field components at the sphere boundary. Analogous expressions are true also inside the sphere, with the spherical Hankel functions of the first kind $h_n^{(1)}(k_2r)$ replaced by the spherical Bessel functions $j_n(k_1r)$, where $k_1 = \sqrt{\mu_1\varepsilon_1}\omega/c$, and vice versa.

Now using the equation of continuity of electric and magnetic field tangential components on the sphere surface, unknown coefficients α_{nm} and β_{nm} can be found:

$$\beta_{nm}^{(2)} = -\frac{\varepsilon_1\frac{d}{dz_2}[z_2j_n(z_2)]j_n(z_1) - \varepsilon_2\frac{d}{dz_1}[z_1j_n(z_1)]j_n(z_2)}{\varepsilon_1\frac{d}{dz_2}[z_2h_n^{(1)}(z_2)]j_n(z_1) - \varepsilon_2\frac{d}{dz_1}[z_1j_n(z_1)]h_n^{(1)}(z_2)}b_{nm}$$
$$= -q_nb_{nm}, \qquad (6.21)$$

$$\alpha_{nm}^{(2)} = -\frac{\mu_1\frac{d}{dz_2}[z_2j_n(z_2)]j_n(z_1) - \mu_2\frac{d}{dz_1}[z_1j_n(z_1)]j_n(z_2)}{\mu_1\frac{d}{dz_2}[z_2h_n^{(1)}(z_2)]j_n(z_1) - \mu_2\frac{d}{dz_1}[z_1j_n(z_1)]h_n^{(1)}(z_2)}a_{nm}$$
$$= -p_na_{nm}. \qquad (6.22)$$

Here, $z_{1,2} = \sqrt{\varepsilon_{1,2}\mu_{1,2}}k_0a = k_{1,2}a$. The coefficients p_n and q_n are the Mie reflection coefficients (Mie, 1908) for spherical waves of different polarizations (TE and TM, correspondingly).

The solution obtained is a generalization of the Mie solution and can be applied to solve many important problems. This solution has special significance in nano-optics since nanoparticles can interact with a separate atom effectively.

If one put $r' \to \infty$ in the solution found, then the dipole field near the sphere will be approximately equal to a plane wave, $\mathbf{E} = k^2e^{ikr'}\left(\mathbf{d}_0 - \mathbf{n}\left(\mathbf{n}\mathbf{d}_0\right)\right)/r'$, where $\mathbf{n}$ is a unit vector from the center of the sphere to the dipole position, and the solution found can be used to solve the problem of plane wave diffraction on a sphere, that is, the Mie problem (Mie, 1908; Bohren and Huffman, 1983).

The problems of light radiation by a dipole source inside a sphere (Klimov et al., 1996a) and by sources of higher multipolarity

(Klimov et al., 1996a, 1996b; Klimov and Letokhov, 2000) have a perfectly analogous solution. The problem of a dipole source of light inside the sphere is especially important for a description of optical properties of nanoparticles with embedded fluorophores, because such systems can be easily prepared.

6.2 Optical Resonances in Spherical Particles of an Arbitrary Size

The expressions stated above for electric and magnetic fields scattered by a sphere are remarkable and allow describing many different effects. On the other hand, these expressions are rather complex and often their application is complicated in specific situations.

As said above, separate terms of Eqs. (6.20) and (6.19) correspond to electric (TM, q_n coefficients) or magnetic (TE, p_n coefficients) spherical waves. Moreover, often only one term plays the dominant role in complete expressions for electromagnetic fields. In this case, one should speak about resonances.

Resonances arise when one of the denominators in the Mie coefficients becomes small or even goes to zero. In the case of a sphere, the Mie coefficients depend on frequency and the sphere radius, and the dependency of frequency arises also due to dependency of electric permittivity or magnetic permeability on frequency implicitly, and for spheres made of different material resonances arise under different conditions. According to the approach presented in Chapter 5, it is more convenient to consider electric (or magnetic) permittivity as an independent variable since in this case the description of resonances becomes universal.

In a nonmagnetic case, the Mie coefficients depend on four dimensionless parameters: $n = 1, 2, 3, \ldots, \varepsilon_1, \varepsilon_2, ka = \omega a/c$. In Fig. 6.2, the dependence of the Mie coefficients' absolute value q_n for $n = 1$ (dipole case, TM waves) on electric permittivity and on the size parameter of the sphere, $x = \omega a/c$ is shown. Black color is for the Mie coefficient equal to one, that is, for resonance, and white color is for the Mie coefficient equal to zero, that is, for the absence of resonance.

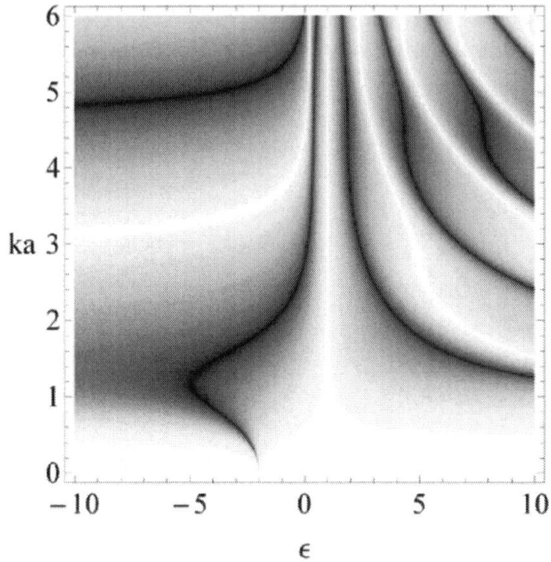

Figure 6.2 The dependence of the Mie coefficient q_1 (TM waves) absolute values on the sphere electric permittivity ε and the size parameter ka. The black color corresponds to the Mie coefficient equal to one, that is, to the resonance, while the white color corresponds to the Mie coefficient equal to zero (absence of resonance).

This figure shows that in the case concerned, the existence of many (more precisely, infinitely many) resonances is possible. Resonances for the positive ε_1 correspond to so-called "whispering gallery" modes (WGMs) when light travels inside a sphere and cannot leave it due to the effect of total internal reflection. More precisely, it is an almost total internal reflection since a part of energy still radiates due to nonzero curvature of a sphere. To enable the existence of WGMs in nanospheres, one should provide very large positive values of electric permittivity, which cannot be provided practically. Therefore, WGMs exist in spheres of radii of several or even tens of micrometers. In such "big" spheres, radiative losses are very small and the quality factor can attain the values of the order of 10^{10} and more (Braginsky and Il'chenko, 1987; Braginsky et al., 1989; Collot et al., 1993). Very interesting effects occur also in nanospheres with rather high permittivities. For example, nanospheres made of silicon, having permittivity of about 12 in mid-infrared (IR), present a very strong magnetic response, which usually is not expected for dielectric nanoparticles (García-Etxarri et al., 2011).

Physically, in a dielectric microsphere the origin of WGMs is related to the effect of almost total internal reflection at its surface, and mode frequency is defined roughly by the Bohr quantization condition:

$$2\pi a \approx N\lambda_1 = N\frac{2\pi c}{\omega\sqrt{\varepsilon_1}}, \quad N = 1, 2, 3, \ldots,$$

where λ_1 is a wavelength in a dielectric microsphere.

Precise equations defining the position of resonances can be obtained by equating the denominators q_n and p_n to zero (see Eqs. (6.21) and (6.22)):

$$\frac{\left[J_{n+1/2}(z_1)\right]'}{J_{n+1/2}(z_1)} = \frac{1}{2z_2}\left(\sqrt{\frac{\varepsilon_1}{\varepsilon_2}} - \sqrt{\frac{\varepsilon_2}{\varepsilon_1}}\right) + \sqrt{\frac{\varepsilon_1}{\varepsilon_2}}\frac{\left[H^{(1)}_{n+1/2}(z_2)\right]'}{H^{(1)}_{n+1/2}(z_2)} \quad \text{(TM case)}, \qquad (6.23)$$

$$\frac{\left[J_{n+1/2}(z_1)\right]'}{J_{n+1/2}(z_1)} = \sqrt{\frac{\varepsilon_2}{\varepsilon_1}}\frac{\left[H^{(1)}_{n+1/2}(z_2)\right]'}{H^{(1)}_{n+1/2}(z_2)} \quad \text{(TE case)}, \qquad (6.24)$$

where $z_{1,2} = \sqrt{\varepsilon_{1,2}\mu_{1,2}}k_0 a$.

In the case of WGMs, an approximately whole number of wavelengths fall along a circumference. Using this fact, Eqs. (6.23) and (6.24) can be simplified with the help of the Debye asymptotic expansions for the Hankel functions and asymptotic expansions in the transition region for the Bessel functions (Abramowitz and Stegun, 1972). As a result, we obtain the following equations instead of Eqs. (6.23) and (6.24):

$$\frac{Ai'\left(-2^{1/3}t\right)}{Ai\left(-2^{1/3}t\right)} = \left(\frac{\nu}{2}\right)^{1/3}\sqrt{\frac{\varepsilon_1}{\varepsilon_2}}\, sh\alpha\left(1 - ie^{-2\nu(\alpha - th\alpha)}\right) \text{(TM case)}, \qquad (6.25)$$

$$\frac{Ai'\left(-2^{1/3}t\right)}{Ai\left(-2^{1/3}t\right)} = \left(\frac{\nu}{2}\right)^{1/3}\sqrt{\frac{\varepsilon_2}{\varepsilon_1}}\, sh\alpha\left(1 - ie^{-2\nu(\alpha - th\alpha)}\right) \text{(TE case)}, \qquad (6.26)$$

Here, $\nu = n + 1/2$, $t = (z_1 - \nu)/\nu^{1/3}$, $ch\alpha = \sqrt{\varepsilon_1/\varepsilon_2}(\nu/z_1)$.

In the case of large n, this equation can be solved by iterations over t. As a result, the expressions for resonance frequencies take the form (Vainshtein, 1966; Klimov et al., 1999a):

$$\left(\frac{\omega a}{c}\right)_{\text{res}} = \frac{1}{\sqrt{\varepsilon_1}}\left(j_{n+1/2} - \frac{\varepsilon_2}{\varepsilon_1}\sqrt{\frac{\varepsilon_1}{\varepsilon_1 - \varepsilon_2}}\right) \text{ (TE case)}, \quad (6.27)$$

$$\left(\frac{\omega a}{c}\right)_{\text{res}} = \frac{1}{\sqrt{\varepsilon_1}}\left(j_{n+1/2} - \sqrt{\frac{\varepsilon_1}{\varepsilon_1 - \varepsilon_2}}\right) \text{ (TE case)}, \quad (6.28)$$

where $j_{n+1/2}$ is one of the roots of the Bessel function $J_{n+1/2}$.

The use of the first nontrivial root of the Bessel function in Eq. (6.27) or Eq. (6.28) results in the mode of "whispering gallery" with angular dependence defined by the spherical harmonic $Y_{nm}(\vartheta, \varphi)$ and no zeros in a radial direction inside the microsphere. By using the second root, the first zero appears in a radial direction inside a microsphere, and so on. This circumstance determines a convenient classification of WGMs by three numbers: the number of zeroes l in a radial direction inside a microsphere, the order of the spherical Bessel function n, and the azimuth quantum number m. In this sense, WGMs are similar to wavefunctions of an atom. Let us note that "whispering gallery" and other modes with a different azimuth quantum number m have the same frequencies as in the case of an atom with a spherically symmetric potential.

For estimations of radiation linewidth Γ_{res} or quality factor Q of main WGMs' resonances, the formula resulting from the iterations Eq. (6.25) and Eq. (6.26) may be used for TM resonances (Vainshtein, 1966):

$$\frac{\Gamma_{\text{res}}}{\omega_{\text{res}}} = \frac{1}{Q_{\text{TM}}} \approx \frac{2}{j_{n+1/2}}\frac{\varepsilon_2}{\varepsilon_1}\sqrt{\frac{\varepsilon_1}{\varepsilon_1 - \varepsilon_2}}e^{-2T_{\text{TM}}},$$
$$T_{\text{TM}} = \nu\left(\operatorname{arch}\sqrt{\frac{\varepsilon_1}{\varepsilon_2}} - \sqrt{1 - \frac{\varepsilon_2}{\varepsilon_1}}\right) - t_0\nu^{1/3}\sqrt{1 - \frac{\varepsilon_2}{\varepsilon_1} + \frac{\varepsilon_2}{\varepsilon_1} + \ldots} \quad (6.29)$$

and the formula:

$$\frac{\Gamma_{\text{res}}}{\omega_{\text{res}}} = \frac{1}{Q_{\text{TE}}} \approx \frac{2}{j_{n+1/2}}\sqrt{\frac{\varepsilon_1}{\varepsilon_1 - \varepsilon_2}}e^{-2T_{\text{TE}}},$$
$$T_{\text{TE}} = \nu\left(\operatorname{arch}\sqrt{\frac{\varepsilon_1}{\varepsilon_2}} - \sqrt{1 - \frac{\varepsilon_2}{\varepsilon_1}}\right) - t_0\nu^{1/3}\sqrt{1 - \frac{\varepsilon_2}{\varepsilon_1} + 1 + \ldots} \quad (6.30)$$

may be used for TE resonances. In Eqs. (6.29) and (6.30) $\nu = n + 1/2, t_0 = (j_{n+1/2} - \nu)/\nu^{1/3}$. For sufficiently large microspheres ($ka > 100$), radiation linewidths and quality factors Eqs. (6.29) and (6.30) become negligibly small in comparison with other types of losses (absorption in substance, scattering on surface roughnesses, etc.).

In the vicinity of the resonance, the Mie reflection coefficients take the simple form:

$$q_n \approx -i\frac{\text{Im}(\Omega_{\text{res}})}{\omega - \Omega_{\text{res}}}, \tag{6.31}$$

where Ω_{res} is the complex frequency characterizing the resonance mode $\Omega_{\text{res}} = \omega_{\text{res}} - i\Gamma_{\text{res}}^{(n)}/2$ and $\Gamma_{\text{res}}^{(n)}$ characterizes the resonance mode width and the quality factor of the resonator with this mode.

Analogous expressions are true also for TE modes in the vicinity of TE resonance:

$$p_n \approx -i\frac{\text{Im}(\Omega_{\text{res}})}{\omega - \Omega_{\text{res}}}. \tag{6.32}$$

Let us note again that the expressions stated above describe WGMs' resonances arising when electric permittivity values are *positive* and sphere radii are relatively big (of the order of the wavelength or bigger). It is very important to note that the imaginary parts of Ω_{res} in Eqs. (6.31) and (6.32) are *negative* in contrast to the imaginary parts of electric permittivities of real materials.

If the real part of the sphere electric permittivity is *negative,* resonances are possible for arbitrary small sizes of the sphere. These are just localized plasmon resonances. To find the resonance values of electric permittivity $\varepsilon_{\text{res},n}$, Eq. (6.23) must be expanded in the series over the small parameter k_0a and then the approximated value $\varepsilon_1 = \varepsilon_{\text{res},n}(k_0a)$ is to be found, satisfying the equation with the required accuracy. In particular, for $n = 1, 2,$ and 3 these expressions for $\varepsilon_n(k_0a)$ have the form:

$$\begin{aligned}
\frac{\varepsilon_{\text{res, 1}}}{\varepsilon_2} &= -2 - \frac{12}{5}(k_2a)^2 - 2i(k_2a)^3 + \dots, \\
\frac{\varepsilon_{\text{res, 2}}}{\varepsilon_2} &= -\frac{3}{2} - \frac{5}{14}(k_2a)^2 - \frac{65}{392}(k_2a)^4 - \frac{i}{12}(k_2a)^5 + \dots, \\
\frac{\varepsilon_{\text{res, 3}}}{\varepsilon_2} &= -\frac{4}{3} - \frac{56}{405}(k_2a)^2 - \frac{11788}{601425}(k_2a)^4 \\
&\quad - \frac{469672}{95954625}(k_2a)^6 - \frac{4i}{2025}(k_2a)^7 + \dots,
\end{aligned} \tag{6.33}$$

These eigenvalues will correspond to dipole, quadrupole, and octupole plasmon oscillations in the nanosphere, correspondingly. It is interesting that the imaginary parts of resonance electric permittivity (which are related to radiation losses) are proportional to $(ka)^{2n+1}$ and become less and less pronounced for higher multipole radiation. It is also important to have in mind that according to the general theory (see Chapter 5), the imaginary parts of $\varepsilon_{\text{res},n}$ in Eq. (6.33) are negative in contrast to the imaginary parts of electric permittivities of real materials.

Let us stress again that the plasmon oscillations satisfying the conditions in Eq. (6.33) have place for arbitrary small spheres and are of great significance for different applications.

The case of the spheres made of materials with nonzero dielectric and magnetic permittivities (metamaterials) is also of great interest and will be considered in Chapter 11.

6.3 Optical Properties of a Spherical Particle of an Arbitrary Size

After the classical problem of sphere influence on dipole radiation or plane wave propagation has been solved, practically all interesting optical characteristics of the sphere can be found.

When a plane wave is incident on a sphere, the influence of the sphere can be described by scattering and absorption cross sections. The sum of the scattering and absorption cross sections defines an extinction cross section, that is, attenuation cross section.

The scattering cross section, as its name implies, allows characterizing the part of the energy flow being incident on a sphere, which scatters in various directions after the interaction with the sphere. More formally, the total scattering cross section is defined by the ratio between radiation power and incident energy flow. In the case when a plane wave is incident on a sphere, the scattering cross section is independent neither on its direction nor on polarization and is described by the expression:

$$\sigma_{\text{sca}} = \frac{2\pi}{k_2^2} \sum_{n=1}^{\infty} (2n+1)\left(|q_n|^2 + |p_n|^2\right), \tag{6.34}$$

where the coefficients q_n, p_n are given by the expressions in Eqs. (6.21) and (6.22).

The absorption cross section describes the energy absorbed in a sphere and is expressed by the ratio between the power absorbed and energy flow in an incident wave. In the case when the plane wave is incident on the sphere, the absorption cross section is also independent neither on its direction nor on polarization and is described by the expression:

$$\sigma_{\text{abs}} = \frac{2\pi}{k_2^2} \sum_{n=1}^{\infty} (2n+1) \left(\frac{1}{2} - \left| q_n - \frac{1}{2} \right|^2 - \left| p_n - \frac{1}{2} \right|^2 \right). \quad (6.35)$$

The extinction cross section determines the attenuation of a plane wave passing through a region where many spheres are placed and is equal to the sum of scattering and absorption cross sections and may be described by the formula:

$$\sigma_{\text{ext}} = \frac{2\pi}{k_2^2} \sum_{n=1}^{\infty} (2n+1) \text{Re} \{q_n + p_n\}. \quad (6.36)$$

In Figs. 6.3–6.5, different cross sections of a gold nanoparticle are shown, which is placed in vacuum and water. The last case is important for medical and biological applications (see Chapter 13). Figure 6.3 implies that the cross section of a particle in vacuum is maximal for particles of 50 nm radius at the wavelength of 520 nm order. In the case when a nanoparticle is placed in water (Fig. 6.4), the whole situation is generally similar, but the absorption maximum takes place for a particle of the radius of 20 nm order. Figure 6.5 shows that for the smallest particles (of a radius less than 50 nm), absorption prevails, while light scattering prevails for larger particles. It is these diagrams that explain the colors of colloid solutions discussed in Chapter 2.

The maxima on the diagrams of nanoparticles' scattering and absorption are determined by plasmon resonances in the nanoparticles. However, analyzing the dependencies of cross sections on one or the other parameter, one cannot make any conclusions on what actually happens inside a nanoparticle in the region of plasmon resonance. This question has been investigated by Bashevoy et al. (2005), who found a complex spatial structure of energy flows in metal nanoparticles near plasmon resonance.

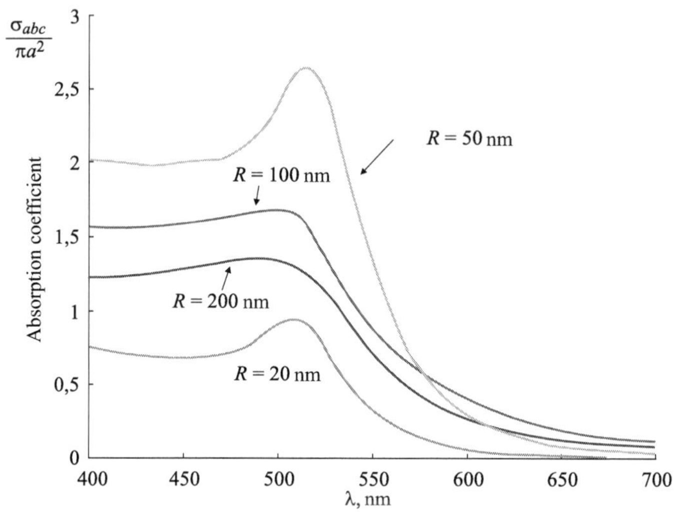

Figure 6.3 The dependence of the absorption cross section of a spherical gold particle on the radiation wavelength for different particle radii ($a = 20$, 50, 100, 200 nm). The particle is in vacuum.

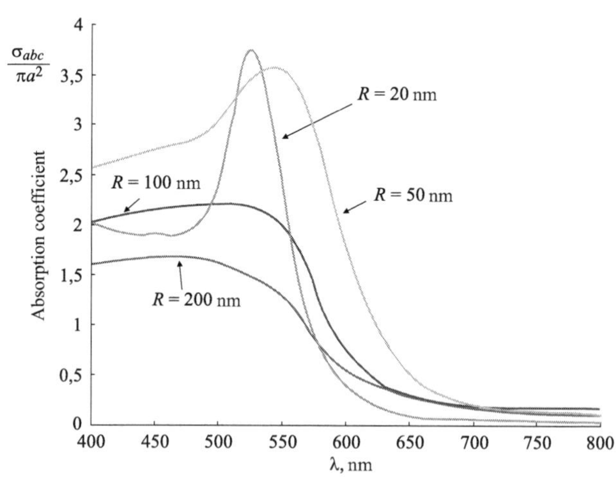

Figure 6.4 The dependence of a spherical gold particle absorption cross section on the radiation wavelength for different particle radii ($a = 20$, 50, 100, 200 nm). The particle is in water.

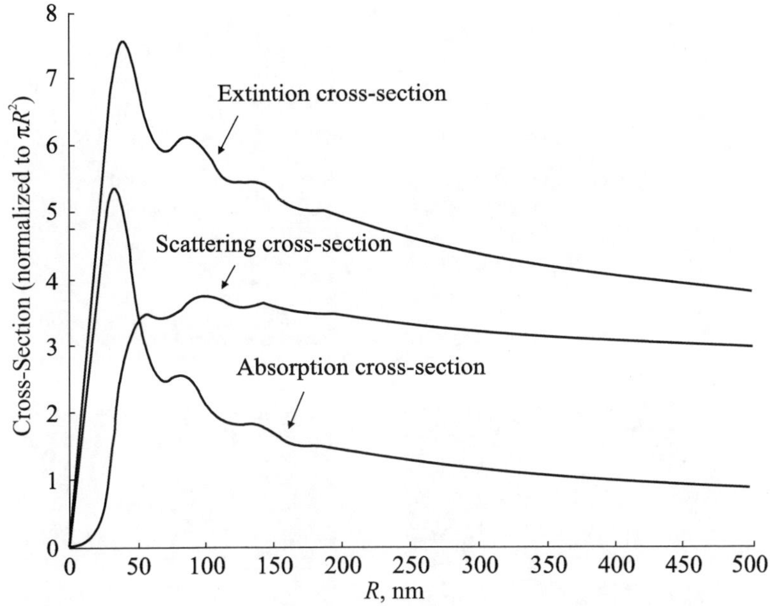

Figure 6.5 The dependence of spherical gold particle absorption, scattering, and extinction cross sections on its size for a fixed radiation wavelength, $\lambda = 532$ nm. The particle is in water.

In Fig. 6.6, distributions of the energy flow are shown when a plane wave is incident from the left side on a metal sphere of the radius 20 times less than the wavelength for different values of permittivity. In Fig. 6.6a, the scattering and absorption cross sections are much less than the geometrical cross section ($\sigma_{\text{abs}}/(\pi a^2) = 0.47$, $\sigma_{\text{sca}}/(\pi a^2) = 0.03$), the particle is almost invisible, and almost all lines of the energy flow envelope the nanoparticle. In the case represented in Fig. 6.6b, the absorption cross section increases due to losses decrease and approach to the plasmon resonance ($\sigma_{\text{abs}}/(\pi a^2) = 3.6$, $\sigma_{\text{sca}}/(\pi a^2) = 0.24$), and many lines of the energy flow enter the nanoparticle both from the front and from the rear. When the scattering and absorption cross sections increase even more, the energy flow lines form vortex structures around the nanoparticle. This situation is represented in Fig. 6.6c and Fig. 6.6d for silver nanoparticles at the wavelength of 354 nm (where

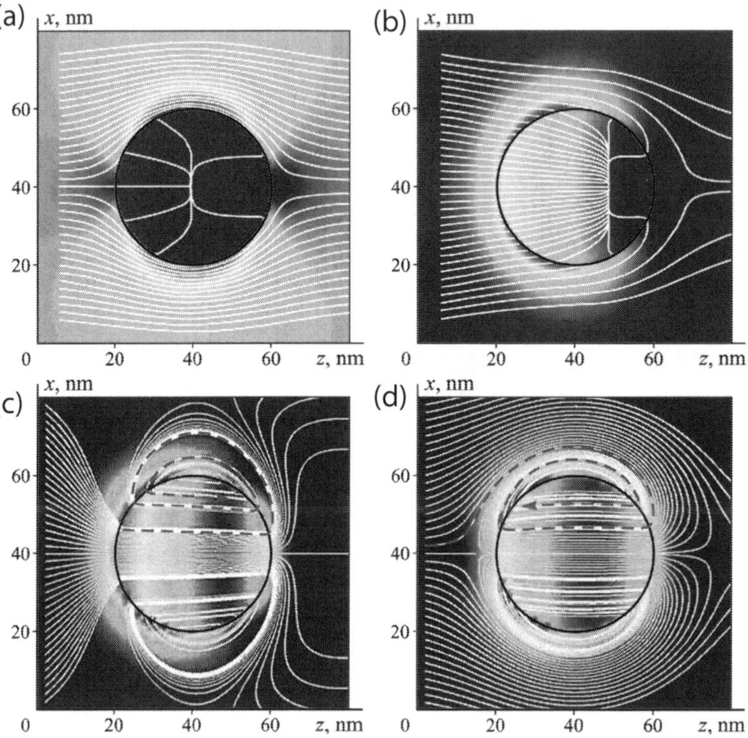

Figure 6.6 Spatial distribution of the energy flow near a spherical nanoparticle of the 20 nm radius ($\lambda/a = 20$) in the plane containing the propagation direction (from left to right) and the vector of incident wave polarization. The colors denote the absolute values of the Umov–Poynting vector, and the white lines denote the energy flow. (a) $\varepsilon = -2 + 10i$, $\lambda =$ 400 nm; (b) $\varepsilon = -2 + i$, $\lambda = 400$ nm; (c) $\varepsilon = -2 + 0.28i$ is the electric permittivity of the sphere for $\lambda = 354$ nm (Lide, 2000); and (d) $\varepsilon = -2.71 + i0.25$, $\lambda = 367$ nm (Lide, 2000). The red dashed lines show the structure of plasmon vortices (Bashevoy et al., 2005).

$\varepsilon = -2.0 + i0.28$, $\sigma_{\text{abs}}/(\pi a^2) = 5.8$, $\sigma_{\text{sca}}/(\pi a^2) = 1.8$) and of 367 nm (where $\varepsilon = -2.71 + i0.25$, $\sigma_{\text{abs}}/(\pi a^2) = 4.1$, $\sigma_{\text{sca}}/(\pi a^2) = 2.0$).

The existence of vortex structures gives an obvious graphical interpretation of the fact that the absorption cross section can be much larger than the geometric cross section since when a vortex is being formed, the lines of energy flow pass through the particle several times and intensify the interaction between the light and the substance, resulting in the large absorption cross section.

In the case when not a plane but spatially, strongly nonhomogeneous wave is incident on a sphere, the cross-section notion is not useful and each problem has its own characteristics. For example, in an important case when an atom or a molecule radiates near a sphere, the sphere influence is turned into the modification of radiation linewidth and to its frequency shift.

As in the case of a plane wave, the influence of a spherical particle on the radiation of an atom or a molecule can be characterized with regard to different problem statements. For example, when we are interested in photons actually radiated by an atom, we should know the radiation linewidth or the radiation part of the decay rate.

And when we are interested in the sphere influence on the lifetime of an exited atom or a molecule, we should consider the photons both flown away from the sphere and absorbed in it. This characteristic is analogous to the extinction cross section.

In contrast to the plane wave incident on a sphere, these characteristics depend both on the atom's or molecule's position relative to the sphere and on the dipole moment orientation. Using the expression for a field reflected from the sphere, Eq. (6.20), and the approach represented in Appendix A1, expressions for the total radiation linewidths and frequency shifts of the atom or the molecule situated near a spherical particle can be obtained (Klimov et al., 1996b).

For the frequency shift of a spectral line of an atom placed outside a sphere, we have the following expression:

$$\left(\frac{\omega-\omega_0}{\gamma_0}\right) = -\frac{3}{4}\mathrm{Im}\left\{\cos^2\psi\sum_{n=1}^{\infty}n(n+1)(2n+1)q_n\left(\frac{h_n^{(1)}(z')}{z'}\right)^2\right.$$
$$+\sin^2\psi\sum_{n=1}^{\infty}\left(n+\frac{1}{2}\right)\left[p_n\left(h_n^{(1)}(z')\right)^2\right.$$
$$\left.\left.+q_n\left(\frac{d(z'h_n^{(1)}(z'))}{z'dz'}\right)^2\right]\right\}, \tag{6.37}$$

while the total linewidth is described by the expression:

$$\left(\frac{\gamma}{\gamma_0}\right) = 1 - \frac{3}{2}\mathrm{Re}\left\{\cos^2\psi \sum_{n=1}^{\infty} n(n+1)(2n+1)q_n \left(\frac{h_n^{(1)}(z')}{z'}\right)^2 + \sin^2\psi \sum_{n=1}^{\infty}\left(n+\frac{1}{2}\right)\left[p_n\left(h_n^{(1)}(z')\right)^2 + q_n\left(\frac{d(z'h_n^{(1)}(z'))}{z'dz'}\right)^2\right]\right\}. \quad (6.38)$$

In these expressions, γ_0 is the linewidth of the atom radiation in the absence of a sphere, ψ is the angle between the orientation of the atom dipole moment and the radial direction, $z' = k_2 r$ is a parameter characterizing the distance between the atom and the sphere center, and p_n and q_n are given by Eqs. (6.21) and (6.22).

If we are interested in photons flying into infinity, the so-called radiative part of the linewidth can be described by the expression (Klimov et al., 1996b):

$$\left(\frac{\gamma}{\gamma_0}\right)_{\mathrm{rad}}^{E1} = \frac{3}{2}\sum_{n=1}^{\infty} n\left(n+1\right)\left(2n+1\right)\frac{\left|j_n\left(k_2 r\right) - q_n h_n^{(1)}\left(k_2 r\right)\right|^2}{\left(k_2 r\right)^2}, \quad (6.39)$$

for the radial orientation of the dipole moment and by the expression:

$$\left(\frac{\gamma}{\gamma_0}\right)_{\tan}^{E1} = \frac{3}{2}\sum_{n=1}^{\infty}\left(n+\frac{1}{2}\right)\left|j_n\left(z'\right) - p_n h_n^{(1)}\left(z'\right)z'\right|^2 + \frac{3}{2}\sum_{n=1}^{\infty}\left(n+\frac{1}{2}\right)\left|\frac{d\left\{z'\left(j_n\left(z'\right) - q_n h_n^{(1)}\left(z'\right)z'\right)\right\}}{z'dz'}\right|^2, \quad (6.40)$$

for the tangential orientation of the dipole moment relative to the sphere surface.

In Fig. 6.7a, the radiative part of the decay rate of an atom placed near the surface of a metal spherical particle is represented as functions of the sphere size. There is a peak in the region of small sizes, which corresponds to plasmon excitation in a nanosphere (see the next section). When the sizes increase, the periodical system of maxima appears, which correspond to surface plasmon

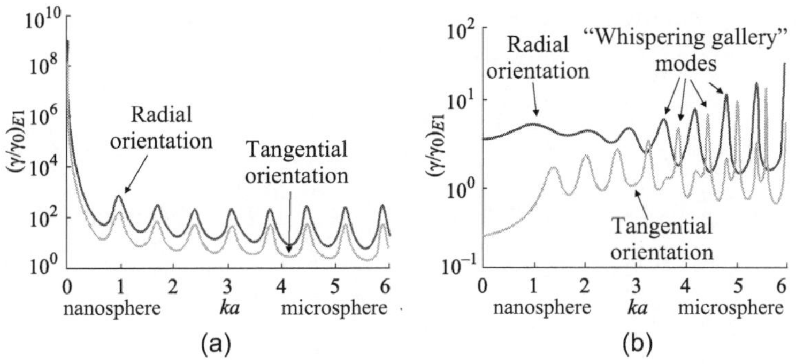

Figure 6.7 The dependence of the radiative decay rate for an atom placed near the surface of a metal nanoparticle with $\varepsilon = -2$, $\mu = 1.05$ (a) and near the dielectric particle with $\varepsilon = 4$, $\mu = 1.05$ (b).

propagation along the sphere boundary. Positions of these peaks are approximately determined by the Bohr quantization rule: $k_{\mathrm{sp}}a = n + 1/2$, where k_{sp} is the surface plasmon wavenumber (see Eq. (4.5)). The WGMs do not arise in this case because there are no propagating waves inside the metal sphere. In Fig. 6.7b, the radiative part of the decay rate of an atom placed near a dielectric sphere is shown. In this case, there are no resonance modes for nanoparticles but there are WGMs at relatively large sizes on the contrary.

Analogous expressions can be obtained also for radiation of the magnetic dipole, electric quadrupole, etc., placed near the sphere (Klimov and Letokhov, 1996a, 1996b; Klimov and Letokhov, 2000; Klimov and Letokhov, 2005).

6.4 The Quasi-Static Theory of Optical Properties of Spherical Nanoparticles

In this case, there is no necessity to use the full Maxwell theory, since the retardation is negligibly small. Instead, one can restrict himself to solution of electrostatic equations within the framework of a long-wavelength approximation.

In particular, the eigenvalues ε_n and the eigenfunctions $\mathbf{e}_n$ of the "ε-method" can be found by solving the equations:

$$\begin{aligned} \operatorname{div}\left(\varepsilon\left(\mathbf{r}\right)\mathbf{e}_n\right) &= 0, \\ \operatorname{rot}\mathbf{e}_n &= 0, \end{aligned} \qquad (6.41)$$

which may be reduced to the solution of Laplace equations by substituting $\mathbf{e}_n = -\nabla\Phi_n$:

$$\begin{aligned} \Delta\Phi_n^{\mathrm{in}} &= 0, \\ \Delta\Phi_n^{\mathrm{out}} &= 0, \\ \varepsilon_n \left.\frac{\partial\Phi_n^{\mathrm{in}}}{\partial\mathbf{n}}\right|_S &= \left.\frac{\partial\Phi_n^{\mathrm{out}}}{\partial\mathbf{n}}\right|_S . \end{aligned} \qquad (6.42)$$

In Eq. (6.42), Φ_n^{in}, Φ_n^{out} are the potentials of the eigenfunctions both inside and outside the particle and $\partial\Phi_n/\partial\mathbf{n}|_S$ denotes the formal derivative on the particle boundary.

In the case of a sphere of radius a, the boundary problem Eq. (6.42) has a very simple solution:

$$\varepsilon_n = -\frac{n+1}{n}, \qquad \Phi = \begin{cases} \left(\dfrac{r}{a}\right)^n Y_{nm}\left(\theta,\varphi\right), & r < a, \\ \left(\dfrac{a}{r}\right)^{n+1} Y_{nm}\left(\theta,\varphi\right), & r > a, \end{cases} \qquad (6.43)$$

which resembles the solution of the Schrödinger equation for a hydrogen atom.

In Fig. 6.8, radial wavefunctions for the simplest dipole ($n = 1$) plasmon oscillation of a nanosphere and the $2p$ state of a hydrogen atom are shown. As one can see from the picture, the wavefunctions are very similar. This permits us to refer to plasmon oscillations in a sphere and other nanoparticles as plasmonic atoms. Later, we will show that this analogy is deeper than it can appear at first sight, and plasmonic molecules can be formed from plasmonic atoms (Chapter 10). Let us also note that the solution of Eq. (6.43) agrees with the expansion of the exact solution over powers of ka in Eq. (6.33).

Knowing the general solution for nanosphere eigenfunctions in Eq. (6.43), the fields appearing when the sphere is placed in an arbitrary external field can be found with the help of the general theory represented in Chapter 5.

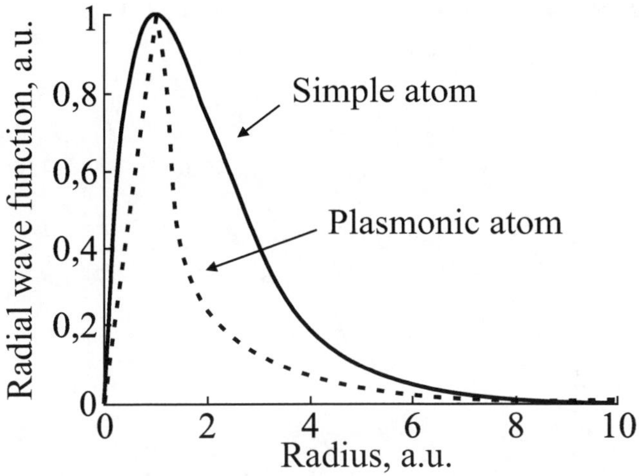

Figure 6.8 The comparison of the radial wavefunctions in a hydrogen atom and in a spherical particle.

In a *homogeneous* electric field, only dipole resonances with $n = 1$ can be excited in a sphere. In this case, the scattered field outside the particle is the dipole field with the dipole momentum:

$$\mathbf{d} = \varepsilon_2 \alpha_1 \mathbf{E}_0, \tag{6.44}$$

where

$$\alpha_1 = \frac{\varepsilon(\omega) - \varepsilon_2}{\varepsilon(\omega) + 2\varepsilon_2} a^3 \tag{6.45}$$

is the electric dipole polarizability of the sphere and $\varepsilon(\omega)$ is the electric permittivity of the sphere substance (we suppose that the sphere is placed in a medium with electric permittivity ε_2).

Knowing the dipolar polarizability of the sphere, the scattering and absorption cross sections of a sphere made of any substance can be found with the help of the general expressions from Chapter 5:

$$\begin{aligned} \sigma_{\text{abs}} &= 4\pi k_2 \text{Im} \alpha_1, \\ \sigma_{\text{sca}} &= \frac{8\pi}{3} k_2^4 |\alpha_1|^2. \end{aligned} \tag{6.46}$$

Using the general expressions of Eqs. (6.36) and (6.34) and expanding them in the small parameter *ka* series, more precise expressions can be found:

$$\sigma_{\text{abs}} \approx \frac{12\pi k_2 a^3 \varepsilon''(\omega)\,\varepsilon_2}{|\varepsilon(\omega)+2\varepsilon_2|^2}\left\{1+\frac{1}{90}\frac{k_2^2 a^2}{|\varepsilon(\omega)+2\varepsilon_2|^2}\right.$$
$$\times\left[\varepsilon(\omega)^4+8|\varepsilon(\omega)|^2\varepsilon'(\omega)\,\varepsilon_2+134|\varepsilon(\omega)|^2\varepsilon_2^2\right.$$
$$\left.+16\left(\varepsilon'(\omega)\right)^2\varepsilon_2^2+104\varepsilon'(\omega)\,\varepsilon_2^3-344\varepsilon_2^4\right]$$
$$\left.+\frac{5}{18}k_0^2 a^2\left|\frac{\varepsilon(\omega)+2\varepsilon_2}{2\varepsilon(\omega)+3\varepsilon_2}\right|^2-4k_2^3 a^3\frac{\varepsilon''(\omega)\,\varepsilon_2}{|\varepsilon(\omega)+2\varepsilon_2|^2}+O\left(k_0^4\right)\right\}, \tag{6.47}$$

$$\sigma_{\text{sca}} \approx \frac{8}{3}\pi k_2^4 a^6\left|\frac{\varepsilon(\omega)-\varepsilon_2}{\varepsilon(\omega)+2\varepsilon_2}\right|^2$$
$$\times\left[1+\frac{6}{5}k_2^2 a^2\frac{|\varepsilon(\omega)|^2-4\varepsilon_2^2}{|\varepsilon(\omega)+2\varepsilon_2|^2}-4k_2^3 a^3\frac{\varepsilon''(\omega)\,\varepsilon_2}{|\varepsilon(\omega)+2\varepsilon_2|^2}+O\left(k_0^4\right)\right], \tag{6.48}$$

in which the main terms coincide with the quasi-static solutions in Eq. (6.46), naturally. As seen from these expressions, as the sphere size decreases, the absorption processes always prevail over scattering ones, and thus in this case the extinction cross section is defined only by the absorption cross section. Let us also note that this circumstance is rather general, since according to Eq. (6.46) the scattering cross section is proportional to the squared volume of the nanoparticle, while the absorption cross section is linearly proportional to the volume.

However, if the radiation source (an atom or a molecule) is situated near the nanosphere, its field is highly nonuniform near the sphere and plasmon oscillations of any multipolarity can be excited in principle.

When an atom is situated near the nanosphere, the relative change of the total decay rate of the atom with the radial or tangential orientation of the dipole momentum can be described by the expressions:

$$\left(\frac{\gamma}{\gamma_{0,2}}\right)_{\text{radial}}^{\text{total}} \underset{ka\to 0}{\longrightarrow} \frac{3}{2\left(k_2 r\right)^3}\operatorname{Im}\sum_{n=0}^{\infty}(n+1)^2\left(\frac{a}{r}\right)^{2n+1}\frac{\alpha_n}{a^{2n+1}}$$
$$+\operatorname{Re}\left(1+\frac{2\alpha_1}{r^3}\right)^2+O\left(\frac{1}{k_2 a}\right), \tag{6.49}$$

$$\left(\frac{\gamma}{\gamma_{0,2}}\right)^{\text{total}}_{\text{tangential}} \underset{ka\to 0}{\longrightarrow} \frac{3}{4\,(k_2 r)^3}\mathrm{Im}\sum_{n=0}^{\infty} n\,(n+1)\left(\frac{a}{r}\right)^{2n+1}\frac{\alpha_n}{a^{2n+1}}$$
$$+\mathrm{Re}\left(1-\frac{\alpha_1}{r^3}\right)^2 + O\left(\frac{1}{k_2 a}\right), \quad (6.50)$$

where

$$\alpha_n = a^{2n+1}\frac{\varepsilon\,(\omega)-\varepsilon_2}{\varepsilon\,(\omega)+\varepsilon_2\,(n+1)/n} \quad (6.51)$$

are the multipolar polarizabilities of the order n, which generalize the notion of dipolar polarizability for $n = 1$.

A relative change of the radiative decay rate, that is, of the decay rate related with photons outgoing to infinity, can be described by the following expressions:

$$\left(\frac{\gamma}{\gamma_{0,2}}\right)^{\text{radiative}}_{\text{radial}} \underset{ka\to 0}{\longrightarrow} \left|1+\frac{2\,(\varepsilon(\omega)-\varepsilon_2)}{\varepsilon(\omega)+2\varepsilon_2}\frac{a^3}{r^3}\right|^2 + O\left((ka)^2\right), \quad (6.52)$$

$$\left(\frac{\gamma}{\gamma_{0,2}}\right)^{\text{radiative}}_{\text{tangential}} \underset{ka\to 0}{\longrightarrow} \left|1-\frac{(\varepsilon(\omega)-\varepsilon_2)}{\varepsilon(\omega)+2\varepsilon_2}\frac{a^3}{r^3}\right|^2 + O\left((ka)^2\right). \quad (6.53)$$

Nonradiative rates (i.e., the energy of an excited atom that converts to heat) are the differences between Eq. (6.49), Eq. (6.50) and Eq. (6.52), Eq. (6.53) and are determined by the first terms of Eqs. (6.49) and (6.50) mainly. The comparison of Eq. (6.49), Eq. (6.50) with Eq. (6.52), Eq. (6.53) shows that the total decay rate contains the component that increases infinitely (as $1/(ka)^3$) when the radius of the lossy sphere goes to zero. Moreover, as is seen from Eqs. (6.49) and (6.50), the series become divergent for an atom placed near the surface ($r \to a$), implying that for a sphere with losses the rate of nonradiative processes tends to infinity when the radiating atom approaches the surface. This fact is called radiation quenching and represents the general rule according to which in the contexts of local theories of electric permittivity the rate of nonradiative decays increases inversely proportional to the cubed distance Δ from the surface with losses.

$$\left(\frac{\gamma}{\gamma_0}\right)^{\text{nonradiative}} \sim \mathrm{Im}\frac{\varepsilon\,(\omega)-\varepsilon_2}{\varepsilon\,(\omega)+\varepsilon_2}\frac{1}{(k_2\Delta)^3}. \quad (6.54)$$

Often the quenching phenomenon makes difficulties for the investigation of radiation processes of atoms and molecules near metal particles and surfaces and for the creation of optical nanodevices. Therefore exact calculation, control, and minimization of nonradiative processes are the most important tasks for the development of nano-optics.

In the case of a perfectly conductive nanosphere ($\varepsilon(\omega) \to -\infty$), only radiative decays take place, for which the following expressions are true correspondingly:

$$\left(\frac{\gamma}{\gamma_0}\right)_{\text{radial}}^{\text{radiative}} \underset{ka\to 0}{\longrightarrow} \left|1+2\left(\frac{a}{r}\right)^3\right|^2 + O\left((ka)^2\right), \tag{6.55}$$

$$\left(\frac{\gamma}{\gamma_0}\right)_{\text{tangential}}^{\text{radiative}} \underset{ka\to 0}{\longrightarrow} \left|1-\left(\frac{a}{r}\right)^3\right|^2 + O\left((ka)^2\right). \tag{6.56}$$

If an atom is placed in the immediate vicinity of a sphere surface, then instead of Eqs. (6.55) and (6.56) the following important expressions can be obtained:

$$\left(\frac{\gamma}{\gamma_0}\right)_{\text{radial}}^{\text{radiative}} = 9, \tag{6.57}$$

$$\left(\frac{\gamma}{\gamma_0}\right)_{\text{tangential}}^{\text{radiative}} = 0. \tag{6.58}$$

Vanishing of the decay rate of an atom with a dipole moment of tangential orientation relative to a surface is a general rule, determined by the fact that the dipole moment induced on a perfectly conducting surface is exactly equal to the atom dipole moment but has the opposite orientation. As a result, atoms and molecules with a dipole moment of tangential orientation relative to the surface radiate very weakly. This situation is also true for any material, but suppression of the decay rate becomes less intense. In general case, the dipole oriented normally relative to a surface radiates more intensively than the dipole oriented tangentially.

Eqs. (6.47), (6.48), and Eqs. (6.52), (6.53) are true for most of the practically important cases. However, there is a region of parameters where these expressions are inapplicable (Klimov, 2002; Tribelsky and Luk'yanchuk, 2006). The matter concerns a specific case for which the following conditions are true:

$$\varepsilon(\omega) + \varepsilon_2(n+1)/n << (ka)^2 << 1. \tag{6.59}$$

It is a case of so-called weakly absorbing metals. In this case, the quasi-static approximation is incorrect, the formulas Eqs. (6.45), (6.46), (6.52), and (6.53) are inapplicable, and the full Maxwell theory should be used both for scattering cross sections and for decay rates of atoms and molecules (see Eqs. (6.36) and (6.38)).

In Fig. 6.9, the extinction cross sections of a spherical particle of radius a are shown in the limiting case of a lossless sphere. It is assumed that the electric permittivity is described by the Drude's dispersion law $\varepsilon(\omega) = 1 - \omega_{\mathrm{pl}}^2/\omega^2$. In this case, the dipole resonance arises for $\varepsilon = -2$, $\omega = \omega_{\mathrm{pl}}/\sqrt{3}$ and the quadrupole resonance arises for $\varepsilon = -3/2$, $\omega = \omega_{\mathrm{pl}}\sqrt{2}/\sqrt{5}$. As is seen from the picture,

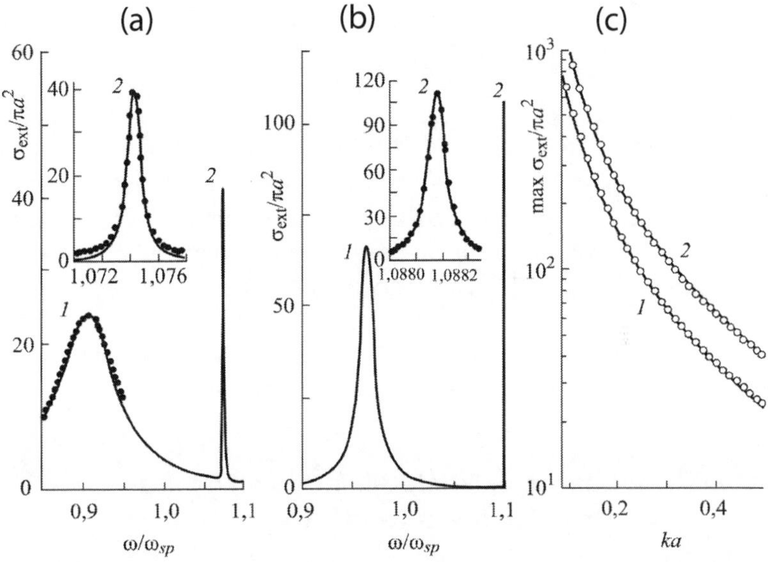

Figure 6.9 The normalized extinction cross section of a lossless spherical particle of radius a as a function of the frequency for $ka = 0.5$ (a) and for $ka = 0.3$ (b) for the Drude dispersion law $\varepsilon(\omega) = 1 - \omega_{\mathrm{pl}}^2/\omega^2$. $\omega_{\mathrm{sp}} = \omega_{\mathrm{pl}}/\sqrt{3}$ is the frequency of the dipole plasmon oscillation of the sphere. The numbers 1 and 2 denote the dipole and quadrupole resonances, correspondingly. The structure of the quadrupole resonance is shown in the insets. The solid line is calculations for Mie's exact theory, and the dotted line is the resonance approximation. (c) The dependence of the extinction cross section maxima for dipole and quadrupole resonances on the sphere radius ka (Tribelsky and Luk'yanchuk, 2006).

for a small particle the quadrupole resonance becomes dominant at corresponding frequencies. Naturally, if Eqs. (6.46) and (6.47) were used, this resonance would not appear at all. However, if real materials are taken into consideration, the situation might change. For example, for gold the imaginary part of electric permittivity at the frequency of quadrupole transition is large enough, and the quadrupole resonance usually cannot be detected (see, e.g., Fig. 6.3). However, for other materials (Na, K, Al), this anomalous situation can be quite observable.

In the case of an exact electric dipole plasmon resonance $\varepsilon(\omega) + 2\varepsilon_2 = 0$, the following expressions for spontaneous decay rates can be obtained using the general expressions of Eqs. (6.39) and (6.40):

$$\left(\frac{\gamma}{\gamma_{0,2}}\right)^{\text{radiative}}_{\text{radial}} \underset{ka\to 0}{\longrightarrow} \frac{225}{(ka)^4}\left(\frac{a}{r}\right)^6 \frac{1}{\varepsilon_2^2\left(\mu_1+5\mu_2\right)^2}, \tag{6.60}$$

$$\left(\frac{\gamma}{\gamma_{0,2}}\right)^{\text{radiative}}_{\text{tangential}} \underset{ka\to 0}{\longrightarrow} \frac{225}{4\,(ka)^4}\left(\frac{a}{r}\right)^6 \frac{1}{\varepsilon_2^2\left(\mu_1+5\mu_2\right)^2}. \tag{6.61}$$

An important feature of these expressions is that in the case of an exact resonance in nanospheres ($ka \to 0$), an unlimited increase of spontaneous decay rates occurs for any orientation of the dipole moment of an atom or a molecule. A perfectly analogous situation takes place also for the resonances of higher multipolarities, where the singularity just increases as the orbital number n does.

6.5 The Influence of Nonlocal Effects on Optical Properties of Spherical Particles

The whole analysis above supposed implicitly that electric permittivity and magnetic permeability are scalar constants. To be more precise, all the formulas stated above are true for a case when strong frequency dispersion is absent, that is, when the frequency scale of material dispersion is large in comparison with the linewidth of an atom radiation (narrow linewidth approximation). Usually, this case occurs in practice.

However, when considering the problems of a small *spatial* scale (an atom near an interface or a nanobody), it is necessary to take into account the spatial dispersion, that is, the dependence of electric induction **D** on electric field gradients or, that is the same, the dependence of electric permittivity on the wavevector ((Landau and Lifshitz, 1984); see also Chapter 3).

Specific spatial scales a^{NL}, where the spatial dispersion and nonlocal effects appear, are defined by the relationship between frequency and the Fermi rate in metal:

$$a^{\mathrm{NL}} \sim v_{\mathrm{F}}/\omega, \tag{6.62}$$

since in normal metals the Fermi velocity is of the order of 10^8 cm/s (see Table 3.2) and optic frequencies are of the order of 10^{15} s^{-1}, than $a^{NL} \approx 1$ nm. Thus, in nanoparticles of several nanometers size, spatial dispersion will play an important role.

In the presence of spatial dispersion, electric permittivity is a tensor, even in an isotropic medium. If the medium not only is isotropic but also has an inversion center, the tensor ε_{ik} can be composed only of the components of the vector **k** and the unit tensor δ_{ik}. The general form of a tensor of this kind can be written as:

$$\varepsilon_{ik}(\omega, \mathbf{k}) = \varepsilon_t(\omega, k)\left(\delta_{ik} - \frac{k_i k_k}{k^2}\right) + \varepsilon_l(\omega, k)\frac{k_i k_k}{k^2}, \tag{6.63}$$

where $\varepsilon_l(\omega, k)$ and $\varepsilon_t(\omega, k)$ depend on the wavevector absolute value only (and on ω). If the electric field strength **E** is directed along the wavevector, the induction $\mathbf{D} = \varepsilon_l \mathbf{E}$, and if $\mathbf{E} \perp \mathbf{k}$, $\mathbf{D} = \varepsilon_t \mathbf{E}$. Correspondingly, the values $\varepsilon_l(\omega, k)$ and $\varepsilon_t(\omega, k)$ are called the longitudinal and TE permittivities. Thus, the presence of spatial dispersion results in propagation both of longitudinal and of transverse modes in a medium, characterized by the dispersion laws

$$\begin{aligned} k_t^2 &= \left(\frac{\omega}{c}\right)^2 \varepsilon_t(\omega, k_t), \\ \varepsilon_l(\omega, k_l) &= 0, \end{aligned} \tag{6.64}$$

correspondingly.

The transverse modes with the spatial structure described by the following expressions correspond to the transverse dispersion law:

$$\begin{aligned} \mathbf{M}(\mathbf{r}) &= \mathrm{rot}\left(\mathbf{r} Y_{nm}(\theta, \varphi)\, j_n(k_t r)\right), \\ \mathbf{N} &= \frac{1}{k_t}\mathrm{rot}\left(\mathbf{M}(\mathbf{r})\right), \end{aligned} \tag{6.65}$$

where the vector harmonics **M** and **N** can describe the TE and magnetic fields depending on polarization. These are the expressions that were used to solve the problem of a dipole radiation near a sphere with no spatial dispersion (Section 6.1).

The structure of longitudinal modes that do not appear in the absence of spatial dispersion is defined by the expression:

$$\begin{aligned}\mathbf{E}_l(\mathbf{r}) &= \mathbf{L}(\mathbf{r}) = \mathrm{grad}\left(Y_{nm}(\theta, \varphi)\, j_n(k_l r)\right),\\ \mathbf{H}_l(\mathbf{r}) &= 0.\end{aligned} \tag{6.66}$$

Usual boundary conditions of tangential components of electric and magnetic fields continuity are insufficient for determination of the longitudinal modes $\mathbf{E}_l(\mathbf{r})$ in nanoparticles, and additional boundary conditions are required, allowing electromagnetic fields in nanoparticles to be found uniquely. Now, there is no common approach to this problem, and the spatial dispersion in nanoparticles is taken into account within the frameworks of different models. The discussion of the boundary conditions problem in the presence of spatial dispersion can be found, for example, in (Agranovich and Ginsburg, 1984).

The numerical investigation of nonradiative decay rates of an atom near a nanosphere, taking into account nonlocal effects, was carried out by Ekardt and Penzar (1986). In this paper, the following expression for nonradiative decay rates was used:

$$\begin{aligned}\gamma^{\mathrm{total}} &= \frac{3}{2(k_2 r)^3}\sum_{l=1}^{\infty}\frac{\mathrm{Im}\alpha_l^{\mathrm{NL}}(\omega)}{a^{2l+1}}\left(\frac{a}{r}\right)^{2l+1}(l+1)\\ &\times\left[(l+1)\cos^2\theta + \frac{l}{2}\sin^2\theta\right],\end{aligned} \tag{6.67}$$

where θ is the angle between the radius and the atom dipole moment orientation. Eq. (6.67) is obtained from Eqs. (6.49) and (6.50) by retention of the main terms inversely proportional to r^3 and where local multipolar polarizabilities, Eq. (6.51), are substituted by their nonlocal analogs $\alpha_l \to \alpha_l^{\mathrm{NL}}$, which can be calculated by the numerical method elaborated by Ekardt (1985) within the context of the time-dependent density functional method (TDDFT).

In Fig. 6.10, the dependencies of nonradiative decay rates, Eq. (6.67), on dimensionless radiation frequency are shown. It is seen from the pictures that nonlocal effects result both in essential

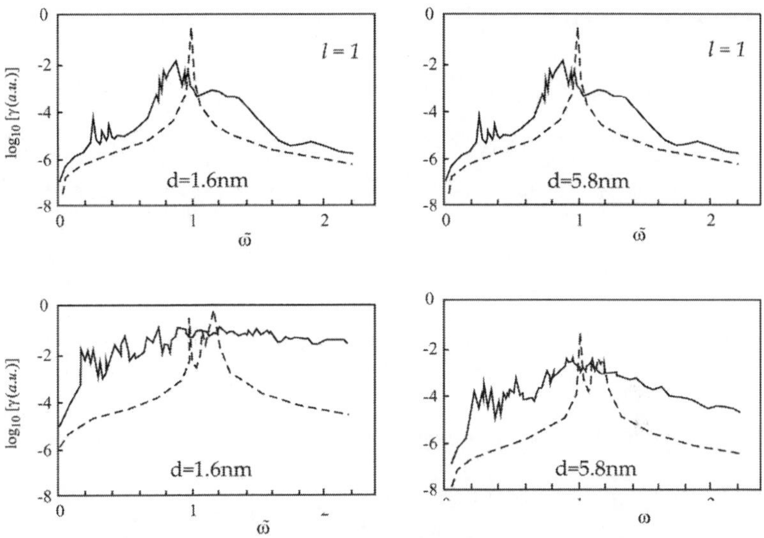

Figure 6.10 The dependence of the nonradiative decay rate of an atom near a sodium sphere on the dimensionless frequency $\tilde{\omega} = \omega/\left(\omega_{\text{pl}}/\sqrt{3}\right)$ obtained within the TDDFT numerical approach, taking nonlocal effects into account (solid lines). The dashed lines correspond to calculations according to classical local theory, Eq. (6.49). The dipole momentum of the atom has radial orientation, and the atom is situated at distance d from the surface of a sodium sphere of 9.5 nm radius. In the upper row, only dipole contributions are taken into account, while contributions of all modes are shown in the bottom row (Ekardt and Penzar, 1986).

increase of nonradiative decay rates in comparison with the local case (the dashed line) and in appearance of a set of additional resonances.

An analogous approach was used by Leung (1990), Hider and Leung (2002) for development of a simple *analytical* model of influence of permittivity spatial dispersion on the nonradiative decay rate of an atom near a nanosphere. In this paper, they use the expression analogous to Eq. (6.67) for the total decay rate and for frequency shifts. However, in contrast to the approach of Ekardt and Penzar (1986) they replaced local susceptibilities, Eq. (6.51), by analytical expressions for nonlocal susceptibilities proposed by Fuchs and Claro (1987). If the expressions for relative nonradiative decay rates and frequency shifts are represented as:

$$\left(\frac{\gamma}{\gamma_{\text{vac}}}\right) = 1 + \frac{3}{2\left(k_2 r\right)^3}\operatorname{Im}\left(G\right), \quad \left(\frac{\Delta\omega}{\gamma_{\text{vac}}}\right) = -\frac{3}{4\left(k_2 r\right)^3}\operatorname{Re}\left(G\right), \tag{6.68}$$

where r is the distance between the sphere center and the atom, the normalized part of the reflected Green function takes the form:

$$G_{\text{radial}} = \sum_l \left(l+1\right)^2 \frac{\alpha_l^{\text{NL}}}{a^{2l+1}} \left(\frac{a}{r}\right)^{2l+1}, \tag{6.69}$$

$$G_{\text{tangential}} = \frac{1}{2}\sum_l l\left(l+1\right) \frac{\alpha_l^{\text{NL}}}{a^{2l+1}} \left(\frac{a}{r}\right)^{2l+1}, \tag{6.70}$$

where nonlocal polarizabilities

$$\alpha_l^{\text{NL}} = a^{2l+1}\frac{l\left(\xi_l - 1\right)}{l\left(\xi_l + 1\right) + 1}, \tag{6.71}$$

are obtained from local expressions, Eq. (6.51), by replacement of the electric permittivity ε by the "effective electric permittivity" ξ_l. In turn, in the nonlocal theory the "effective electric permittivity" ξ_l is defined as (Fuchs and Claro, 1987):

$$\frac{1}{\xi_l} = \left(2l+1\right) a F_l = \frac{2}{\pi}\left(2l+1\right) a \int_0^\infty dk \frac{j_l^2\left(ka\right)}{\varepsilon\left(k,\omega\right)}. \tag{6.72}$$

In Eq. (6.72), $\varepsilon(\text{k},\omega)$ is the electric permittivity of a homogeneous substance with spatial dispersion taken into account. In the case of the hydrodynamic model of permittivity:

$$\varepsilon\left(k,\omega\right) = 1 - \frac{\omega_{\text{pl}}^2}{\omega\left(\omega + i\Gamma\right) - \beta^2 k^2}, \tag{6.73}$$

where $\beta^2 = 3/5 v_{\text{F}}^2$, v_{F} is the electrons Fermi rate in metal, and the effective electric permittivity takes the form:

$$\xi_l = \left[\frac{1}{\varepsilon\left(\omega\right)} + \left(2l+1\right)\left[\frac{a\omega_{\text{pl}}}{\beta u}\right]^2 I_{l+1/2}\left(u\right) K_{l+1/2}\left(u\right)\right]^{-1}, \tag{6.74}$$

where $u = a\sqrt{\omega_{\text{pl}}^2 - \omega\left(\omega + i\Gamma\right)}/\beta$ and $\varepsilon\left(\omega\right) = 1 - \omega_{\text{pl}}^2/\omega\left(\omega + i\Gamma\right)$ are the local dielectric functions (Drude function). In Fig. 6.11, the dependencies of nonradiative decay rates on dimensionless

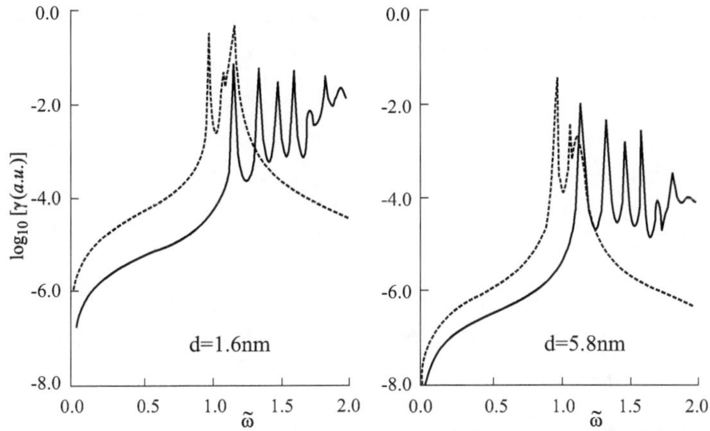

Figure 6.11 The dependence of the nonradiative decay rate of an atom near a sodium sphere of 9.5 nm radius on the dimensionless frequency $\tilde{\omega} = \omega/\left(\omega_{\mathrm{pl}}/\sqrt{3}\right)$ obtained within analytical theory, Eq. (6.71) (solid lines). The dashed lines correspond to calculations according to classical local theory, Eq. (6.49). The dipole momentum of an atom has radial orientation, and the atom is situated at the distance d from the sphere surface (Leung, 1990).

frequency $\tilde{\omega} = \omega/\left(\omega_{\mathrm{pl}}/\sqrt{3}\right)$ near a nonlocal nanosphere are shown (Leung, 1990). As is seen from the pictures within the context of this theory, nonlocal effects result in a *decrease* of nonradiative rates in the region of $\tilde{\omega} < 1$, that is, at frequencies lower than the frequency of the localized plasmon, and in an *increase* for $\tilde{\omega} > 1$.

In fact, Ekardt and Penzar (1986), Leung, (1990), Hider and Leung (2002) have investigated nonlocal optical properties of a nanoparticle, taking into account longitudinal electric fields only. More detailed investigation of the spatial dispersion influence on optical properties of nanoparticles with transverse electromagnetic wave and retardation effects taken into account was carried out by Ruppin (1975), Agarwal and Neil (1983).

As in Section 6.1 of this paper, the transverse fields are represented as a series over transverse vector spherical harmonics, Eq. (6.65), with the transverse wavevector:

$$k_t^2 = \left(\frac{\omega}{c}\right)^2 \varepsilon_t(\omega), \tag{6.75}$$

while the longitudinal fields are represented as a series over longitudinal harmonics, Eq. (6.66), with the longitudinal wavevector:

$$k_l^2 = \frac{5}{3v_F^2}\left(\omega^2 - i\Gamma\omega - \omega_{pl}^2\right), \tag{6.76}$$

which is the solution of the dispersion equation:

$$\varepsilon_l(k,\omega) = 1 - \frac{\omega_{pl}^2}{\omega(\omega + i\Gamma) - \beta^2 k^2} = 0. \tag{6.77}$$

The procedure of the vector spherical harmonics matching on the sphere surface includes not only usual continuity conditions for the tangential components of **E** and **H** but also an additional continuity condition for the *normal* component of the electric field (Ruppin, 1975). This statement of the problem results in modification of the Mie coefficients q_n, see Eq. (6.21), for the TM waves:

$$q_n^{NL} = \frac{\varepsilon_t\left[z_2 j_n(z_2)\right]' j_n(z_t) - \varepsilon_2\left[z_t j_n(z_t)\right]' j_n(z_2) - (\varepsilon_t - \varepsilon_2)A}{\varepsilon_t\left[z_2 h_n^{(1)}(z_2)\right]' j_n^{(1)}(z_t) - \varepsilon_2\left[z_t j_n^{(1)}(z_t)\right]' h_n^{(1)}(z_2) - (\varepsilon_t - \varepsilon_2)B},$$

$$A = \frac{j_n(z_2) j_n(z_t) n(n+1) j_n(z_l)}{z_l j_n'(z_l)},$$

$$B = \frac{h_n^{(1)}(z_2) j_n(z_t) n(n+1) j_n(z_l)}{z_l j_n'(z_l)}, \tag{6.78}$$

where z_t, z_l, z_2 = $k_t a$, $k_l a$, $k_2 a$, correspondingly. For TE modes, the coefficients p_n, Eq. (6.22), do not change within this approach.

In the case of the absence of spatial dispersion (the limit $\beta^2 = \frac{3}{5}v_F^2 \to 0$) k_l, $z_l \to \infty$, and Eq. (6.78) transforms into the Mie coefficients, Eq. (6.21).

As usual, the denominator of Eq. (6.78) determines the condition for plasmon resonance existence in a spherical particle with nonlocal effects taken into account. In the case of nanoparticles (z_t, $z_2 \to 0$), the denominator of Eq. (6.78) simplifies to the form:

$$(\varepsilon_t(\omega) - \varepsilon_2) + \frac{k_l(\omega) a j_n'(k_l(\omega) a)}{n(n+1) j_n(k_l(\omega) a)}\left(n\varepsilon_t(\omega) + \varepsilon_2(n+1)\right) = 0. \tag{6.79}$$

This dispersion equation determines the spectrum of localized plasmons in a spherical nanoparticle with the nonlocal effects taken

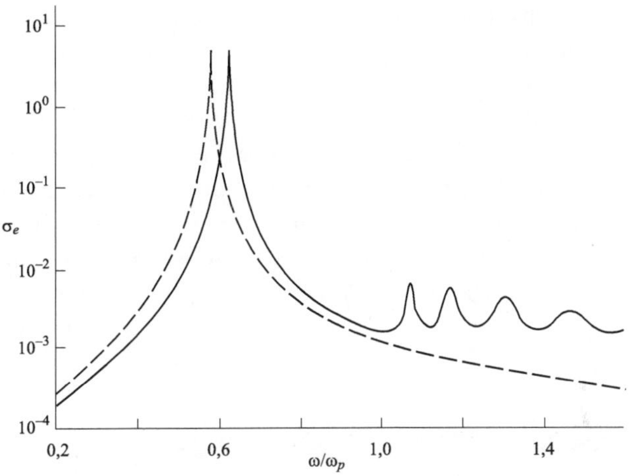

Figure 6.12 The extinction coefficients of a sodium sphere of the 1.5 nm radius. The solid curve is the nonlocal theory, and the dashed line corresponds to calculations within the local Mie theory (Ruppin, 1975).

into account. In the local limit, $k_l a \to \infty$ and this equation reduces to the standard equation of dipole plasmon resonance for *transverse* modes:

$$n\varepsilon_t + \varepsilon_2 (n+1) = 0. \tag{6.80}$$

In small nanoparticles where Eq. (6.79) is true, the spatial dispersion results in transverse and longitudinal plasmons interaction, and the plasmon oscillations have a hybrid character, and their spectrum has lines corresponding both to transverse and longitudinal oscillations.

Ruppin (1975) used the modified Mie coefficients, Eq. (6.78), for calculation of scattering and absorption cross sections of nanoparticles, substituting them into Eqs. (6.34) and (6.35). In Fig. 6.12, the results of calculation of extinction coefficients within this approach are shown. As seen from the picture, taking nonlocal effects into account results in the shift of the dipole plasmon resonance to the region of higher frequency and in the appearance of additional resonances above the plasmon frequency.

Agarwal and Neil (1983) used the nonlocal coefficients q_n^{NL}, Eq. (6.78), for calculation of radiative decay rates, Eqs. (6.39) and (6.40), where instead of the Mie coefficient q_n its nonlocal generalization q_n^{NL} was used for TM modes.

The resonance positions and characteristics of decay rates obtained by Agarwal and Neil (1983), Leung (1990), Ekardt and Penzar (1986) are not in good agreement, and thus additional theoretical and experimental investigations of nonlocal effects in spherical nanoparticles are required. Nevertheless, the general conclusion about the nonlocal effects' importance for nanoparticles of a size less than 10 nm remains valid.

6.6 Optical Properties of Layered Spherical Particles

Until now, we have considered a simple spherical particle. However, the approach represented in Section 6.1 can be applied to any layered spherical system (often systems of this kind are called nanomatryoshkas or nanoshells). As in the case of a simple spherical particle, here the main point is the expansions of the dipole source field or a plane wave over the vector spherical harmonics, which are pinned to the sphere center, Eq. (6.18). After that, in each layer the field is to be found as an expansion over vector spherical harmonics, Eqs. (6.19) and (6.20). Applying the continuity condition for tangential fields' components at all interfaces, all unknown coefficients can be found, and thus the problem of the plane or spherical wave diffraction on a layered particle can be solved and the scattering and absorption cross sections and rates of spontaneous decays of an atom placed near a layered nanoparticle can be found.

6.6.1 *Optical Properties of Layered Nanoparticles in a Homogeneous Field*

In the case of nanoparticles, the situation becomes simpler again since quasi-static approximation can be used. In the case of a layered geometry, shown in Fig. 6.13, the solution of the Laplace equation in different regions has the form:

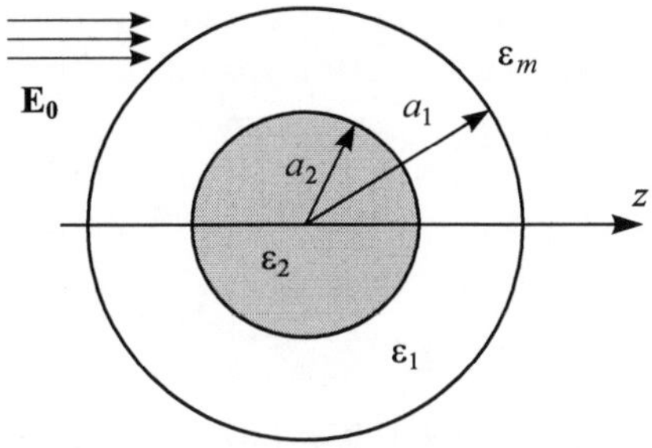

Figure 6.13 The geometry of a layered particle in a homogeneous electric field.

$$\varphi_2 = -E_0 \frac{9\varepsilon_1\varepsilon_m}{(\varepsilon_1 + 2\varepsilon_m)(\varepsilon_2 + 2\varepsilon_1) + 2\beta(\varepsilon_1 - \varepsilon_m)(\varepsilon_2 - \varepsilon_1)} r\cos\theta, \quad r < a_2,$$

$$\varphi_1 = -E_0 \frac{3\left((\varepsilon_2 + 2\varepsilon_1) - (\varepsilon_2 - \varepsilon_1)(a_2/r)^3\right)\varepsilon_m}{(\varepsilon_1 + 2\varepsilon_m)(\varepsilon_2 + 2\varepsilon_1) + 2\beta(\varepsilon_1 - \varepsilon_m)(\varepsilon_2 - \varepsilon_1)} r\cos\theta, \quad a_2 < r < a_1,$$

$$\varphi_0 = -E_0 \left[1 - \frac{(\varepsilon_1 - \varepsilon_m)(\varepsilon_2 + 2\varepsilon_1) + \beta(\varepsilon_m + 2\varepsilon_1)(\varepsilon_2 - \varepsilon_1)}{(\varepsilon_1 + 2\varepsilon_m)(\varepsilon_2 + 2\varepsilon_1) + 2\beta(\varepsilon_1 - \varepsilon_m)(\varepsilon_2 - \varepsilon_1)} \left(\frac{a_1}{r}\right)^3\right] \times r\cos\theta, \quad r > a_1, \tag{6.81}$$

where $\beta = (a_2/a_1)^3$ is the ratio between the volume of the particle core and its full volume.

It is seen from this solution that in the external domain ($r > a_1$) the field consists of the homogeneous excitation field $\varphi_0 = -E_0 z$ and the scattered field, which coincides with the field of a point dipole.

So, the dipole polarizability of this three-layered system has the form:

$$\alpha = a_1^3 \frac{(\varepsilon_1 - \varepsilon_m)(\varepsilon_2 + 2\varepsilon_1) + \beta(\varepsilon_m + 2\varepsilon_1)(\varepsilon_2 - \varepsilon_1)}{(\varepsilon_1 + 2\varepsilon_m)(\varepsilon_2 + 2\varepsilon_1) + 2\beta(\varepsilon_1 - \varepsilon_m)(\varepsilon_2 - \varepsilon_1)}, \tag{6.82}$$

and for $\varepsilon_2 = \varepsilon_1$ transforms into the expression for the polarizability of a simple sphere. In turn, knowing the polarizability, the scattering and absorption cross sections can be found using the formulas, Eq. (6.46).

It is interesting that the polarizability of a homogeneous spherical particle becomes equal to zero if the particle is coated by a shell having such a radius and permittivity that the numerator of Eq. (6.82) is equal to zero. This is equivalent to the condition fulfillment:

$$\left(\frac{a_2}{a_1}\right)^3 \frac{(\varepsilon_2 - \varepsilon_1)}{(\varepsilon_2 + 2\varepsilon_1)} = \frac{(\varepsilon_m - \varepsilon_1)}{(\varepsilon_m + 2\varepsilon_1)}. \tag{6.83}$$

Practically, zero polarizability implies that the nanoparticle coated by the shell is invisible.

Certainly, this fact is true only for small particles (nanoparticles). To make a big object invisible, much more complex constructions made on the basis of modern metamaterials are required. The problem of invisible coatings ("invisibility cloak") is considered in Chapter 13, Section 13.8, in detail.

When the denominator of Eq. (6.82) is close to zero, a plasmon resonance appears in the system. This resonance is of particular importance for nanoshells, which are dielectric nanoparticles coated by a metal layer (normally, silver or gold) (Prodan et al., 2003). The importance of the resonances in nanoshells is due to the fact that their wavelengths can vary within wide limits (between 500 nm and 10,000 nm) as the ratio between internal and external radii of the nanoshell changes. In a particular case when $\varepsilon_2 = 1$, $\varepsilon_m = 1$ (an empty nanoshell placed in vacuum), localized plasmons exist if permittivity of the shell satisfies the condition:

$$\varepsilon_1 = -\frac{5 + 4\beta}{4(1-\beta)} \pm \frac{3\sqrt{1+8\beta}}{4(1-\beta)}. \tag{6.84}$$

In the case of the Drude dispersion law $\varepsilon_1(\omega) = 1 - \omega_{\text{pl}}^2/\omega^2$, the resonant permittivities, Eq. (6.84), correspond to plasmon frequencies:

$$\omega_{\pm}^2 = \frac{\omega_{\text{pl}}^2}{2}\left[1 \pm \frac{\sqrt{1+8\beta}}{3}\right]. \tag{6.85}$$

The plasmon resonances in Eqs. (6.84) and (6.85) describe the nanoshell dipole symmetric or antisymmetric oscillations. In nanoshells, the plasmon resonances with higher multipole moments are possible as they are in any spherical symmetric systems. In the case of an arbitrary multipolar resonance in the nanoshell of ε_2, $\varepsilon_m = 1$, the plasmon frequencies can be written as:

$$\omega_{l,\pm}^2 = \frac{\omega_{\rm pl}^2}{2}\left[1 \pm \frac{\sqrt{1+4l\,(l+1)\,(a_2/a_1)^{2l+1}}}{2l+1}\right], \quad l = 1, 2, 3\ \ldots \tag{6.86}$$

Naturally, for $l = 1$ (dipole oscillations) Eq. (6.86) reduces to Eq. (6.85).

6.6.2 *Spontaneous Emission of an Atom in the Presence of Layered Spherical Particles*

If a layered spherical particle is in the field of a spherical wave (dipole field), all the fields characteristics can be found analytically (Klimov and Letokhov, 1999; Tomas, 2001). In particular, in the case of an atom placed in the center of a layered spherical system shown in Fig. 6.13, the expression for the total decay rate may be written by the following formulas:

$$\left(\frac{\gamma}{\gamma_{0,2}}\right) = 1 - \mathrm{Re}\,(G), \tag{6.87}$$

$$G = \frac{\varepsilon_1 h_1^{(R)}(z_1)\left(z_2 h_1^{(1)}(z_2)\right)' - \varepsilon_2 h_1^{(1)}(z_2)\left(z_1 h_1^{(R)}(z_1)\right)'}{\varepsilon_1 h_1^{(R)}(z_1)\,(z_2 j_1(z_2))' - \varepsilon_2 j_1(z_2)\left(z_1 h_1^{(R)}(z_1)\right)'}, \tag{6.88}$$

$$h_1^{(R)}(z) = h_1^{(1)}(z) - R h_1^{(2)}(z),$$

Here, γ_{02} is the decay rate in the homogeneous medium 2, $z_1 = k_1 a_2$, $z_2 = k_2 a_2$, and R is the coefficient of the reflection at the external boundary ($r = a_1$):

$$R = \frac{\varepsilon_1 h_1^{(1)}(z_3)\left(z_4 h_1^{(1)}(z_4)\right)' - \varepsilon_m h_1^{(1)}(z_4)\left(z_3 h_1^{(1)}(z_3)\right)'}{\varepsilon_1 h_1^{(2)}(z_3)\left(z_4 h_1^{(1)}(z_4)\right)' - \varepsilon_m h_1^{(1)}(z_4)\left(z_3 h_1^{(2)}(z_3)\right)'}, \tag{6.89}$$

where $z_3 = k_1 a_1$, $z_4 = k_m a_1$.

In the case of $\varepsilon_m = \varepsilon_1$, the reflection coefficient tends to zero ($R = 0$), and Eq. (6.87) reduces to the expression for the total decay rate of an atom situated inside a spherical particle or cavity.

In the case of nanoparticles, the quasi-static theory can be used again, and the electric potential of the "atom dipole + nanoparticle" system can be found in the form of:

$$\begin{aligned}
\varphi_2 &= \left(\frac{d_0}{\varepsilon_2 r^2} + t_2 r\right)\cos\theta, \quad r < a_2, \\
\varphi_1 &= \left(\frac{u_1}{r^2} + t_1 r\right)\cos\theta, \quad a_2 < r < a_1, \\
\varphi_0 &= \frac{d_{tot}}{\varepsilon_m r^2}\cos\theta, \quad r > a_1,
\end{aligned} \tag{6.90}$$

where in the first equation the term $(d_0/\varepsilon_2 r^2)\cos\theta$ describes the dipole field of the emission of the atom or molecule in the center of the cavity, and d_{total} describes the total dipole moment of the system. The unknown coefficients t_2, t_1, u_1, and d_{total} can be found from the continuity condition of potential and normal components of electric induction on all interfaces.

In particular,

$$d_{\text{total}} = \frac{9\varepsilon_1\varepsilon_m d_0}{(\varepsilon_1 + 2\varepsilon_m)(\varepsilon_2 + 2\varepsilon_1) + 2\beta(\varepsilon_1 - \varepsilon_m)(\varepsilon_2 - \varepsilon_1)}, \tag{6.91}$$

and thus the following expression for the radiative decay rate of an atom placed in the center of a spherically layered system (Fig. 6.13) is obtained (see Appendix A1):

$$\begin{aligned}
\left(\frac{\gamma}{\gamma_{vac}}\right) &= \sqrt{\varepsilon_m}\left|\frac{d_{\text{total}}}{d_0}\right|^2 \\
&= \sqrt{\varepsilon_m}\left|\frac{9\varepsilon_1\varepsilon_m}{(\varepsilon_1 + 2\varepsilon_m)(\varepsilon_2 + 2\varepsilon_1) + 2\beta(\varepsilon_1 - \varepsilon_m)(\varepsilon_2 - \varepsilon_1)}\right|^2,
\end{aligned} \tag{6.92}$$

where γ_{vac} is the decay rate of an atom in vacuum. Certainly, Eq. (6.92) is applicable when the core and the environment are transparent, $\text{Im}(\varepsilon_2) = \text{Im}(\varepsilon_m) = 0$.

Usually, zeros of the denominator in Eq. (6.92) determine the position of resonances. Comparing of the denominators in Eqs. (6.82) and (6.92) confirms their identity obviously, and therefore the plasmon resonances of polarizability and the rates of atom radiative decay are equal. This also implies that the dipole source in the center of the system excites the same plasmon modes as the homogeneous external field does.

References

Abramowitz, M., Stegun, I.A. (1972) *Handbook of Mathematical Functions* (National Bureau of Standards).

Agarwal, G.S., Neil, S.V. (1983) Effect of hydrodynamics scattering of the metal on surface plasmons and surface enhanced phenomena in spherical geometries, *Phys. Rev. B*, **28**, 487–493.

Agranovich, V.M., Ginzburg, V.L. (1984) *Crystal Optics with Spatial Scattering and Excitons* (Springer-Verlag, Berlin).

Bashevoy, M.V., Fedotov, V.A., Zheludev, N.I. (2005) Optical whirlpool on an absorbing metallic nanoparticle, *Opt. Express*, **13**, 8372–8379 (http://www.opticsinfobase.org/oe/abstract.cfm?id=85768).

Braginsky, V.B., Il'chenko, V.S. (1987) Properties of optical dielectric microresonators, *Sov. Phys. Dokl.*, **32**, 306–307.

Braginsky, V.B., Gorodetsky, M.L., Il'chenko, V.S. (1989) Quality-factor and nonlinear properties of optical whispering-gallery modes, *Phys. Lett. A*, **137**, 393–397.

Bohren, C.F., Huffman, D.R. (1983) *Absorption and Scattering of Light by Small Particles* (Wiley, New York).

Chew, H. (1988) Radiation and lifetimes of atoms inside dielectric particles, *Phys. Rev. A*, **38**, 3410–3416.

Chew, H. (1987) Transition rates of atoms near spherical surfaces, *J. Chem. Phys.*, **87**, 1355.

Chew, H., McNulty, P.J., Kerker, M. (1976) Model for Raman and fluorescent scattering by molecules embedded in small particles, *Phys. Rev. A*, **13**, 396–404.

Collot, L., Lefevre-Seguin, V., Brune, M., Raimond, J.M., Haroche, S. (1993) Very high-Q whispering-gallery mode resonances observed on fused silica microspheres, *Europhys. Lett.*, **23**, 327.

Dutta Gupta S., Agarwal, G.S. (1995) Strong coupling cavity physics in microspheres with whispering gallery modes, *Opt. Commun.*, **115**, 597–605.

Ekardt, W., Penzar, Z. (1986) Nonradiative lifetime of excited states near a small metal particle, *Phys. Rev. B*, **34**, 8444–8448.

Ekardt, W. (1985) Collective multipole excitations in small metal particles: critical angular momentum l^{cr} for the existence of collective surface modes, *Phys. Rev. B*, **32**, 1961–1970.

Fuchs, R., Claro, F. (1987) Multipolar response of small metallic spheres: nonlocal theory *Phys. Rev. B*, **35**, 3722–3727.

Hider, M.H., Leung, P.T. (2002) Nonlocal electrodynamic modeling of fluorescence characteristics for molecules in a spherical cavity, *Phys. Rev. B*, **66**, 195106.

García-Etxarri A., Gómez-Medina R., Froufe-Pérez L.S., López C., Chantada L., Scheffold F., Aizpurua J., Nieto-Vesperinas M., Sáenz J.J. (2011) Strong magnetic response of submicron silicon particles in the infrared, *Opt. Express*, **19**, 4815–4826.

Jackson, J.D. (1975) *Classical Electrodynamics* (Wiley, New York).

Klimov, V.V. (2002) Spontaneous emission of an excited atom placed near a "left-handed" sphere, *Opt. Commun.*, **211**, 183–196.

Klimov, V.V., Ducloy, M., Letokhov, V.S. (1996a) Spontaneous emission rate and level shift of atom inside dielectric microsphere, *J. Mod. Opt.*, **43**, 549–563.

Klimov, V.V., Ducloy, M., Letokhov, V.S. (1996b) Radiative frequency shift and line width of an atom dipole in the vicinity of a dielectric microsphere, *J. Mod. Opt.*, **43**, 2251–2267.

Klimov, V.V., Ducloy, M., Letokhov, V.S. (1999a) Strong interaction of two-level atom with whispering gallery modes of dielectric microsphere: quantum consideration, *Phys. Rev. A*, **59**, 2996–3014.

Klimov, V.V., Ducloy, M., Letokhov, V.S. (1999b) Quasi orbital motion of ultracold excited atomic dipole near dielectric microsphere, *Eur. Phys. J. D*, **5**, 345–350.

Klimov, V.V., Letokhov, V.S. (1996a) Quadrupole radiation of atom in the vicinity of a dielectric microsphere, *Phys. Rev. A*, **54**, 4408–4423.

Klimov, V.V., Letokhov, V.S. (1996b) Increase of spontaneous quadrupole transition rate in the vicinity of a dielectric microsphere, *Opt. Commun.*, **122**, 155–162.

Klimov, V.V., Letokhov, V.S. (1999) The enhancement and inhibition of the spontaneous-emission rates in nanobubbles, *Chem. Phys. Lett.*, **301**, 441–448.

Klimov, V.V., Letokhov, V.S. (2000) Effect of the curvature of nanostructures on radiative multipole transition rates, comments on atomic and molecular physics, *Comments Mod. Phys.*, **2D**, 15–27.

Klimov, V.V. and Letokhov, V.S. (2005) Electric and magnetic dipole transitions of an atom in the presence of spherical dielectric interface, *Laser Phys.*, **15**, 61–73.

Klimov, V.V. (2003) Spontaneous atomic radiation in the presence of nanobodies, *Phys.-Uspekhi*, **46**(9), 13–18.

Landau, L.D., Lifshitz, E.M. (1984) *Electrodynamics of Continuous Media*, 2nd ed. (Pergamon Press, Oxford).

Leung, P.T. (1990) Decay of molecules at spherical surfaces: nonlocal effects, *Phys. Rev. B*, **42**, 7622–7625.

Lide, R., ed. (2000) *Handbook of Chemistry and Physics* (CRC Press, New York).

Mie, G. (1908) Beitrage zur Optik truber Medien, speziell kolloidaler Metallosungen, *Ann. Phys. IV*, **25**, 377.

Prodan, E., Radloff, C., Halas, N.J., Nordlander, P. (2003) A hybridization model for the plasmon response of complex nanostructures, *Science*, **302**, 419.

Ruppin, R. (1975) Optical properties of small metallic spheres, *Phys. Rev. B*, **11**, 2871–2876.

Tomas, M.S. (2001) Local-field corrections to the decay rate of excited molecules in absorbing cavities: the Onsager model, *Phys. Rev. A*, **63**, 053811.

Tribelsky, M.I., Luk'yanchuk, B.S. (2006) Anomalous light scattering by small particles, *Phys. Rev. Lett.*, **97**, 263902.

Vainshtein, L.A. (1988) *Elektromagnitnye Volny (Electromagnetic Waves)*, 2nd ed. (Radio i Svyaz', Moscow) (in Russian).

Vainshtein, L.A. (1966) *Open Resonators and Open Waveguides* (Sovietskoe Radio) (in Russian).

Chapter 7

Plasmonic Properties of Nanospheroids

Nanoparticles in the shape of oblate or prolate spheroids, which are obtained as a result of the ellipse rotation about the short axis or the long axis, correspondingly (Fig. 7.1), have an even greater significance for optical applications than spherical particles have. On the one hand, understanding of the optical properties of spheroid-shaped nanoparticles allows estimating of the influence of small deviations from sphericity on various applications. On the other hand, great attention is paid to spheroidal nanoparticles that differ essentially from spherical particles (nanoneedles, nanorods, nanodiscs) since the plasmon resonances in particles of this kind are rich in structure, which allows their usage in many applications.

For the moment, the methods of nanorod synthesis are developed in detail (Yu et al., 1997; Pérez-Juste et al., 2004; Lisiecki and Pileni, 2003; Wang et al., 2006; Mohamed et al., 2000), the images of nanorods and the bar charts describing the distribution of particles over the aspect ratios are shown. Certainly, the shape of these particles is not perfectly spheroidal, and their sizes differ slightly from each other. Nevertheless, the spheroidal approximation appears to be adequate enough since it allows one to describe analytically the properties of such particles more or less correctly. In

Nanoplasmonics
Vasily Klimov

ISBN 978-981-4267-16-8 (Hardcover), 978-981-4267-42-7 (eBook)
www.panstanford.com

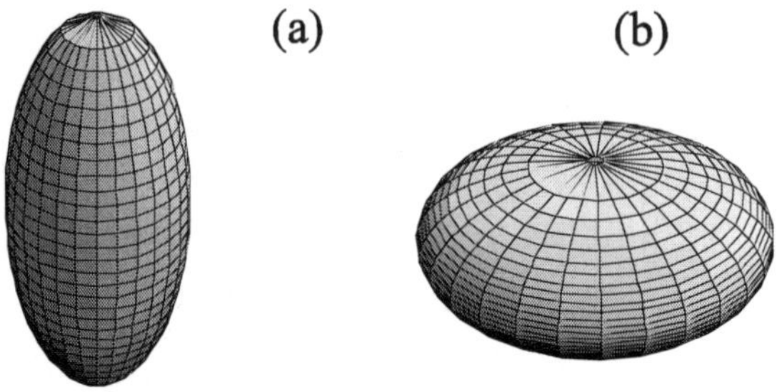

Figure 7.1 The geometry of a prolate spheroidal nanoparticle (a) and of an oblate spheroidal particle (b).

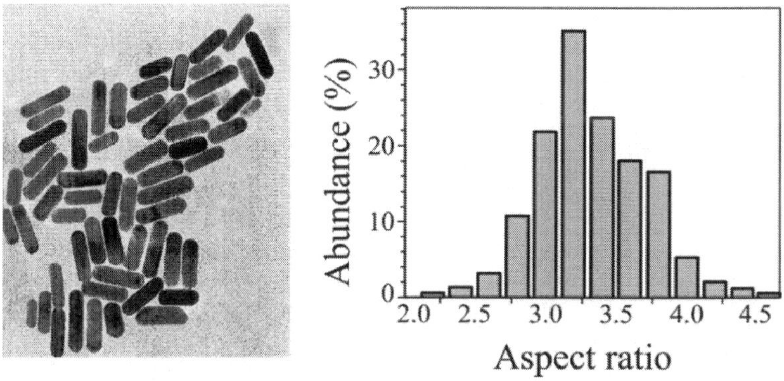

Figure 7.2 TEM image of gold nanorods in solution and their distribution over the sizes (Mohamed et al., 2000). *Abbreviation*: TEM, transmission electron microscopy.

the last section of this chapter, shapes of nanoparticles more similar to nanorods will be considered.

In the case of perfectly conducting spheroids, the analytical description in the form of a series of spheroidal functions is possible even in the case of point sources of light (atoms, molecules, or quantum dots). Thus, Hatcher and Leitner (1954), Belkina (1957a) have solved the problem of the field of the electric dipole placed on the symmetry axis of a perfectly conducting spheroid and oriented parallel to it. Belkina (1957b) has also found a solution for the field of the magnetic dipole placed on the symmetry axis of a perfectly conducting spheroid, too, but oriented normally to it. These solutions are overwhelmingly important but indirectly relevant to nanoplasmonics since there are no plasmon oscillations in perfectly conducting particles.

The case of spheroidal particles made of real materials (dielectric or metal) is even more complicated, and now only the solutions for diffraction of plane electromagnetic waves on spheroidal particles of an arbitrary material are known (Asano and Yamamoto, 1975; Farafonov, 1983; Voshchinnikov and Farafonov, 1993). In these studies, the solution of the diffraction problem was reduced to the solution of an infinite system of linear equations with rather complicated coefficients. For specific calculations of the optical properties of spheroidal nanoparticles, the effective calculation of spheroidal functions is important, and for this purpose effective algorithms have been developed (Li et al., 1998; Li et al., 2002; Falloon, 2001). Spheroidal shells can be also treated with the help of analogous methods (Farafonov, 2001).

Nevertheless, all these analytical and semianalytical methods are extremely complex, and the study of nanoparticles' plasmon properties is rather difficult when these methods are used. Fortunately, in the case of nanoparticles the retardation can be neglected, and the quasi-static "ε-method" can be used for determination of optical (plasmon) properties of nanoparticles (Chapter 5). Therefore, a great part of the current chapter is devoted to the description of plasmon properties of nanospheroids in the quasi-static approximation. Let us note that the simplest (dipole and quadrupole) properties of plasmon oscillations in spheroidal nanoparticles can be found as the limiting case of plasmon oscillations in ellipsoidal nanoparticles (see Chapter 8).

7.1 Plasmon Resonances in Spheroidal Nanoparticles (Quasi-Static Approximation)

7.1.1 *Prolate Spheroids*

To find the plasmon modes of a prolate spheroidal nanoparticle in the context of the "ε-method" (see Chapter 5), it is natural to make use of prolate spheroidal coordinates, where the z axis coincides

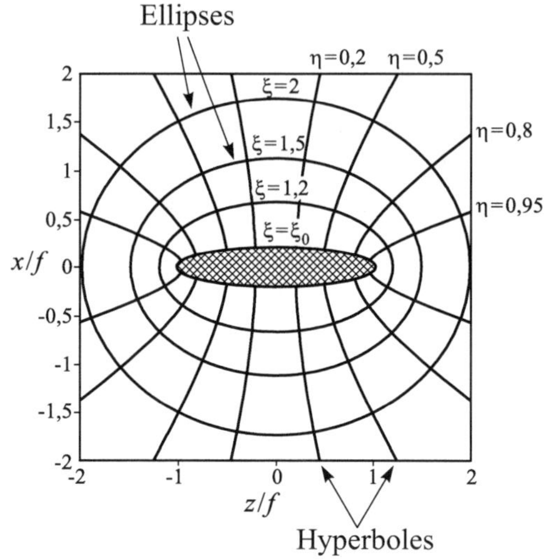

Figure 7.3 The coordinates of a prolate spheroid.

with the symmetry axis of the particle (Fig. 7.3). The relation between the spheroidal coordinates $1 < \xi < \infty$, $-1 < \eta < 1$, $0 < \varphi < 2\pi$ and the Cartesian coordinates x, y, z has the following form (Smythe, 1969):

$$x = f\sqrt{(1-\eta^2)(\xi^2-1)}\cos\varphi$$
$$y = f\sqrt{(1-\eta^2)(\xi^2-1)}\sin\varphi$$
$$z = f\xi\eta, \tag{7.1}$$

where f is the half distance between the foci of an ellipse. In spheroidal coordinates, the variables of the Laplace equation can be separated and its arbitrary solution takes the form:

$$\varphi = P_n^m(\eta)\begin{bmatrix} Q_n^m(\xi) \\ P_n^m(\xi) \end{bmatrix}\begin{bmatrix} \cos m\varphi \\ \sin m\varphi \end{bmatrix}, \tag{7.2}$$

where $n \geq m$ are arbitrary integer numbers.

In Eq. (7.2) and further, $P_n^m(\xi)$, $P_n^m(\eta)$, and $Q_n^m(\xi)$ are associated Legendre functions of the first and second kinds, correspondingly (Abramowitz and Stegun, 1964). Herewith, the associated polynomials of the η argument are defined on the branch cut $-1 < \eta < 1$, while the associated polynomials of the $\xi > 1$ argument have the branch cut from $-\infty$ to 1. The surface $\xi = \xi_0 > 1$ describes the surface of the nanoparticle under consideration. The ξ_0 parameter is expressed through the semiaxes $c > a$ of the ellipsoid forming the spheroid by the expression:

$$\xi_0 = c/\sqrt{c^2 - a^2}. \tag{7.3}$$

The usual form of the expressions for the Legendre functions of $-1 < \eta < 1$ is:

$$\begin{aligned}
&P_1(\eta) = \eta \qquad Q_1(\eta) = \frac{1}{2}\eta \ln\frac{1+\eta}{1-\eta} - 1 \\
&P_1^1(\eta) = \sqrt{1-\eta^2} \qquad Q_1^1(\eta) = \sqrt{1-\eta^2}\left(\frac{1}{2}\ln\frac{1+\eta}{1-\eta} + \frac{\eta}{1-\eta^2}\right) \\
&P_2(\eta) = \frac{1}{2}\left(3\eta^2 - 1\right) \qquad Q_2(\eta) = \frac{1}{4}\left(3\eta^2 - 1\right)\ln\frac{1+\eta}{1-\eta} - \frac{3\eta}{2} \\
&P_2^1(\eta) = 3\eta\sqrt{1-\eta^2} \quad Q_2^1(\eta) = \sqrt{1-\eta^2}\left(\frac{3}{2}\eta\ln\frac{1+\eta}{1-\eta} + \frac{3\eta^2-2}{1-\eta^2}\right) \\
&P_2^2(\eta) = 3\left(1-\eta^2\right) \quad Q_2^2(\eta) = \left(1-\eta^2\right)\left(\frac{3}{2}\ln\frac{1+\eta}{1-\eta} + \frac{5\eta-3\eta^3}{\left(1-\eta^2\right)^2}\right).
\end{aligned} \tag{7.4}$$

For $\xi > 1$, the expressions for the Legendre functions are obtained from Eq. (7.4) by replacements $\eta \to \xi$, $\sqrt{1-\eta^2} \to \sqrt{\xi^2 - 1}$ and, furthermore, by replacements $1 - \eta \to \xi - 1$ in logarithmic terms.

The associated Legendre polynomials $P_n^m(\xi)$ increase at $\xi \to \infty$, and therefore the solution of the Laplace equation outside a particle $(\xi > \xi_0)$ should be sought as follows:

$$\tilde{\varphi}^{(2)} = \sum_{n=0}^{\infty}\sum_{m=0}^{n} P_n^m(\eta)\, Q_n^m(\xi)\left[\alpha_{nm}^{(2)}\cos m\varphi + \beta_{nm}^{(2)}\sin m\varphi\right]. \tag{7.5}$$

On the other hand, the associated Legendre polynomials $Q_n^m(\xi)$ have singularities for $\xi = 1$ (on the line between the foci), and inside the spheroid $(1 < \xi < \xi_0)$ the solution finiteness requirement determines the necessity to seek the solution in the form of:

$$\tilde{\varphi}^{(1)} = \sum_{n=0}^{\infty}\sum_{m=0}^{n} P_n^m(\eta)\, P_n^m(\xi)\left[\alpha_{nm}^{(1)}\cos m\varphi + \beta_{nm}^{(1)}\sin m\varphi\right]. \tag{7.6}$$

The continuity condition of the field tangential components and normal induction components results in the following system of equations:

$$\alpha_{nm}^{(1)} P_n^m(\xi_0) = \alpha_{nm}^{(2)} Q_n^m(\xi_0)$$
$$\varepsilon \alpha_{nm}^{(1)} \left(P_n^m(\xi_0)\right)' = \varepsilon_{\mathrm{H}} \alpha_{nm}^{(2)} \left(Q_n^m(\xi_0)\right)' \quad (7.7)$$

where ε_{H} is the electric permittivity of the environment and ε is the electric permittivity of the particle regarded as a spectral parameter. The analogous system is true also for the determination of β coefficients.

The nontrivial solution of the system is possible only in the case of plasmon resonance, that is, the case when the electric permittivity of the particle $\varepsilon = \varepsilon_n^m$ has the form

$$\frac{\varepsilon_n^m}{\varepsilon_{\mathrm{H}}} = \frac{\left(Q_n^m(\xi_0)\right)' P_n^m(\xi_0)}{\left(P_n^m(\xi_0)\right)' Q_n^m(\xi_0)} \quad (7.8)$$

In the most important case of $n = 1$ (dipole plasmon), the expression for resonance electric permittivity has the form

$$\varepsilon_1^0 = \frac{\xi_0\left(-2\xi_0 + \left(\xi_0^2 - 1\right)\ln\left((\xi_0+1)/(\xi_0-1)\right)\right)}{\left(\xi_0^2 - 1\right)\left(-2 + \xi_0 \ln\left((\xi_0+1)/(\xi_0-1)\right)\right)}\varepsilon_{\mathrm{H}} \quad (7.9)$$

$$\varepsilon_1^1 = \frac{4 - 2\xi_0^2 + \xi_0\left(\xi_0^2 - 1\right)\ln\left((\xi_0+1)/(\xi_0+1)\right)}{\xi_0\left(-2\xi_0 + \left(\xi_0^2 - 1\right)\ln\left((\xi_0+1)/(\xi_0-1)\right)\right)}\varepsilon_{\mathrm{H}} \quad (7.10)$$

where ξ_0 is determined by Eq. (7.3).

Let us note that the dipole moment of the modes with the index $m = 0$ is along the z axis, while the dipole moments of the modes with the index $m = 1$ are in the plane perpendicular to the symmetry axis.

In Fig. 7.4, the dependencies of resonance values of electric permittivity, Eq. (7.8), on the aspect ratio a/c of the nanoparticle are shown. As is seen from these pictures, first of all, the values of the resonance permittivity lie in the range $(-\infty,-1)$ and tend to $-\infty$ for axially symmetric modes $(m = 0)$ in strongly elongated particles. In this case, the resonance permittivities of the other modes tend to -1. In the case of the Drude model for electric permittivity, it corresponds to the resonance frequency tending to $\omega \to 0$ or to the frequency of a surface plasmon, $\omega \to \omega_{\mathrm{pl}}/\sqrt{2}$, correspondingly.

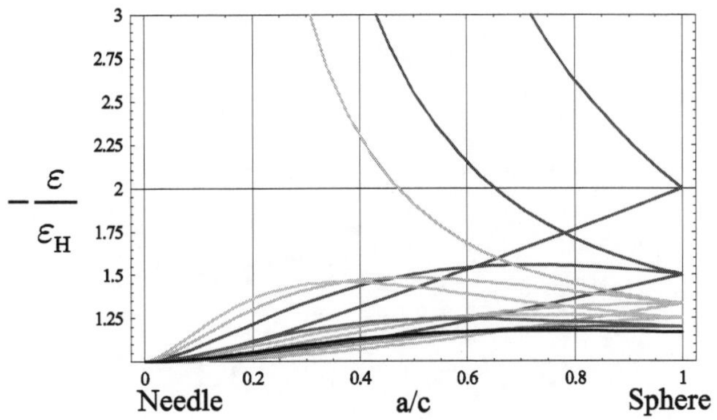

Figure 7.4 The dependencies of the resonance values of the electric permittivity for a prolate spheroidal nanoparticle on the aspect ratio.

In the case of $a/c \to 1$, that is, in the spherical limit, the dependency on the azimuthal number m disappears and the eigenvalues of electric permittivity go to the corresponding values for the sphere:

$$\varepsilon_n^m \underset{c/a \to 1}{\longrightarrow} -\frac{n+1}{n}; n = 1, 2, 3 \ldots \quad (7.11)$$

Plasmon wavefunction corresponding to Eq. (7.8) will take the form

$$\varphi_n^m = P_n^m(\eta)\, P_n^m(\xi)\, Q_n^m(\xi_0) \begin{bmatrix} \cos m\varphi \\ \sin m\varphi \end{bmatrix}, \xi < \xi_0$$

$$\varphi_n^m = P_n^m(\eta)\, P_n^m(\xi_0)\, Q_n^m(\xi) \begin{bmatrix} \cos m\varphi \\ \sin m\varphi \end{bmatrix}, \xi > \xi_0. \quad (7.12)$$

In Fig. 7.5, the plasmon frequencies corresponding to Eq. (7.9) for real metals (Au, Ag, and Cu), are shown.

In Fig. 7.6, the surface charge distributions are shown for the plasmon modes with $n = 3$. As seen from these pictures, generally these surface charge distributions are similar to the surface charge distributions in a sphere. Knowing eigenvalues of electric permittivity and corresponding potentials, the solution of any problem related to a nanospheroid in an arbitrary external field can be found.

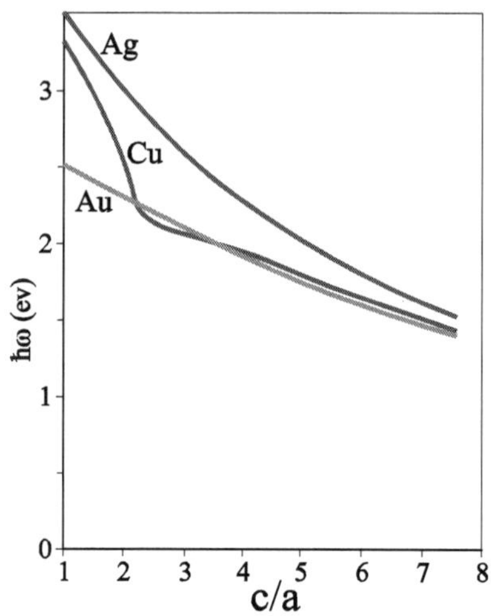

Figure 7.5 The dependencies of the plasmon frequencies of a prolate spheroidal particle of different materials on the aspect ratio (Gersten and Nitzan, 1980).

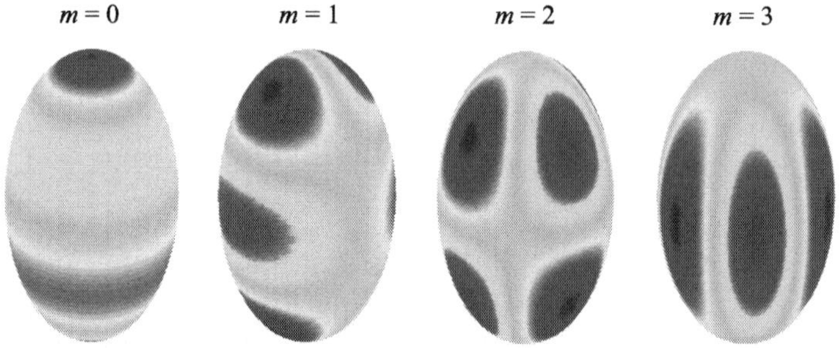

Figure 7.6 The surface charge of the plasmon modes $n = 3, m = 0, 1, 2, 3$ of a prolate spheroidal nanoparticle with the aspect ratio $c/a = 1.81$.

7.1.2 *Oblate Spheroids*

To find the plasmon modes of an oblate spheroidal nanoparticle in the context of the "ε-method" (see Chapter 5), it is natural to make use of the oblate spheroid coordinates $0 < \xi < \infty$, $-1 < \eta < 1$, $0 < \varphi < 2\pi$, where the z axis coincides with the symmetry axis of the particle. The relation between oblate spheroidal and Cartesian coordinates has the following form (Smythe, 1969):

$$x = f\sqrt{(1-\eta^2)(\xi^2+1)}\cos\varphi$$
$$y = f\sqrt{(1-\eta^2)(\xi^2+1)}\sin\varphi$$
$$z = f\xi\eta. \tag{7.13}$$

Within these spheroidal coordinates, the variables in the Laplace equation can be separated again, and an arbitrary solution will have the following form:

$$\varphi = P_n^m(\eta)\begin{bmatrix} Q_n^m(i\xi) \\ P_n^m(i\xi) \end{bmatrix}\begin{bmatrix} \cos m\varphi \\ \sin m\varphi \end{bmatrix}, \tag{7.14}$$

where $n \geq m$ are arbitrary integers.

As in the case of prolate spheroids in Eq. (7.14), $P_n^m(\xi)$, $P_n^m(\eta)$, and $Q_n^m(\xi)$ are the associated Legendre polynomials of the first and second kinds, correspondingly (Abramowitz and Stegun, 1964). Hereby, the associated polynomials of the η argument are defined on the branch cut $-1 < \eta < 1$, while the associated polynomials of the $i\xi$ argument have the branch cut from $-\infty$ to 1. The surface $\xi = \xi_0 > 0$ describes the surface of the nanoparticle considered. ξ_0 is expressed through the semiaxes of the ellipse forming the spheroid by the formula:

$$\xi_0 = c/\sqrt{a^2-c^2}. \tag{7.15}$$

For convenience, let us mention also the expression for some associated Legendre functions of imaginary arguments, which take place in the case of an oblate spheroid (Smythe, 1969):

$$P_1(i\xi) = i\xi \qquad Q_1(i\xi) = \xi arcctg\xi - 1$$
$$P_1^1(i\xi) = i\sqrt{1+\xi^2} \qquad Q_1^1(i\xi) = \sqrt{1+\xi^2}\left(arcctg\xi - \frac{\xi}{\xi^2+1}\right)$$
$$P_2(i\xi) = -\frac{1}{2}\left(3\xi^2+1\right) \qquad Q_2(i\xi) = \frac{i}{2}\left(\left(3\xi^2+1\right)arcctg\xi - 3\xi\right)$$
$$P_2^1(i\xi) = -3\xi\sqrt{1+\xi^2}$$
$$Q_2^1(i\xi) = i\sqrt{1+\xi^2}\left(3\xi arcctg\xi - \frac{3\xi^2+2}{\xi^2+1}\right)$$
$$P_2^1(i\xi) = -3\left(\xi^2+1\right)$$
$$Q_2^2(i\xi) = i\left(1+\xi^2\right)\left(3arcctg\xi - \frac{5\xi+3\xi^3}{\left(\xi^2+1\right)^2}\right) \tag{7.16}$$

The Legendre associated polynomials $P_n^m(i\xi)$ increase at infinity, and the solution of the Laplace equation outside the nanoparticle $(\xi > \xi_0)$ should be sought as follows:

$$\tilde{\varphi}^{(2)} = \sum_{n=0}^{\infty}\sum_{m=0}^{n} P_n^m(\eta)\, Q_n^m(i\xi)\left[\alpha_{nm}^{(2)}\cos m\varphi + \beta_{nm}^{(2)}\sin m\varphi\right] \quad (7.17)$$

On the other hand, the Legendre associated polynomials $Q_n^m(i\xi)$ have singularities for $\xi = 0$ (on the line between the foci), and therefore inside the spheroid $0 < \xi < \xi_0$ the finiteness requirement determines the necessity to seek the solution in the form of:

$$\tilde{\varphi}^{(1)} = \sum_{n=0}^{\infty}\sum_{m=0}^{n} P_n^m(\eta)\, P_n^m(i\xi)\left[\alpha_{nm}^{(1)}\cos m\varphi + \beta_{nm}^{(1)}\sin m\varphi\right]. \quad (7.18)$$

The continuity condition of the electric field tangential components and induction normal components results in the following system of equations:

$$\begin{aligned} \alpha_{nm}^{(1)} P_n^m(i\xi_0) &= \alpha_{nm}^{(2)} Q_n^m(i\xi_0) \\ \varepsilon\alpha_{nm}^{(1)}\left(P_n^m(i\xi_0)\right)' &= \varepsilon_H \alpha_{nm}^{(2)}\left(Q_n^m(i\xi_0)\right)' \end{aligned} \quad (7.19)$$

The analogous system is true also for the β coefficients' determination.

The nontrivial solution of the system takes place only in the case of plasmon resonance $\varepsilon = \varepsilon_n^m$, that is, when the permittivity of the particle has the form:

$$\frac{\varepsilon_n^m}{\varepsilon_{\mathrm{H}}} = \frac{\left(Q_n^m(i\xi_0)\right)' P_n^m(i\xi_0)}{\left(P_n^m(i\xi_0)\right)' Q_n^m(i\xi_0)} \quad (7.20)$$

Let us note that Eq. (7.20) differs from Eq. (7.8) only by the replacement $\xi_0 \to i\xi_0$; however, this replacement results in quite different values of the resonance dielectric permittivities ε_n^m.

In the most important case of the dipolar $(n = 1)$ plasmon resonance in oblate spheroids, the resonance dielectric permittivities have the form:

$$\varepsilon_1^0 = \frac{\xi_0\left(-\xi_0 + \left(1+\xi_0^2\right) arcctg\xi_0\right)}{\left(1+\xi_0^2\right)\left(\xi arcctg\xi_0 - 1\right)}\varepsilon_{\mathrm{H}} \quad (7.21)$$

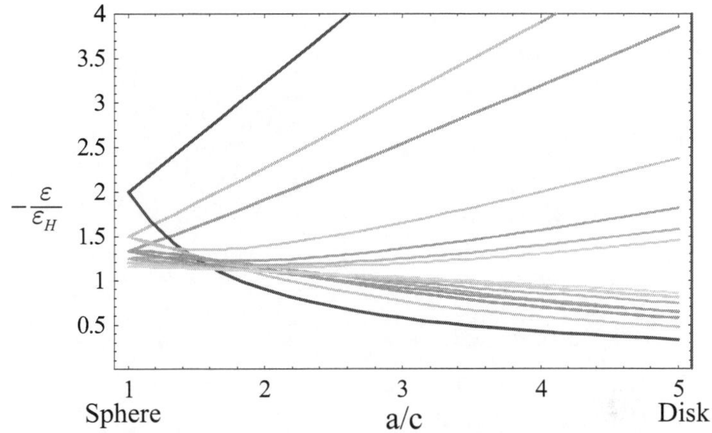

Figure 7.7 The dependencies of the electric permittivity resonance values for an oblate nanoparticle on the aspect ratio a/c.

$$\varepsilon_1^1 = \frac{-2 - \xi_0^2 + \xi_0\left(1 + \xi_0^2\right) arcctg\xi_0}{\xi_0\left(-\xi_0 + \left(1 + \xi_0^2\right) arcctg\xi_0\right)}\varepsilon_{\mathrm{H}} \tag{7.22}$$

where ξ_0 is described by Eq. (7.15).

Let us note that the dipole moment of the plasmon modes with the $m = 0$ index is along the z axis, while the dipole moment of the modes with the $m = 1$ is in the plane normal to the symmetry axis.

In Fig. 7.7, the dependencies of resonance values of electric permittivity, Eq. (7.8), on the aspect ratio a/c are shown. As seen from the pictures, in the case of $a/c \to 1$, that is, in the spherical limit, the dependence on the azimuthal number m disappears and the electric permittivity eigenvalues go to corresponding values for a nanosphere:

$$\frac{\varepsilon_n^m}{\varepsilon_H} \underset{a/c \to 1}{\longrightarrow} -\frac{n+1}{n}; n = 1, 2, 3 \ldots \tag{7.23}$$

In the case of strongly oblate particles ($a/c \to \infty$), values of the resonance dielectric permittivities with even $n + m$ tend to infinity, always staying smaller than -1. On the other hand, the values of the electric permittivities with odd $n + m$ in a spherical case ($c/a = 1$) are smaller than -1, while by increasing of oblateness their values tend to 0 and thus become greater than -1. This phenomenon was not observed in prolate spheroids, where the resonance dielectric

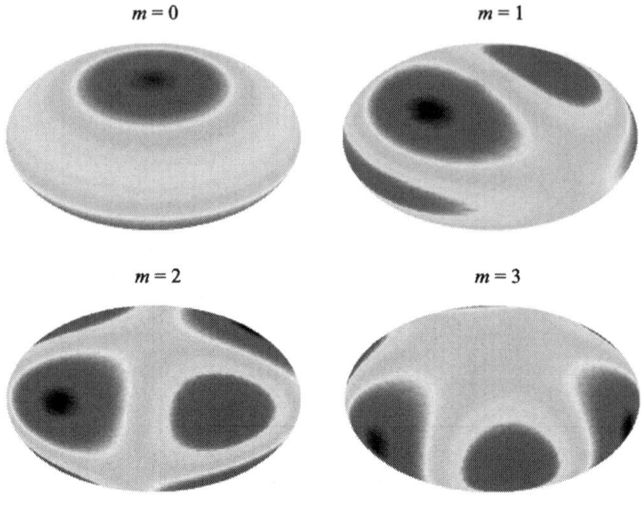

Figure 7.8 The surface charge of some plasmon modes $n = 3, m = 0, 1, 2, 3$ of an oblate spheroidal nanoparticle with the aspect ratio $c/a = 0.57$.

permittivity value is always smaller than -1, but it is typical for more complex particles, such as a cluster of two spheres (Chapter 10) or a nanocube and an octahedron (Chapter 9), for example. Effects of this kind may form a basis for the methods of detection of nanoparticles' shape and, specifically, ellipticity of oblate spheroids.

In case of the Drude model for electric permittivity in strongly oblate particles ($a/c \to \infty$), the plasmon frequency tends to 0 (for even $n + m$) or to the frequency of a bulk plasmon $\omega \to \omega_{\text{pl}}$ (odd $n + m$).

The equation corresponding to plasmon wavefunctions of Eq. (7.20) have the form:

$$\varphi_n^m = P_n^m(\eta)\, P_n^m(i\xi)\, Q_n^m(i\xi_0) \begin{bmatrix} \cos m\varphi \\ \sin m\varphi \end{bmatrix}, \xi < \xi_0$$

$$\varphi_n^m = P_n^m(\eta)\, P_n^m(i\xi_0)\, Q_n^m(i\xi) \begin{bmatrix} \cos m\varphi \\ \sin m\varphi \end{bmatrix}, \xi > \xi_0. \qquad (7.24)$$

In Fig. 7.8, the surface charge distributions for the modes with $n = 3$ are shown. As seen from the pictures, these distributions are similar qualitatively to the distributions of plasmon modes surface charges in a sphere.

7.2 Optical Properties of Spheroids

Knowing the eigenvalues of electric permittivity and corresponding potentials, one can find the solution of any problem related to a nanospheroid in an arbitrary external field. One of the usual problems of this kind is the problem of optical radiation scattering by nanoparticles placed in a solid-state matrix or in a colloidal solution. This class of problems is characterized by nanoparticle polarizability and by scattering and absorption cross sections. Strongly nonhomogeneous fields, for example, a molecule radiation field, play a significant role in nano-optics and nanoplasmonics. For these cases, such characteristics as the spontaneous decay rate and the radiation frequency shift are important. Often, the spontaneous emission rate is related to the same modes as scattering cross sections do, while the frequency shift is sensible to a greater number of plasmon modes.

7.2.1 *Spheroids' Polarizability*

To find spheroids' polarizability, the plasmon modes dipole moments are to be calculated when excited by a homogeneous field. The spheroids (and ellipsoids; see Chapter 8) have an interesting peculiarity: only modes with $n = 1$ can be excited by a homogeneous field and have a nonzero dipole moment. As a result, the quasi-static problem of spheroidal particle excitation by a homogeneous field can be solved in the form:

$$\mathbf{E} = \mathbf{E}^0 + \sum_m \mathbf{e}_1^m \frac{(\varepsilon(\omega) - \varepsilon_{\mathrm{H}})}{(\varepsilon_1^m - \varepsilon(\omega))} \frac{\int\limits_{V+} \mathbf{e}_1^m \mathbf{E}^0 dV}{\int\limits_{V+} (\mathbf{e}_1^m)^2 dV}$$

$$= \sum_m \mathbf{e}_1^m \frac{(\varepsilon_1^m - \varepsilon_{\mathrm{H}})}{(\varepsilon_1^m - \varepsilon(\omega))} \frac{\int\limits_{V+} \mathbf{e}_1^m \mathbf{E}^0 dV}{\int\limits_{V+} (\mathbf{e}_1^m)^2 dV}, \quad \mathbf{e}_1^m = -\nabla \varphi_1^m, \quad (7.25)$$

where ε_1^m and φ_1^m are given by Eqs. (7.21), (7.22), and (7.24) for an oblate spheroid and by Eqs. (7.9), (7.10), and (7.6) for a prolate spheroid correspondigly, $\varepsilon(\omega)$ describes the permittivity of the particle's specific material, and ε_{H} is the dielectric of the medium surrounding the particle.

Taking into account the fact that the plasmon potentials of the modes with $n = 1$ are linear in the coordinates inside the particle, the following expression is obtained instead of Eq. (7.25) for the electric field inside a spheroid:

$$E_x = E_x^0 \frac{\varepsilon_1^1 - \varepsilon_\mathrm{H}}{\varepsilon_1^1 - \varepsilon(\omega)}, \quad E_y = E_y^0 \frac{\varepsilon_1^1 - \varepsilon_\mathrm{H}}{\varepsilon_1^1 - \varepsilon(\omega)}, \quad E_z = E_z^0 \frac{\varepsilon_1^0 - \varepsilon_\mathrm{H}}{\varepsilon_1^0 - \varepsilon(\omega)}. \tag{7.26}$$

Remembering that the particle dipole moment is defined by the difference of the particle and environment polarizations in the particle volume, that is, by the relation:

$$\mathbf{d} = \frac{\varepsilon(\omega) - \varepsilon_\mathrm{H}}{4\pi} \int_V \mathbf{E} dV = \frac{\varepsilon(\omega) - \varepsilon_\mathrm{H}}{4\pi} \mathbf{E} V = \varepsilon_\mathrm{H} \overleftrightarrow{\alpha} \mathbf{E}_0, \tag{7.27}$$

we obtain the final expressions for the polarizability of prolate and oblate spheroids:

$$\alpha_x = \frac{ca^2}{3} \frac{(\varepsilon(\omega) - \varepsilon_\mathrm{H})(\varepsilon_1^1/\varepsilon_\mathrm{H} - 1)}{\varepsilon_1^1 - \varepsilon(\omega)} = \frac{ca^2}{3} \frac{(\varepsilon(\omega) - \varepsilon_\mathrm{H})}{\varepsilon_\mathrm{H} + L_1(\varepsilon(\omega) - \varepsilon_\mathrm{H})}$$
$$\alpha_z = \frac{ca^2}{3} \frac{(\varepsilon(\omega) - \varepsilon_\mathrm{H})(\varepsilon_1^0/\varepsilon_\mathrm{H} - 1)}{\varepsilon_1^0 - \varepsilon(\omega)} = \frac{ca^2}{3} \frac{(\varepsilon(\omega) - \varepsilon_\mathrm{H})}{\varepsilon_\mathrm{H} + L_0(\varepsilon(\omega) - \varepsilon_\mathrm{H})} \tag{7.28}$$

where in the right part the polarizabilities are expressed through the geometric factor L that is related to the eigenvalues of electric permittivity by the formula:

$$L_m = \frac{1}{1 - \varepsilon_1^m/\varepsilon_H} \tag{7.29}$$

In Fig. 7.9, the relation between the geometric factor and the resonance electric permittivities of the lowest dipole modes of spheroids ($n = 1$) is shown. The electric fields' polarizations are shown by the arrows. Let us note that for prolate spheroids this curve describes the longitudinal modes for 0 (*cylinder*) $< L < 1/3$ (*sphere*) and the transverse modes for 1/3 (*sphere*) $< L < 1/2$ (*cylinder*). For oblate spheroids, this curve describes the longitudinal modes for 1/3 (*sphere*) $< L < 1$ (*disk*) and the transverse modes for 0 (*disk*) $< L < 1/3$ (*sphere*).

Substituting Eqs. (7.9) and (7.10) in Eq. (7.29) for a prolate spheroid and Eqs. (7.21) and (7.22) in Eq. (7.29) for an oblate

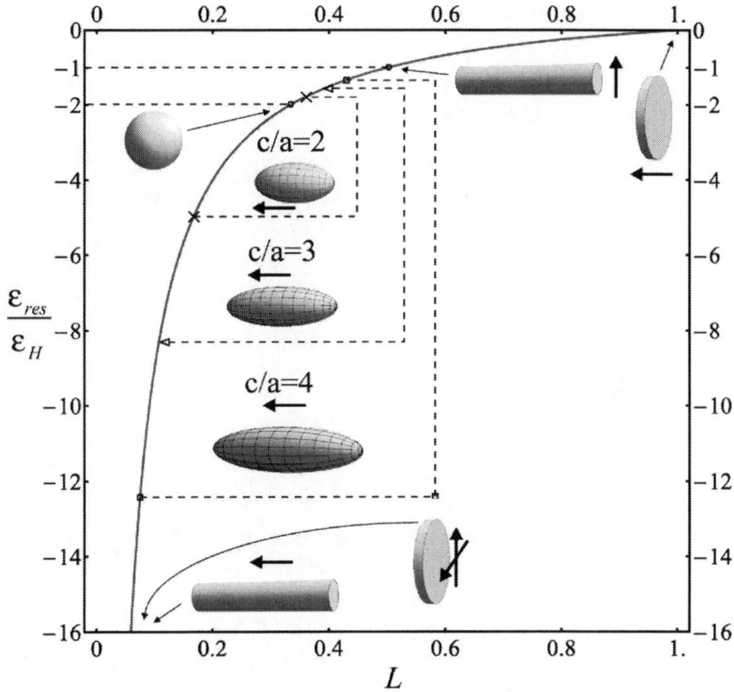

Figure 7.9 The relation between the geometric factor L and the resonance electric permittivities of the lowest dipole modes of spheroids. The electric field polarizations are shown by the arrows.

spheroid, the depolarization coefficients can be presented as

$$L_0 = \left(\xi_0^2 - 1\right)\left(\frac{\xi_0}{2}\ln\left(\frac{\xi_0 + 1}{\xi_0 - 1}\right) - 1\right), \xi_0^2 = \frac{c^2}{c^2 - a^2}$$

$$L_1 = \frac{1}{2}\left(1 - L_0\right) \qquad (7.30)$$

for prolate spheroids ($a = b < c$) and as

$$L_0 = \left(1 + \xi_0^2\right)\left(1 - \xi_0 arcctg\xi_0\right), \xi_0^2 = \frac{c^2}{a^2 - c^2},$$

$$L_1 = \frac{1}{2}\left(1 - L_0\right) \qquad (7.31)$$

for oblate spheroids ($a = b > c$).

When a spheroid is close to being a sphere, for any case we have

$$L_0 \approx \frac{1}{3} - \frac{2}{15}\frac{c^2 - a^2}{c^2} + \dots \tag{7.32}$$

In the case of plasmon resonance, the enhancement of the polarization and, consequently, local electric fields occurs. This enhancement is one of the most important features of nano-optics and nanoplasmonics. The enhancement of local electric fields is especially great when the resonance occurs in prolate particles or near the spikes. In the case of prolate spheroids, the factor of longitudinal field enhancement near the spheroid surface is especially great and has the form

$$\frac{E^{\text{out}}}{E_0} = \frac{\varepsilon}{\varepsilon_{\text{H}}}\frac{\varepsilon_1^0 - \varepsilon_{\text{H}}}{\varepsilon_1^0 - \varepsilon} = \frac{\varepsilon}{(1 - L_0)\,\varepsilon_{\text{H}} + L_0\varepsilon} \tag{7.33}$$

Near the resonance in strongly prolate spheroids $\varepsilon_1^0 - \varepsilon \approx i\varepsilon''$ and $\left|\varepsilon_1^0\right| \sim \varepsilon_{\text{H}}/L_0 \gg 1$, the enhancement factor takes the form:

$$\left|\frac{E^{\text{out}}}{E_0}\right| \approx \frac{\varepsilon_{\text{H}}}{L_0^2\varepsilon''}, \tag{7.34}$$

which implies that the field enhancement increases both as the geometric factor L decreases (the lightning rod effect) and as the losses in the nanoparticle ε'' are reduced. Since many phenomena are proportional to the square (fluorescence and luminescence) or even to the biquadrate (Raman scattering) of the electric field, it is seen that prolate spheroidal particles can play an important role in most diverse applications of nanoplasmonics and especially in nanodevices based on detection and excitation of single molecules.

7.2.2 *Spheroids' Scattering and Absorption Cross Sections*

Knowing the polarizabilities of spheroidal particles, one can find the scattering and absorption cross sections from the general formulas (see Chapter 5). There is a small difficulty that spheroids' polarizability is not an isotropic tensor in contrast to the case of a sphere and the cross sections depend on radiation polarization.

Let the R matrix describe the transition from the system of coordinates related to the spheroids axes (x, y, z) to the laboratory system of coordinates (x', y', z') where the z'-axis is the direction of incident light propagation:

$$R_{ij} = \mathbf{e}_i \cdot \mathbf{e}'_j, \tag{7.35}$$

where $\mathbf{e}_i$, $\mathbf{e}'_j$ are unit vectors along the axes of reference in the spheroid system and in laboratory system of coordinates.

Now, the absorption and scattering cross sections of the light, polarized along the x'-axis of the laboratory system of coordinates, will have the form (Bohren and Huffman, 1983):

$$\begin{aligned} \sigma^{\text{abs}} &= 4\pi k_{\text{H}} \text{Im} \left(\alpha_x R_{11}^2 + \alpha_x R_{21}^2 + \alpha_z R_{31}^2 \right) \\ \sigma^{\text{sc}} &= \frac{8\pi}{3} k_{\text{H}}^4 \left(|\alpha_x|^2 R_{11}^2 + |\alpha_x|^2 R_{21}^2 + |\alpha_z|^2 R_{31}^2 \right), \end{aligned} \tag{7.36}$$

where $k_{\text{H}} = \sqrt{\varepsilon_{\text{H}}}\omega/c$ is the wavevector in the particle environment and the polarizabilities α_x, α_z are defined by the expressions in Eq. (7.28). Analogous expressions are true also for the other (y) polarization.

In the case of chaotically oriented spheroidal particles, the averaged values of the squares of direction cosines are equal:

$$\left\langle R_{ij}^2 \right\rangle = 1/3 \tag{7.37}$$

and for any polarization of incident light the absorption cross section has the form:

$$\begin{aligned} \sigma^{\text{abs}} &= 4\pi k_H \text{Im} \left(\alpha_x \frac{2}{3} + \alpha_z \frac{1}{3} \right), \\ \sigma^{\text{sc}} &= \frac{8\pi}{3} k_{\text{H}}^4 \left(|\alpha_x|^2 \frac{2}{3} + |\alpha_z|^2 \frac{1}{3} \right) \end{aligned} \tag{7.38}$$

The polarizabilities, Eq. (7.28), and the cross sections, Eq. (7.36), correspondingly, have the maxima when the electric permittivity of the substance for some frequencies turns out to be close to the resonance electric permittivity $\varepsilon = \varepsilon' + i\varepsilon'' \approx \varepsilon_n^m$.

In the case of limiting forms of nanoparticles (spheres, disks, and needles), the expressions for the orientation averaged cross sections of scattering and absorption have the form

$$\begin{aligned}
\left\langle\sigma^{\mathrm{abs}}\right\rangle_{\text{sphere}} &= \frac{k_{\mathrm{H}}V}{3}\left[\frac{27\varepsilon_{\mathrm{H}}}{(\varepsilon'+2\varepsilon_{\mathrm{H}})^2+\varepsilon''^2}\right]\varepsilon'' \\
\left\langle\sigma^{\mathrm{abs}}\right\rangle_{\text{needle}} &= \frac{k_{\mathrm{H}}V}{3}\left[\frac{8\varepsilon_{\mathrm{H}}}{(\varepsilon'+\varepsilon_{\mathrm{H}})^2+\varepsilon''^2}+\frac{1}{\varepsilon_{\mathrm{H}}}\right]\varepsilon'' \\
\left\langle\sigma^{\mathrm{abs}}\right\rangle_{\text{disk}} &= \frac{k_{\mathrm{H}}V}{3}\left[\frac{\varepsilon_{\mathrm{H}}}{\varepsilon'^2+\varepsilon''^2}+\frac{2}{\varepsilon_{\mathrm{H}}}\right]\varepsilon'' \\
\left\langle\sigma^{\mathrm{sc}}\right\rangle_{\text{sphere}} &= \frac{k_{\mathrm{H}}^4V^2}{18\pi}\left|\varepsilon-\varepsilon_{\mathrm{H}}\right|^2\left[\frac{27}{(\varepsilon'+2\varepsilon_{\mathrm{H}})^2+\varepsilon''^2}\right] \\
\left\langle\sigma^{\mathrm{sc}}\right\rangle_{\text{needle}} &= \frac{k_{\mathrm{H}}^4V^2}{18\pi}\left|\varepsilon-\varepsilon_{\mathrm{H}}\right|^2\left[\frac{8}{(\varepsilon'+\varepsilon_{\mathrm{H}})^2+\varepsilon''^2}+\frac{1}{\varepsilon_{\mathrm{H}}^2}\right] \\
\left\langle\sigma^{\mathrm{sc}}\right\rangle_{\text{disk}} &= \frac{k_{\mathrm{H}}^4V^2}{18\pi}\left|\varepsilon-\varepsilon_{\mathrm{H}}\right|^2\left[\frac{1}{\varepsilon'^2+\varepsilon''^2}+\frac{2}{\varepsilon_{\mathrm{H}}^2}\right],
\end{aligned} \tag{7.39}$$

where V is the volume of the spheroidal particle, $\varepsilon = \varepsilon' + i\varepsilon''$ is the electric permittivity of the nanoparticle, and ε_{H} is the electric permittivity of the environment substance. The wavevector $\mathbf{k}$ in Eq. (7.39) is also related to the medium with the electric permittivity ε_{H}, $k_{\mathrm{H}} = \sqrt{\varepsilon_{\mathrm{H}}}\omega/c$.

7.2.3 *Rates of Spontaneous Emission of Molecules in the Vicinity of a Spheroid*

As has been said repeatedly, the problems related to atoms' and molecules' emission in the vicinity of nanoparticles and the control of these processes have great significance in nano-optics, regarded as an important direction of high-end nanotechnologies (systems of nanosystems). In the quasi-static approximation both the spontaneous emission rate and the frequency shift of this emission may be calculated (see Appendix A1).

A remarkable feature of spheroidal particles is that for both these characteristics an analytical expression can be found. In the case of arbitrary relative positions of a molecule and a nanoparticle, formulas are rather complicated and can be found in (Klimov et al., 2002b). These expressions can be also obtained as a limiting case of an atom radiation in the vicinity of a triaxial nanoellipsoid (Guzatov and Klimov, 2005; Chapter 8).

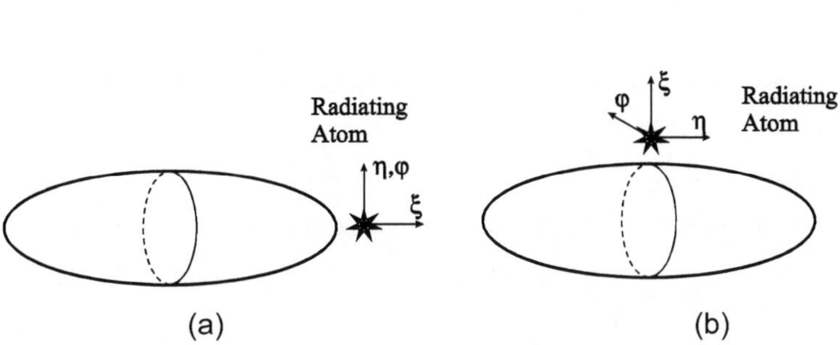

Figure 7.10 The geometry of the problem of molecule emission in the vicinity of a spheroidal nanoparticle.

Here, we will consider only some particular cases of spontaneous emission in the vicinity of a nanospheroid, which are important for the general understanding of the radiation processes near a nanospheroid. For simplicity, we will consider that a nanoparticle and an atom (a molecule) are placed in vacuum ($\varepsilon_H = 1$).

7.2.3.1 Prolate spheroids

Physically, interesting are the cases when an excited atom or a molecule is placed on a spheroid symmetry axis ($x = y = 0$) (Fig. 7.10a) or on its equator ($z = 0$) (Fig. 7.10b).

In the case of an atom placed on the spheroid symmetry axis z, the expressions for spontaneous emission rates take the form:

$$\left(\frac{\gamma}{\gamma_0}\right)_z = \left|1 + G_{10}\frac{d}{d\xi}Q_1(\xi)\right|^2\Bigg|_{\xi=z/f}, \tag{7.40}$$

for the z-orientation of the dipole moment and

$$\left(\frac{\gamma}{\gamma_0}\right)_\perp = \left|1 + G_{11}\frac{d}{d\xi}Q_1(\xi)\right|^2\Bigg|_{\xi=z/f} \tag{7.41}$$

for the dipole moment perpendicular to the symmetry axis.

In the case of an atom placed in the plane $z = 0$, correspondingly, we have:

$$\left(\frac{\gamma}{\gamma_0}\right)_r = \left|1 + G_{11}\frac{d}{d\xi}Q_1^1(\xi)\frac{\sqrt{\xi^2-1}}{\xi}\right|^2\Bigg|_{\xi=\sqrt{1+\left(x^2+y^2\right)/f^2}} \tag{7.42}$$

for radial orientation of the dipole moment in the plane x-y,

$$\left(\frac{\gamma}{\gamma_0}\right)_z = \left|1 + G_{10}\frac{Q_1(\xi)}{\xi}\right|^2\Bigg|_{\xi=\sqrt{1+\left(x^2+y^2\right)/f^2}}, \tag{7.43}$$

for z-orientation of the dipole moment, and

$$\left(\frac{\gamma}{\gamma_0}\right)_\varphi = \left|1 + G_{11}\frac{d}{d\xi}Q_1(\xi)\right|^2\Bigg|_{\xi=\sqrt{1+\left(x^2+y^2\right)/f^2}}, \tag{7.44}$$

for φ-orientation of the dipole moment.

In these expressions:

$$G_{10} = \frac{(\varepsilon - \varepsilon_{\mathrm{H}})}{\varepsilon_1^0 - \varepsilon}\frac{\xi_0}{Q_1(\xi_0)}$$

$$= -\frac{2(\varepsilon - \varepsilon_{\mathrm{H}})\xi_0(\xi_0^2 - 1)}{2\left(\varepsilon_{\mathrm{H}}\xi_0^2 - \varepsilon(\xi_0^2 - 1)\right) + (\varepsilon - \varepsilon_{\mathrm{H}})\xi_0(\xi_0^2 - 1)\ln\left((\xi_0+1)/(\xi_0-1)\right)}$$

$$G_{11} = \frac{(\varepsilon - \varepsilon_{\mathrm{H}})}{\varepsilon_1^1 - \varepsilon}\frac{P_1^1(\xi_0)}{Q_1^1(\xi_0)}$$

$$= \frac{2(\varepsilon - \varepsilon_{\mathrm{H}})\xi_0(\xi_0^2 - 1)}{2\left(\varepsilon_{\mathrm{H}}\xi_0^2 - \varepsilon\xi_0^2 - 2\varepsilon_{\mathrm{H}}\right) + (\varepsilon - \varepsilon_{\mathrm{H}})\xi_0(\xi_0^2 - 1)\ln\left((\xi_0+1)/(\xi_0-1)\right)},$$

and x, y, z are the coordinates of the atom position outside the spheroid.

The most interesting feature of the decay rate for the φ orientation of the dipole moment is its independence of the position along a surface of any spheroid.

Further simplification of the results of Eqs. (7.40)–(7.44) is possible when the atom is considered to be placed in the immediate vicinity of the spheroid surface ($\xi \to \xi_0$). In this case, we obtain the following expressions for the atom placed on the prolate spheroid pole:

$$\left(\frac{\gamma}{\gamma_0}\right)_z = \left|\frac{2\varepsilon}{\left(\xi_0(\varepsilon - 1)(\xi_0^2 - 1)\ln\frac{\xi_0+1}{\xi_0-1} + 2\left(\varepsilon + \xi_0^2 - \xi_0^2\varepsilon\right)\right)}\right|^2 \tag{7.45}$$

for the z-orientation of the dipole moment and

$$\left(\frac{\gamma}{\gamma_0}\right)_\perp = \left(\frac{\gamma}{\gamma_0}\right)_{\varphi,\eta=1} = \left|\frac{4}{\xi_0(\varepsilon - 1)(\xi_0^2 - 1)\ln\frac{\xi_0+1}{\xi_0-1} + 2\left(\xi_0^2 - \xi_0^2\varepsilon - 2\right)}\right|^2 \tag{7.46}$$

for the dipole moment normal to the symmetry axis.

In the case of an atom placed on the equator, we have, correspondingly:

$$\left(\frac{\gamma}{\gamma_0}\right)_r = \left|\frac{4\varepsilon}{\xi_0(\varepsilon-1)(\xi_0^2-1)\ln\frac{\xi_0+1}{\xi_0-1}+2(\xi_0^2-\xi_0^2\varepsilon-2)}\right|^2 \quad (7.47)$$

for the radial orientation of the dipole moment,

$$\left(\frac{\gamma}{\gamma_0}\right)_z = \left|\frac{2}{\xi_0(\varepsilon-1)(\xi_0^2-1)\ln\frac{\xi_0+1}{\xi_0-1}+2(\varepsilon+\xi_0^2-\xi_0^2\varepsilon)}\right|^2 \quad (7.48)$$

for the z-orientation of the dipole moment, and

$$\left(\frac{\gamma}{\gamma_0}\right)_\varphi = \left|\frac{4}{\xi_0(\varepsilon-1)(\xi_0^2-1)\ln\frac{\xi_0+1}{\xi_0-1}+2(\xi_0^2-\xi_0^2\varepsilon-2)}\right|^2 \quad (7.49)$$

for the φ-orientation of the dipole moment.

These expressions display the regularity: the decay rates in Eq. (7.45) exceed the decay rates in Eq. (7.48) by ε^2 times for any form of the spheroid. The analogous relation is true also for Eq. (7.47) and Eqs. (7.49), (7.46). These features are related to the boundary conditions on the spheroid surface: the tangential components of the field are continuous, while the electric field components normal to the surface differ by the factor ε.

In Fig. 7.11, the dependence of spontaneous emission rates on the aspect ratio a/c for the dipoles oriented parallel to the axis of the spheroid made of silver ($\varepsilon = -15.37 + i0.231$, $\lambda = 632.8$ nm (Shu and Hansma, 2001)) and gold ($\varepsilon = -8.37 + i1.16$, $\lambda = 600$ nm (Hass and Hadley, 1963)) is shown. An essential (10^4–10^6 times) enhancement of the spontaneous decay rate for the spheroid of $a/c \approx 0.2$ for Ag and $a/c \approx 0.35$ for Au is seen from the picture. It means that nanoparticles of this geometry can be used for the detection and selection of different molecules. It is also seen from the picture that the highest emission enhancement occurs in the case of an atom placed in the vicinity of the prolate spheroid pole (tip). This phenomenon is related to the electric field enhancement (in this case, the fields of vacuum fluctuations) in the vicinity of the tip (see Eq. (7.34)). There is no such enhancement of fields in the vicinity of the equator, and thus the enhancement of the spontaneous decay rate is minor.

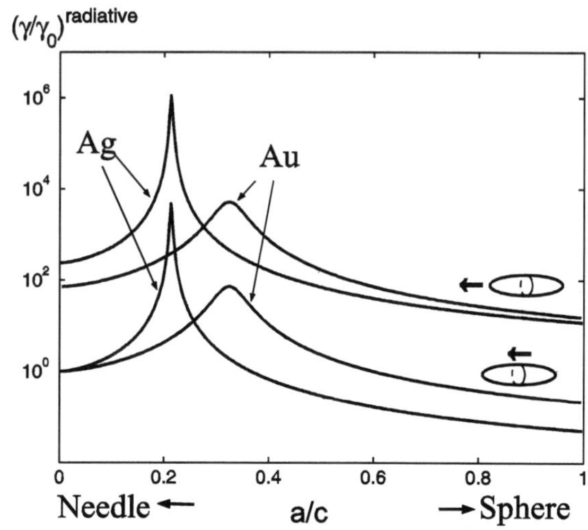

Figure 7.11 The dependence of the spontaneous transitions rates in the vicinity of gold and silver spheroids (Au, $\varepsilon = -8.37 + i1.16$, $\lambda = 600$ nm; Ag, $\varepsilon = -15.37 + i0.231$, $\lambda = 632.8$ nm) on the spheroid axes ratio a/c. The dipole is oriented parallel to the spheroid axis.

Since in metal nanoparticles the conductivity is nonzero, all optical processes are accompanied by the radiation absorption and consequent heating of nanoparticles. This is also true in the case of the emission of an atom or a molecule placed in the vicinity of a nanoparticle. The rate of emitted energy absorption in the nanoparticle is characterized by the nonradiative decay rate (see Appendix A1). It is essential that when the emitting atom is placed at small distances from the particle surface, nonradiative losses always dominate over the radiation losses (see the general expression Eq. (6.54)), and this is to be taken into account in any investigation of processes within the context of nano-optics and nanoplasmonics. In Fig. 7.12, the dependencies of the radiative and nonradiative parts of decay rates of an atom on its distance from the surface of a prolate silver nanospheroid are shown in the case of plasmon resonance at the wavelength of atom emission. It is seen from the picture that when the dipole moment is perpendicular to the axis, the nonradiative decay rates begin to dominate at short (of the order of several nanometers) distances from the surface. However, in the

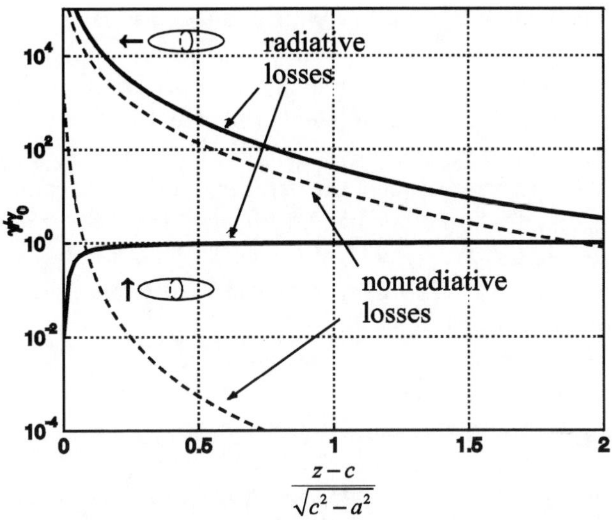

Figure 7.12 The dependencies of the radiative and nonradiative decay rates of an atom on its distance from the surface of a prolate silver nanospheroid in the case of plasmon resonance ($\xi_0 = 1.0236$, $a/c = 0.213$, $f = \sqrt{c^2 - a^2} = 0.1\lambda$). It is assumed that the atom is placed on the spheroid axis near its pole and radiates at the wavelength $\lambda = 632.8$ nm, at which the spheroid permittivity is equal to $\varepsilon = -15.37 + i0.231$.

case of the z-orientation of the dipole momentum, the radiative part of the decay rate dominates down to distances of the order of 1 nm. This is determined both by the small imaginary part of the silver electric permittivity and by the fact that absorption takes place only in the vicinity of the vertex, that is, in the region that is essentially smaller than the whole nanoparticle volume.

7.2.3.2 Oblate spheroids

The problem of atom or molecule emission in the presence of an oblate spheroid can be solved by a perfect analogy with the case of a prolate spheroid, and all results can be obtained by the replacement $\xi_0\ (\xi_0 > 1) \Rightarrow i\xi_0\ (\xi_0 > 0)$ in the expressions of the previous section.

7.3 Plasmon Oscillations in Spheroidal Shells

Both in natural and in technological processes, the shape of synthesized nanoparticles is similar to spheroidal ones but the structure is layered. For metal nanoparticles, this is determined by the oxide formation on the surface and other components' precipitation. On the other hand, artificially formed nanoparticles of layered shape have great significance. In Fig. 7.13, the process of particle synthesis is shown, with a dielectric core consisting of Fe_2O_3 (hematite) and a shell consisting of gold.

These artificially formed layered nanoparticles (nanoshells, nanorice) have great practical value since their plasmon resonances can be tuned into a wide range of frequencies.

If the core and shell of a particle can be described with the help of confocal spheroids, their plasmon properties can be described in a perfect analogy with the description in Section 7.1. In particular, in the case of a prolate layered structure (Fig. 7.14), the field outside the particle can be described by the expression:

$$\tilde{\varphi} = \sum_{n=0}^{\infty} \sum_{m=0}^{n} P_n^m(\eta)\, Q_n^m(\xi)\, \alpha_n^m e^{im\varphi}, \tag{7.50}$$

the field in the shell by the expression:

$$\tilde{\varphi} = \sum_{n=0}^{\infty} \sum_{m=0}^{n} P_n^m(\eta) \left[\beta_n^m Q_n^m(\xi) + \gamma_n^m P_n^m(\xi)\right] e^{im\varphi}, \tag{7.51}$$

and, finally, the field inside the particle is described by the expression:

$$\tilde{\varphi} = \sum_{n=0}^{\infty} \sum_{m=0}^{n} P_n^m(\eta)\, P_n^m(\xi)\, \delta_n^m e^{im\varphi} \tag{7.52}$$

In Eqs. (7.50)–(7.52), α, β, γ, δ are some coefficients to be determined from boundary conditions.

From the continuity of the potentials and the normal components of electric induction on the interfaces between the core and the shell and between the shell and the environment, the resonance values of electric permittivity can be found, for which free plasmon oscillations take place. In our case, the electric permittivity values

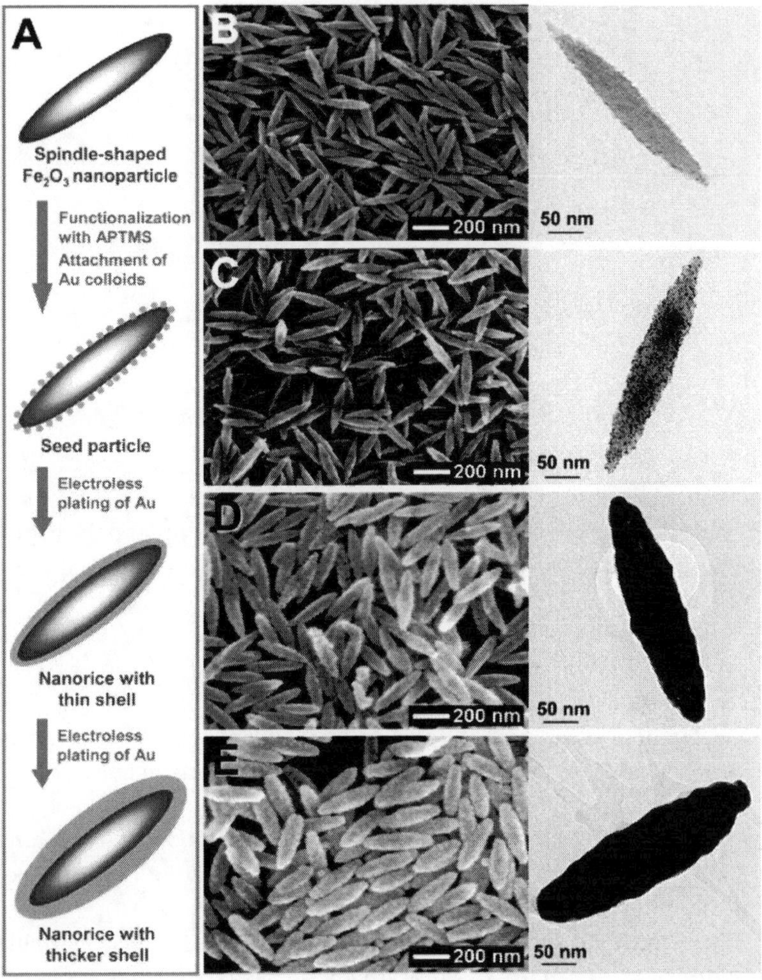

Figure 7.13 (A) The scheme of the synthesis of nanoparticles with a hematite core (Fe_2O_3) coated by a gold shell (nanorice). (B) The image of initial hematite crystals (the longitudinal size is 340 ± 20 nm; the transverse size is 54 ± 4 nm). (C) The image of hematite nanocrystals coated with gold seed nanoparticles of the 2 nm diameter. (D) The image of hematite crystals coated by a thin gold film (13.1 ± 1.1 nm) electrolytically. (E) The image of hematite crystals coated by a rather thick gold film (27.5 ± 1.7 nm) electrolytically (Wang et al., 2006).

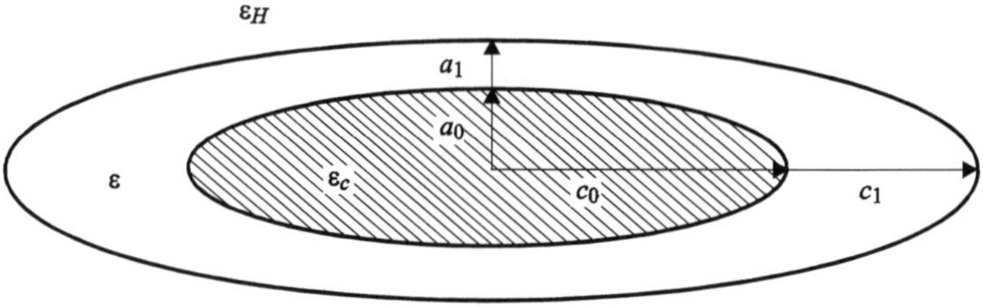

Figure 7.14 The geometry of a spheroid with a shell.

are governed by the quadratic equation:

$$\varepsilon^2 Q_1 P_1 \left(Q_0' P_1' - Q_1' P_0'\right) + \varepsilon \left(\varepsilon_{\mathrm{H}} P_0 Q_1' \left(Q_1 P_0' - Q_0' P_1\right) + \varepsilon_{\mathrm{C}} P_0' Q_1 \left(Q_1' P_0 - Q_0 P_1'\right)\right) + \varepsilon_{\mathrm{H}} \varepsilon_{\mathrm{C}} Q_1' P_0' \left(Q_0 P_1 - Q_1 P_0\right) = 0 \quad (7.53)$$

where the indices n, m in the Legendre functions are omitted for simplicity, and the indices 0 and 1 denote that these are functions of the argument $\xi_0 = c_0/\sqrt{c_0^2 - a_0^2}$ or $\xi_1 = c_1/\sqrt{c_1^2 - a_1^2}$, and the prime denotes the derivative of the function with respect to its argument, as usual.

It is important that this equation as a function of ε (the case of a metal shell) have two solutions for each pair of the indices n, m. Mathematically, it is related to the fact that the plasmonic shell has two boundaries at $\xi = \xi_0$ and $\xi = \xi_1$. From a physical point of view, such a shell can be treated as a hybrid of a dielectric cavity in a solid metal with the boundary $\xi = \xi_0$ and a metal nanoparticle in a host medium with the boundary $\xi = \xi_1$ (an analogous reasoning is applicable for any shell). As a result, there will be two types of plasmon modes in the system.

In Fig. 7.15, the dependencies of resonance values of dipole modes ($n = 1$) electric permittivity on the ratio between the longer and the shorter semiaxes of the dielectric core of $\varepsilon_{\mathrm{C}} = 9.5$ are shown, while in Fig. 7.16 the plasmon frequencies are shown that correspond to these values and were calculated with the help of the Drude model, $\varepsilon(\omega) = 1 - \omega_{\mathrm{pl}}^2/\omega^2$. Moreover, in Fig. 7.16, the plasmon frequencies of a solid metal spheroid in vacuum and of a dielectric cavity ($\varepsilon_{\mathrm{C}} = 9.5$, hematite) in a solid metal are shown.

These additional dashed curves help to understand the physical meaning of plasmon oscillations in a spheroidal metal shell. In

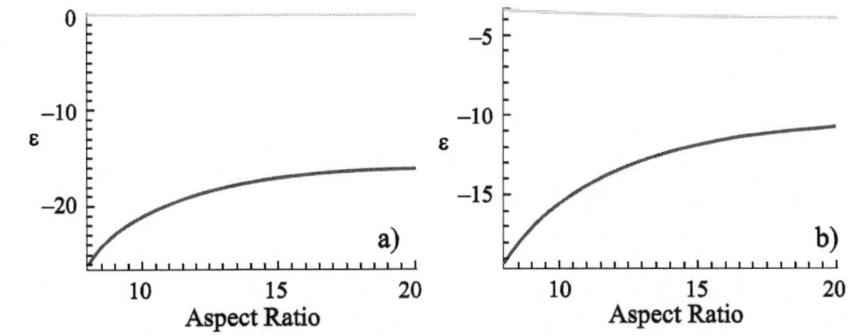

Figure 7.15 The dependencies of the resonance values of dipole modes permittivity ($n = 1$) on the aspect ratio c_0/a_0 of the dielectric core with $\varepsilon_C = 9.5$. (a) Longitudinal modes ($m = 0$) and (b) transverse modes ($m = \pm 1$). The aspect ratio of the shell is fixed ($c_1/a_1 = 4.575$), and the particle is in vacuum, $\varepsilon_H = 1$.

particular, it is clear from Fig. 7.16 that the high-frequency branches of longitudinal plasmon oscillations are basically related to the oscillations at the interface between the core and the shell, while the low-frequency branch is related to the oscillations on the outer surface of the metal shell. In the case of transverse oscillations, the

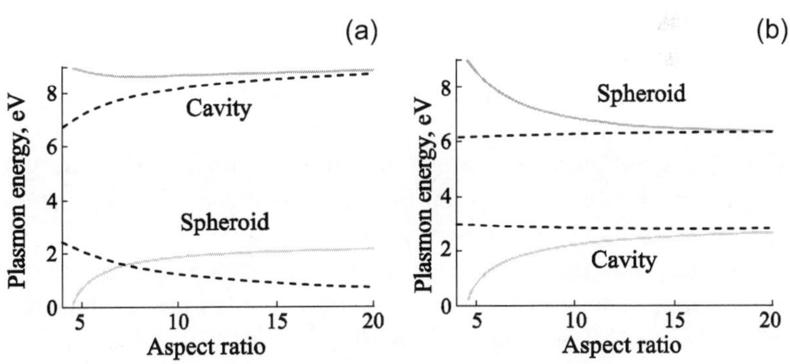

Figure 7.16 The dependencies of dipole modes plasmon frequencies ($n = 1$) on the aspect ratio c_0/a_0 of the dielectric core with $\varepsilon_C = 9.5$. (a) Longitudinal modes ($m = 0$) and (b) transverse modes ($m = \pm 1$). The plasmon frequencies of the metal spheroid in vacuum and dielectric cavity ($\varepsilon_C = 9.5$) in bulk metal are shown by the black dashed lines. The aspect ratio of the shell is fixed ($c_1/a_1 = 4.575$). The particle is in vacuum, $\varepsilon_H = 1$, and the bulk plasmon frequency of the metal is $\omega_{pl} = 8.95$ eV.

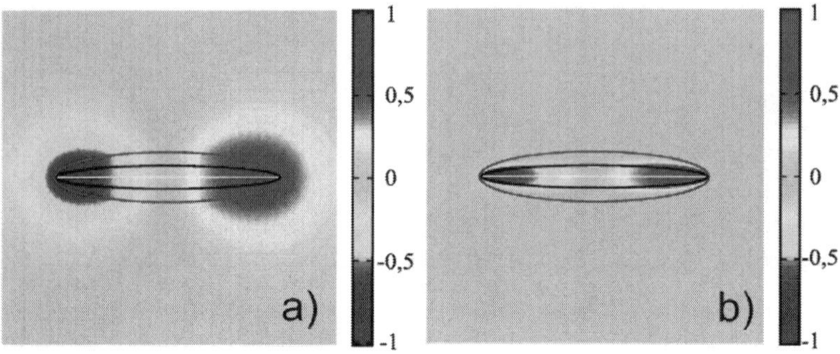

Figure 7.17 The spatial distribution of potential of longitudinal dipole modes ($n = 1, m = 0$, see Fig. 7.16a) for a dielectric core with $\varepsilon_C = 9.5$ and the aspect ratio $c_0/a_0 = 10$. (a) The low-frequency branch $\varepsilon = -21$; (b) the high-frequency branch $\varepsilon = -0.07$. The aspect ratio of the shell is fixed ($c_1/a_1 = 4.575$), and the particle is in vacuum.

situation is exactly the opposite: the high-frequency branches of plasmon oscillations are basically related to the oscillations on the outer surface of the metal shell, while the low-frequency branch is related to the oscillations on the internal surface. It is seen from the pictures that the interaction of these modes weakens with an increase of the dielectric core aspect ratio. This is obvious since at that the confocal core volume decreases. On the other hand, with the decrease of the core aspect ratio, that is, with the decrease of the shell thickness, the interaction of the surface charges on the external and internal sides of the shell becomes stronger, resulting in the "repulsion" of the plasmon frequencies.

In Fig. 7.17, the spatial distribution of the potential of both longitudinal modes ($n = 1, m = 0$, see Fig. 7.16a) for a dielectric core of $\varepsilon_C = 9.5$ and aspect ratio 10 is shown. It is seen from the picture that in the case of the low-frequency branch (Fig. 7.17a) there are strong fields also outside the nanospheroid. It is implied by the fact that basically this mode is related to the nanoparticle external surface oscillations (Fig. 7.16a). And, vice versa, the high-frequency mode (Fig. 7.17b) is basically concentrated in the spheroid dielectric core since it is similar to the mode originating on the shell internal surface.

Essentially, these fields are analogous to those originating in the gap between metal nanoparticles (Chapter 10). In contrast to the latter, in nanospheroidal shells the strong field also exists outside nanoparticles. The controlled enhancement of the external fields by the low-frequency longitudinal mode (Fig. 7.17a) can be used in many practical applications. In particular, each nanoparticle of this kind can be used as a separate nanosensor of biomolecules such as proteins and DNA and other biological samples.

After the resonance dielectric permittivities have been found, the optical properties of nanoparticles of this kind can be found. In particular, the polarizabilities of a spheroidal shell have the form:

$$\alpha_z = \frac{a_1^2 c_1}{3} \frac{(\varepsilon - \varepsilon_H)\left[\varepsilon + (\varepsilon_C - \varepsilon)\left(L_0^{(0)} - fL_0^{(1)}\right)\right] + f\varepsilon(\varepsilon_C - \varepsilon)}{\left(\left[\varepsilon + (\varepsilon_C - \varepsilon)\left(L_0^{(0)} - fL_0^{(1)}\right)\right]\left[\varepsilon_H + (\varepsilon - \varepsilon_H) L_0^{(1)}\right] + fL_0^{(1)}\varepsilon(\varepsilon_C - \varepsilon)\right)}$$

$$\alpha_x = \alpha_y = \frac{a_1^2 c_1}{3} \frac{(\varepsilon - \varepsilon_H)\left[\varepsilon + (\varepsilon_C - \varepsilon)\left(L_1^{(0)} - fL_1^{(1)}\right)\right] + f\varepsilon(\varepsilon_C - \varepsilon)}{\left(\left[\varepsilon + (\varepsilon_C - \varepsilon)\left(L_1^{(0)} - fL_1^{(1)}\right)\right]\left[\varepsilon_H + (\varepsilon - \varepsilon_H) L_1^{(1)}\right] + fL_1^{(1)}\varepsilon(\varepsilon_C - \varepsilon)\right)} \tag{7.54}$$

In these expressions, $L^{(0)}$, $L^{(1)}$ are geometric factors for solid spheroids described by the equations $\xi = \xi_0$ and $\xi = \xi_1$ (see Eq. (7.30) for the prolate spheroid and Eq. (7.31) for the oblate spheroid) and $f = a_0^2 c_0 / a_1^2 c_1$ is the fraction of the total particle volume occupied by the inner core. Let us note that the roots in the denominator of Eq. (7.54) coincide with the roots in Eq. (7.53) for $n = 1$ and $m = 0$ or ± 1, correspondingly, since only for these values of indices the plasmon modes have the dipole moments contributing to dipole polarizability, Eq. (7.54).

Knowing the polarizabilities, all the absorption and scattering cross sections can be found from the formulas given in Section 7.2, in turn.

7.4 The Effect of Retardation in Nanoparticles of Spheroidal and Related Shapes

The main part of this chapter is devoted to the presentation of nanospheroids' properties in the quasi-static approximation. The quasi-static approximation describes nanoparticles of less than

50 nm rather well. For greater nanoparticles, the quasi-static approximation provides only a qualitative description. Certainly, one can do sophisticated calculations with spheroidal functions or direct numerical calculations in each specific case. However, for quick estimation it is convenient to have some working semianalytical formulas of limited accuracy.

In particular, in the case of the Mie theory for a sphere of radius a the electric polarizability on a sphere of an arbitrary size can be presented in the following form:

$$\alpha^E = i\frac{3q_1}{2k_H^3} \tag{7.55}$$

where q_1 is the Mie coefficient defined in Eq. (6.21) and k_H is the wavenumber of the surrounding medium with the dielectric constant ε_H. Assuming that the size of the sphere is small ($ka << 1$), it is easy to find an expression for polarizability with the accuracy to the terms of the order ka^4:

$$\alpha^E(\lambda) = a^3 \frac{1 + (\varepsilon(\lambda) + \varepsilon_H)\, x^2/10 + O(x^4)}{\left(1 + \frac{3\varepsilon_H}{(\varepsilon(\lambda) - \varepsilon_H)}\right) - \frac{1}{10}(\varepsilon(\lambda) + 10\varepsilon_H)\, x^2 - i\frac{2}{3}(\varepsilon_H)^{3/2} x^3 + O(x^4)} \tag{7.56}$$

where $x = ka$ is the size parameter and $\varepsilon(\lambda)$ is the permittivity of the sphere. This formula describes finite size spherical nanoparticles better than the purely quasi-static approximation with $x = ka = 0$.

It is desirable to obtain analogous expressions for spheroids and other nanoparticles. For spheroids (and even for ellipsoids), the resonance denominator of polarizability can be found analytically with an accuracy up to terms of the order ka^4. Indeed, for a term with an imaginary unit (radiative losses) there is a general expression through the quasi-static solution (Chapter 5). The terms of the order ka^2 can be also found analytically, for example, using the solution for plane-wave scattering by ellipsoid (Stevenson, 1953a, 1953b). However, within these approaches it is difficult to find the corrections for the numerator. Moreover, there is no analytical solution for nanoparticles slightly different from spheroidal and ellipsoidal ones.

Therefore (Kuwata et al., 2003; Prescott and Mulvaney, 2006) proposed approximation expressions analogous to Eq. (7.55) and having parameters that take into account both the effects of retardation and more complex nanoparticles' shape more or less universally. The dependencies of these parameters on system sizes can be found once with the help of numerical calculations.

Kuwata et al. (2003) proposed to describe the basic resonance term of the longitudinal polarizability of any prolate particle with the help of the expression:

$$\alpha = \frac{V}{4\pi}\frac{1}{\left(L + \varepsilon_{\mathrm{H}}/\left(\varepsilon - \varepsilon_{\mathrm{H}}\right) + A\varepsilon_{\mathrm{H}}x^2 + B\left(\varepsilon_{\mathrm{H}}x^2\right)^2 - i\frac{4\pi^2\varepsilon_{\mathrm{H}}^{3/2}}{3}\frac{V}{\lambda_0^3}\right)} \tag{7.57}$$

where the parameters A and B are to be found from some or other consideration. Generally, these parameters are dependent on permittivity ε (in order to consider the material influence) and on the geometrical factor L (in order to take the shape into account; see Eq. (7.30)). In the case of small losses in the nanoparticle substance and small radiative losses the resonance condition will have the form:

$$L + \varepsilon_{\mathrm{H}}/\left(\varepsilon' - \varepsilon_{\mathrm{H}}\right) + A\varepsilon_{\mathrm{H}}x^2 + B\left(\varepsilon_{\mathrm{H}}x^2\right)^2 = 0 \tag{7.58}$$

where ε' is the real part of the particle electric permittivity.

To determine the parameters A and B, Kuwata et al. (2003) proposed to compare Eq. (7.58) to the resonances in spheroidal particles of gold and silver calculated within the finite difference time domain (FDTD) numerical simulation. In a remarkable manner, the coefficients A and B do not depend on a material practically and are described by the expressions:

$$A\,(L) = 0.4865L - 1.046L^2 + 0.8481L^3$$

$$B\,(L) = 0.01909L + 0.1999L^2 + 0.6077L^3 \tag{7.59}$$

The same formulas may be used for the description of nanorods, that is, cylinders with spherical or spheroidal caps. The determination of L is an independent problem, and in fact it is equivalent to finding of corresponding resonant values of basic longitudinal mode permittivity in quasi-static approximation.

Prescott and Mulvaney (2006) approached this problem from the other side. They proposed to parameterize the extinction coefficient in the form:

$$Q_{\mathrm{ext}}\,(\lambda) = \frac{2\pi V\varepsilon_{\mathrm{H}}^{3/2}}{GL^2\lambda}\frac{\varepsilon''}{\{\varepsilon' - \varepsilon_{\mathrm{res}}\}^2 + \varepsilon''^2} \tag{7.60}$$

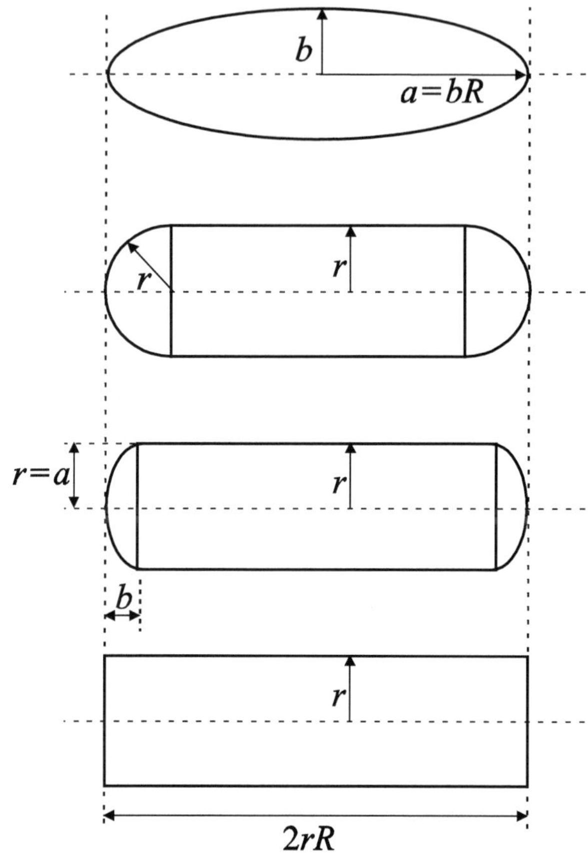

Figure 7.18 The geometry of prolate nanoparticles used for modeling of real nanoparticles (spheroid, cylinder with spherical caps [spherocylinder], cylinder with spheroidal ends [eccentricity e] and with plane ends). In all the cases, the aspect ratio R is the same (Prescott and Mulvaney, 2006).

where G is the area of the particle's geometrical cross section normal to the direction of incident light (the area of geometric shadow), V is the volume of the particle, λ is the emission wavelength in vacuum, $L = 1/(1 - \varepsilon_{\text{res}}/\varepsilon_{\text{H}})$ is the particle's geometric factor, and ε_{res} is the effective value of the resonance electric permittivity of the basic longitudinal mode. Having compared Eq. (7.60) with the results of the discrete dipole approximation (DDA) simulation of light-scattering processes, they found the dependencies of the resonance

Table 7.1 The absolute values of the resonance electric permittivity values of longitudinal resonance $|\varepsilon_{res}|$ for different cap geometries and aspect ratios (R)

R	Spherocylinder	$e = 0.5$	$e = 0.8$	$e = 0.95$	Cylinder
1	2.2	2.3	2.4	2.8	3.5
2	5.4	5.6	6.0	6.5	7.4
3	9.8	10.0	10.5	11.1	11.9
4	15.0	15.3	15.8	16.6	17.6
5	21.1	21.4	22.0	22.8	23.9
6	28.1	28.4	29.1	30.0	31.4
7	35.9	36.3	37.1	38.1	39.6
8	44.8	45.2	46.0	47.1	48.7

dielectric permittivities and geometric factors on the systems' sizes for nanospheroids and nanorods of different shapes (Fig. 7.18).

The results of these calculations are presented in Tables 7.1 and 7.2 for a cylinder of radius 10 nm and for a different aspect ratio R, that is, for the ratio between the radius and the half-length of the particle. The values of geometric factors for cylinders of radii 5, 15, and 20 nm were also found (Prescott and Mulvaney, 2006).

Thus, the results of Kuwata et al. (2003), Prescott and Mulvaney (2006) allow one to estimate conditions of the origin of fundamental longitudinal plasmon resonances in elongated nanoparticles with a cylindrical symmetry. These approaches can be naturally applied to other (e.g., transversal) modes of spheroids and nanorods and even to more complicated nanoparticles, because in fact plasmonic

Table 7.2 The summary of parameters describing the quadratic approximation of the resonance electric permittivity values of longitudinal resonance from the aspect ratio r for a cylinder of the radius of 10 nm, $-\varepsilon_{res} = aR^2 + bR + c$

Cylinder type	a	b	c
Spherocylinder	0.4532	2.0100	−0.3072
$e = 0.5$	0.4495	2.0856	−0.2903
$e = 0.8$	0.4420	2.2417	−0.2489
$e = 0.95$	0.4378	2.3786	0.0144
Cylinder	0.4473	2.4262	0.6937

oscillations can be described with a limited number of scalar parameters (see Eq. (5.23)), which can be approximated with the help of numerical simulations.

References

Abramowitz, M., Stegun, I.A., eds. (1964) *Handbook of Mathematical Functions* (NBS).

Asano, S., Yamamoto, G. (1975) Light scattering by a spheroidal particle, *Appl. Opt.*, **14**, 29–49.

Belkina, M.G. (1957a) Diffraction of electromagnetic waves on the disk, in *Diffraction of Electromagnetic Waves on Some Bodies of Revolution* (Sovetskoe Radio, Moscow), p. 148.

Belkina, M.G. (1957b) Characteristics of radiation of prolate ellipsoid of revolution, in *Diffraction of Electromagnetic Waves on Some Bodies of Revolution* (Sovetskoe Radio, Moscow), p. 126.

Bohren, C.F., Huffman, D.R. (1983) *Absorption and Scattering of Light by Small Particles* (Wiley).

Falloon, P.E. (2001) *Theory and Computation of Spheroidal Harmonics with General Arguments,* thesis (University of Western Australia, Department of Physics).

Farafonov, V.G. (1983) The scattering of a plane electromagnetic wave by a dielectric spheroid, *Differ. Equations,* **19**, 1765.

Farafonov, V.G. (2001) Light scattering by axisymmetric multilayer particles *Opt. Spectrosc.*, **91**, 92.

Gersten, J., Nitzan, A. (1980) Electromagnetic theory of enhanced Raman scattering by molecules adsorbed on rough surfaces, *J. Chem. Phys.*, **73**(7), 3023.

Guzatov, D.V., Klimov, V.V. (2005) Radiative decay engineering by triaxial nanoellipsoid, *Chem. Phys. Lett.*, **412**, 341–346.

Hass, G., Hadley, L. (1963) Optical properties of metals, in *American Institute of Physics HandBook*, ed. D.E. Gray (McGraw-Hill, New York), p. 6.

Hatcher, E.C., Jr., Leitner, A. (1954) Radiation from a point dipole located at the tip of a prolate spheroid, *J. Appl. Phys.*, **25**, 1250.

Klimov, V.V., Ducloy, M., Letokhov, V.S. (2002a) A model of an apertureless scanning microscope with a prolate nanospheroid as a tip and an excited molecule as an object, *Chem. Phys. Lett.*, **358**, 192–198.

Klimov V.V., Ducloy, M., Letokhov, V.S. (2002b) Spontaneous emission of an atom placed near a prolate nanospheroid, *Eur. Phys. J.*, **D20**, 133–148.

Kuwata, H., Tamaru, H., Esumi, K., Miyano, K. (2003) Resonant light scattering from metal nanoparticles: practical analysis beyond Rayleigh approximation, *Appl. Phys. Lett.*, **83**, 4625.

Li, L.W., Leong, M.S., Yeo, T.S., Kooi, P.S., Tan, K.Y. (1998) Computations of spheroidal harmonics with complex arguments: a review with an algorithm, *Phys. Rev. E*, **58**, 6792–6806.

Li, L.-W., Kang, X.-K., Leong, M.-S. (2002) *Spheroidal Wave Functions in Electromagnetic Theory* (John Wiley & Sons).

Lisiecki, I., Pileni, M.P. (2003) Cobalt nanocrystals: influence of the synthesis on the nanocrystal self-organization, *Langmuir*, **19**, 9486.

Mohamed, M.B., Volkov, V., Link, S., El-Sayed, M.A. (2000) The "lightning" gold nanorods: fluorescence enhancement of over a million compared to the gold metal, *Chem. Phys. Lett.*, **317**, 517–523.

Pérez-Juste, J., Liz-Marzán, L.M., Carnie, S., Chan, D.Y.C., Mulvaney, P. (2004) Electric-field-directed growth for gold nanorods, *Adv. Funct. Mater.*, **14**, 571.

Prescott, S.W., Mulvaney, P. (2006) Gold nanorod extinction spectra, *J. Appl. Phys.*, **99**, 123504.

Prescott, S.W., Mulvaney, P. (2006) Erratum: "Gold nanorod extinction spectra" [J. Appl. Phys. 99, 123504 (2006)], *J. Appl. Phys.*, **103**, 119901.

Shu, Q.Q., Hansma, P.K. (2001) Fluorescent apparent quantum yields for excited molecules near dielectric interfaces, *Thin Solid Films*, **384**, 76–84.

Smythe, W.R. (1969) *Static and Dynamic Electricity*, 3rd ed. (McGraw-Hill, New York).

Stevenson, A.F. (1953a) Solution of electromagnetic scattering problems as power series in the ratio (dimension of scatterer)/wavelength, *J. Appl. Phys.*, **24**, 1134–1142.

Stevenson, A.F. (1953b) Electromagnetic scattering by an ellipsoid in the third approximation, *J. Appl. Phys.*, **24**, 1143–1151.

Voshchinnikov, N.V., Farafonov, V.G. (1993) Optical properties of spheroidal particles, *Astrophys. Space Sci.*, **204**, 19–86.

Wang, H., Brandl, D.W., Le, F., Nordlander, P., Halas, N.J. (2006) Nanorice: a hybrid plasmonic nanostructure, *Nano Lett.*, **6**, 827–832.

Yu, Y.Y., Chang, S.S., Lee, C.L., Wang, C.R.C. (1997) Gold nanorods: electrochemical synthesis and optical properties, *J. Phys. Chem. B*, **101**, 6661.

Chapter 8

Optical Properties of a Three-Axial Nanoellipsoid

In the present chapter, the optical properties of a three-axial nanoellipsoid and their influence on the radiation of atoms and molecules placed in the vicinity of such an ellipsoid are discussed. This problem is very important since the three-axial ellipsoid generalizes both a sphere and spheroids. On the other hand, even quite spherical particles have, in fact, asymmetries, which have a considerable influence on their optical properties because all resonance degenerations are removed.

The quasi-static solution of the problem of a homogeneous and isotropic nanoellipsoid placed in a uniform field is well-known (Stratton, 1941). The quasi-static solution of the problem of a nanoellipsoid made of a homogeneous, but anisotropic, dielectric in a uniform electric field was also found (Jones, 1945). Stevenson (1953a, 1953b) found the first wavenumber corrections to the quasi-static solution. Currently, his solution is the most well-known analytical solution for diffraction of a plane wave by a small homogeneous ellipsoid. The solution of the quasi-static problem of a three-axial nanoellipsoid excited by highly inhomogeneous optical fields, such as radiation fields, of an atom or a molecule is also very

Nanoplasmonics
Vasily Klimov

ISBN 978-981-4267-16-8 (Hardcover), 978-981-4267-42-7 (eBook)
www.panstanford.com

important (Guzatov and Klimov, 2005). A solution of this problem may have a lot of applications in diffraction (scattering) theory; it can be used for construction of Green's function in ellipsoidal domains, for a solution of the problem of spontaneous radiation properties of an atom in the vicinity of a three-axial ellipsoid, etc. There is an analytic expression for an induced dipole moment of a three-axial nanoellipsoid excited by a point dipole (an atom), and there is a solution of the problem of the spontaneous emission rate of an atom placed in the vicinity of a nanoellipsoid.

8.1 The General Solution of the Quasi-Static Problem of Plasmon Oscillations in a Three-Axial Nanoellipsoid in the Context of the "ε-Method"

As it was shown in Chapter 5, to find solutions within the ε-method, it is necessary to solve the homogeneous problem of potential theory with the corresponding boundary conditions on the body surface:

$$\begin{aligned} \mathbf{E} &= -\nabla\varphi, \\ \Delta\varphi &= 0, \end{aligned} \tag{8.1}$$

where ∇ is the gradient and Δ is the Laplace operator.

In the case of a three-axial ellipsoid, it is convenient to solve the equations in Eq. (8.1) in the ellipsoidal coordinate system (Hobson, 1931). Let the three-axial nanoellipsoid have semiaxes $a_1 > a_2 > a_3$ directed along each of the axis $x_1,\ x_2,\ x_3$ of the Cartesian coordinate system (Fig. 8.1). Introducing the notions $h_1 = \sqrt{a_2^2 - a_3^2}$, $h_2 = \sqrt{a_1^2 - a_3^2}$ and $h_3 = \sqrt{a_1^2 - a_2^2}$, the relation between the ellipsoidal coordinates $\rho,\ \mu,\ \nu$ and the Cartesian coordinates may be written in the form (Hobson, 1931):

$$\begin{aligned} x_1 &= \frac{1}{h_2 h_3}\rho\mu\nu, \\ x_2 &= \frac{1}{h_1 h_3}\sqrt{\rho^2 - h_3^2}\sqrt{\mu^2 - h_3^2}\sqrt{h_3^2 - \nu^2}, \\ x_3 &= \frac{1}{h_1 h_2}\sqrt{\rho^2 - h_2^2}\sqrt{h_2^2 - \mu^2}\sqrt{h_2^2 - \nu^2}, \end{aligned} \tag{8.2}$$

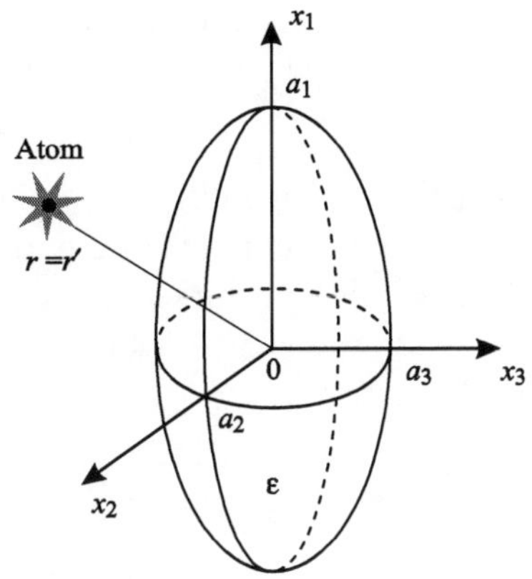

Figure 8.1 The geometry of a three-axial ellipsoid.

where $0 \leq \nu^2 \leq h_3^2 \leq \mu^2 \leq h_2^2 \leq \rho^2 < \infty$. The expression $\rho = a_1$ defines the surface of the ellipsoidal nanoparticle $\sum_{\alpha=1}^{3} \frac{x_\alpha^2}{a_\alpha^2} = 1$.

In what follows, for convenience we will use the Greek letters $\alpha, \beta, \gamma, \ldots$ as indices, understanding that they take the values $1, 2, 3, \ldots$.

An arbitrary solution of the Laplace equation, Eq. (8.1), inside and outside an ellipsoid may be written as the expansion of the potential φ over the Lame function (Hobson, 1931):

$$\varphi = \begin{cases} \sum_{n=1}^{\infty} \sum_{m=1}^{2n+1} A_n^m E_n^m(\rho) E_n^m(\mu) E_n^m(\nu), & \rho \leq a_1, \\ \sum_{n=1}^{\infty} \sum_{m=1}^{2n+1} B_n^m F_n^m(\rho) E_n^m(\mu) E_n^m(\nu), & \rho > a_1, \end{cases} \tag{8.3}$$

where A_n^m and B_n^m are the expansion coefficients. The external Lame functions $F_n^m(\rho)$ are expressed through the internal ones $E_n^m(\rho)$ through the relation (Hobson, 1931):

$$F_n^m(\rho) = (2n+1) E_n^m(\rho) \int_{\rho}^{\infty} \frac{du}{\left(E_n^m(u)\right)^2 \sqrt{u^2 - h_2^2}\sqrt{u^2 - h_3^2}} \tag{8.4}$$

The functions defined in such a way are regular both inside and outside an ellipsoid and vanish at infinity. It is important to note that, in fact, Lame functions are the functions of four variables:

$$E_n^m(\rho) = E_n^m(\rho; a_1, a_2, a_3),$$
$$F_n^m(\rho) = F_n^m(\rho; a_1, a_2, a_3), \tag{8.5}$$

and their calculation is rather difficult.

Using the boundary condition for the potential and the normal component of the electric induction on the ellipsoid surface, we obtain the following system of equations for the expansion coefficients A_n^m and B_n^m:

$$A_n^m E_n^m(a_1) = B_n^m F_n^m(a_1),$$
$$\varepsilon A_n^m E'^m_n(a_1) = B_n^m F'^m_n(a_1). \tag{8.6}$$

In Eq. (8.6), the prime denotes the differentiation of the function with respect to its first argument. We denoted the dielectric constant of the nanoellipsoid substance by ε, and we take the electric permittivity of the medium, in which the ellipsoid is submerged, as equal to one. From the condition of consistency of equations in Eq. (8.6), the expressions for the electric permittivity eigenvalues and corresponding plasmon wavefunctions can be found easily:

$$\varepsilon_n^m(a_1, a_2, a_3) = \frac{E_n^m(a_1; a_1, a_2, a_3)\, F'^m_n(a_1; a_1, a_2, a_3)}{E'^m_n(a_1; a_1, a_2, a_3)\, F_n^m(a_1; a_1, a_2, a_3)}, \tag{8.7}$$

$$\varphi = \begin{cases} E_n^m(\rho)\, E_n^m(\mu)\, E_n^m(\nu), & \rho \le a_1 \\ \dfrac{E_n^m(a_1)}{F_n^m(a_1)} F_n^m(\rho)\, E_n^m(\mu)\, E_n^m(\nu), & \rho > a_1 \end{cases} \tag{8.8}$$

Equations (8.7) and (8.8) exhaust the solution to the problem of the plasmon oscillations in a three-axial nanoellipsoid.

As it is shown in Chapter 5, knowing the eigenfunctions for the potential and the corresponding eigenvalues and dielectric constant, one can solve any problem concerning a three-axial nanoellipsoid placed in the electromagnetic field of an arbitrary source and, in particular, find the plasmon frequencies' values. Nevertheless, the solution obtained is rather formal since the effective ways of calculation of the Lame functions depending on four parameters have been not elaborated yet.

Nevertheless, in the next section the method will be described that, at least partially, allows calculation of the plasmon oscillations in nanoparticles of ellipsoidal shape. This method is based on the usage of Cartesian coordinates rather than ellipsoidal ones.

8.2 Explicit Expressions for Plasmon Modes of a Nanoellipsoid in the Cartesian Coordinates

To find optical properties of ellipsoidal nanoparticles in the context of the ε-method constructively, it is convenient to use the integral formulation for plasmon eigenoscillations (Guzatov et al., 2010):

$$\varphi_n(\mathbf{r}) = \frac{\varepsilon_n - 1}{4\pi} \int_V d\mathbf{r}' \left(\nabla' \varphi_n(\mathbf{r}') \nabla\right) \frac{1}{|\mathbf{r} - \mathbf{r}'|} \tag{8.9}$$

where the integral is taken over the ellipsoid volume.

To solve this integral equation, it is convenient to use the usual Cartesian coordinates instead of the ellipsoidal ones, used in the previous section. This possibility is determined by the fact that ellipsoidal functions inside a particle may be represented as Niven functions, which have the general form (Whittaker and Watson, 1963):

$$\Phi(x_1, x_2, x_3) = \left\{ \begin{matrix} & x_1 \; x_1x_2 & \\ 1 & x_2 \; x_2x_3 \; x_1x_2x_3 \\ & x_3 \; x_3x_1 & \end{matrix} \right\}$$
$$\Theta_1(x_1, x_2, x_3)\,\Theta_2(x_1, x_2, x_3)\ldots\Theta_p(x_1, x_2, x_3),$$
$$\Theta_p(x_1, x_2, x_3) = \frac{x_1^2}{a_1^2 + \theta_p} + \frac{x_2^2}{a_2^2 + \theta_p} + \frac{x_3^2}{a_3^2 + \theta_p} - 1 \tag{8.10}$$

where θ_p is a set of parameters, the choice of which allows $\Phi(x_1, x_2, x_3)$ to satisfy the Laplace equation in the finite region of the space (inside the ellipsoid).

Substituting Eq. (8.10) in Eq. (8.9), it is possible to determine the electric permittivity eigenvalues and corresponding eigenpotentials (wavefunctions) inside the nanoellipsoid. To obtain the potential values outside the particle, the found eigenfunctions and eigenvalues are to be substituted in the right part of Eq. (8.9). After that, the potential at an arbitrary point of space can be found by integration.

The general approach to the construction of Eq. (8.9) solution in the form of Eq. (8.10), that is, to finding of the plasmon oscillations' eigenfunctions and eigenvalues, is as follows. Let the eigenfunction, Eq. (8.10), of the potential in the problem concerning a three-axial ellipsoid have the following general form:

$$\varphi_n(\mathbf{r}) = \sum_{n\geq p+q+s\geq 0} A_{pqs} x_1^p x_2^q x_3^s = A_{000} + A_{100}x_1 + \ldots + A_{n00}x_1^n + \ldots, \tag{8.11}$$

where $p, q, s = 0, 1, 2, \ldots$. Substituting Eq. (8.11) in Eq. (8.9), we obtain the system of equations for the coefficients A_{pqs}:

$$\sum_{p+q+s=0}^{n} A_{pqs} x_1^p x_2^q x_3^s = \frac{\varepsilon_n - 1}{4\pi} \sum_{p+q+s=0}^{n} A_{pqs} \left(p\frac{\partial}{\partial x_1} \int_V d\mathbf{r}' \frac{x'^{p-1}_1 x'^q_2 x'^s_3}{|\mathbf{r}-\mathbf{r}'|} + q\frac{\partial}{\partial x_2} \int_V d\mathbf{r}' \frac{x'^p_1 x'^{q-1}_2 x'^s_3}{|\mathbf{r}-\mathbf{r}'|} + s\frac{\partial}{\partial x_3} \int_V d\mathbf{r}' \frac{x'^p_1 x'^q_2 x'^{s-1}_3}{|\mathbf{r}-\mathbf{r}'|} \right) \tag{8.12}$$

A remarkable feature of an ellipsoid is the fact that the integrals arising in Eq. (8.12) are the polynomials over the Cartesian coordinates of the power $p + q + s + 2$, and its derivative may be present as:

$$p\frac{\partial}{\partial x_1} \int_V d\mathbf{r}' \frac{x'^{p-1}_1 x'^q_2 x'^s_3}{|\mathbf{r}-\mathbf{r}'|} = \sum_{p'+q'+s'=0}^{p'+q'+s'=p+q+s} B_{p'q's'} x_1^{p'} x_2^{q'} x_3^{s'},$$
$$q\frac{\partial}{\partial x_2} \int_V d\mathbf{r}' \frac{x'^p_1 x'^{q-1}_2 x'^s_3}{|\mathbf{r}-\mathbf{r}'|} = \sum_{p'+q'+s'=0}^{p'+q'+s'=p+q+s} C_{p'q's'} x_1^{p'} x_2^{q'} x_3^{s'},$$
$$s\frac{\partial}{\partial x_3} \int_V d\mathbf{r}' \frac{x'^p_1 x'^q_2 x'^{s-1}_3}{|\mathbf{r}-\mathbf{r}'|} = \sum_{p'+q'+s'=0}^{p'+q'+s'=p+q+s} D_{p'q's'} x_1^{p'} x_2^{q'} x_3^{s'}, \tag{8.13}$$

where $p', q', s' = 0, 1, 2, \ldots$. Substituting Eq. (8.13) in Eq. (8.12) and equating the coefficients with the same power $x_1^p x_2^q x_3^s$ on the left and on the right, we obtain the homogeneous system of linear equations:

$$
\begin{aligned}
A_{000} &= \frac{\varepsilon_n - 1}{4\pi} A_{000} \left(B_{000} + C_{000} + D_{000}\right), \\
A_{100} &= \frac{\varepsilon_n - 1}{4\pi} \left(A_{100} + A_{010} + A_{001}\right) \left(B_{100} + C_{100} + D_{100}\right), \\
A_{010} &= \frac{\varepsilon_n - 1}{4\pi} \left(A_{100} + A_{010} + A_{001}\right) \left(B_{010} + C_{010} + D_{010}\right), \\
A_{001} &= \frac{\varepsilon_n - 1}{4\pi} \left(A_{100} + A_{010} + A_{001}\right) \left(B_{001} + C_{001} + D_{001}\right), \\
&\vdots
\end{aligned} \tag{8.14}
$$

The nontrivial solution of Eq. (8.14) provides the electric permittivity resonance values and the plasmon oscillation wavefunctions. Let us note, there are $2n + 1$ independent plasmon oscillations for the polynomial Eq. (8.11) of the order n, that is, for polynomials with $A_{p+q+s=n} \neq 0$.

Consequently, it is convenient to denote eigenfunctions and eigenvalues as φ_n^m, ε_n^m, where $m = 1 \ldots 2n + 1$. At that, the lower index characterizes the highest power of the polynomial φ_n^m. Let us note, this concept agrees with spherical harmonics definitions and confirms the similarity of the spherical harmonics $r^n Y_n^m(\theta, \varphi)$ and the ellipsoidal harmonics $\varphi_n^m(x_1, x_2, x_3)$.

The plasmon modes' potentials outside the ellipsoid g_n^m may be found with the help of integration:

$$
g_n^m(\mathbf{r}) = \frac{\varepsilon_n^m - 1}{4\pi} \int_V d\mathbf{r}' \left(\nabla' \phi_n^m(\mathbf{r}') \nabla\right) \frac{1}{|\mathbf{r} - \mathbf{r}'|} \tag{8.15}
$$

Let us consider a specific case of localized plasmon oscillations in a nanoellipsoid.

The plasmon modes with the potential inside the particle being a linear function of coordinates

$$
\varphi_1^\alpha = x_\alpha, \ \alpha = 1, 2, 3 \tag{8.16}
$$

and the corresponding three eigenvalues

$$
\varepsilon_1^\alpha = 1 - \left(\frac{a_1 a_2 a_3}{2} I_\alpha\right)^{-1}, \tag{8.17}
$$

where

$$
I_\alpha = \int_0^\infty \frac{du}{\left(u + a_\alpha^2\right) R(u)}, \qquad R(u) = \prod_{\alpha=1}^{3} \left(u + a_\alpha^2\right)^{1/2}, \tag{8.18}
$$

are the simplest and the most investigated. These modes have a dipole moment, and therefore they are the most important in applications.

The potentials of these plasmon modes outside the particle have the form:

$$g_1^{\alpha}(\mathbf{r}) = \frac{I_\alpha(u_0)}{I_\alpha} x_\alpha, \tag{8.19}$$

where

$$I_\alpha(u_0) = \int_{u_0}^{\infty} \frac{du}{\left(u + a_\alpha^2\right) R(u)}, \quad R(u) = \prod_{\alpha=1}^{3} \left(u + a_\alpha^2\right)^{1/2} \tag{8.20}$$

and u_0 is the positive root of the equation $\sum_{\alpha=1}^{3} \frac{x_\alpha^2}{u+a_\alpha^2} = 1$.

Modes with the potentials quadratic $(n = 2)$ over the coordinates inside the ellipsoid are more complicated. Two of them have the following form:

$$\varphi_2^1 = \left(\sum_{\beta=1}^{3} \frac{x_\beta^2}{\Lambda' - a_\beta^2} + 1\right),$$

$$\varphi_2^2 = \left(\sum_{\beta=1}^{3} \frac{x_\beta^2}{\Lambda'' - a_\beta^2} + 1\right), \tag{8.21}$$

where

$$\left.\begin{matrix} \Lambda' \\ \Lambda'' \end{matrix}\right\} = \frac{1}{3}\left\{ \pm \left[\sum_{\alpha=1}^{3} \left(a_\alpha^4 - \frac{a_1^2 a_2^2 a_3^2}{a_\alpha^2}\right)\right]^{1/2} + \sum_{\alpha=1}^{3} a_\alpha^2 \right\}. \tag{8.22}$$

The corresponding to Eq. (8.21) resonant permittivity values have the form:

$$\varepsilon_2^1 = 1 + \left(\frac{a_1 a_2 a_3}{2} \sum_{\alpha=1}^{3} \frac{a_\alpha^2 I_\alpha}{\Lambda' - a_\alpha^2}\right)^{-1},$$

$$\varepsilon_2^2 = 1 + \left(\frac{a_1 a_2 a_3}{2} \sum_{\alpha=1}^{3} \frac{a_\alpha^2 I_\alpha}{\Lambda'' - a_\alpha^2}\right)^{-1}, \tag{8.23}$$

where I_α are defined above (see Eq. (8.18)).

Three other eigenfunctions with $n = 2$ have the form:

$$\phi_2^{2+\alpha} = \frac{x_1 x_2 x_3}{x_\alpha}, \quad \alpha = 1, 2, 3 \tag{8.24}$$

The corresponding resonance electric permittivities have the form:

$$\varepsilon_2^3 = 1 - \left(\frac{a_1 a_2 a_3}{2}\left(a_1^2 + a_2^2\right) I_{12}\right)^{-1},$$
$$\varepsilon_2^4 = 1 - \left(\frac{a_1 a_2 a_3}{2}\left(a_2^2 + a_3^2\right) I_{23}\right)^{-1},$$
$$\varepsilon_2^5 = 1 - \left(\frac{a_1 a_2 a_3}{2}\left(a_3^2 + a_1^2\right) I_{31}\right)^{-1}, \qquad (8.25)$$

where the integrals $I_{\alpha\beta}$ are defined by the relations:

$$I_{\alpha\beta} = \int_0^\infty \frac{du}{\left(u + a_\alpha^2\right)\left(u + a_\beta^2\right) R\left(u\right)}, \; R\left(u\right) = \prod_{\alpha=1}^{3}\left(u + a_\alpha^2\right)^{1/2} \qquad (8.26)$$

Using the eigenfunctions of the internal problem Eqs. (8.21) and (8.24), with the help of Eq. (8.9) we obtain the following expressions for eigenfunctions outside the ellipsoid:

$$g_2^1\left(\mathbf{r}\right) = \frac{\sum\limits_{\beta=1}^{3} \frac{a_\beta^2 I_\beta(u_0)}{\Lambda' - a_\beta^2}\left(\sum\limits_{\alpha=1}^{3} \frac{x_\alpha^2}{\Lambda' - a_\alpha^2} + 1\right) + \frac{2u_0}{R(u_0)} \sum\limits_{\alpha=1}^{3} \frac{x_\alpha^2}{\left(\Lambda' - a_\alpha^2\right)\left(u_0 + a_\alpha^2\right)}}{\sum\limits_{\beta=1}^{3} \frac{a_\beta^2 I_\beta}{\Lambda' - a_\beta^2}} \qquad (8.27)$$

$$g_2^2\left(\mathbf{r}\right) = \frac{\sum\limits_{\beta=1}^{3} \frac{a_\beta^2 I_\beta(u_0)}{\Lambda'' - a_\beta^2}\left(\sum\limits_{\alpha=1}^{3} \frac{x_\alpha^2}{\Lambda'' - a_\alpha^2} + 1\right) + \frac{2u_0}{R(u_0)} \sum\limits_{\alpha=1}^{3} \frac{x_\alpha^2}{\left(\Lambda'' - a_\alpha^2\right)\left(u_0 + a_\alpha^2\right)}}{\sum\limits_{\beta=1}^{3} \frac{a_\beta^2 I_\beta}{\Lambda'' - a_\beta^2}} \qquad (8.28)$$

$$g_2^3\left(\mathbf{r}\right) = \frac{I_{23}\left(u_0\right)}{I_{23}} x_2 x_3 \qquad (8.29)$$

$$g_2^4\left(\mathbf{r}\right) = \frac{I_{31}\left(u_0\right)}{I_{31}} x_3 x_1 \qquad (8.30)$$

$$g_2^5\left(\mathbf{r}\right) = \frac{I_{12}\left(u_0\right)}{I_{12}} x_1 x_2 \qquad (8.31)$$

where the integrals $I_{\alpha\beta}\left(u_0\right)$ are defined by the relation:

$$I_{\alpha\beta}(u_0) = \int_{u_0}^{\infty} \frac{du}{\left(u + a_\alpha^2\right)\left(u + a_\beta^2\right) R(u)} \tag{8.32}$$

and u_0 is again the positive root of the equation $\sum_{\alpha=1}^{3} \frac{x_\alpha^2}{u+a_\alpha^2} = 1$.

The modes φ_2^m, $m = 1 \ldots 5$ have not a dipole but quadrupole moment. Therefore, these modes will not be excited by a homogeneous external field, and sometimes they are called "dark" modes. The quadrupole modes φ_2^m, $m = 1 \ldots 5$ are sensitive to the field gradients, and therefore they can be excited even in regions where electric fields, but not their gradients, are equal to zero. Such complicated fields are frequent in nano-optics and singular optics.

Plasmon oscillations with the potentials φ_n^m, $n = 3$, $m = 1 \ldots 7$ have even more complicated properties. Six of the eigenfunctions have the form:

$$\varphi_3^{2\alpha-1} = x_\alpha \left(\sum_{\gamma=1}^{3} \frac{x_\gamma^2}{\Lambda'_\alpha - a_\gamma^2} + 1 \right),$$

$$\varphi_3^{2\alpha} = x_\alpha \left(\sum_{\gamma=1}^{3} \frac{x_\gamma^2}{\Lambda''_\alpha - a_\gamma^2} + 1 \right), \alpha = 1 \ldots 3, \tag{8.33}$$

where Λ'_α and Λ''_α have the form:

$$\left.\begin{array}{l} \Lambda'_\alpha \\ \Lambda''_\alpha \end{array}\right\} = \frac{1}{5} \left\{ \pm \left[4 \sum_{\beta=1}^{3} a_\beta^4 - 3a_\alpha^4 - (a_1 a_2 a_3)^2 \left(\sum_{\beta=1}^{3} \frac{1}{a_\beta^2} + \frac{6}{a_\alpha^2} \right) \right]^{1/2} \right.$$

$$\left. +2 \sum_{\beta=1}^{3} a_\beta^2 - a_\alpha^2 \right\} \tag{8.34}$$

The corresponding eigenvalues may also be written in analytical form:

$$\varepsilon_3^{2\alpha-1} = 1 + \left\{ \frac{a_1 a_2 a_3}{4} \left(\Lambda'_\alpha + 2a_\alpha^2 \right) \sum_{\beta=1}^{3} \frac{\left(1 + 2\delta_{\alpha\beta}\right) a_\beta^2 I_{\alpha\beta}}{\Lambda'_\alpha - a_\beta^2} \right\}^{-1},$$

$$\varepsilon_3^{2\alpha} = 1 + \left\{ \frac{a_1 a_2 a_3}{4} \left(\Lambda''_\alpha + 2a_\alpha^2 \right) \sum_{\beta=1}^{3} \frac{\left(1 + 2\delta_{\alpha\beta}\right) a_\beta^2 I_{\alpha\beta}}{\Lambda''_\alpha - a_\beta^2} \right\}^{-1}, \tag{8.35}$$

where $I_{\alpha\beta}$ is defined above (see Eq. (8.26)).

The seventh eigenfunction has a rather simple form:

$$\phi_3^7 = x_1 x_2 x_3 \tag{8.36}$$

The corresponding eigenvalue has the form:

$$\varepsilon_3^7 = 1 - \left\{ \frac{(a_1 a_2 a_3)^3}{2} \left(\sum_{\alpha=1}^{3} a_\alpha^{-2} \right) I_{123} \right\}^{-1} \tag{8.37}$$

where

$$I_{123} = \int_0^\infty \frac{du}{(R(u))^3} \tag{8.38}$$

Plasmon potentials outside the particle are obtained from φ_3^m, $m = 1 \ldots 7$ with the help of Eq. (8.9) and have the form ($\alpha = 1,2,3$):

$$g_3^{2\alpha-1}(\mathbf{r}) = x_\alpha \left\{ \sum_{\gamma=1}^{3} \frac{(1+2\delta_{\alpha\gamma})\, a_\gamma^2 I_{\alpha\gamma}(u_0)}{\Lambda'_\alpha - a_\gamma^2} \left(\sum_{\beta=1}^{3} \frac{x_\beta^2}{\Lambda'_\alpha - a_\beta^2} + 1 \right) \right.$$
$$\left. + \frac{2u_0}{R(u_0)\left(\Lambda'_\alpha + 2a_\alpha^2\right)\left(u_0 + a_\alpha^2\right)} \left(\sum_{\beta=1}^{3} \frac{\left(2a_\alpha^2 + a_\beta^2\right) x_\beta^2}{\left(\Lambda'_\alpha - a_\beta^2\right)\left(u_0 + a_\beta^2\right)} + 1 \right) \right\}$$
$$\times \left\{ \sum_{\gamma=1}^{3} \frac{(1+2\delta_{\alpha\gamma})\, a_\gamma^2 I_{\alpha\gamma}}{\Lambda'_\alpha - a_\gamma^2} \right\}^{-1}, \tag{8.39}$$

$$g_3^{2\alpha}(\mathbf{r}) = x_\alpha \left\{ \sum_{\gamma=1}^{3} \frac{(1+2\delta_{\alpha\gamma})\, a_\gamma^2 I_{\alpha\gamma}(u_0)}{\Lambda''_\alpha - a_\gamma^2} \left(\sum_{\beta=1}^{3} \frac{x_\beta^2}{\Lambda''_\alpha - a_\beta^2} + 1 \right) \right.$$
$$\left. + \frac{2u_0}{R(u_0)\left(\Lambda''_\alpha + 2a_\alpha^2\right)\left(u_0 + a_\alpha^2\right)} \left(\sum_{\beta=1}^{3} \frac{\left(2a_\alpha^2 + a_\beta^2\right) x_\beta^2}{\left(\Lambda''_\alpha - a_\beta^2\right)\left(u_0 + a_\beta^2\right)} + 1 \right) \right\}$$
$$\times \left\{ \sum_{\gamma=1}^{3} \frac{(1+2\delta_{\alpha\gamma})\, a_\gamma^2 I_{\alpha\gamma}}{\Lambda''_\alpha - a_\gamma^2} \right\}^{-1}, \tag{8.40}$$

$$g_3^7(\mathbf{r}) = \frac{I_{123}(u_0)}{I_{123}} x_1 x_2 x_3, \tag{8.41}$$

where

$$I_{\alpha\beta\gamma}\left(u_0\right)=\int\limits_{u_0}^{\infty}\frac{du}{\left(u+a_\alpha^2\right)\left(u+a_\beta^2\right)\left(u+a_\gamma^2\right)R\left(u\right)} \tag{8.42}$$

and u_0 is the positive root of the equation $\sum\limits_{\alpha=1}^{3}\frac{x_\alpha^2}{u+a_\alpha^2}=1$.

The modes φ_3^m, $m = 1\ldots 7$ also have no dipole moment, and therefore they can be excited only by fields strongly nonhomogeneous at nanoscale.

Eigenfunctions of a higher order ($n > 3$) and corresponding eigenvalues of electric permittivity can be found quite analogously. One may state that it is possible to find analytical expressions for the eigenfunctions and eigenvalues right up to $n = 10$, for which the Lame functions are known (Maier, 2008). For $n = 4$, explicit expressions for the resonance electric permittivities and the corresponding plasmon modes' potentials were calculated by (Guzatov et al., 2010).

In Fig. 8.2, the surface charge distribution (in relative units) of a three-axial ellipsoid's different modes is shown in a false color. The ellipsoid's semiaxes ratio is $a_2/a_1 = 0.6$ and $a_3/a_1 = 0.4$. The warm colors correspond to the positive charge, and the cold colors correspond to the negative charge. As seen from Fig. 8.2, as the index of the mode n increases, the number of maxima and minima of the surface charges increases (compare Fig. 8.2a–c) and corresponds to the transition from the "dipole" distribution (Fig. 8.2a) to "quadrupole" (Fig. 8.2b) and then to "octupole" (Fig. 8.2c). It should also be mentioned that as the ellipsoid topology changes, for example, to a sphere ($a_1 = a_2 = a_3$), the charge distribution corresponds completely to the charge distribution on the sphere that can be obtained with the help of spherical harmonics of the corresponding degree and order.

In Fig. 8.3, the eigenvalues of the three-axial ellipsoid electric permittivity are shown as the function of one of the semiaxes ratio a_3/a_1, while the other ratio is fixed, $a_2/a_1 = 0.6$. As seen, the solutions obtained have a nontrivial character. There are nonmonotonic functions with maxima and minima as can be seen from Fig. 8.3b ($n = 2$) and Fig. 8.3c ($n = 3$). For some values a_3/a_1 (a_2/a_1 is fixed), the eigenvalues corresponding to different ε_n^m can

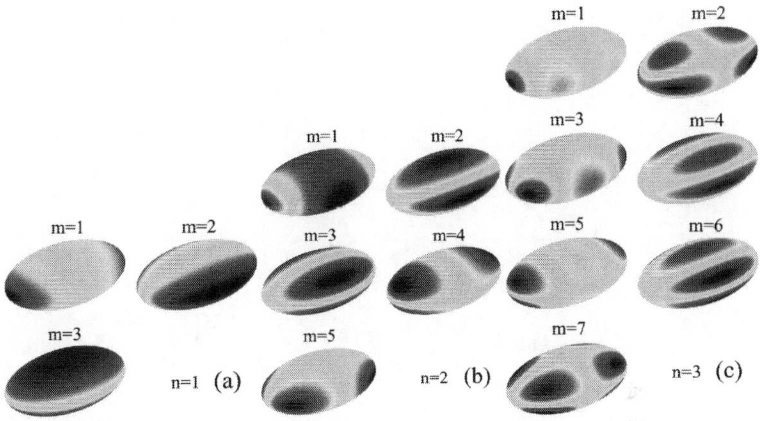

Figure 8.2 The surface charge distribution (arbitrary units) of different plasmon modes of a three-axial ellipsoid with the semiaxes ratio $a_2/a_1 = 0.6$ and $a_3/a_1 = 0.4$. The warm and cold colors correspond to the positive and negative signs of the charge, respectively. (a) "Dipole" modes ($n = 1$), (b) "quadrupole" modes ($n = 2$), and (c) "octupole" modes ($n = 3$).

coincide with each other. As it is well seen from Fig. 8.3 in the point $a_3/a_1 = 0.6$ ($a_2 = a_3 < a_1$), the following coincidences take place:

$$\varepsilon_1^2 = \varepsilon_1^3, \varepsilon_2^2 = \varepsilon_2^4, \ \varepsilon_2^3 = \varepsilon_2^5 \text{ and } \varepsilon_3^2 = \varepsilon_3^7, \ \varepsilon_3^3 = \varepsilon_3^5, \ \varepsilon_3^4 = \varepsilon_3^6$$

while in the point $a_3/a_1 = 1$ ($a_1 = a_3 > a_2$) we have:

$$\varepsilon_1^1 = \varepsilon_1^3, \quad \varepsilon_2^1 = \varepsilon_2^5, \ \varepsilon_2^3 = \varepsilon_2^4 \text{ and } \varepsilon_3^1 = \varepsilon_3^5, \ \varepsilon_3^2 = \varepsilon_3^6, \ \varepsilon_3^3 = \varepsilon_3^7.$$

From this figure it is also seen that the resonance electric permittivity of a three-axial ellipsoid takes all values in the region $[0, -\infty]$. This fact allows one to choose such a shape and a material of an ellipsoid that provide an effective resonance interaction with an arbitrary molecule within both its absorption and its emission bands. This property can be used for the construction of bright, artificial fluorophores on the basis of a three-axial nanoellipsoid or more complicated nanoparticles (see Chapter 13, Section 13.6).

Thus, in this section explicit expressions for the plasmon modes of ellipsoids of the "dipole," "quadrupole," and "octupole" types were presented. With the help of these expressions, many effects having practical significance may be described. In particular, knowing the electric permittivities and using one or another law of ellipsoid

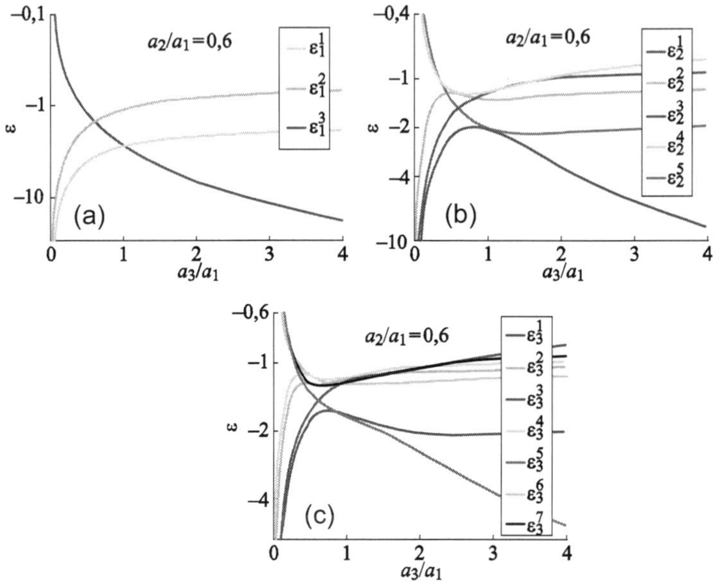

Figure 8.3 The eigenvalues of the three-axial ellipsoid electric permittivity, depending on the semiaxes ratio a_3/a_1 ($a_2/a_1 = 0.6$). (a) $n = 1$, (b) $n = 2$, and (c) $n = 3$.

substance dispersion, explicit expressions for plasmon resonance frequencies can be found. The present approaches may be also used for finding of plasmon modes in layered ellipsoidal particles.

8.3 Plasmon Resonances in an Ellipsoid of a Finite Size (Taking the Effect of Retardation into Account)

In previous sections, properties of plasmon oscillations in a nano-ellipsoid were considered in the quasi-static approximation, that is, neglecting the effects of retardation during light propagation along the particle. Size-dependent corrections to quasi-static resonance permittivity values have a great significance since they allow determining of both the position of the resonance peak and its width more precisely. For a three-axial nanoellipsoid, some first wavenumber corrections to the resonance electric permittivity can be found. These corrections will be stated below.

As shown in Chapter 5, the solution to the vector problem of electromagnetic field diffraction on a body of an arbitrary size with the electric permittivity ε can be presented in the form:

$$\mathbf{E} = \mathbf{E}^i + \sum_p A_p \mathbf{e}_p, \tag{8.43}$$

where $\mathbf{E}^i$ is the source electric field and $\mathbf{e}_p$ is the electric field of the pth plasmon mode. The coefficients A_p have the form:

$$A_p = \left(\frac{1-\varepsilon}{\varepsilon - \varepsilon_p}\right) \frac{\int_V d\mathbf{r}\, (\mathbf{E}^i \mathbf{e}_p)}{\int_V d\mathbf{r}\, (\mathbf{e}_p)^2} = \left(\frac{1-\varepsilon}{\varepsilon - \varepsilon_p}\right) B_p, \tag{8.44}$$

where ε_p is the resonance electric permittivity of the plasmon mode. In Eq. (8.44), the integral is taken over the particle volume V.

In the case of nanobodies, that is, in the case when the body size is much less than the typical wavelength, all the values entering Eqs. (8.43) and (8.44) may be expanded into the series of the powers $k_0 = \omega/c \to 0$, that is, it can be written as:

$$\begin{aligned}
\mathbf{E} &= \sum_n (ik_0)^n \mathbf{E}_n, & \mathbf{E}^i &= \sum_n (ik_0)^n \mathbf{E}^i_n, \\
\mathbf{e}_p &= \sum_n (ik_0)^n \mathbf{e}_{p,n}, & A_p &= \sum_n (ik_0)^n A_{p,n}, \\
B_p &= \sum_n (ik_0)^n B_{p,n}, & \varepsilon_p &= \sum_n (ik_0)^n \varepsilon_{p,n}.
\end{aligned} \tag{8.45}$$

Substituting Eq. (8.45) in Eq. (8.43) and equating the coefficients with the same powers k_0^n on the left and on the right, we obtain an infinite number of equations of the type:

$$\begin{aligned}
\mathbf{E}_0 &= \mathbf{E}^i_0 + \sum_p A_{p,0}\mathbf{e}_{p,0}, \\
\mathbf{E}_1 &= \mathbf{E}^i_1 + \sum_p \left(A_{p,0}\mathbf{e}_{p,1} + A_{p,1}\mathbf{e}_{p,0}\right), \\
\mathbf{E}_2 &= \mathbf{E}^i_2 + \sum_p \left(A_{p,0}\mathbf{e}_{p,2} + A_{p,1}\mathbf{e}_{p,1} + A_{p,2}\mathbf{e}_{p,0}\right),
\end{aligned} \tag{8.46}$$

where

$$A_{p,0} = (1-\varepsilon)\frac{B_{p,0}}{\varepsilon - \varepsilon_{p,0}},$$
$$A_{p,1} = (1-\varepsilon)\left(\frac{B_{p,1}}{\varepsilon - \varepsilon_{p,0}} + \frac{\varepsilon_{p,1}B_{p,0}}{\left(\varepsilon - \varepsilon_{p,0}\right)^2}\right),$$
$$A_{p,2} = (1-\varepsilon)\left(\frac{B_{p,2}}{\varepsilon - \varepsilon_{p,0}} + \frac{\varepsilon_{p,1}B_{p,1}}{\left(\varepsilon - \varepsilon_{p,0}\right)^2} + \frac{\varepsilon_{p,1}^2 B_{p,0}}{\left(\varepsilon - \varepsilon_{p,0}\right)^3} + \frac{\varepsilon_{p,2}B_{p,0}}{\left(\varepsilon - \varepsilon_{p,0}\right)^2}\right). \tag{8.47}$$

In turn, the $B_{p,n}$ coefficients can be expressed through $\mathbf{e}_{p,n}$ and $\mathbf{E}_n^i$. As a result, we have a system of equations that can be solved successively if a zero-order quasi-static solution is known.

The best-known solution to the problem of plane electromagnetic wave diffraction on a three-axial ellipsoid in long wave approximation was obtained by Stevenson (1953a, 1953b). He found the first three terms ($n = 0, 1, 2$) of the scattered field expansion over the wavenumber $k_0 \to 0$. Comparing Stevenson's solution with Eqs. (8.46) and (8.47), the corrections to the resonance electric permittivity ε_1^α, $\alpha = 1, 2, 3$ (up to k_0^3) can be found as:

$$\varepsilon_1^m = \varepsilon_{1,0}^m + (ik_0)\,\varepsilon_{1,1}^m + (ik_0)^2\,\varepsilon_{1,2}^m + (ik_0)^3\,\varepsilon_{1,3}^m \ldots, \tag{8.48}$$

where $\varepsilon_{1,0}^\alpha$ is the expression for the resonance electric permittivity $\varepsilon_{1.0}^\alpha = 1 - \left(\frac{a_1a_2a_3}{2}I_\alpha\right)^{-1}$ found earlier (see Eq. (8.17)). The k_0a corrections of the first order are equal to zero due to the ellipsoid symmetry, $\varepsilon_{1,1}^\alpha = 0$. And, finally, the corrections to the resonance electric permittivity of the second order have the form:

$$\varepsilon_{1,2}^\alpha = \frac{a_1a_2a_3}{4}\left(1-\varepsilon_{1,0}^\alpha\right)^2\left\{-\frac{1}{5}a_\alpha^2\left(-I_\alpha + a_\alpha^2 I_{\alpha\alpha}\right)\right.$$
$$\left. + \frac{1}{2}\left(I - \frac{1}{5}\sum_\beta a_\beta^2 I_\beta + a_\alpha^2\left(I_\alpha - \frac{1}{5}\sum_\beta a_\beta^2 I_{\alpha\beta}\right)\right)\right\}, \tag{8.49}$$

where $\alpha, \beta = 1, 2, 3$ and the integrals I_α and $I_{\alpha\beta}$ are determined above (see Eqs. (8.18) and (8.32)) and $I = \int\limits_0^\infty du \prod\limits_\alpha \left(u + a_\alpha^2\right)^{-1/2}$.

In the case of a sphere ($a_1 = a_2 = a_3$), the corrections in Eq. (8.49) agree with results of the expansion of the Mie solution over powers of k (see Eq. (6.33)). From Eq. (8.49), the corrections of the second order may be obtained also in the case of spheroids.

The correction of the third order is universal and can be calculated for any particle (see Chapter 5, Eq. (5.32)). In the case of ellipsoids, the correction of the third order takes the form:

$$\varepsilon_{1,3}^{\alpha} = \frac{2}{9} a_1 a_2 a_3 \left(\varepsilon_{1,0}^{\alpha} - 1\right)^2. \tag{8.50}$$

Thus, the expressions in Eqs. (8.48), (8.49), and (8.50) allow one to find the resonance values of the ellipsoid electric permittivity with an accuracy to the third order of $k_0 a$, inclusive. Physically, the quadratic correction to the resonance electric permittivity is a real value, resulting in a frequency shift of the dipole plasmon resonances. In turn, the cubic corrections are related to the emission and take an imaginary value, resulting in plasmon line broadening.

In Fig. 8.4, the resonance electric permittivity of a three-axial ellipsoid is shown, taking into account the quadratic correction Eq. (8.40) as a function of the semiaxes ratio a_1/a_3 for the fixed value $a_2/a_3 = 0.6$.

As seen, taking into account the quadratic correction (in the picture $k_0 a_3 = 0.1$) moves the dependency down to the lower values of the resonance electric permittivity. At that, the value of the resonance shift is always negative, and the corresponding red shift of the plasmon frequency grows as the ellipsoid volume increases. The corrections stated above were obtained under the assumption that the particle is in the medium with $\varepsilon_{\mathrm{H}} = 1$ (vacuum). If the electric permittivity of the environment $\varepsilon_{\mathrm{H}} \neq 1$, then in Eqs. (8.48), (8.49), and (8.50) $\varepsilon_{1,q}^{\alpha}$ should be replaced by $\varepsilon_{1,q}^{\alpha}/\varepsilon_{\mathrm{H}}$. Moreover, the environment wavenumber $k_{\mathrm{H}} = \sqrt{\varepsilon_{\mathrm{H}}}\omega/c$ is to be used instead of $k_0 = \omega/c$.

8.4 Optical Properties of a Nanoellipsoid in a Homogeneous External Field

Knowing the eigenfunctions and eigenvalues, one can also find solutions to many problems of practical interest. In particular, if an ellipsoid with the electric permittivity ε and the semiaxes a_1, a_2, a_3 is in a homogeneous external field, only plasmons φ_1^m, $m = 1 \ldots 3$

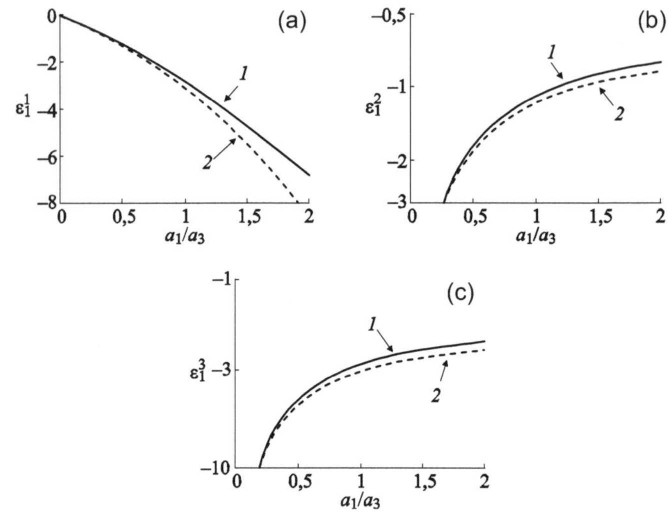

Figure 8.4 The resonance dielectric permittivity of a three-axial ellipsoid with the retardation taken into account as a function of the semiaxes ratio a_1/a_3 ($a_2/a_3 = 0.6$). (a) ε_1^1, $\varphi_1^1 \sim x_1$, (b) ε_1^2, $\varphi_1^2 \sim x_2$, and (c) ε_1^3, $\varphi_1^3 \sim x_3$. The solid lines (1) correspond to the quasi-static approximation, while the dashed lines (2) take the quadratic corrections in Eq. (8.49) into account. The absolute value of the third semiaxis length was determined by $k_0 a_3 = 0.1$, that is, $a_3 \approx \lambda/60$.

can be excited, which have a linear potential with respect to the coordinates inside the ellipsoid and, correspondingly, the homogeneous electric field.

At greater distances from the ellipsoid, its field is the field of the dipole arising as a result of ellipsoid polarization. In this case, the polarization tensor is not isotropic and may be written as:

$$\overleftrightarrow{\Pi} = \begin{bmatrix} \Pi_1 & & \\ & \Pi_2 & \\ & & \Pi_3 \end{bmatrix}, \tag{8.51}$$

where, according to Eq. (5.21), the polarizabilities along the axes x_1, x_2, $x_3 (x, y, z)$ have the form:

$$\Pi_i = \frac{a_1 a_2 a_3}{3} \frac{\left(1 - \varepsilon_1^i/\varepsilon_{\mathrm{H}}\right)\left(\varepsilon(\omega) - \varepsilon_{\mathrm{H}}\right)}{\varepsilon(\omega) - \varepsilon_1^i} = \frac{a_1 a_2 a_3}{3} \frac{\varepsilon(\omega) - \varepsilon_{\mathrm{H}}}{\varepsilon_{\mathrm{H}} + L_i\left(\varepsilon(\omega) - \varepsilon_{\mathrm{H}}\right)},$$

$$L_i = \frac{a_1 a_2 a_3}{2} I_i, \tag{8.52}$$

where the integral I_i is determined in Eq. (8.18) and ε_{H} is the electric permittivity of the environment.

The geometric factor L_i of the ellipsoid is related to the electric permittivity values by the formula:

$$L_i = \frac{1}{1 - \varepsilon_1^i/\varepsilon_{\mathrm{H}}}. \tag{8.53}$$

The geometric factor L is related to the depolarization coefficient $\bar{L}$, which is determined by how much the external field inside the spheroid is being reduced by the ellipsoid polarization:

$$E_i = E_i^0 - 4\pi \bar{L}_i P_i, \tag{8.54}$$

where E_i, E_i^0 are the fields inside the spheroid and in the absence of the ellipsoid, correspondingly, and $\boldsymbol{P}$ is the polarization vector. The depolarization tensor makes sense only for the particles since for the cavities (bubbles) the polarization P is equal to zero in them. The correlation connecting the geometric factor and the depolarization tensor has the form:

$$\bar{L}_i = \frac{L_i}{\varepsilon_{\mathrm{H}}} \frac{\varepsilon - \varepsilon_{\mathrm{H}}}{\varepsilon - 1}. \tag{8.55}$$

Only when the particle is in free space, the geometric factor becomes equal to the depolarization coefficient.

Eq. (8.52) is obtained in the quasi-static approximation. However, if the corrections taking into account the retardation effects (see Eqs. (8.49) and (8.50)) are used as the resonance electric permittivity in Eq. (8.55), a much more precise expression for polarizability can be found.

The polarizability tensor can be also found for layered ellipsoids having a core with the semiaxes a_1, a_2, a_3 and electric permittivity ε_{C} and a confocal shell with the semiaxes a'_1, a'_2, a'_3 and electric permittivity ε. The confocality of the shells implies that the values of their semiaxes are related by the ratios:

$$a'^2_1 - a_1^2 = a'^2_2 - a_2^2 = a'^2_3 - a_3^2 = t > 0. \tag{8.56}$$

If a layered particle of this kind is in a medium with the permittivity ε_{H}, the components of the polarization tensor have the form:

$$\Pi_i = \frac{a'_1 a'_2 a'_3}{3} \times \frac{(\varepsilon - \varepsilon_{\mathrm{H}})\left[\varepsilon + (\varepsilon_{\mathrm{C}} - \varepsilon)(L_i - f L'_i)\right] + f\varepsilon(\varepsilon_{\mathrm{C}} - \varepsilon)}{\left[\varepsilon + (\varepsilon_{\mathrm{C}} - \varepsilon)(L_i - f L'_i)\right]\left[\varepsilon_{\mathrm{H}} + (\varepsilon - \varepsilon_{\mathrm{H}}) L'_i\right] + f L'_i \varepsilon (\varepsilon_{\mathrm{C}} - \varepsilon)}, \tag{8.57}$$

where $f = a_1a_2a_3/a'_1a'_2a'_3$ is the ratio of the core volume to the total particle volume and L_i, L'_i are geometric factors of the core and the external surface of the nanoparticle. Naturally, for $f = 0$ (there is no core) Eq. (8.57) turns into Eq. (8.52).

As usual, the denominator vanishing in Eq. (8.57) corresponds to the localized plasmon oscillation excitation. If the denominator in Eq. (8.57) is considered a function of the core electric permittivity $\varepsilon_{\rm C}$, then the properties of the corresponding resonance values of the electric permittivity and plasmon frequencies will be qualitatively similar to the properties of a metal ellipsoid without a shell.

If the denominator in Eq. (8.57) is treated as a function of ε, then for each plasmon polarization there will be two modes, one of which is similar to the mode of an ellipsoidal cavity in metal $\left(\varepsilon_i^{\rm cavity} = \varepsilon_{\rm C}/\left(1 - 1/L_i\right)\right)$ and localized on the core surface, and another is similar to the plasmon mode of a solid ellipsoid $\left(\varepsilon_i^{\rm ellipsoid} = \varepsilon_{\rm H}\left(1 - 1/L'_i\right)\right)$ and localized on the external surface of the particle. In all cases, the electric field inside the core is homogeneous.

Knowing the tensor of ellipsoidal nanoparticles polarizability from the general formulas (see Chapter 5), the scattering and absorption cross section can be found:

$$\sigma^{\rm abs} = 4\pi k \,{\rm Im} \sum_{ij} \mathbf{e}_i^* \Pi_{ij} \mathbf{e}_j,$$
$$\sigma^{\rm sc} = 8\pi k^4 \sum_{ijk} \mathbf{e}_i^* \Pi_{ji}^* \Pi_{jk} \mathbf{e}_k,$$
$$i, j, k = 1 \ldots 3, \tag{8.58}$$

where $k = \sqrt{\varepsilon_{\rm H}}\omega/c$ is the wavevector in the medium surrounding the nanoparticle and $\mathbf{e}_k$ is the unit polarization vector.

In Eq. (8.58), both the polarizability tensor of the particle and the light polarization vector are related to the same system of coordinates. Practically, it is convenient to determine the polarization vectors in the laboratory system of coordinates, and the polarizability tensor in the system of coordinates related to its main axes. Let the matrix R describe the transition from the system of coordinates related to the ellipsoid axes (x, y, z) to the laboratory system of coordinates (x', y', z'), where the z' axis is the direction of the incident light propagation:

$$R_{ij} = \mathbf{e}_i \cdot \mathbf{e}'_j. \tag{8.59}$$

In Eq. (8.59), $\mathbf{e}_i$, $\mathbf{e}'_j$ are the unit vectors along the coordinate axes in the ellipsoid system and in the laboratory systems of coordinates, respectively. Then the absorption and scattering cross sections of the x'-polarized light on the ellipsoid will have the form (Bohren and Huffman, 1983):

$$\begin{aligned} \sigma_x^{\text{abs}} &= 4\pi k_{\text{H}} \text{Im}\left(\Pi_x R_{11}^2 + \Pi_y R_{21}^2 + \Pi_z R_{31}^2\right), \\ \sigma_x^{\text{sc}} &= \frac{8\pi}{3} k_{\text{H}}^4 \left(|\Pi_x|^2 R_{11}^2 + \left|\Pi_y\right|^2 R_{21}^2 + |\Pi_z|^2 R_{31}^2\right), \end{aligned} \tag{8.60}$$

where $k_{\text{H}} = \sqrt{\varepsilon_{\text{H}}}\omega/c$ is the wavevector in the medium surrounding the nanoparticle and the polarizabilities Π_x, Π_y, Π_z are determined by the expressions in Eq. (8.52) for a homogeneous ellipsoid and the expressions in Eq. (8.57) for a layered one. Analogous expressions are also true for the other (y') polarization.

In a case of chaotically oriented ellipsoidal particles, averaged values of the square direction cosines are equal:

$$\left\langle R_{ij}^2 \right\rangle = 1/3, \tag{8.61}$$

and for any polarization of the incident light the absorption and scattering cross sections have the form:

$$\begin{aligned} \sigma_x^{\text{abs}} &= \frac{4\pi k_{\text{H}}}{3} \text{Im}\left(\Pi_x + \Pi_y + \Pi_z\right), \\ \sigma_x^{\text{sc}} &= \frac{8\pi}{9} k_{\text{H}}^4 \left(|\Pi_x|^2 + \left|\Pi_y\right|^2 + |\Pi_z|^2\right). \end{aligned} \tag{8.62}$$

Polarizabilities in Eqs. (8.52) and (8.57) and, correspondingly, cross sections in Eqs. (8.60) and (8.62) have maxima when at some frequencies (the plasmon resonance frequencies) the electric permittivity of the ellipsoid becomes close to the resonance electric permittivity $\varepsilon = \varepsilon' + i\varepsilon'' \approx \varepsilon_n^m$. At that, the structure of the electric field in the vicinity of the nanoparticle will be determined by the electric field of the corresponding plasmon mode and will not depend on properties of the nanoparticle material.

8.5 The Influence of a Metal Nanoellipsoid on the Spontaneous Emission of an Atom

In the previous section, the optical properties of an ellipsoid placed in a spatially homogeneous field were considered. This task formulation is crucially important in many applications. However, now situations when the optical fields are changing at distances that are much smaller than the wavelength have greater importance. In particular, fields of this kind arise when considering the problem of molecule emission in the vicinity of an ellipsoidal nanoparticle, which is very important practically. The general solution to the problem of an ellipsoid in an arbitrary external field has the form (Chapter 5):

$$\mathbf{E}(\mathbf{r}) = \mathbf{E}^i(\mathbf{r}) - (\varepsilon - 1)\sum_{n=1}^{\infty} \frac{\mathbf{e}_n(\mathbf{r})}{\varepsilon - \varepsilon_n} \frac{\int_V d\mathbf{r}'\left(\mathbf{E}^i(\mathbf{r}')\,\mathbf{e}_n(\mathbf{r}')\right)}{\int_V d\mathbf{r}\,(\mathbf{e}_n(\mathbf{r}'))^2}, \tag{8.63}$$

where $\mathbf{E}^i(\mathbf{r})$ is the nonhomogeneous exciting field and $\mathbf{e}_n(\mathbf{r}) = -\nabla\varphi_n(\mathbf{r})$ is the electric field of the plasmon modes, and for simplicity we consider the case when the nanoparticle is in vacuum.

In the case of the dipole radiation of an atom or a molecule placed at the point **r'**, the field incident on the particle has the form:

$$\mathbf{E}^i(\mathbf{r}) = \mathbf{E}^d(\mathbf{r}) = -\left(\mathbf{d}_0\nabla'\right)\nabla\frac{1}{|\mathbf{r} - \mathbf{r}'|}, \tag{8.64}$$

where $\mathbf{d}_0$ is the dipole moment of the transition from the excited state of an atom to its ground state and the monochromatic dependency on the time is not shown.

A remarkable feature of an ellipsoid is the fact that the dipole moment $\delta\mathbf{d}$, induced in it by the atom near field, Eq. (8.64), can be calculated in the explicit form (Guzatov and Klimov, 2005):

$$\delta\mathbf{d} = \overleftrightarrow{\Pi}\cdot\left\langle\mathbf{E}^d\right\rangle, \tag{8.65}$$

where the dipole field averaged over the ellipsoid volume has the form:

$$\left\langle\mathbf{E}^d\right\rangle = \frac{1}{V}\int_V d\mathbf{r}\mathbf{E}^d(\mathbf{r}) = \frac{1}{V}\left(\mathbf{d}_0\nabla'\right)\nabla'\int_V \frac{d\mathbf{r}}{|\mathbf{r} - \mathbf{r}'|}, \tag{8.66}$$

and $\overset{\leftrightarrow}{\Pi}$ is the polarizability tensor of the three-axial ellipsoid with the main axes directed along the Cartesian axes of coordinates (Eq. (8.52)).

Calculating the integral in Eq. (8.66), we obtain the explicit expressions for the induced dipole moment:

$$\delta\mathbf{d} = -\frac{3}{2}\sum_{\alpha=1}^{3} d_{0,\alpha}\Pi_{\alpha\alpha}I_{\alpha}\left(u'_0\right)\mathbf{n}_{\alpha} + \frac{3}{R\left(u'_0\right)}\sum_{\alpha=1}^{3}\frac{\Pi_{\alpha\alpha}x'_{\alpha}\mathbf{n}_{\alpha}}{u'_0+a_{\alpha}^2}\sum_{\beta=1}^{3}\frac{d_{0,\beta}x'_{\beta}}{u'_0+a_{\beta}^2}\left(\sum_{\gamma=1}^{3}\frac{x'^2_{\gamma}}{\left(u'_0+a_{\gamma}^2\right)^2}\right)^{-1}, \tag{8.67}$$

where $\mathbf{n}_{\alpha}$ are the unit vector along the axes of coordinates α, u'_0 is the positive root of the equation $\sum_{\alpha=1}^{3}\frac{x'^2_{\alpha}}{u'+a_{\alpha}^2} = 1$, x'_{α} are the coordinates of the atom position, and the components of the polarizability tensor $\Pi_{\alpha\alpha}$ are given by Eq. (8.52). Substituting Eq. (8.67) in the general expression for the radiative decay rate in the vicinity of nanobodies (see Appendix A1), we obtain the general expression for the radiative decay rate of an atom or a molecule placed near a three-axial nanoellipsoid:

$$\left(\frac{\gamma}{\gamma_0}\right)^{\text{radiative}} = \frac{|\mathbf{d}_{\text{total}}|^2}{|\mathbf{d}_0|^2}, \tag{8.68}$$

where $\mathbf{d}_{\text{total}} = \mathbf{d}_0 + \delta\mathbf{d}$ is the total dipole moment of the system "atom + ellipsoid" and γ_0 is the decay rate in the absence of a nanoparticle. In some particular cases of a spheroid (a needle, a disk, etc.) and a sphere, the expressions for the decay rates in these cases can be derived from Eqs. (8.67) and (8.68).

In a case of an atom placed on one of the ellipsoid's main axes, the expression for the decay rates can be simplified essentially. In a particular case of an atom having the dipole moment $\mathbf{d}_0 = d_0\mathbf{n}_{\alpha}$ and placed at the point x'_{β} of the x_{β}-axis (outside the ellipsoid), we obtain the following expression for the radiative decay rate $(\alpha, \beta = 1,\ 2,\ 3)$ (Guzatov and Klimov, 2005):

$$\frac{\gamma_{\text{rad}}}{\gamma_0} = \left|1 - 3\Pi_{\alpha\alpha}\left(\frac{1}{2}I_{\alpha}\left(x'^2_{\beta} - a_{\beta}^2\right) - \frac{\delta_{\beta\alpha}}{R\left(x'^2_{\alpha} - a_{\alpha}^2\right)}\right)\right|^2, \tag{8.69}$$

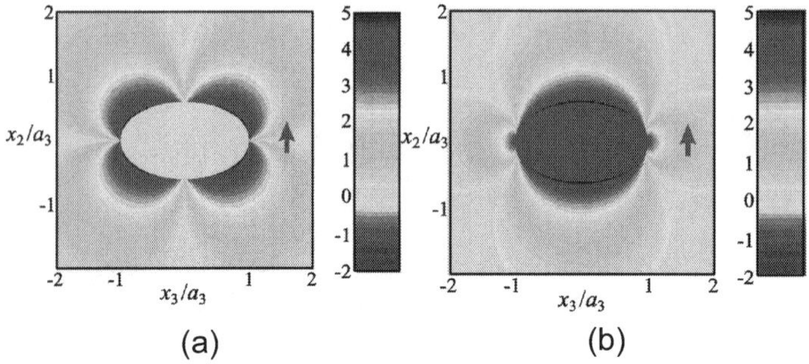

Figure 8.5 The distribution of the relative radiative decay rate of an atom placed in the vicinity of a silver ellipsoid ($\varepsilon = -15.37 + i0.231$, $\lambda = 632.8$ nm) as a function of its position in the plane $x_1 = 0$. (a) $a_2/a_3 = 0.6, a_1/a_3 = 0.105$ and (b) $a_2/a_3 = 0.6$, $a_1/a_3 = 0.046$. The atom dipole moment is oriented along the x_2-axis. The scale of the pseudocolor is logarithmic.

where $\delta_{\beta\alpha}$ is the Kronecker's delta symbol. If an atom having the dipole moment $\mathbf{d}_0 = d_0\mathbf{n}_\alpha$ is placed inside the ellipsoid, the decay rate does *not* depend on the atom position inside the ellipsoid and is defined by the expression (Guzatov and Klimov, 2005):

$$\frac{\gamma_{\text{rad}}}{\gamma_0} = \left|1 - \frac{3}{2}\Pi_{\alpha\alpha} I_\alpha\right|^2, \tag{8.70}$$

which depends on the atom dipole moment orientation only.

Let us note here that when the electric permittivity of the host medium is equal to ε_{H}, then according to the theory present in Appendix A1, it is necessary to make the replacements $\varepsilon \to \varepsilon/\varepsilon_{\text{H}}$ and $\gamma_0 \to \gamma_0\sqrt{\varepsilon_{\text{H}}}$.

In Figs. 8.5 and 8.6, spatial distributions of radiative decay rates of an atom placed near a silver nanoellipsoid are shown for two geometries. The geometry (a), when the ellipsoid has the aspect ratios $a_1/a_3 = 0.1048$ and $a_2/a_3 = 0.6$, corresponds to the plasmon resonance with the polarization along the x_1-axis, while the geometry (b), when the ellipsoid has the aspect ratios $a_1/a_3 = 0.0457$ and $a_2/a_3 = 0.6$, corresponds to the plasmon resonance with polarization along the x_2-axis.

In Fig. 8.5, the radiative decay rate as a function of the atom position in the plane $x_1 = 0$ in the vicinity of a silver nanoellipsoid ($\varepsilon = -15.37 + i0.231$ at $\lambda = 632.8$; (Shu and Hansma, 2001)) is

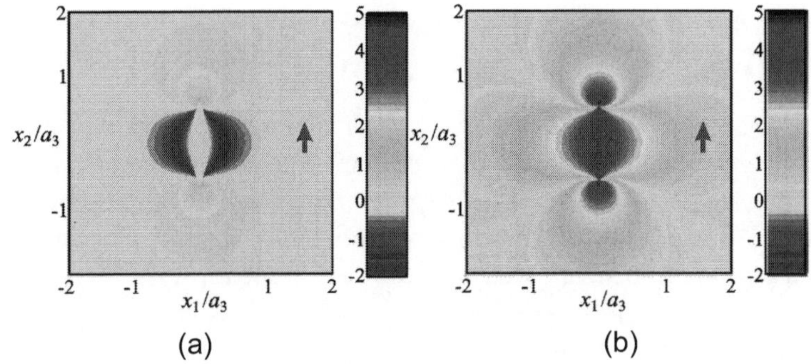

Figure 8.6 The distribution of the relative radiative decay rate of an atom placed in the vicinity of a silver ellipsoid ($\varepsilon = -15.37 + i0.231$, $\lambda = 632.8$ nm) as a function of its position in the plane $x_3 = 0$. The dipole moment is oriented along the x_2-axis. The scale of the pseudocolor is logarithmic.

shown. In Fig. 8.5a, four regions of the atom decay rate enhancement are seen. When the atom is situated in these regions, the plasmon oscillations with x_1-polarizations are excited. However, when the ellipsoid is compressed along the x_1-axis (Fig. 8.5b), the excitation of the other plasmon resonance (with polarization along the x_2-direction) occurs, resulting in a different spatial picture of the decay rate. Again, there are four regions of the enhancement of the radiative decay rate, but in contrast to Fig. 8.5a these regions have different geometry.

In Fig. 8.6, the radiative decay rate as a function of the atom position is shown for the same ellipsoids. However, now the atom is considered to be placed at different points of the plane $x_3 = 0$. As seen from Fig. 8.6a, there are regions of the decay rate inhibition situated symmetrically in the space to the left and to the right of the ellipsoid. However, if the ellipsoid is squeezed along the x_1-axis, a quite different situation is observed (Fig. 8.6b) and the decay enhancement regions appear instead of the regions of its inhibition. As in the case of Fig. 8.5, it is related to the change of the plasmon mode, which is in resonance with the atom emission.

Thus, the spontaneous decay radiative rate of an atom placed in the vicinity of a three-axial ellipsoid depends on the ellipsoid geometry, the atom position, and its dipole moment orientation in a complicated manner. This means that a three-axial metal ellipsoid can be used for effective optical field control at the nanoscale (see also Chapter 13, Section 13.6).

Another important characteristic of an ellipsoid is the nonradiative decay rate of an atom placed near a three-axial nanoellipsoid, which is related to the Joule loss rate of the electromagnetic energy inside the ellipsoid. According to Appendix A1, the nonradiative decay rate can be estimated with the help of the formula:

$$\left(\frac{\gamma}{\gamma_0}\right)^{\text{total}} = \frac{3}{2}\text{Im}\left(\frac{\mathbf{d}_0\delta\mathbf{E}(\mathbf{r}')}{k^3} + \ldots\right), \tag{8.71}$$

where $\mathbf{d}_0$ is the dipole moment value of the emitting atom and $\delta\mathbf{E}(\mathbf{r}')$ is the field reflected by the ellipsoid (the second term in the right part of Eq. (8.63). Substituting the expression for $\delta\mathbf{E}(\mathbf{r}')$ in Eq. (8.71), we obtain the expression for the nonradiative decay rate of an atom or a molecule placed in the vicinity of a metal nanoellipsoid:

$$\left(\frac{\gamma}{\gamma_0}\right)^{\text{nonrad}} = \frac{6\pi}{d_0^2 k_0^3}\sum_{n=1}^{\infty}\sum_{m=1}^{2n+1}\text{Im}\left(\frac{\varepsilon - 1}{\varepsilon - \varepsilon_n^m}\right)\frac{\left[(\mathbf{d}_0\nabla')\, g_n^m(\mathbf{r}')\right]^2}{\left(1 - \varepsilon_n^m\right) N_n^m}, \tag{8.72}$$

where $N_n^m = \int_V d\mathbf{r}\left(\mathbf{e}_n^m(\mathbf{r})\right)^2$ is the plasmon modes' normalization and $g_n^m(\mathbf{r}')$ is the plasmon modes' potentials outside the particle taken at the point of the atom location.

Unlike the radiative decay rate (which is related to the dipole plasmon modes only), the nonradiative decay rates depend on *all* plasmon modes in a nanoparticle, and when moving closer to the ellipsoid surface, the higher ellipsoidal harmonics become more and more significant for the calculations of the nonradiative decay rates. As a good approximation, Eq. (8.72) can be used, in which only the expressions for the plasmon modes with $n = 1,2,3$ (see Section 8.2) are taken into account. The approximation thus obtained, certainly, is not correct in the immediate vicinity of the ellipsoid surface. In this case the asymptotic may be used that is applicable at the distances on which the ellipsoid surface can be approximated by the plane:

$$\left(\frac{\gamma}{\gamma_0}\right)^{\text{nonrad}}_{\text{tang}} = \frac{3}{16\left(k_0\Delta'\right)^3}\text{Im}\frac{\varepsilon - 1}{\varepsilon + 1}; \quad \left(\frac{\gamma}{\gamma_0}\right)^{\text{nonrad}}_{\text{norm}} = 2\left(\frac{\gamma}{\gamma_0}\right)^{\text{nonrad}}_{\text{tang}} \tag{8.73}$$

where Δ' is the distance between the atom and the surface. In the case of an ellipsoid:

$$\Delta' = v'_0 \sqrt{\sum_{\alpha=1}^{3} \frac{x'^2_\alpha}{\left(v'_0 + a^2_\alpha\right)^2}}, \tag{8.74}$$

where v'_0 is the positive root of the equation:

$$\sum_{\alpha=1}^{3} \frac{x'^2_\alpha a^2_\alpha}{\left(v'_0 + a^2_\alpha\right)^2} = 1. \tag{8.75}$$

The indices "*tang*" and "*norm*" in Eq. (8.73) denote the tangential and normal orientations of the dipole moment relative to the ellipsoid surface.

References

Bohren, C.F., Huffman, D.R. (1983) Absorption and Scattering of Light by Small Particles (JohnWiley-Interscience, New York).

Guzatov, D.V., Klimov, V.V. (2005) Radiative decay engineering by triaxial nanoellipsoids, *Chem. Phys. Lett.*, **412**, 341–346.

Guzatov, D.V., Klimov, V.V., Pikhota M. Yu. (2010) Plasmonic oscillations in ellipsoid nanoparticles: beyond dipole approximation, *Laser Phys.*, **20**, 85–99.

Hobson, E.W. (1931) *Theory of Spherical and Ellipsoidal Harmonics* (Cambridge University Press, Cambridge).

Jones, R.C. (1945) A generalization of the dielectric ellipsoid problem, *Phys. Rev.*, **68**, 93–96.

Maier, R.S. (2008) Lame polynomials, hyperelliptic reductions and Lame band structure, *Philos. Trans. R. Soc. A: Math, Phys. and Engineering Sciences*, **366**, 1115.

Shu, Q.Q., Hansma, P.K. (2001) Fluorescent apparent quantum yields for excited molecules near dielectric interfaces, *Thin Solid Films*, **384**, 76–84.

Stevenson, A.F. (1953a) Solution of electromagnetic scattering problems as power series in the ratio (dimension of scatterer)/wavelength, *J. Appl. Phys.*, **24**, 1134–1142.

Stevenson, A.F. (1953b) Electromagnetic scattering by an ellipsoid in the third approximation, *J. Appl. Phys.*, **24**, 1143–1151.

Stratton, J.A. (1941) *Electromagnetic Theory* (McGraw-Hill, New York).

Whittaker, E.T., Watson, G.N. (1963) *A Course of Modern Analysis*, Vol. 2 (State Physics and Mathematics Publishing House).

Chapter 9

Localized Plasmons in Polyhedral Nanoparticles

In previous chapters, plasmon properties of nanoparticles were considered, having forms coinciding with coordinate surfaces of different systems of coordinates and describable analytically to one extent or another. Certainly, the diversity of nanoparticles is not restricted to these forms. Moreover, often nanoparticles have forms of a polyhedron, related to the crystal structure and other factors.

Recently, monocrystal polyhedral nanoparticles were synthesized with very small size dispersion (Tao et al., 2006). Images obtained with the help of an electron microscope show that the nanoparticles have clear-cut faces with sharp vertices and edges. Small nanoparticles (under 80 nm) usually have nanocube shapes. When particles grow in size (by increase of synthesis time), the formation of nanoparticles in the form of cubes with the truncated edges, cuboctahedrons, and, finally, octahedrons of 300 nm size takes place.

Thus, in the present chapter optical properties of polyhedron nanoparticles will be considered. At that, we will pay a special attention to regular polyhedrons (Platonic solids). Platonic solids have great significance in most diverse fields of human activity, from

Nanoplasmonics
Vasily Klimov

ISBN 978-981-4267-16-8 (Hardcover), 978-981-4267-42-7 (eBook)
www.panstanford.com

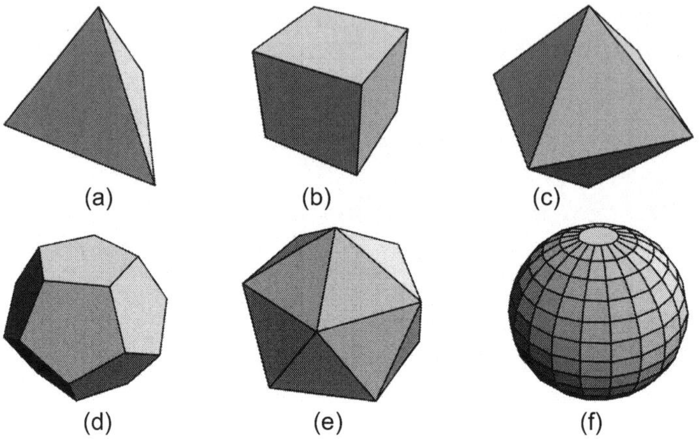

Figure 9.1 Platonic solids (regular polyhedra). (a) Tetrahedron, (b) cube (hexahedron), (c) octahedron, (d) dodecahedron, (e) icosahedron, and (f) sphere.

music to cosmology. These shapes are shown in Fig. 9.1, and some geometric parameters of Platonic solids are presented in Table 9.1 (a is the edge length). Despite a simple and symmetric shape such nanoparticles can be investigated with the help of numerical methods mainly.

9.1 Optical Properties of Dielectric Particles in the Form of Regular Polyhedra (Platonic Solids)

A considerable number of investigations were devoted to optical properties of dielectric nanoparticles, that is, nanoparticles having positive electric permittivity. In this case, plasmon resonances cannot be observed directly. Nevertheless, according to general results of Chapter 5, the polarizabilities of dielectric bodies are determined by the plasmon modes with a nonzero dipole electric moment.

Generally speaking, the polarizabilities of nonspherical bodies are tensor quantities. However, for Platonic solids of high symmetry (see Fig. 9.1 and Table 9.1), the polarizabilities are scalar quantities, and in this respect these bodies are similar to spheres.

Table 9.1 Geometric characteristics of regular polyhedra

Regular polyhedron	Edge number and type	Circumscribed sphere radius	Inscribed sphere radius	Surface area	Volume
Tetrahedron	4 regular triangles	$\frac{a}{4}\sqrt{6}$	$\frac{a}{12}\sqrt{6}$	$a^2\sqrt{3}$	$\frac{a^3}{12}\sqrt{2}$
Cube	6 squares	$\frac{a}{2}\sqrt{3}$	$\frac{a}{2}$	$6a^2$	a^3
Octahedron	8 equilateral triangles	$\frac{a}{2}\sqrt{2}$	$\frac{a}{6}\sqrt{6}$	$2a^2\sqrt{3}$	$\frac{a^3}{3}\sqrt{2}$
Dodecahedron	12 regular pentagons	$\frac{a}{4}\left(1+\sqrt{5}\right)\sqrt{3}$	$\frac{a}{4}\sqrt{10+22/\sqrt{5}}$	$3a^2\sqrt{5\left(5+2\sqrt{5}\right)}$	$\frac{a^3}{4}\left(15+7\sqrt{5}\right)$
Icosahedron	20 equilateral triangles	$\frac{a}{4}\sqrt{2\left(5+\sqrt{5}\right)}$	$\frac{a}{4\sqrt{3}}\left(3+\sqrt{5}\right)$	$5a^2\sqrt{3}$	$\frac{5a^3}{12}\left(3+\sqrt{5}\right)$

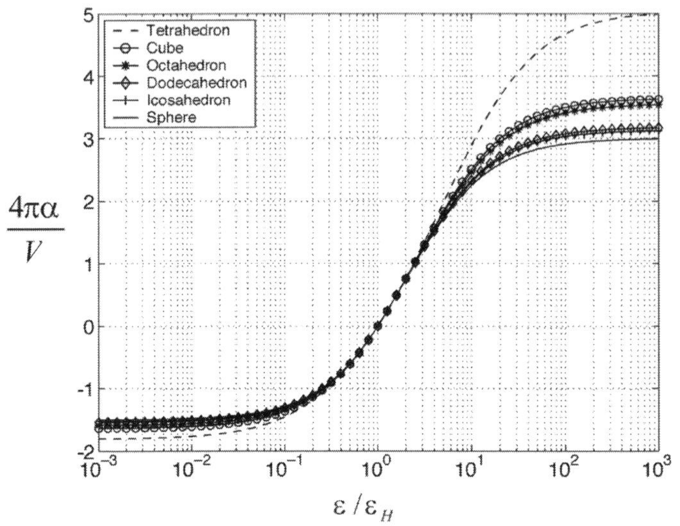

Figure 9.2 The dependencies of different Platonic solids' polarizabilities on their relative electric permittivity $\tau = \varepsilon/\varepsilon_H$. © 2004 IEEE. Reprinted with permission from Sihvola et al. (2004).

However, the absolute values of polarizabilities depend on the shape essentially.

The analysis of Platonic solids' polarizability was carried out by Avelin and Sihvola (2002), Sihvola et al. (2004) with the help of the method of surface integral equations (see Chapter 5, Eq. (5.44)). Shown in Fig. 9.2 are the dependencies of polarizabilities of different Platonic solids on the relative electric permittivity $\tau = \varepsilon/\varepsilon_H$, where ε_H is the electric permittivity of the environment.

As follows from these graphs, for any electric permittivity values six platonic bodies may be arranged in the following order of increasing polarizability: sphere $\rightarrow$ icosahedron $\rightarrow$ dodecahedron $\rightarrow$ octahedron $\rightarrow$ cube $\rightarrow$ tetrahedron.

Like many other cases, a sphere has the minimal value of polarizability. An icosahedron has the polarizability most similar to that of a sphere; a tetrahedron has the highest polarizability of all the bodies considered.

There is a strong correlation between the polarizability and the solid angle at the vertex of a polyhedron, that is, the sharper is the vertex angle, the higher is the polyhedron polarizability. A tetrahedron has the sharpest vertices, and the solid angle at its vertices is minimal; an octahedron has sharper vertices than a cube does, and a dodecahedron's vertices are sharper than those of an icosahedron. Consequently, an octahedron's polarizability is higher than that of a cube, and a dodecahedron's polarizability is higher than that of an icosahedron.

The number of edges is also a good polarizability indicator. A tetrahedron (4 edges) has the maximal polarizability, a cube and an octahedron (12 edges) are in the middle, while a dodecahedron and an icosahedron (30 edges) have minimal polarizability similar to that of a sphere, which can be considered as having an infinite number of edges.

The dependencies shown in Fig. 9.2 are rather smooth and can be well approximated. In cases of a tetrahedron, cube, octahedron, dodecahedron, and icosahedron, we have the following Pade approximations of polarizabilities, correspondingly (Sihvola et al., 2004):

$$\alpha = V\frac{\alpha_\infty}{4\pi}(\tau - 1)\frac{\tau^3 + 7.65667\tau^2 + 8.50919\tau - \alpha_0}{\tau^4 + 14.1983\tau^3 + 44.9182\tau^2 + 30.2668\tau + \alpha_\infty}, \tag{9.1}$$

$$\alpha = V\frac{\alpha_\infty}{4\pi}(\tau - 1)\frac{\tau^3 + 4.83981\tau^2 + 5.54742\tau - \alpha_0}{\tau^4 + 8.0341\tau^3 + 19/3534\tau^2 + 15.4349\tau + \alpha_\infty}, \tag{9.2}$$

$$\alpha = V\frac{\alpha_\infty}{4\pi}(\tau - 1)\frac{\tau^3 + 5.13936\tau^2 + 5.86506\tau - \alpha_0}{\tau^4 + 8.26227\tau^3 + 19.8267\tau^2 + 156191\tau + \alpha_\infty}, \tag{9.3}$$

$$\alpha = V\frac{\alpha_\infty}{4\pi}(\tau - 1)\frac{\tau^2 + 2.342101\tau - 2.56842\alpha_0/\alpha_\infty}{\tau^3 + 4.72932\tau^2 + 6.53464\tau + 2.56842}, \tag{9.4}$$

$$\alpha = V\frac{\alpha_\infty}{4\pi}(\tau - 1)\frac{\tau^2 + 3.04968\tau - 4.0896\alpha_0/\alpha_\infty}{\tau^3 + 5.29169\tau^2 + 8.52687\tau + 4.0896}. \tag{9.5}$$

In these expressions, α_∞, α_0 are constants characterizing the polarizabilities for the infinitely high and zero electric permittivities

Table 9.2 The values of constants characterizing the polarizabilities for the infinitely high and zero electric permittivities of the nanoparticle material

Polyhedron	α_∞	α_0
Tetrahedron	5.0285	−1.8063
Cube	3.6442	−1.6383
Octahedron	3.5507	−1.5871
Dodecahedron	3.1779	−1.5422
Icosahedron	3.1304	−1.5236
Sphere	3	−1.5

of the nanoparticle substance. The values of these important parameters are presented in Table 9.2.

These values are typical and enormously important for the description of nanoparticles' plasmon properties. Indeed, according to Eq. (5.23), in Chapter 5, the polarizability of any body is expressed through the eigenfunctions and eigenvalues of plasmon oscillations in the nanoparticle:

$$\alpha_{\mu\nu} = \frac{\varepsilon - \varepsilon_{\rm H}}{4\pi} \sum_m C^m_{\mu\nu} \frac{\varepsilon_m/\varepsilon_{\rm H} - 1}{\varepsilon_m - \varepsilon} \quad . \tag{9.6}$$

In the case of Platonic solids, the dipole polarizability is isotropic:

$$\begin{aligned} \alpha_{\mu\nu} &= \alpha\delta_{\mu\nu}, \\ \alpha &= \frac{\varepsilon - \varepsilon_{\rm H}}{4\pi} \sum_m C^m \frac{\varepsilon_m/\varepsilon_{\rm H} - 1}{\varepsilon_m - \varepsilon}, \end{aligned} \tag{9.7}$$

where $C^m = C^m_{\mu\mu}/3$.

In the case of infinitely low and infinitely high electric permittivities, we obtain from Eq. (9.7):

$$\begin{aligned} \frac{\alpha_\infty}{4\pi} &= -\frac{1}{4\pi} \sum_m C^m \left(\varepsilon_m/\varepsilon_{\rm H} - 1\right) = \frac{V}{4\pi} - \frac{1}{4\pi} \sum_m C^m \varepsilon_m/\varepsilon_{\rm H}, \\ \frac{\alpha_0}{4\pi} &= -\frac{1}{4\pi} \sum_m C^m \left(1 - \frac{\varepsilon_{\rm H}}{\varepsilon_m}\right) = -\frac{V}{4\pi} + \frac{1}{4\pi} \sum_m C^m \frac{\varepsilon_{\rm H}}{\varepsilon_m}. \end{aligned} \tag{9.8}$$

In the case when α_∞ and α_0 are known (see Table 9.2), sum rules are obtained that can be used for determination of correctness of numerical calculations of the nanoparticles' plasmon properties.

9.2 Properties of Localized Plasmons in Nanoparticles of a Complex Form

9.2.1 *A Nanocube and Related Geometries*

A particular interest in plasmon properties of nanocubes arose when the methods of the synthesis of silver and gold nanocubes had been successfully developed (Sun and Xia, 2002) (see also Fig. 2.18, Chapter 2).

For the first time, the theoretical analysis of plasmon oscillations in nanoparticles of cubic form was carried out by Fuchs (1975). In his calculations, he used the method of surface integral equations in its simplest form (Chapter 5). The results of the calculations for the principal modes, that is, for the modes with the maximal values of the dipole moment, are shown in Table 9.3.

According to the general theory, the sum of all the coefficients C_n is to be equal to 1, $\sum C^n = 1$. The sum of six components C_n stated in the table is equal to 0.96, and so only 4% fall to the share of others' resonances having nonzero dipole moments. Moreover, only the first two resonances among these six have significant amplitudes, so it can be expected that the nanocube optical response will be determined by these two resonances basically.

Shown in Fig. 9.3 are the dependencies of cube polarizabilities, calculated from Eqs. (9.2) and (9.7), on relative permittivity. It is clear from these pictures that in the region of positive values of electric permittivity, the polarizability, Eq. (9.7), calculated with the

Table 9.3 The resonance values of the permittivity for the cube main plasmonic modes and the corresponding values of these modes' strengths (Fuchs, 1975)

n	C_n	ε_n
1	0.44	−3.68
2	0.24	−2.37
3	0.04	−1.9
4	0.05	−1.27
5	0.1	−0.78
6	0.09	−0.42

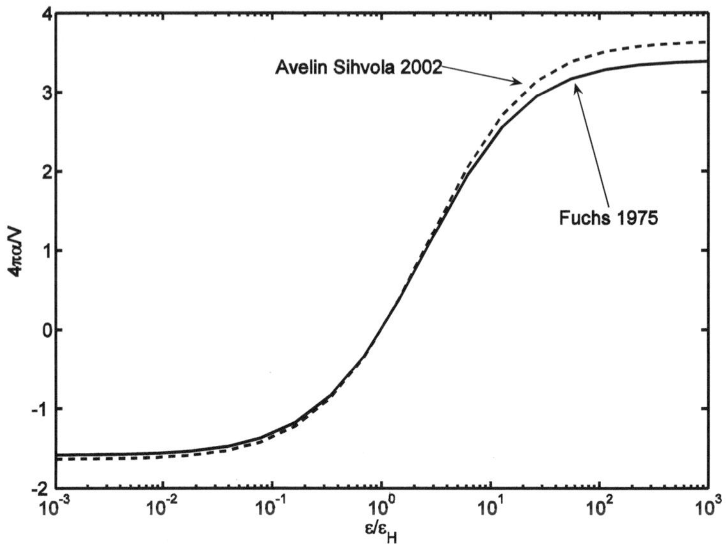

Figure 9.3 The dependency of the cube polarizability calculated with the help of the approximating formula, Eq. (9.2), and with the help of Eq. (9.7), in which the values of Table 9.3 are used.

help of the data presented in Table 9.3, is in rather good agreement with the calculation of the dielectric cube polarizability, stated in the previous section. It confirms the correctness of the results present in Table 9.3 once more.

The distribution of the surface charge corresponding to these modes is shown schematically in Fig. 9.4, where only 1/8 of a cube is shown. The dipole moment of all modes is directed along the z axis, and the origin of the coordinate system lies in the center of a cube. These surface modes have the symmetry of a cube, and thus they are symmetric about the rotation by the angle multiple of $\pi/2$, around the z axis and antisymmetric under the reflection across the plane xy. This symmetry allows one to determine the surface charge distribution of the other parts of the cube.

The extinction cross section for a silver nanocube were investigated by Gonzalez and Noguez (2007), Noguez (2007) with the help of the discrete dipole approximation (DDA) numerical method (see Appendix A2), allowing one to take retardation effects

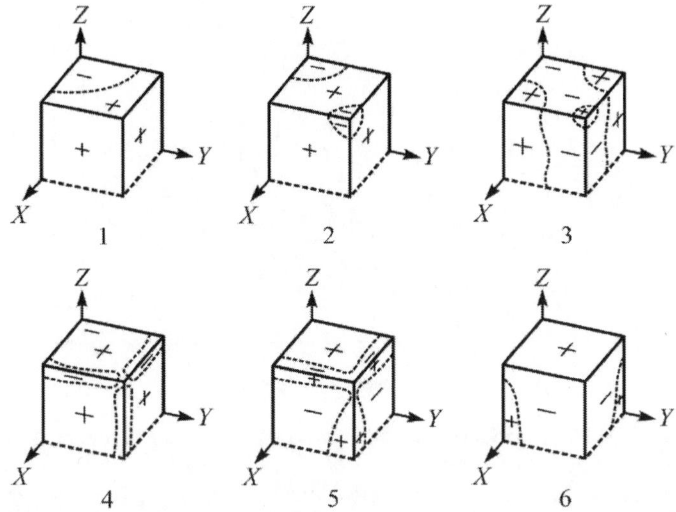

Figure 9.4 The scheme of the surface charge distribution of the main plasmon modes in a nanocube (Fuchs, 1975).

into account. In practice, within the context of this method a nanoparticle is divided into a great number of smaller cubes, which are approximated by the dipoles. In Fig. 9.5, the normalized extinction cross section for a silver nanocube of a volume equal to the volume of a sphere of 2.2 nm radius is shown; at that 10^5 dipoles were used for discretization of this particle.

For a nanocube in vacuum only two primary plasmon resonances are conspicuous, corresponding to $n = 1$ and 2 in Table 9.3. If a silver nanoparticle is placed in a silica matrix, then all six principal resonances stated in Table 9.3 become apparent (the red solid line in Fig. 9.5).

Since a nanocube has sharp edges and vertices, questions of the accuracy of numerical calculations of its plasmon properties arise. To make these questions clear, it is reasonable to consider sequentially real and hypothetical nanobodies having a smooth surface and tending in the limit to the perfect cube. The optical properties consideration by a continuous change of nanoparticles' shape is important not just from a mathematical point of view; nanoparticles' changes of such kind occur in practice of nanoparticle

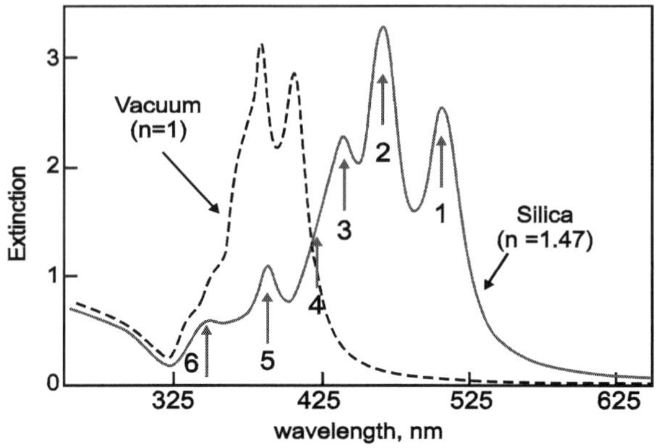

Figure 9.5 The normalized extinction cross section of a silver nanocube placed in vacuum and in silica and having a volume equal to the volume of a nanosphere of 2.2 nm radius. Six main resonances are shown by the arrows. The calculation was carried out with the help of the DDA method by approximating the cube by 10^5 dipoles (Noguez, 2007).

crystallization or melting frequently. Thus, the investigation of optical spectra dynamics in such processes may be an important instrument of control over them.

Recently, investigations of nanoparticles' forms' sequences were carried out Gonzalez and Noguez (2007), Noguez (2007), where within the DDA method the cube spectra, its several truncated versions, an icosahedron, and a sphere were analyzed. At that, by "truncated cubes" we mean figures obtained by truncation of eight vertices by a plane normal to the line joining the center and the

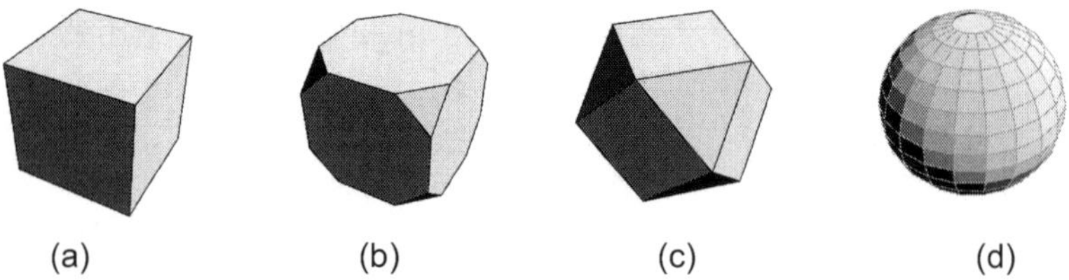

Figure 9.6 The illustration of the process of a cube transition to a sphere through truncation of vertices. (a) Cube, (b) cube with a truncated vertex ($r = 1/4$), (c) cube with a truncated vertex ($r = 1/2$), or cuboctahedron, and (d) sphere.

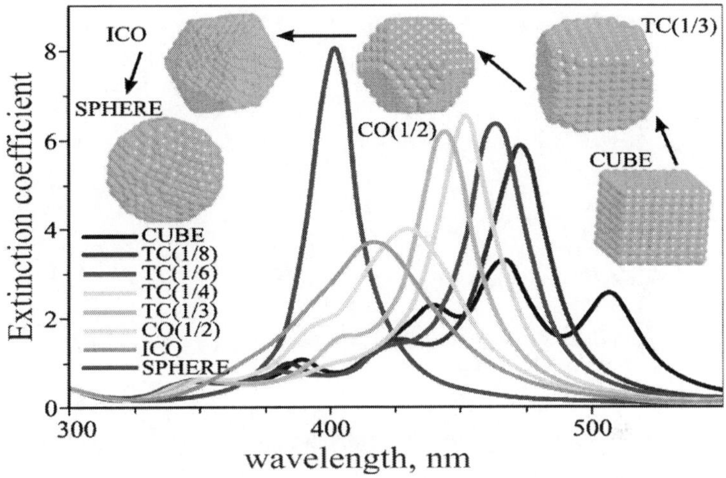

Figure 9.7 The extinction cross section depending on the wavelength of a cube, truncated cubes, an icosahedron, and a sphere. Nanoparticles are placed in a silica matrix (the refractive index is $n = 1.47$) and have a volume equal to the volume of a sphere of radius $r = 2.2$ nm (Noguez, 2007).

vertex (Fig. 9.6). The figures obtained are characterized by the value r, $0 < r < 1/2$, which determines the distance $\delta d = r \cdot a$ from the vertex to the truncating plane along the main diagonal of the cube with the side a. $r = 0$ corresponds to an unaltered cube, while $r = 1/2$ corresponds to a cuboctahedron (Fig. 9.6c). All the truncated cubes have six octagonal and eight trigonal faces, while a cuboctahedron is formed by 6 square and 8 trigonal faces.

Shown in Fig. 9.7 are results of the normalized extinction cross sections calculation for such a nanoparticles' sequence, with an icosahedron added to the sequence, which is rather allied to a sphere but cannot be obtained by the truncation of the cube vertices.

It is clear from this figure that even small changes of the cube shape result in substantial changes of its extinction cross section. Generally, the results of Gonzalez and Noguez (2007), Noguez (2007) show that an increase in the edge number results in the growth of the influence of one plasmon mode only (structurally analogous to the dipole plasmon mode in a sphere) and the shift of its frequency toward shorter wavelengths.

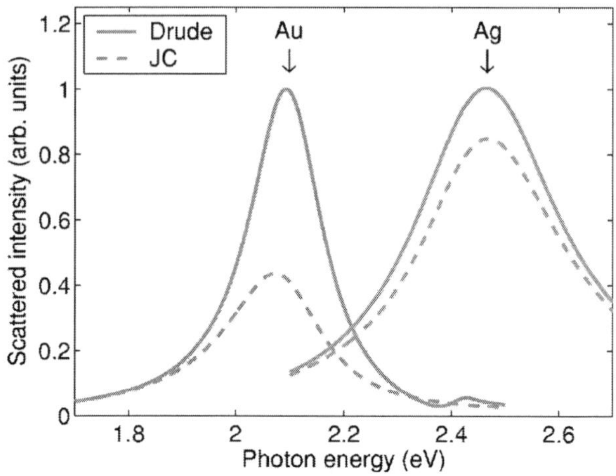

Figure 9.8 The extinction cross sections for a 50 nm cube with smoothed edges as a function of energy (quasi-static approximation). The solid curve corresponds to the Drude model of electric permittivity, and the dotted line corresponds to the experimental data of Johnson and Christy (1972), Hohenester and Krenn (2005).

The influence of nanocube smoothening on its plasmon properties was also investigated within the context of the method of surface integral equations (Chapter 5) without taking the retardation effect into account (Hohenester and Krenn, 2005). In this work, nanocubes with smoothed edges corresponding to really synthesized nanoparticles were explored. In Fig. 9.8 (Hohenester and Krenn, 2005), the scattering cross section of a cube with rounded edges made of gold or silver is shown. Here, either the Drude model (the solid line) or experimental data (Johnson and Christy, 1972) was used as the dielectric function. The comparison of the results obtained making use of these dielectric functions shows that in the case of gold there is a substantial discrepancy between the Drude approximation and experimental values for the electric permittivity (Johnson and Christy, 1972). It is due to the fact that the damping related to interband transitions is disregarded in the Drude model. On the contrary, in the case of silver the agreement between the Drude model and experimental data (Johnson and Christy, 1972) is good enough.

Since the method of surface integral equations includes only the dielectric function of a cube, the difference between the curves in Fig. 9.8 is related, namely, to the differences in Drude's dielectric functions and the dielectric functions of Johnson and Christy (1972) (see Chapter 3, Figs. 3.1 and 3.2).

Let us note that the spectra in Fig. 9.8 are shifted essentially relative to the spectra in Fig. 9.5 (the case of a nanocube in vacuum), testifying the limited accuracy of numerical calculations again.

Only a small number of objects of a shape more or less similar to an ideal cube is considered in (Gonzalez and Noguez, 2007; Noguez, 2007; Hohenester and Krenn, 2005). So the investigation of the continuous sequence of smooth surfaces tending to a cube in the limiting case is of great interest. Such an investigation was carried out by Klimov (2008), Pikhota (2008), where the plasmon spectra of superellipsoidal particles with the surface described by the equation:

$$|x|^n + |y|^n + |z|^n = a^n, n > 0, a > 0, \qquad (9.9)$$

were investigated in the context of the ε-method. For $n = 2$, this equation describes a sphere; for $n \to \infty$ it describes a cube. The parameterization, Eq. (9.9), is of great importance since all the surfaces described by it are smooth, and this removes many questions in numerical investigations of the optical properties of nanobodies. The electric permittivity spectra of particles with the surface described by Eq. (9.9) were found with the help of the boundary element method (BEM) (Chapter 5) and are shown in Fig. 9.9 as functions of the parameter n, characterizing the nanoparticle surface.

From this picture, one can see how the electric permittivity eigenvalues are being split due to the symmetry change and turning into the cubes' eigenvalues, gradually. In particular, it is clear that the triply degenerate dipole modes of a sphere with $\varepsilon = -2$ do not split and turn into the triply degenerate strong dipole modes of a cube with $\varepsilon \approx -4.6$. A remarkable feature of the spectra in this picture is that the plasmon spectrum changes essentially for $n = 2.5$, where the branches with $\varepsilon > -1$ are formed, which are typical not for a sphere but for a cube. Actually, for this value of n the phase transition takes place in plasmon spectra. An analogous transition

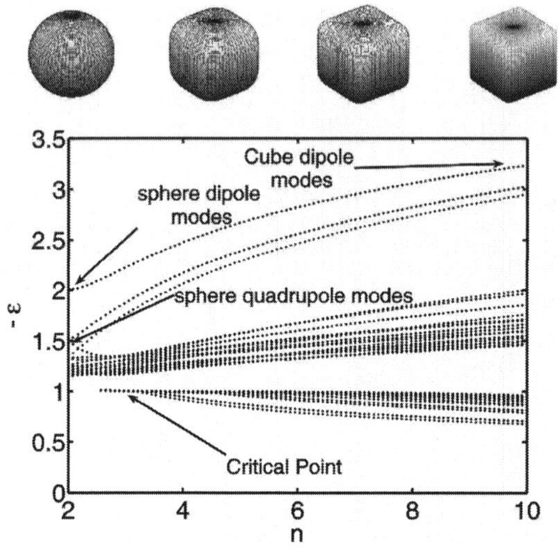

Figure 9.9 The electric permittivity spectra depending on the shape of nanoparticles of cube symmetry (see Eq. (9.9)).

also takes place in other particles of a complex shape (see Figs. 7.7 and 10.7). The spectra obtained have great significance since they allow controlling of the crystallization or melting of nanoparticles by purely spectroscopic means.

Eq. (9.9) is remarkable not merely because it allows one to describe the continuous transition from a sphere to a cube and vice versa but also because by changing the parameter from $n = 2$ to $n = 1$, the transition from a sphere to an octahedron occurs. The dependency of the resonance electric permittivity of the parameter n in this region is shown in Fig. 9.10, from which it is seen how the dipole modes of a sphere with $\varepsilon = -2$ (the upper curve) go over into the octahedron dipole modes continuously. It is very important to note that the resonance electric permittivity of this mode tends to the value that exceeds substantially the maximum in the absolute magnitude value of the resonance electric permittivity in a cube.

Thus, there is a range of methods of calculating optical spectra of cubes, cuboctahedrons, and octahedrons. It is highly important that these spectra are dependent on the particles' geometry

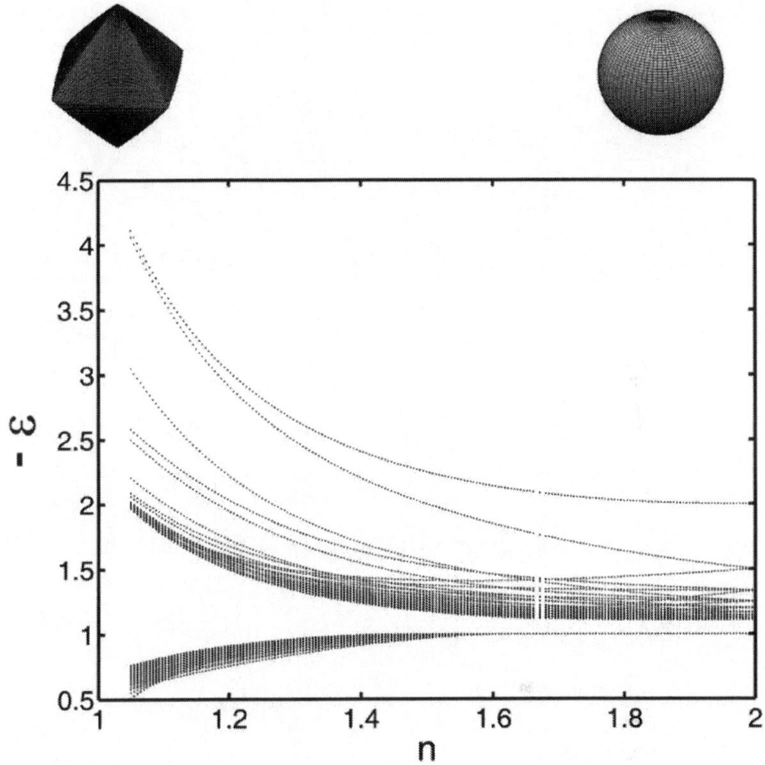

Figure 9.10 The electric permittivity spectra depending on the shape of nanoparticles of octahedron symmetry.

substantially. Spectral peaks of large particles are shifted to the red region as a result of the retardation effects increase. If the latter are considered correctly, it is possible to say that generally theoretical and experimental spectra descriptions are in good mutual agreement. However, it should be kept in mind that this agreement is related considerably to a high absorption in materials that smooth spectra detail. The spectra shown in Figs. 9.9 and 9.10 are not related to absorption and thus appear to be more fundamental. Nevertheless, the purely mathematical explanation of the spectra peculiarities of particles of this kind and more complex ones is still ahead.

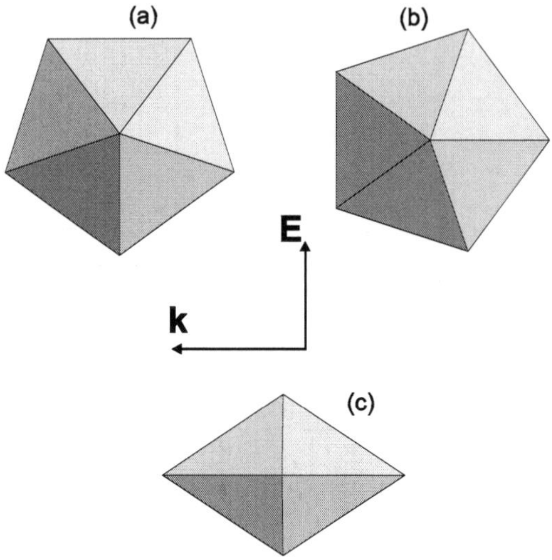

Figure 9.11 The decahedron geometry.

9.2.2 *A Decahedron and Related Geometries*

Decahedrons and pentagonal bipyramids, obtained as a result of some synthesis technologies (Wang, 2000; Yacaman et al., 2001; Kuo et al., 2004; Wei et al., 2005; Nilius et al., 2000), are also one of the important geometries of metal nanoparticles. It is known that metal nanoparticles of several nanometers' size can take very different shapes depending on their sizes, composition, and thermodynamic conditions (Baletto et al., 2002; Baletto and Ferrando, 2005). Regular decahedrons consist of 10 trigonal faces, which form two pentagons. A decahedron is an asymmetric figure in the sense that its polarizability is a tensor, in contrast to the polarizability of Platonic solids and a sphere, and depends on the polarization of the electric field incident on it. In Fig. 9.11, we can see three different orientations of a decahedron with respect to the electric field directed vertically. In Fig. 9.11a, the field is directed along one of the central edges, while in Fig. 9.11b the field is directed along one of the outside edges. In Fig. 9.11c, the field is directed normally to the pyramid base.

From DDA analysis of a silver decahedron carried out by Gonzalez and Noguez (2007), it follows that substantial anisotropy is present for the polarizations parallel and normal to the pyramid base. When the electric field is directed parallel to the pentagonal base, the corresponding extinction spectrum has a very wide maximum with a half-width of 90 nm at the wavelength of 403 nm. On the other hand, when the electric field is normal to the pentagon, there is a relatively narrow maximum at the wavelength of 343 nm in the spectrum. For the parallel polarization, the spectra are almost the same, excepting the region near the maximum, where moderate differences are noticeable.

On the other hand, the maximum of the spectrum averaged with respect to the nanoparticles' orientations is situated at the wavelength of 410 nm and has a half-width of 90 nm. It testifies that the dominant contribution to polarizability and the extinction cross section arises from nanoparticles with the pentagon base parallel to the field. Generally speaking, nanoparticles of the decahedron geometry have many plasmon resonances. However, in the considered case of silver only three of them are observed. It is determined by the fact that all resonances are close to each other, and due to the broadening related to the losses inside particles, the corresponding maxima merge.

In the case when the nanoparticles' size lies between 1 nm and 5 nm, regular decahedrons are almost not synthesized. In this range of sizes, Marks decahedrons and rounded decahedrons usually occur (Fig. 9.12).

The Marks decahedrons (Marks, 1994) are very stable (Fig. 9.12b,c). In an ultraclean environment and under the condition of a weak interaction with the substrate, a Marks decahedron becomes dominant. The limiting case of the Marks decahedron is decahedrons in the form of a star (Fig. 9.12d). In addition to the Marks decahedron, small nanoparticles grow in the form of round decahedrons often (Fig. 9.12e,f,g). The geometry of particles of this kind is obtained from the geometry of a regular decahedron by truncation of the sharpest vertices. Nanoparticles of this kind originate when colloidal methods of nanoparticle growth are used (see Chapter 2, "Nanochemistry Methods"). The investigation of

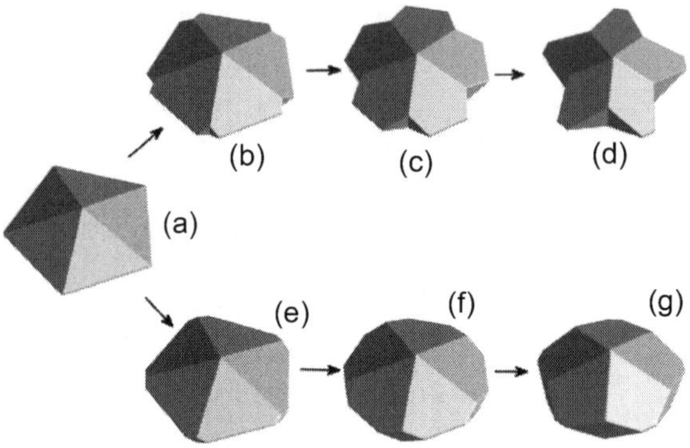

Figure 9.12 Nanoparticles of the decahedral symmetry. (a) A regular decahedron, (b, c) a Marks decahedron, (d) a star, and (e, f, g) a rounded decahedron.

optical properties of such nanoparticles was carried out by Gonzalez and Noguez (2007).

9.2.3 *A Tetrahedron, a Trigonal Prism, and Related Geometries*

A tetrahedron, a truncated tetrahedron, and a trigonal prism are other important polyhedrons. These geometries, investigated both theoretically and experimentally, are shown in Fig. 9.13. The significance of these geometries lies in the fact that nanoparticles of a similar form can be obtained with the help of nanosphere lithography and nanochemistry methods (see Chapter 2).

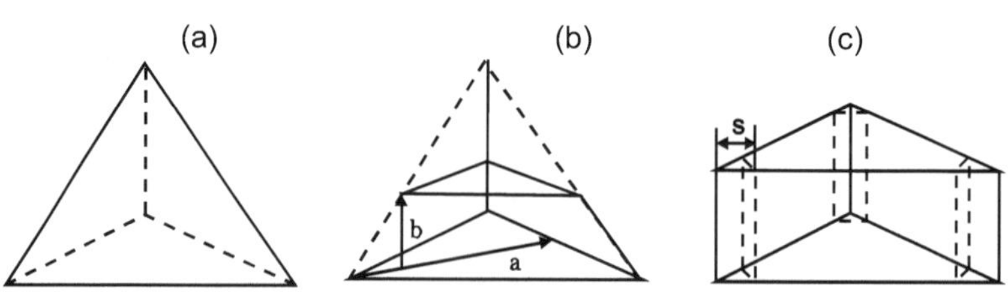

Figure 9.13 The geometries allied to a tetrahedron's geometry. (a) A regular tetrahedron, (b) a truncated tetrahedron, and (c) a trigonal prism.

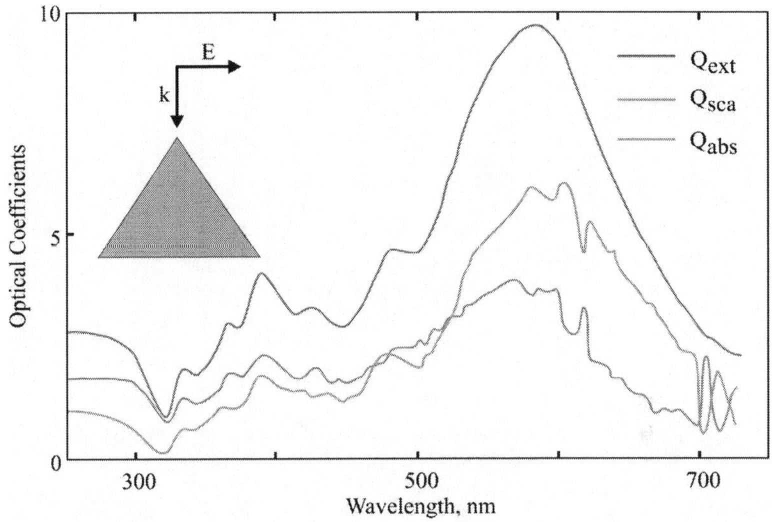

Figure 9.14 The efficiencies of scattering (green line), absorption (blue line), and extinction (red line) for a silver tetrahedron with a side of 164 nm (65,000 dipoles (Sosa et al., 2003)).

Sosa et al. (2003) have investigated the optical properties of a silver tetrahedron in the case of a field polarization parallel to one of its edges with the help of the DDA method. Results of the calculations in the case of a tetrahedron with an edge of 164 nm and with the usage of 65,000 point dipoles are shown in Fig. 9.14. In this case, the scattering and absorption occur in a wide range of wavelength. For $\lambda < 500$ nm, the absorption effects are a little bigger than the scattering effects, and for $\lambda > 500$ nm the situation is reverse. In this case, the plasmon resonances' structure is very rich, and that is related to the fact that a tetrahedron has the sharpest vertices of all Platonic solids. This peculiarity becomes apparent in the case of a dielectric tetrahedron, which has the maximal (dipole) polarizability of all Platonic solids. Let us note that the resonances in the region of wavelengths more than 700 nm are related to a bad convergence of the DDA method and have no physical significance.

The calculations of tetrahedral nanoparticles' properties were carried out also by Yang et al. (1995), and the results of this

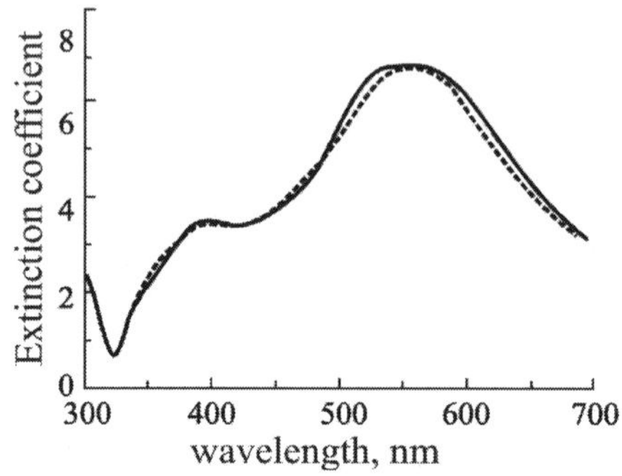

Figure 9.15 The extinction coefficient for a silver tetrahedron obtained with the help of the DDA simulation as a function of wavelength. The solid and dotted lines correspond to 2,321 and 1,622 dipoles, respectively. The edge length is 57.7 nm (Yang et al., 1995).

calculation are shown in Fig. 9.15 for a tetrahedron of silver with the dielectric function from (Hagemann et al., 1975). Generally, these results are in agreement with the results of Sosa et al. (2003) (see Fig. 9.14). At that, we can see from the picture that the results for 1,622 dipoles and for 2,321 dipoles differ a little. Nevertheless, investigations of the tetrahedron optical properties appear to be expedient in the context of other numerical methods. This is of prime importance because the DDA description of quadrupole plasmon oscillations and those of a higher multipolarity, which also contribute to different optical processes, is not quite adequate.

The optical properties of a truncated tetrahedron were investigated by Jensen et al. (1999a), Jensen et al. (1999b), Kelly et al. (2003). Let us note that nanoparticles of the shape similar to a truncated tetrahedron are obtained as a result of the nanoparticles' synthesis by the method of nanosphere lithography (Chapter 2), and thus the results have an immediate practical significance.

More detailed investigations of the truncated tetrahedron plasmon resonances, both theoretical and experimental, were carried out by Jensen et al. (1999b).

The influence of the environment electric permittivity ε_H on the optical properties of a truncated tetrahedron was carried out by Kelly et al. (2003) within the DDA approach. The environment's influence on the locations of the plasmon resonance can be understood easily in the case of an arbitrary particle for which the precise condition of the dipole plasmon resonance is given by (see Chapter 5):

$$\mathrm{Re}\left(\varepsilon\left(\omega\right)\right) - \tilde{\varepsilon}_{\mathrm{res}}\varepsilon_{\mathrm{H}} = 0, \qquad (9.10)$$

where $\varepsilon\left(\omega\right)$ and ε_{H} are the electric permittivity of a metal and a solvent, in which the nanoparticle is situated, correspondingly, and $\tilde{\varepsilon}_{\mathrm{res}}$ is the relative value of the resonance permittivity found within the ε-method. For a nanosphere, $\tilde{\varepsilon}_{\mathrm{res}} = -2$. In the case of the Drude model, the electric permittivity of the metal is equal to:

$$\varepsilon\left(\omega\right) = 1 - \frac{\omega_{\mathrm{pl}}^2}{\omega^2} \qquad (9.11)$$

(where ω_{pl} is the bulk plasmon frequency) and the plasmon resonance takes place for wavelength

$$\lambda_{\mathrm{p}} = \lambda_{\mathrm{pl}}\sqrt{1 - \tilde{\varepsilon}_{\mathrm{res}}\varepsilon_{\mathrm{H}}} \approx \lambda_{\mathrm{pl}}\sqrt{1 - \tilde{\varepsilon}_{\mathrm{res}}}\left\{1 + \frac{n_{\mathrm{H}} - 1}{1 - \tilde{\varepsilon}_{\mathrm{res}}}\right\}, \qquad (9.12)$$

where $n_{\mathrm{H}} = \sqrt{\varepsilon_{\mathrm{H}}}$ is the environment (dissolvent) refractive index and it is assumed that $n_{\mathrm{H}} \approx 1$. Results of DDA modeling and corresponding experimental data are shown in Fig. 9.16 (Kelly et al., 2003). From the figure, we can see that both the theory and the experiment give the linear dependencies of the resonance position on the environment refractive index. Unfortunately, the theory and the experiment demonstrate different slopes of the lines, and some additional investigations are required for the final solution of this important problem.

Development of synthesis technology of silver nanoparticles in the form of trigonal prisms with atomic smooth faces was a significant result of colloidal chemistry (Jin et al., 2001). The investigation of the optical properties of such nanoparticles was carried out with the help of DDA modeling by Kelly et al. (2003). Both regular trigonal prisms and prisms with somehow truncated angles were considered (see Fig. 9.13 and the insertion in Fig. 9.17). All calculations were carried out for a case when nanoparticles are situated in water. Shown in Fig. 9.17 are the spectra averaged with respect to the nanoparticles' orientations. Regular prisms had bases in the form of regular triangles with a side of 100 nm and a height of

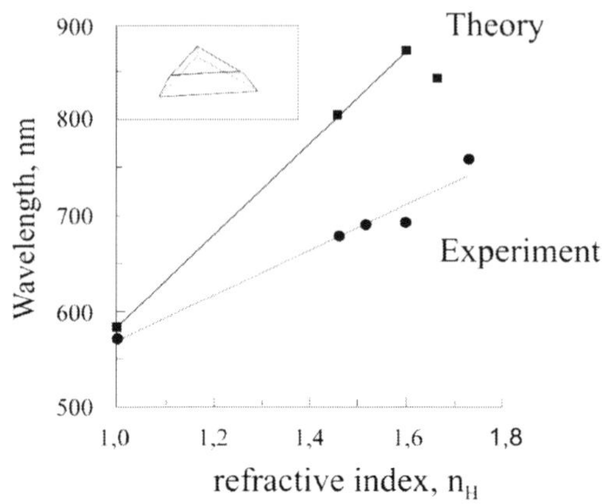

Figure 9.16 The DDA simulation of the environment permittivity influence on the position of the truncated tetrahedron plasmon resonance. The experimental data (Malinsky et al., 2001) were obtained by matching of the substrate ($n_{\text{substrate}}$) and solvent (n_{H}) refractive indices, so the truncated tetrahedron may be regarded as placed in a homogeneous space. Reprinted with permission from Kelly et al. (2003). Copyright 2003 American Chemical Society.

16 nm, which are similar to values measured experimentally. DDA modeling was performed by the prism partition into cubes with edges of 2 nm. This partition was sufficient for the convergence of the method.

The analysis of Fig. 9.17 shows that the extinction spectra have three peaks. In the case of a regular trigonal prism the main peak is situated at 770 nm, the weaker one is at 460 nm, and a small, but very sharp, peak is situated at 335 nm. The main peak (in the red region of the spectrum) turns out to be very sensitive to the form of prism angles, and when the size of a truncated angle (a snip) increases, a substantial shift of this peak to the blue region occurs. From the physical point of view this effect is quite clear since by such a truncation the particle is becoming more similar to a sphere, and this fact corresponds to increase of plasmon oscillations' frequency.

For better comprehension of the physical nature of extinction spectra in Fig. 9.18a, results of DDA modeling for different polarizations of the incident radiation relative to the prism major axes are shown. In all cases, two peaks are noticeable in the

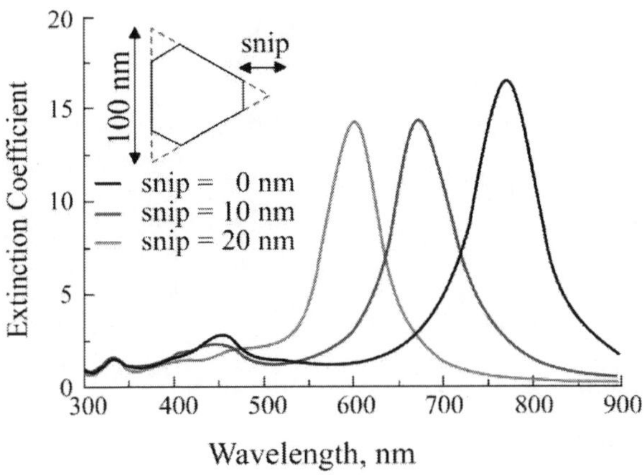

Figure 9.17 The extinction spectra of trigonal prisms with a side of 100 nm and end truncation at distances of 0 nm, 10 nm, and 20 nm averaged with respect to orientations. In the insertion, the geometry of a prism with truncated angles is shown. The prism height is 16 nm. During DDA calculations, a cubic grid with a cell size of 2 nm was used. In the case of a regular trigonal prism, 68,704 dipoles were used. Reprinted with permission from Kelly et al. (2003). Copyright 2003 American Chemical Society.

extinction spectrum, and the spectra in directions parallel to the prism base (*y*- and *z*-polarizations) turn out to be very similar to each other and to have the peaks at 460 nm and 670 nm. In the case of the light polarization perpendicular to the prism base (along the shorter axis), the maxima take place at 335 nm and 430 nm.

In Fig. 9.18b, results of the comparison of averaged curves from Fig. 9.18a and the experimental data are shown. The agreement is good and confirms that a trigonal prism with 10 nm truncated vertices describes the averaged synthesized particle well. Let us note that though there are four peaks in polarization-resolved spectra, the peak at the wavelength of 400 nm disappears as a result of the averaging.

For comprehension of the physical nature of plasmon resonances noticeable in Figs. 9.17 and 9.18, dipoles' distributions for different resonance (460 nm and 770 nm) modes are shown in Fig. 9.19 in the case of a regular trigonal prism. In the case of a mode with the resonance at 770 nm, the polarization

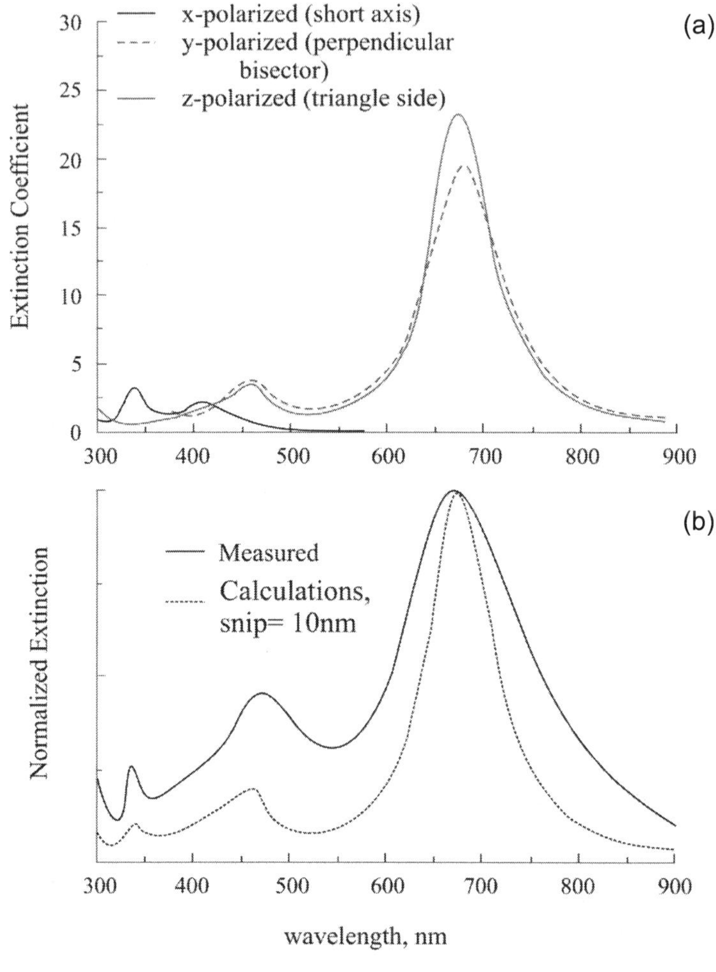

Figure 9.18 The extinction spectra of silver trigonal nanoprisms with a side of 100 nm, a height of 16 nm, and a snip of 10 nm. (a) The calculated spectra for different polarizations and (b) the spectrum averaged with respect to polarization and the measured extinction value. Reprinted with permission from Kelly et al. (2003). Copyright 2003 American Chemical Society.

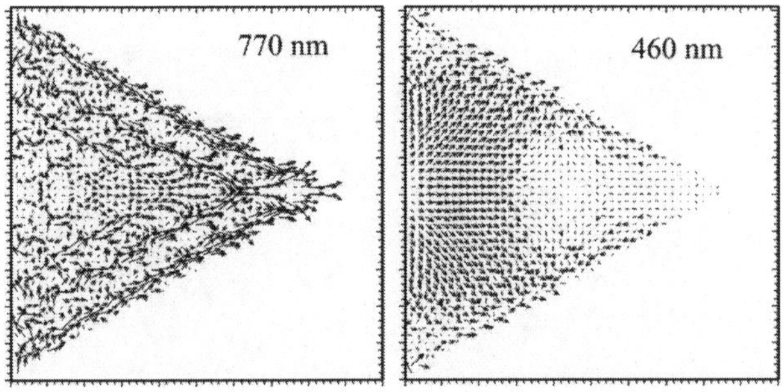

Figure 9.19 The polarization distribution for the dipole (on the left) and quadrupole (on the right) resonances at wavelengths of 770 nm and 460 nm, correspondingly, in a silver trigonal prism with the same parameters as in Fig. 9.18 (no snips). Reprinted with permission from Kelly et al. (2003). Copyright 2003 American Chemical Society.

distribution is similar to the polarization distribution in the dipole mode of a sphere and has a big horizontal (along the field) dipole moment. In the case of the resonance at 460 nm, the polarization distribution is entirely different, and only half of the grid cells have a noticeable polarization. This distribution has a quadrupole structure. The conclusion may be made from the stated analysis that the resonances at 770 nm and 460 nm have dipole and quadrupole natures, respectively. In Fig. 9.20, the electric field absolute value distribution is shown. Here, we see that the maximal field enhancement (more than 20 times) takes place for the dipole resonance at 770 nm near vertices, while for the quadrupole resonance at 460 nm the maximal field enhancement occurs at the prism lateral faces. Another typical feature that can be observed in Fig. 9.20 is that the field distribution decreases very quickly for the quadrupole resonance when moving away from the particle surface. This also has a simple explanation since the dipole near field decreases inversely proportional to the second power of the distance and the quadrupole near field inversely proportional to the fourth power of the distance from the particle (Jackson, 1975).

Quite recently, properties of not ensembles of trigonal prisms but of separate nanoparticles were investigated successfully (Sherry et al., 2006). Shown in Fig. 9.21 are results of the comparison

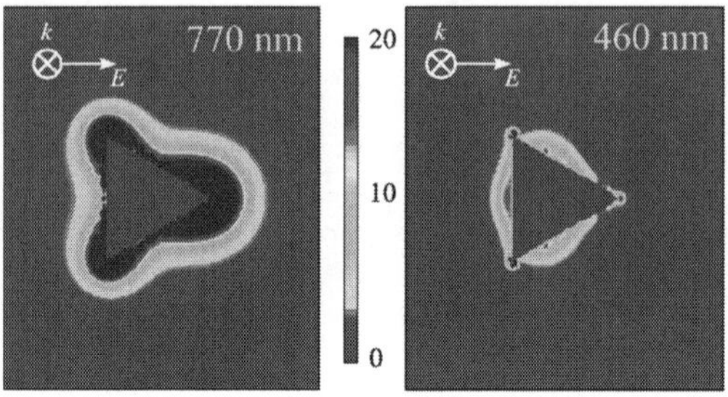

Figure 9.20 The electric field distribution in the plane normal to the shorter prism axis and passing through its bisecting point. The light is incident on the prism parallel to its shorter axis and has the horizontal polarization. On the left, $\lambda = 770$ nm. On the right, $\lambda = 460$ nm. The edge length of the triangle lying in the base is 100 nm, and the prism height is 16 nm (Kelly et al., 2003).

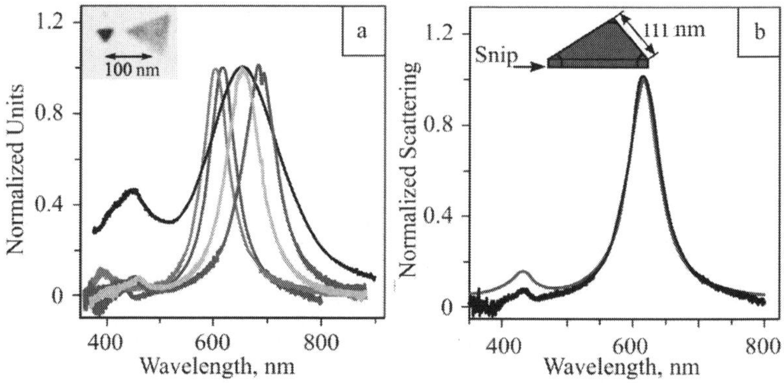

Figure 9.21 The spectra of localized plasmons of a silver trigonal prism. (a) The extinction spectrum of the nanoprisms' ensemble in an aqueous solution (the black curve) and the spectra of the separate particles' scattering (the colored curves). Typical TEM images of the nanoprisms are shown in the insertion. (b) Results of DDA simulation (the red curve) and of experimental investigation (the black curve) of a single trigonal prism placed in an inert nitrogen environment ($n = 1.000297$). The theoretical spectra were calculated for a nanoprism with an edge of 111 nm, a height of 10 nm, and a snip of 15 nm (Sherry et al., 2006).

of the trigonal prisms' optical properties, obtained by three different methods:

1) ensemble extinction spectroscopy;
2) the spectroscopy of Rayleigh scattering (microscopy of the dark field of separate nanoprisms); and
3) the modeling of separate nanoprisms with the help of the DDA method.

In Fig. 9.21a, the ensemble extinction (scattering + absorption) spectra in the aqueous solution ($n = 1.33$) of trigonal silver nanoprisms are compared with the scattering spectra of the four separately selected different nanoprisms also situated in the aqueous solution. In the insertion, typical transmission electron microscopy (TEM) images of nanoprisms are shown. In Fig. 9.21b, the scattering spectra of trigonal nanoprisms in the nitrogen atmosphere ($n = 1.000297$) are compared with the results of the DDA modeling (see Appendix A2). These spectra show the strong dipole resonance at 620 nm and the weaker quadrupole resonance at 433 nm. Let us note that nanoprisms' parameters used in the modeling were not measured experimentally but were selected to fit the experimental data.

9.3 Conclusions

By now, great theoretical and experimental material concerning properties of nanoparticles of the form of different polyhedrons is accumulated. Main theoretical results are related to the DDA modeling method, which is not free from defects concerned with convergence, and thus the lattices of tens and hundreds of thousands of dipoles are required. Therefore, the usage of other numerical and analytical methods (the ε-method) remains desirable. Nevertheless, main quantitative conclusions for the nanoparticles in the form of polyhedrons may be made:

1) As the particle sizes increase, the spectra, as a rule, shift to the red region (resonance values of electric permittivity become more negative);

2) As the environment refractive index increases, the spectra shift in the red region;
3) Sharper faces and edges provoke an increase of resonances number, which, though, are not always noticeable in the spectra due to losses inside polyhedral particles substance; and
4) For plasmon modes having a strong dipole moment local fields arise in the vicinity of sharp vertices.

References

Avelin, J., Sihvola, A. (2002) Polarizability of polyhedral dielectric scatterers, *Microwave Opt. Technol. Lett.*, **32**, 60–64.

Baletto, F., Ferrando, R. (2005) Structural properties of nanoclusters: energetic, thermodynamic, and kinetic effects, *Rev. Mod. Phys.*, **77**, 371.

Baletto, F., Ferrando, R., Fortunelli, A., Mottet, C. (2002) Crossover among structural motifs in transition and noble-metal clusters, *J. Chem. Phys.*, **116**, 3856.

Fuchs, R. (1975) Theory of optical properties of ionic crystal cubes, *Phys. Rev. B*, **11**, 1732.

Gonzalez, L., Noguez, C. (2007) Influence of morphology on the optical properties of metal nanoparticles, *J. Comput. Theor. Nanosci.*, **4**, 231–238.

Hagemann, H.J., Gudat, W., Kunz, C. (1975) Optical constants from the far infrared to the x-ray region: Mg, Al, Cu, Ag, Au, Bi, C, and Al_2O_3, *J. Opt. Soc. Am.*, **65**, 742–744.

Hohenester, U., Krenn, J. (2005) Surface plasmon resonances of single and coupled metallic nanoparticles: a boundary integral method approach, *Phys. Rev. B*, **72**, 195429.

Jackson, J.D. (1975) *Classical Electrodynamics*, 2nd ed. (Wiley & Sons, New York).

Jensen, T.R., Schatz, G.C., Van Duyne, R.P. (1999a) Nanosphere lithography: surface plasmon resonance spectrum of a periodic array of silver nanoparticles by ultraviolet-visible extinction spectroscopy and electrodynamic modeling, *J. Phys. Chem. B*, **103**, 2394–2401.

Jensen, T., Kelly, L., Lazarides, A., Schatz, G.C. (1999b) Electrodynamics of noble metal nanoparticles and nanoparticle clusters, *J. Cluster Sci.*, **10**, 295–317.

Jin, R., Cao, Y., Mirkin, C.A., Kelly, K.L., Schatz, G.C., Zheng, J.-G. (2001) Photoinduced conversion of silver nanospheres to nanoprisms, *Science*, **294**, 1901.

Johnson, P.B., Christy, R.W. (1972) *Phys. Rev. B*, **6**, 4370–4379.

Kelly, K.L., Coronado, E., Zhao, L.L., Schatz, G.C. (2003) The optical properties of metal nanoparticles: the influence of size, shape, and dielectric environment, *J. Phys. Chem. B*, **107**, 668–677.

Klimov, V.V. (2008) Nanoplasmonics, *Phys. - Uspekhi*, **178**, 875–879.

Kuo, C.-H., Chiang, T.-F., Chen, L.-J., Huang, M.H. (2004) Synthesis of highly faceted pentagonal- and hexagonal-shaped gold nanoparticles with controlled sizes by sodium dodecyl sulfate, *Langmuir*, **20**, 7820.

Malinsky, M.D., Kelly, K.L., Schatz, G.C., Van Duyne, R.P. (2001) Nanosphere lithography: effect of substrate on the localized surface plasmon resonance spectrum of silver nanoparticles, *J. Phys. Chem. B*, **105**, 2343–2350.

Marks, L.D. (1994) Experimental studies of small particle structures, *Rep. Prog. Phys.*, **57**, 603.

Nilius, N., Ernst, N., Freund, H.-J. (2000) Photon emission spectroscopy of individual oxide-supported silver clusters in a scanning tunneling microscope, *Phys. Rev. Lett.*, **84**, 3994–3997.

Noguez, C. (2007) Surface plasmons on metal nanoparticles: the influence of shape and physical environment, Invited feature article, *J. Phys. Chem. C*, **111**, 3806–3819.

Pikhota, M.Y. (2008) *Plasmon Resonances in Nanoparticles of Cubic Morphology*, Graduation theses (Moscow Institute of Physics and Technology).

Sherry, L.J., Jin, R., Mirkin, C.A., Schatz, G.C., Van Duyne, R.P. (2006) Localized surface plasmon resonance spectroscopy of single silver triangular nanoprisms, *Nano Lett.*, **6**, 2060–2065.

Sihvola, A., Yla-Oijala, P., Jarvenpaa, S., Avelin, J. (2004) Polarizabilities of platonic solids, *IEEE Trans. Antennas Propag.*, **52**, 2226–2233.

Sosa, I., Noguez C., Barrera, R.G. (2003) Optical properties of metal nanoparticles with arbitrary shapes, http://arxiv.org/abs/cond-mat/0304216.

Sun, Y., Xia, Y. (2002) Shape-controlled synthesis of gold and silver nanoparticles, *Science*, **298**, 2176.

Tao, A., Sinsermsuksakul, P., Yang, P. (2006) Polyhedral silver nanocrystals with distinct scattering signatures, *Angew. Chem., Int. Ed.*, **45**, 4597–4601.

Wang, Z.L. (2000) Transmission electron microscopy of shape-controlled nanocrystals and their assemblies, *J. Phys. Chem. B*, **104**, 1153–1175.

Wei, G., Zhou, H., Liu, Z., Song, Y., Wang, L., Sun, L., Li, Z. (2005) One-step synthesis of silver nanoparticles, nanorods, and nanowires on the surface of DNA network, *J. Phys. Chem. B*, **109**, 8738–8743.

Yacaman, M.J., Ascencio, J.A., Liu, H.B., Gardea-Torresdey, J. (2001) Structure shape and stability of nanometric sized particles, *J. Vac. Sci. Technol. B*, **19**, 1091.

Yang, W.-H., Schatz, G.C., Van Duyne, R.P. (1995) Discrete dipole approximation for calculating extinction and Raman intensities for small particles with arbitrary shapes, *J. Chem. Phys.*, **103**, 869.

Chapter 10

Localized Plasmons in Nanoparticle Clusters

In previous chapters, optical properties of isolated plasmon nanoparticles were considered. These properties turn out to be very interesting and to have a range of significant applications. It was shown that they depend substantially on the nanoparticles' shapes. However, in all typical cases the plasmon localization region in one isolated particle coincides with its volume, more or less. Plasmon wavefunctions of single particles and the spectrum of their oscillations are similar to the atoms' wavefunctions in general, and thus they can be referred to as "plasmon atoms."

Practically, isolated (noninteracting) nanoparticles occur in space, the atmosphere, and oceans and also when carrying out specific experiments. When the particles' concentration increases, the interaction between them becomes stronger, and when the distance between them becomes as small as the order of their sizes or even smaller, it is more expedient to talk about not separate nanoparticles but nanoparticles' clusters. On the other hand, the main tendency of modern nanotechnologies is the necessity to place as many objects as possible in a given volume (high-density integration), and as a result these elements occur at a near distance

Nanoplasmonics
Vasily Klimov

ISBN 978-981-4267-16-8 (Hardcover), 978-981-4267-42-7 (eBook)
www.panstanford.com

and the interaction between them becomes strong. In this case, it is also expedient to talk about clusters. Third, in nature, the processes of particles' self-organization occur frequently, and as a result, separate particles form more or less determinate structures, which should be described in terms of clusters and arrays (lattices) of nanoparticles.

Fundamentally and practically, the significance of plasmon properties of nanoparticle clusters is related to the fact that in clusters, contrary to separate particles, there are at least two dimensional parameters, nanoparticles' size and the distance between them. A change in each parameter results in a substantial change of the nanoparticles' plasmon properties and allows one to control effectively their optical properties and the spectra of plasmon-polariton (photon-polariton) oscillations and to provide effective interaction of plasmon particle clusters with atoms, molecules, and other nanosized devices. Above that, in the gap between the nanoparticles strong local fields arise, which can exceed the incident fields by several orders of magnitude and are also significant for many applications. Practical applications of the nanoparticles' plasmon properties will be also considered in Chapter 13.

Clusters of two nanoparticles have the most fundamental significance, and thus the present chapter is devoted mainly to investigation of plasmon properties of two-nanoparticle clusters. Certain attention will be paid also to linear clusters of several or many nanoparticles.

10.1 The Classification of Plasmon Oscillations in a Cluster on the Basis of Plasmon Oscillations in the Particles Composing It

10.1.1 *The System of Linear Integral Equations Describing Plasmon Oscillations in a Cluster of Nanoparticles*

Plasmon oscillations in a cluster of two nanoparticles of a finite volume are localized plasmon oscillations with a discrete spectrum, and therefore all general approaches stated in Chapter 5 are exactly applicable to them. However, as we have seen in previous chapters, plasmon properties of isolated (single) nanoparticles are complicated sufficiently. It is naturally expected that properties of plasmon oscillations in nanoparticle clusters will be even more complicated. At that, of course, a question arises as to how plasmon oscillations in a cluster of two nanoparticles are related to plasmon oscillations in isolated particles composing the cluster.

As it was shown in Chapter 5, plasmon oscillations in any system can be described by solutions of integral equations for eigenvalues problem of surface charges:

$$\sigma(\mathbf{s}) = \frac{1}{2\pi}\lambda \int\limits_S d^2\mathbf{s}'\sigma(\mathbf{s}')\frac{\mathbf{n}(\mathbf{s})(\mathbf{s}-\mathbf{s}')}{|\mathbf{s}-\mathbf{s}'|^3}, \ \mathbf{s}, \mathbf{s}' \in S, \qquad (10.1)$$

where the integral is taken over the surface of all particles composing the cluster and $\lambda = (\varepsilon-1)/(\varepsilon+1)$ are eigenvalues of the problem determining the resonance values of permittivity and frequency. If a cluster consists, for example, of two nanoparticles with surfaces S_1 and S_2, then its plasmon modes can be conveniently described with the help of two functions $\sigma_1(\mathbf{s})$, $\sigma_2(\mathbf{s})$ defining the surface charges determined on the surfaces S_1 and S_2, correspondingly (see Fig. 10.1), and satisfying the coupled system of two integral equations:

$$\sigma_1(\mathbf{s}) = \frac{1}{2\pi}\lambda\left[\int\limits_{S_1} d^2\mathbf{s}'\sigma_1(\mathbf{s}')\frac{\mathbf{n}(\mathbf{s})(\mathbf{s}-\mathbf{s}')}{|\mathbf{s}-\mathbf{s}'|^3} + \int\limits_{S_2} d^2\mathbf{s}'\sigma_2(\mathbf{s}')\frac{\mathbf{n}(\mathbf{s})(\mathbf{s}-\mathbf{s}'-\mathbf{R}_{12})}{|\mathbf{s}-\mathbf{s}'-\mathbf{R}_{12}|^3}\right],$$

$$\sigma_2(\mathbf{s}) = \frac{1}{2\pi}\lambda\left[\int\limits_{S_1} d^2\mathbf{s}'\sigma_1(\mathbf{s}')\frac{\mathbf{n}(\mathbf{s})(\mathbf{s}-\mathbf{s}'+\mathbf{R}_{12})}{|\mathbf{s}-\mathbf{s}'+\mathbf{R}_{12}|^3} + \int\limits_{S_2} d^2\mathbf{s}'\sigma_2(\mathbf{s}')\frac{\mathbf{n}(\mathbf{s})(\mathbf{s}-\mathbf{s}')}{|\mathbf{s}-\mathbf{s}'|^3}\right], \ \mathbf{s}, \mathbf{s}' \in S_1, S_2 \qquad (10.2)$$

where $\mathbf{R}_{12}$ is the vector directed from the center of one particle to the center of another and the vectors $\mathbf{s}$, $\mathbf{s}'$ are reckoned from the centers of the corresponding particles.

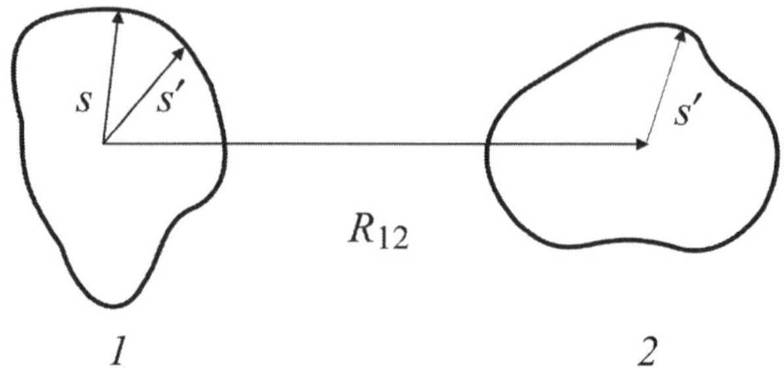

Figure 10.1 The geometry of a two-nanoparticle cluster, used during the integral Eq. (10.2) derivation.

The solution of the system in Eq. (10.2) allows one to find both eigenvalues of the cluster electric permittivity:

$$\lambda = \frac{(\varepsilon - 1)}{(\varepsilon + 1)}, \tag{10.3}$$

and corresponding distributions of the surface charge and then potential sand fields distributions according to known formulas of electrostatics:

$$\varphi_n(\mathbf{r}) = \int \frac{d\mathbf{s}'\sigma_n(\mathbf{s}')}{|\mathbf{r} - \mathbf{s}'|}, \; \mathbf{E}_n = -\nabla\varphi_n(\mathbf{r}). \tag{10.4}$$

10.1.2 *Plasmon Oscillation Properties in Clusters with Large Distances between Nanoparticles: The Point Dipole Model*

If the distance R_{12} between the particles is large in comparison with their sizes, the cross terms in the system Eq. (10.2) are small, and the system of two unbounded integral equations describing plasmon oscillations in isolated nanoparticles can be written instead of Eq. (10.2) in the first approximation:

$$\begin{aligned} \sigma_1(\mathbf{s}) &= \frac{1}{2\pi}\lambda\left[\int_{S_1} d^2\mathbf{s}'\sigma_1(\mathbf{s}')\frac{\mathbf{n}(\mathbf{s})(\mathbf{s}-\mathbf{s}')}{|\mathbf{s}-\mathbf{s}'|^3}\right], \; \mathbf{s}, \mathbf{s}' \in S_1, \\ \sigma_2(\mathbf{s}) &= \frac{1}{2\pi}\lambda\left[\int_{S_2} d^2\mathbf{s}'\sigma_2(\mathbf{s}')\frac{\mathbf{n}(\mathbf{s})(\mathbf{s}-\mathbf{s}')}{|\mathbf{s}-\mathbf{s}'|^3}\right], \; \mathbf{s}, \mathbf{s}' \in S_2. \end{aligned} \tag{10.5}$$

When the distance between the particles becomes shorter, the interaction between them becomes stronger, and the interaction terms in Eq. (10.2) cannot be neglected. Assuming the particles' sizes' smallness in comparison with the distance between them, R_{12}, and using the fact that the full charge on the particles is equal to zero, the following expression can be written instead of Eq. (10.2) by expanding the cross terms in the series:

$$\begin{aligned}\sigma_1(\mathbf{s}) &= \frac{1}{2\pi}\lambda\left[\int\limits_{S_1} d^2\mathbf{s}'\sigma_1(\mathbf{s}')\frac{\mathbf{n}(\mathbf{s})(\mathbf{s}-\mathbf{s}')}{|\mathbf{s}-\mathbf{s}'|^3}+\mathbf{n}(\mathbf{s})\cdot\mathbf{E}_2(\mathbf{R}_{12})\right], \mathbf{s},\mathbf{s}'\in S_1\\ \sigma_2(\mathbf{s}) &= \frac{1}{2\pi}\lambda\left[\mathbf{n}(\mathbf{s})\cdot\mathbf{E}_1(\mathbf{R}_{12})+\int\limits_{S_2} d^2\mathbf{s}'\sigma_2(\mathbf{s}')\frac{\mathbf{n}(\mathbf{s})(\mathbf{s}-\mathbf{s}')}{|\mathbf{s}-\mathbf{s}'|^3}\right], \mathbf{s},\mathbf{s}'\in S_2\end{aligned} \tag{10.6}$$

where

$$\mathbf{E}_i(\mathbf{R}) = -\frac{\mathbf{d}_i}{|\mathbf{R}|^3}+3\frac{(\mathbf{d}_i\cdot\mathbf{R})\mathbf{R}}{|\mathbf{R}|^5},$$

is the electric field of the dipole $\mathbf{d}_i$ and

$$\mathbf{d}_1 = \int\limits_{S_1} d^2\mathbf{s}'\sigma_1(\mathbf{s}')\,\mathbf{s}';\quad \mathbf{d}_2 = \int\limits_{S_2} d^2\mathbf{s}'\sigma_2(\mathbf{s}')\,\mathbf{s}',$$

are the dipole moments of plasmon oscillations in the first and second nanoparticles, respectively.

The origin of corrections related to the nanoparticles' interaction has a simple physical sense: it is simply the charge density on the surface of one nanoparticle induced by a dipole electric field from plasmonic oscillations in another. Since in Eq. (10.6) the correction terms are small for quite long distances between the particles, the solution of Eq. (10.6) can be found within the perturbation theory on the basis of the solution of Eq. (10.5) as in the first approximation.

After a quite evident fact that plasmon modes modification in nanoparticle clusters is related to their electrostatic interaction is ascertained, the model (the point dipole model) can be easily developed, allowing one to estimate changes of plasmon oscillations' spectra.

The essence of this approach is that the particles are replaced by point dipoles with polarizabilities equal to polarizabilities of corresponding particles in a homogeneous field. This replacement is correct since the electric field near the nanoparticles is almost homogeneous for long enough distances between them. After this approximation, the spectra of electric permittivity or plasmon frequencies can be found in a self-consistent manner.

To derive a self-consistent system of equations, let us first consider a case of two nanoparticles and suppose that one of them is situated at the point $\mathbf{r}_1$ and has the dipole moment $\mathbf{d}_1$ and the polarizability α_1. Another particle is situated at the point $\mathbf{r}_2$ and has the dipole moment $\mathbf{d}_2$ and the polarizability α_2. Let us also assume for simplicity that the nanoparticles are located in vacuum. In the first approximation, the polarizabilities of the particles do not depend on the presence of other particles and are determined by one of the plasmon modes existing in them, so the following may be written:

$$\alpha_i = \frac{\xi_i}{\varepsilon - \varepsilon_i}, \tag{10.7}$$

where ξ_i, ε_i characterize the mode of plasmon oscillations in the nanoparticles. For example, in the case of a spherical particle and the dipole resonance in it

$$\alpha_i = a_i^3 \frac{\varepsilon_i - 1}{\varepsilon_i + 2}, \tag{10.8}$$

where a_i, ε_i are the radius and permittivity of the nanosphere, respectively.

In the case of spheroidal or ellipsoidal nanoparticles, the polarizabilities are stated in Chapters 7 and 8. In the general case, polarizabilities are tensor values. Practically, application of the point dipole method is reasonable when the particles composing a cluster can be approximated by spherical or ellipsoidal nanoparticles.

Due to the above-mentioned, the self-consistent system of equations for the dipole moments has the form

$$\begin{aligned}
\mathbf{d}_1 &= \alpha_1 \mathbf{E}_2(\mathbf{r}_1) = \alpha_1 \overleftrightarrow{G}^0(\mathbf{R}_{12})\, \mathbf{d}_2, \\
\mathbf{d}_2 &= \alpha_2 \mathbf{E}_1(\mathbf{r}_2) = \alpha_2 \overleftrightarrow{G}^0(\mathbf{R}_{12})\, \mathbf{d}_1, \\
\overleftrightarrow{G}^0_{ij}(\mathbf{R}) &= -\frac{\delta_{ij}}{|\mathbf{R}|^3} + 3\frac{\mathbf{R}_i \mathbf{R}_j}{|\mathbf{R}|^5},
\end{aligned} \tag{10.9}$$

where i, $j = 1, 2, 3$ stand for indices numbering Cartesian coordinates, $\overleftrightarrow{G}^0$ is the quasi-static Green function of the point dipole source, and $\mathbf{R}_{12}$ is the vector of the distance between the particles.

Excluding, for example, the dipole moment of the second particle from Eq. (10.9), we obtain the vector equation

$$\left[\overleftrightarrow{1} - \alpha_1 \overleftrightarrow{G}^0 (\mathbf{R}_{12}) \alpha_2 \overleftrightarrow{G}^0 (\mathbf{R}_{12})\right] \mathbf{d}_1 = 0, \tag{10.10}$$

which has nonzero solutions (plasmons) when the determinant is equal to zero:

$$\left|\overleftrightarrow{1} - \alpha_1 \overleftrightarrow{G}^0 (\mathbf{R}_{12}) \alpha_2 \overleftrightarrow{G}^0 (\mathbf{R}_{12})\right| = 0. \tag{10.11}$$

In the case of two spherical nanoparticles situated on the z axis at a distance R from each other, we have

$$\begin{vmatrix} 1 - \dfrac{\alpha_1\alpha_2}{R^6} & 0 & 0 \\ 0 & 1 - \dfrac{\alpha_1\alpha_2}{R^6} & 0 \\ 0 & 0 & 1 - 4\dfrac{\alpha_1\alpha_2}{R^6} \end{vmatrix} = 0. \tag{10.12}$$

This system has three pairs of eigenvalues. The polarization (dipole moment orientation) of one is along the line joining the centers of the particles, while the polarizations of other two solutions are perpendicular to this line.

Let us consider, for example, the case of longitudinal polarization, having the following form of the dispersion equation:

$$1 - 4\frac{\alpha_1\alpha_2}{R^6} = 0. \tag{10.13}$$

Substituting the expression for the polarizations α_1, α_2 from Eq. (10.8) to Eq. (10.13), we obtain the dispersion equation for the case of equal spheres:

$$4\left(\frac{a}{R}\right)^6 \left(\frac{\varepsilon - 1}{\varepsilon + 2}\right)^2 = 1, \tag{10.14}$$

which allows one to find the resonance permittivities:

$$\varepsilon = -2\left(1 \pm 3\left(\frac{a}{R}\right)^3\right). \tag{10.15}$$

One solution of Eq. (10.15) corresponds to the case when the dipole moments of both nanospheres are oriented in the same

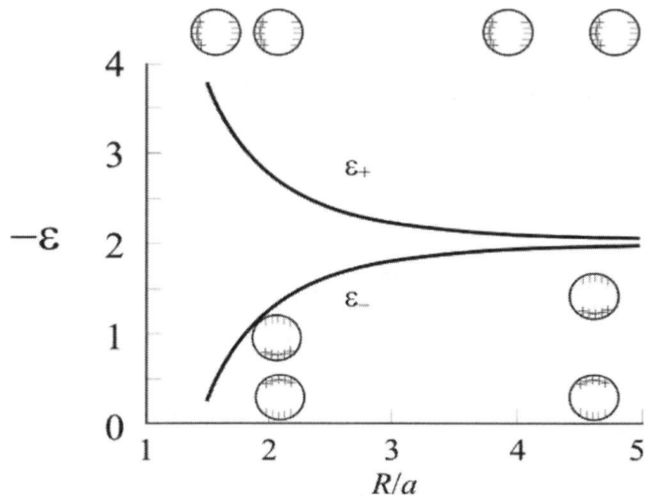

Figure 10.2 The dependencies, Eq. (10.15), of electric permittivity eigenvalues on the distance between the spherical nanoparticles in a cluster. The lower branch corresponds to symmetric plasmon oscillations, while the upper branch describes antisymmetric plasmon oscillations. The same dependencies describe qualitatively the case of any particles spaced rather distantly.

direction along the axis (the sign "+" results in reduction of electric permittivity resonance value), for another solution they are antiparallel (the sign "–" results in increase of the electric permittivity resonance value). The dependencies, Eq. (10.15), are schematically shown in Fig. 10.2. Naturally, these dependencies are true only for $(a/R) \ll 1$.

It should be stressed again that Eq. (10.15) determines resonance values of the cluster electric permittivity with correction to the particles' interaction. Due to this, the polarizability of a cluster with an excited longitudinal mode with parallel orientations of dipoles will have the simple form

$$\alpha = 2a^3 \frac{\varepsilon(\omega) - 1}{\varepsilon(\omega) + 2\left(1 + 3\left(a/R_{12}\right)^3\right)}, \tag{10.16}$$

where $\varepsilon(\omega)$ is electric permittivity of the material of the nanoclusters' particles. Naturally, electric polarizability of a mode with antiparallel orientation of dipoles is equal to zero. On the other hand, the equation

$$\varepsilon(\omega) + 2\left(1 \pm 3\,(a/R)^3\right) = 0, \qquad (10.17)$$

allows one to find plasmon frequencies of the considered cluster modes. The case of transverse cluster oscillations can be analyzed analogously.

The generalization of the considered case to the case of a system of many particles is straightforward, and corrections for eigenvalues of electric permittivity and plasmon oscillations' frequencies can be found by considering the eigenvalues of the homogeneous system of equations:

$$\begin{aligned} &\mathbf{d}_i = \alpha_i \sum_{j \neq i} \overset{\leftrightarrow}{G}{}^0\left(\mathbf{R}_{ij}\right)\mathbf{d}_j \\ &i, j = 1 \ldots N \end{aligned} . \qquad (10.18)$$

From a qualitative point of view, the interaction influence is illustrated in Figs. 10.2 and 10.3. In Fig. 10.2, it is shown that

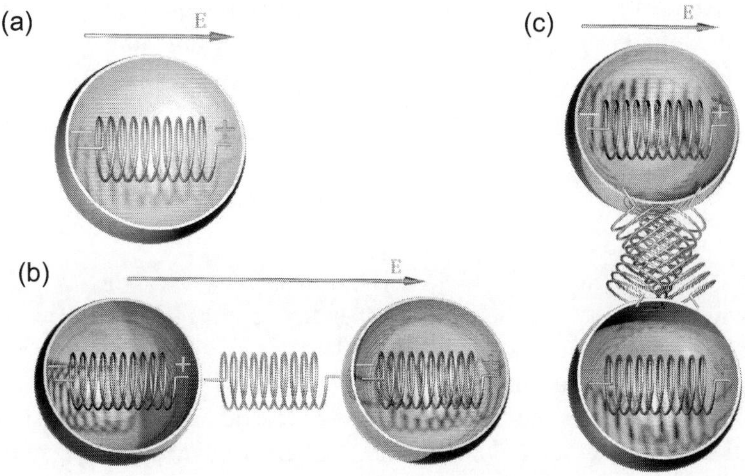

Figure 10.3 The qualitative picture explaining the plasmon frequency shifts during cluster formation. (a) An isolated nanoparticle; (b) a pair of adjacent nanoparticles with the exciting field parallel to the line joining the particles. In this case, the charges' interaction results in reduction of plasmon oscillation frequency; (c) the exciting field perpendicular to the line joining the particles. In this case the charges' interaction results in an increase of plasmon oscillation frequency.

when the particles are approaching each other, the "repulsion" of electric permittivities' resonance values occurs. In the upper part of the picture, the antisymmetric plasmon oscillations are shown, which have high absolute values of resonance electric permittivity, while in the lower part of the picture the symmetric plasmon oscillations are shown. In this picture, charge distributions in noninteracting particles and their change under the interaction influence are shown schematically. Figure 10.3 illustrates the simple physical mechanism allowing one to understand in which plasmon modes the frequency increase and decrease take place, that is, this picture illustrates the curves' "repulsion" shown in Fig. 10.2.

We have considered the dipole model rather thoroughly since this model allows one to estimate relatively simply some properties of clusters made of particles located not so close to each other. This model is also interesting since it allows one to take into account the retardation effects, which can be important in long clusters. To do this, it is sufficient to replace the quasi-static Green functions in Eq. (10.9) by the retarded Green functions (Jackson, 1975):

$$\overleftrightarrow{G}^{0}_{ij}(\mathbf{R}) = \left[k^2(\delta_{ij} - \mathbf{n}_i\mathbf{n}_j)\frac{1}{|\mathbf{R}|} + (3\mathbf{n}_i\mathbf{n}_j - \delta_{ij})\left(\frac{1}{|\mathbf{R}|^3} - \frac{ik}{|\mathbf{R}|^2}\right)\right] e^{ik|\mathbf{R}|}, \tag{10.19}$$

where $k = \omega/c$ is the wavenumber in vacuum, $\mathbf{R}$ is the distance between the particles, and $\mathbf{n} = \mathbf{R}/|\mathbf{R}|$ is the unit vector directed from one particle to another.

However, this model has also severe restrictions. In particular, it does not allow making of any conclusions on the spatial structure of plasmon modes. To find the latter, Eq. (10.6) should be solved with the help of different versions of the perturbation theory or the total system of integral equations in Eq. (10.2). Above that, this model ceases to work when nanoparticles are situated rather closely, when the interaction of the higher multipoles is to be taken into account, and when the structure of plasmon oscillations and spectra can change even qualitatively.

10.1.3 *Weakly and Strongly Localized Plasmon Oscillations in Clusters of Strongly Interacting Nanoparticles*

As the distance between nanoparticles becomes shorter, the interaction becomes stronger, and simple reasoning does not provide answers to the questions of the spectra behavior and the surface charge distribution even qualitatively.

The main question arising in the case of the strong interaction is whether the increase of interaction of cluster nanoparticles is restricted to quantitative change of the eigenvalues and eigenfunctions of isolated nanoparticles or a quite radical modification both of the spectrum and of the eigenfunctions takes place at rather small distances. Further analysis shows that the more complicated second version takes place frequently, and when the particles are approaching each other, brand new oscillations arise together with the modified by interaction states of separate nanoparticles. By analogy with atom and molecular physics, it is possible to say that the bound states of plasmons or "plasmon molecules" emerge. These new states have very interesting properties (strong localization and enhancement of local fields) and will be considered in detail in further sections of the present chapter.

By an increase of the number of particles in a cluster, some phenomena can occur that are absent even qualitatively in a system of noninteracting nanoparticles. For example, in a chain of closely adjacent nanoparticles propagating plasmon modes analogous to one-dimensional surface plasmons (see Chapter 4) can also exist complementary to localized plasmon oscillations qualitatively similar to oscillations in separate nanoparticles and can form the basis for plasmon waveguides. These systems will be also considered below.

10.2 Two-Dimensional Plasmons in a Cluster of Two Nanowires

The simplest plasmon oscillations occur in a cluster of two parallel, circular nanowires. As it was shown in Chapter 5, to find the plasmon modes in the context of the ε-method, the homogeneous (i.e., having

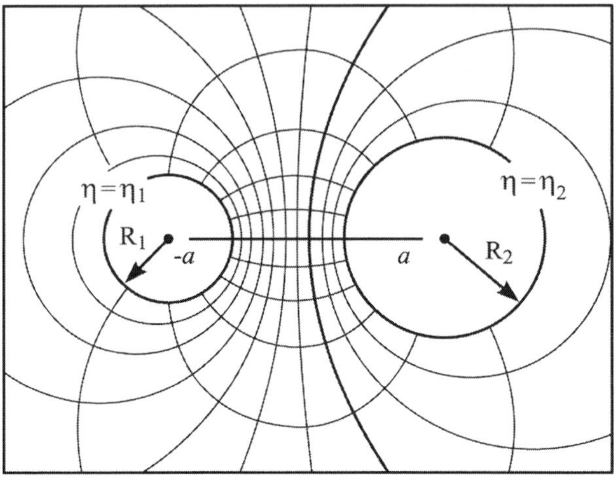

Figure 10.4 The coordinate lines of the bipolar system of coordinates.

no sources) problem of potential theory is to be solved upon the condition of continuity of the potential and the normal component of the electric induction on the nanobody boundary S:

$$\mathbf{E} = -\nabla\Phi; \Delta\Phi = 0,$$

$$[\Phi]_S = 0; \left[\varepsilon\frac{\partial\Phi}{\partial n}\right]_S = 0, \tag{10.20}$$

where ∇ denotes the gradient with respect to the coordinates $\mathbf{r}$ and Δ is the Laplace operator.

In the case of two parallel nanowires, the system Eq. (10.20) can be solved analytically for arbitrary radii and positions of the wires within the bipolar coordinate system ((Morse and Feshbach, 1953); see also Fig. 10.4). Coordinates of the bipolar system $-\infty < \eta < \infty$, $0 < \xi \leq 2\pi$ are related to the Cartesian coordinates by the relations

$$x = a\frac{\sin\xi}{\cosh\,\eta - \cos\xi}; \quad z = a\frac{\sinh\,\eta}{\cosh\,\eta - \cos\xi}, \tag{10.21}$$

where a is a half the distance between the poles of the bipolar system.

Within the bipolar coordinate system, variables can be separated, and any solution of the Laplace equation can be presented as a superposition of $e^{\pm n\eta}e^{\pm in\xi}$, $n = 1, 2, 3\ldots$, and the only problem remaining is to meet the boundary conditions. In the case of equal nanowires, the surface of the wires is determined by the equations $\eta = \pm\eta_0(\eta_0 > 0)$. It corresponds to cylinders of the radius $R = a/\sinh\eta_0$ with the distance between their centers $R_{12} = 2\sqrt{a^2 + R^2}$.

In the case of symmetrical plasmon oscillations ($\varphi(z) = \varphi(-z)$), the resonant dielectric constant takes values

$$\varepsilon^{\text{symm}} = -\tanh(n\eta_0),\ n = 1, 2\ldots \tag{10.22}$$

and corresponding eigenpotentials have the form

$$\varphi^{\text{symm}} = \left\{\begin{array}{ll} e^{n\eta}, & \eta < -\eta_0 \\ \dfrac{\cosh n\eta}{\cosh n\eta_0} e^{-n\eta_0}, & -\eta_0 < \eta < \eta_0 \\ e^{-n\eta}, & \eta > \eta_0 \end{array}\right\} \left\{\begin{array}{c} \cos n\xi \\ \sin n\xi \end{array}\right\}. \tag{10.23}$$

In the case of antisymmetric ($\varphi(z) = -\varphi(-z)$) plasmon oscillations, resonant values of the dielectric constant are equal to

$$\varepsilon^{\text{antisymm}} = -\coth(n\eta_0),\ n = 1, 2, 3\ldots \tag{10.24}$$

and corresponding eigenpotentials have the form

$$\varphi^{\text{antisymm}} = \left\{\begin{array}{ll} e^{n\eta}, & \eta < -\eta_0 \\ -\dfrac{\sinh n\eta}{\sinh n\eta_0} e^{-n\eta_0}, & -\eta_0 < \eta < \eta_0 \\ -e^{-n\eta}, & \eta > \eta_0 \end{array}\right\} \left\{\begin{array}{c} \cos n\xi \\ \sin n\xi \end{array}\right\}. \tag{10.25}$$

In Fig. 10.5, spatial distributions of potentials, Eqs. (10.23) and (10.25), proportional to $\cos n\xi$ are shown for $n = 1$.

From these figures, one can see that these plasmon oscillations are localized in the gap between the nanowires.

One of the most important characteristics of localized plasmons is enhancement of local fields. To find the enhancement factor, one should place their system in the external field. Again, in the case of

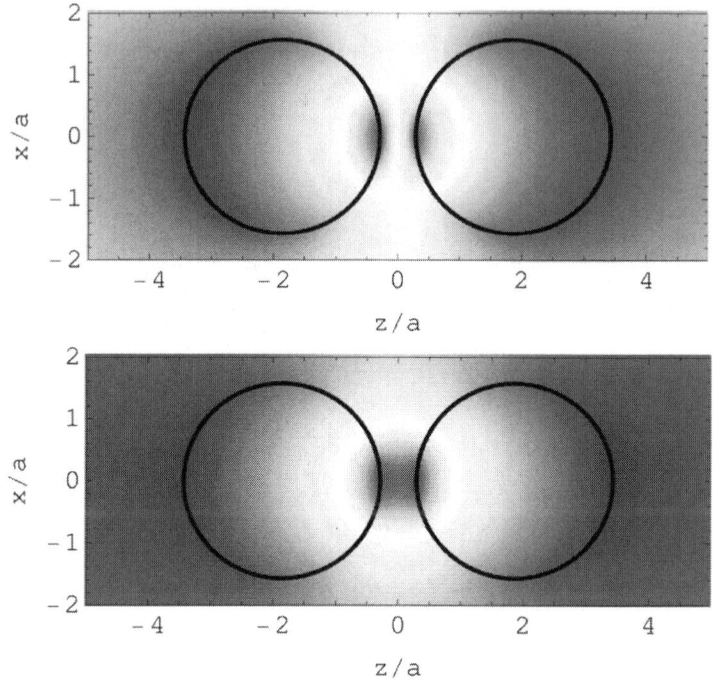

Figure 10.5 The spatial distributions of the eigenpotentials, Eq. (10.25) (top) and Eq. (10.23) (bottom) with $\cos n\xi$ dependence on ξ are shown for $n = 1(\eta_0 = 0.6)$.

two nanowires it is possible to expand the potential of the external electric field in bipolar coordinates:

$$\varphi_0 = -E_0\left(z\cos\varphi + x\sin\varphi\right),$$
$$= -E_0\begin{cases} a\cos\varphi + 2a\sum\limits_{n=1}^{\infty} e^{-n\eta}\cos\left(n\xi - \varphi\right),\ \eta > 0, \\ -a\cos\varphi - 2a\sum\limits_{n=1}^{\infty} e^{n\eta}\cos\left(n\xi + \varphi\right),\ \eta < 0, \end{cases} \tag{10.26}$$

and then to find the solution of Eq. (10.20).

The maximal enhancement of the external electric field occurs when the external field is parallel to the line connecting the wires ($\varphi = 0$ in Eq. (10.26)):

$$\varphi_0 = -E_0 z = -E_0\frac{\eta}{|\eta|}a\left(1 + 2\sum_{n=1}^{\infty} e^{-n|\eta|}\cos\left(n\xi\right)\right). \tag{10.27}$$

In this case of the external field, the solution of Eq. (10.20) has the form

$$\varphi_1^{\text{in}} = E_0 a \left(1 + 2\sum_{n=1}^{\infty} \frac{e^{n(\eta_0+\eta)}\cos(n\xi)}{\varepsilon \sinh(n\eta_0) + \cosh(n\eta_0)}\right),$$

$$\varphi^{\text{out}} = -E_0 a \left(\begin{array}{c} \frac{\eta}{|\eta|}\left(1 + 2\sum\limits_{n=1}^{\infty} e^{-n|\eta|}\cos(n\xi)\right) \\ -2\sum\limits_{n=1}^{\infty} \frac{(\varepsilon-1)\,e^{-n\eta_0}\sinh(n\eta)\cos(n\xi)}{\varepsilon\sinh(n\eta_0)+\cosh(n\eta_0)} \end{array}\right),$$

$$\varphi_2^{\text{in}} = -E_0 a \left(1 + 2\sum_{n=1}^{\infty} \frac{e^{n(\eta_0-\eta)}\cos(n\xi)}{\varepsilon \sinh(n\eta_0) + \cosh(n\eta_0)}\right). \quad (10.28)$$

The strongest enhancement occurs for $n = 1$ resonant dielectric value (see Eq. (10.24) and Fig. 10.5 (top)), and its resonant contribution to the field between the nanowires is

$$\varphi^{\text{out}} \approx 2E_0 a \frac{(\varepsilon-1)\,e^{-\eta_0}\sinh(\eta)\cos(\xi)}{\varepsilon\sinh(\eta_0)+\cosh(\eta_0)}. \quad (10.29)$$

The exact expression for the enhancement factor G of the resonant electric field in the point of one wire closest to another wire can be obtained from Eq. (10.29):

$$G = \left|\frac{E_z}{E_0}\right| = \left|\frac{2(\varepsilon(\omega)-1)}{\varepsilon(\omega)+\coth(\eta_0)}\frac{e^{-\eta_0}}{\sinh(\eta_0)}\cosh(\eta_0)(\cosh(\eta_0)+1)\right| \underset{\eta_0\to 0}{\approx} \frac{4}{\varepsilon''(\omega)\,\eta_0^2} \approx \frac{4}{\varepsilon''(\omega)}\frac{R}{\Delta}, \quad (10.30)$$

where Δ is the gap width between the nanowires of radius R. From this formula, one can see that, indeed, in the gap one can have substantial enhancement of the electric field. This enhancement is inversely proportional to the imaginary part of the permittivity $\varepsilon''(\omega)$ and to the gap width Δ between the nanowires and can be as large as one to three orders of magnitude.

10.3 Plasmons in a Cluster of Two Nanospheres

A wide range of theoretical (Klimov and Guzatov, 2007a, 2007b, 2007c; Guzatov and Klimov, 2005; Claro, 1982; Olivares et al.,

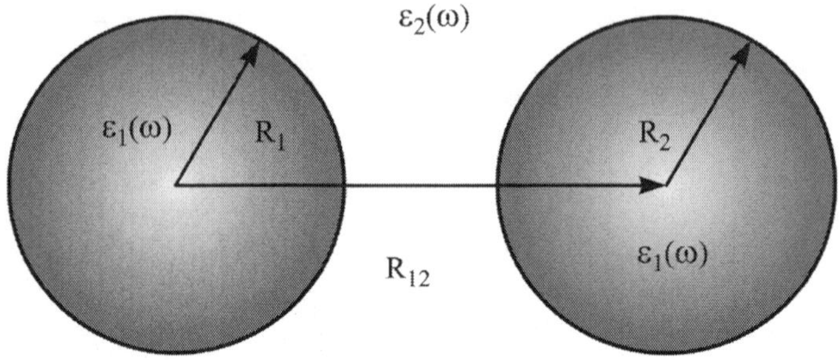

Figure 10.6 The geometry of a cluster of two nanospheres.

1987; Ruppin, 1982, 1989; Blanco and Garcia de Abajo, 2004; Paley et al., 1993; Nordlander et al., 2004; Xu and Kall, 2002; Genov et al., 2004; Chaumet and Dufour, 1998) and experimental (Su et al., 2003; Rechberger et al., 2003; Tamaru et al., 2002; Prikulis et al., 2004) studies is devoted to investigation of optical properties of clusters consisting of two spherical nanoparticles. The case of almost touching nanoparticles is the most interesting since in the interspace between the nanoparticles an especially high enhancement of the local field occurs, resulting in many practical consequences. This case is also the most difficult since the spatial structure of the fields is extremely complex.

In the present section, results of the investigation of optical properties of a cluster consisting of two spherical nanoparticles will be presented (Fig. 10.6). When stating this, we will refer to the research studies by Klimov and Guzatov (2007a, 2007b, 2007c).

In the case of two nanospheres, Eq. (10.20) can be conveniently solved in the bispherical system of coordinates (Fig. 10.4). Coordinates of the bispherical system $-\infty < \eta < \infty, 0 < \xi \leq \pi, 0 \leq \varphi < 2\pi$ are related to the Cartesian coordinates by the relations (Buchholz, 1961; Morse and Feshbach, 1953):

$$x = a\frac{\sin\xi\cos\varphi}{\cosh\ \eta - \cos\xi} \quad y = a\frac{\sin\xi\sin\varphi}{\cosh\ \eta - \cos\xi} \quad z = a\frac{\sinh\ \eta}{\cosh\ \eta - \cos\xi}. \tag{10.31}$$

The equation $\eta = \eta_1 < 0$ defines a sphere of the radius $R_1 = a/|\sinh\ \eta_1|$. Its center is situated in the point $x_1 = y_1 = 0, z_1 = a\coth\ \eta_1$. The second sphere may be defined analogously by the equation $\eta = \eta_2 > 0$. At that, its radius will be equal to $R_2 =$

$a/\sinh\ \eta_2$, and the center will be situated in the point $x_2 = y_2 = 0$, $z_2 = a \coth \eta_2$. The dimensional constant a is half of the distance between the poles of the bispherical system of coordinates. It is defined by the positive root of the equation:

$$R_{12} = \sqrt{R_1^2 + a^2} + \sqrt{R_2^2 + a^2},$$

where $R_{12} = z_2 - z_1$ is an arbitrary distance between the centers of the first and second spheres ($R_{12} \geq R_1 + R_2$). Hereafter in this section, only the case of spheres with equal radii R_0 ($-\eta_1 = \eta_2 = \eta_0$), made of the same material with the electric permittivity ε_1, and placed in a medium with the electric permittivity ε_2 will be considered for simplicity. The generalization to the case of different spheres will be considered in the next section.

In the bispherical system of coordinates, variables can be separated, and the arbitrary solution of the Laplace equation (see Eq. (10.20)) may be written in the form

$$\Phi = \frac{1}{a}\sqrt{\cosh\eta - \cos\xi}\, P_n^m\left(\cos\xi\right) e^{\pm\left(n+\frac{1}{2}\right)\eta \pm im\varphi}, \tag{10.32}$$

where n and m are arbitrary integer numbers and $P_n^m\left(\cos\xi\right)$ are the associated Legendre polynomials.

In the space outside the spheres ($-\eta_0 < \eta < \eta_0$), the most general solution may be sought as

$$\begin{aligned}\Phi = {} & \frac{1}{a}\sqrt{\cosh\ \eta - \cos\xi}\sum_{m=0}^{\infty}\sum_{n=1}^{\infty} P_{m+n-1}^{m}\left(\cos\xi\right)\\ & \times\left\{\left(\alpha_{mn}\frac{\cosh\ \left(\left(m+n-\frac{1}{2}\right)\eta\right)}{\cosh\ \left(\left(m+n-\frac{1}{2}\right)\eta_0\right)} + \gamma_{mn}\frac{\sinh\ \left(\left(m+n-\frac{1}{2}\right)\eta\right)}{\sinh\ \left(\left(m+n-\frac{1}{2}\right)\eta_0\right)}\right)\cos\left(m\varphi\right) +\right.\\ & \left.+\left(\beta_{mn}\frac{\cosh\ \left(\left(m+n-\frac{1}{2}\right)\eta\right)}{\cosh\ \left(\left(m+n-\frac{1}{2}\right)\eta_0\right)} + \delta_{mn}\frac{\sinh\ \left(\left(m+n-\frac{1}{2}\right)\eta\right)}{\sinh\ \left(\left(m+n-\frac{1}{2}\right)\eta_0\right)}\right)\sin\left(m\varphi\right)\right\}\end{aligned} \tag{10.33}$$

while inside the spheres only the solutions decreasing for $|\eta| \to \infty$ can be used according to the potential finiteness requirement. Thus, the potential inside the first ($\eta < -\eta_0 < 0$) and second ($\eta > \eta_0 > 0$)

spheres can be presented as

$$\begin{aligned}\Phi^{(1)} &= \frac{1}{a}\sqrt{\cosh\eta - \cos\xi}\sum_{m=0}^{\infty}\sum_{n=1}^{\infty} e^{(m+n-1/2)\eta} P_{m+n-1}^{m}(\cos\xi) \\ &\times\left\{a_{mn}^{(1)}\cos(m\varphi) + b_{mn}^{(1)}\sin(m\varphi)\right\}, \\ \Phi^{(2)} &= \frac{1}{a}\sqrt{\cosh\ \eta - \cos\xi}\sum_{m=0}^{\infty}\sum_{n=1}^{\infty} e^{-(m+n-1/2)\eta} P_{m+n-1}^{m}(\cos\xi) \\ &\times\left\{a_{mn}^{(2)}\cos(m\varphi) + b_{mn}^{(2)}\sin(m\varphi)\right\}.\end{aligned} \tag{10.34}$$

In Eqs. (10.33) and (10.34) α_{mn}, β_{mn}, γ_{mn}, δ_{mn} and $a_{mn}^{(1)}$, $b_{mn}^{(1)}$, $a_{mn}^{(2)}$, $b_{mn}^{(2)}$ are unknown coefficients, which are to be determined on the basis of the boundary conditions on the surfaces of both spheres.

From the condition of potential continuity on the surface of each nanosphere, we obtain the relation between $a_{mn}^{(1)}$, $a_{mn}^{(2)}$ and α_{mn}, γ_{mn} easily:

$$\begin{aligned} a_{mn}^{(1)} e^{-\left(m+n-\frac{1}{2}\right)\eta_0} &= \alpha_{mn} - \gamma_{mn} \\ a_{mn}^{(2)} e^{-\left(m+n-\frac{1}{2}\right)\eta_0} &= \alpha_{mn} + \gamma_{mn} \end{aligned}. \tag{10.35}$$

The equations for the coefficients $b_{mn}^{(1)}$ and $b_{mn}^{(2)}$ can be obtained from Eq. (10.35) with the help of the replacement $a_{mn}^{(1)} \to b_{mn}^{(1)}, a_{mn}^{(2)} \to b_{mn}^{(2)}$.

The continuity condition of the normal component of electric induction results in an infinite chain of interlocking equations (in contrast to the case of two nanowires; see Section 10.2). In the case of two identical spheres ($-\eta_1 = \eta_2 = \eta_0$), these equations may be written in the following form ($\varepsilon = \varepsilon_1/\varepsilon_2$):

$$\begin{aligned}&(n-1)\left(\varepsilon + \tanh\left((n+m-3/2)\,\eta_0\right)\right)\alpha_{m,n-1} - (2n+2m-1)\cosh\eta_0 \\ &\left(\varepsilon + \tanh\left((n+m-1/2)\,\eta_0\right) - (\varepsilon-1)\frac{\tanh\eta_0}{2n+2m-1}\right)\alpha_{mn} \\ &+ (n+2m)\left(\varepsilon + \tanh\left((n+m+1/2)\,\eta_0\right)\right)\alpha_{m,n+1} = 0\end{aligned} \tag{10.36}$$

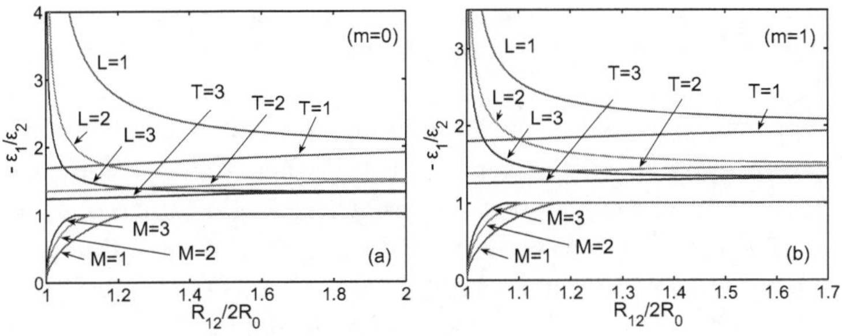

Figure 10.7 The electric permittivity eigenvalues in a cluster of two nanospheres, depending on the distance between them. (a) $m = 0$ and (b) $m = 1$. In the lower part of the pictures, the appearance of new modes (M-modes) on small distances between the nanospheres is distinctively perceptible (Klimov and Guzatov, 2007a).

and

$$(n-1)\left(\varepsilon+\coth\left(\left(n+m-3/2\right)\eta_0\right)\right)\gamma_{m,n-1}-\left(2n+2m-1\right)\cosh\eta_0$$
$$\left(\varepsilon+\coth\left(\left(n+m-1/2\right)\eta_0\right)-\left(\varepsilon-1\right)\frac{\tanh\eta_0}{2n+2m-1}\right)\gamma_{mn}$$
$$+(n+2m)\left(\varepsilon+\coth\left(\left(n+m+1/2\right)\eta_0\right)\right)\gamma_{m,n+1}=0. \tag{10.37}$$

It is essential that the coefficients α_{mn} and γ_{mn} enter in Eqs. (10.36) and (10.37) independently. This means that the resonance electric permittivities of the system of two identical nanospheres (dimer) may be of two independent types, satisfying Eqs. (10.36) and (10.37), correspondingly. At that, the system Eq. (10.36) describes the symmetric oscillations of the electric potential about the plane $z = 0$, and the system Eq. (10.37) describes the antisymmetric ones.

A search of the resonance electric permittivities in the proposed definition corresponds to solution of the so-called generalized eigenvalues problem for the systems of Eqs. (10.36) and (10.37). The generalized problem of eigenvalues (which has the form $\hat{A}x = \lambda\hat{B}x$, where the matrix B is nonnegative) is much more complex than the standard eigenvalues problem (which has the form $\hat{A}x = \lambda x$), and so its solutions describe a larger class of phenomena.

In Fig. 10.7, the dependencies of the eigenvalues (resonance values) of electric permittivity on the distance between nanoparticles composing the cluster for $m = 0$ (Fig. 10.7a) and $m = 1$ (Fig. 10.7b) are shown (Klimov and Guzatov, 2007a, 2007b). First of all, in this picture, the presence of three types of dispersion curves is seen. When the nanospheres approach each other, the modes with resonance electric permittivity tending to $-\infty$ are conditioned by the antisymmetric plasmon oscillations (L-modes) and determined by solutions of the system of equations in Eq. (10.37). The modes with resonance values of electric permittivity tending to a finite limit (T-modes) or to zero (M-modes) are related to symmetric plasmon oscillations and determined by solutions of the system of equations in Eq. (10.36).

As the distance between the nanospheres grows longer, the resonance electric permittivities corresponding to the modes of T- and L-types go over into the well-known values $\varepsilon = -(n+1)/n$ ($n = 1, 2, 3, \ldots$), which correspond to plasmon oscillations of different multipolarity in a single sphere (Chapter 6). The modes of this type are thoroughly studied and can be described with the help of hybridization of separate sphere modes (Nordlander et al., 2004). The modes of M-type were investigated only recently (Klimov and Guzatov, 2005; Klimov and Guzatov, 2007a, 2007b). First, the attention should be paid to the fact that these modes can arise only on the distances not exceeding $R_{12}/2R_0 \leq 1.2$. This circumstance is rather remarkable, and Klimov and Guzatov (2005), Klimov and Guzatov (2007a, 2007b) proposed to call these modes "bound states of plasmon atoms" or "plasmon molecules."

Let us now consider the structure of the wavefunctions (potentials) of plasmon oscillations in a cluster of two identical nanospheres in more detail. In Fig. 10.8, the spatial distribution of the wavefunction (potential) of a plasmon molecule (M-mode) and unbounded states of plasmon atoms (L-, T-modes) in the xz plane are shown.

In the axially symmetric case ($m = 0$, Fig. 10.8a), the spatial structure of antisymmetric (L-modes) and symmetric (T-modes) wavefunctions of unbounded plasmons corresponds generally to the structure of the plasmon wavefunction of isolated spheres. More specifically, the positive charge is situated on one semisphere, while

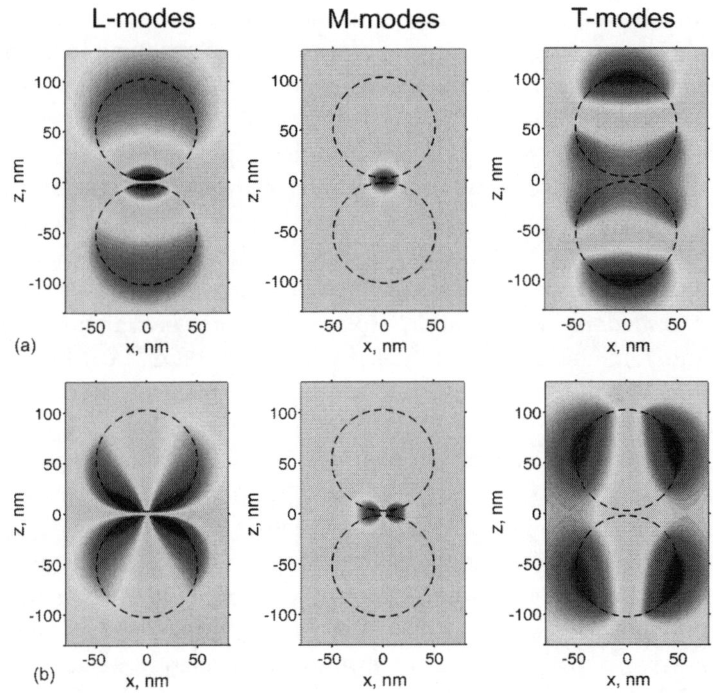

Figure 10.8 The spatial potential distribution in some plasmon modes in a cluster of two nanospheres. (a) $m = 0$ and (b) $m = 1$ (Klimov and Guzatov, 2007b).

the equal electroneutrality due to a negative charge is situated on the opposite part of the sphere. In this case, the interaction of plasmon atoms is restricted to some quantitative charge redistribution on the opposite semispheres.

In the case of *M*-modes ("plasmonic molecules"), the situation is completely different, and both positive and negative charges are localized near the gap between the nanospheres. Far from the gap the plasmon molecule wavefunctions are tending to zero in fact.

In the case of $m = 1$ (Fig. 10.8b), that is, in the case of angle dependency on the wavefunctions of the $\cos\varphi$ or $\sin\varphi$ type, the situation remains perfectly analogous. The only distinction is that dipole moments of unbounded plasmon atoms and plasmon molecules are directed along the x axis.

In general, the wavefunction of the unbounded plasmon atoms (L- and T-modes) remains more or less distributed over the volume of both nanospheres, while the potential of a plasmon molecule (M-mode) is localized strongly near the region between the nanospheres, with the maxima on the sphere surfaces.

As the distance between the spheres grows longer, the localization of plasmon molecules reduces, and when the distance between the spheres is critical, the plasmon molecules disappear, while the unbounded plasmonic states do not experience any changes.

Distinctions in localization of plasmon M-modes and unbounded L- and T-modes determine their fundamental difference relatively to applying of the external field. Unbounded plasmon modes have the polarizability of the order of the nanosphere volume $\alpha \sim R_0^3$ and interact with homogeneous external fields of corresponding orientation and symmetry effectively, and vice versa, plasmon molecules have relatively small polarizability $\alpha \sim \Delta^3$, where Δ is the width of the gap between the spheres. Due to this, M-modes are excited by a homogeneous optic field weakly in comparison with unbounded states of plasmons. On the other hand, M-modes interact effectively with strongly nonhomogeneous fields, which are localized near the interspace between the spheres. Fields of this kind arise during the emission of atoms and molecules situated near the gap. This circumstance makes plasmon molecules extremely prospective for construction of nanosensors and elements of nanodevices sensitive to single molecules. It is quite possible that surface-enhanced Raman scattering (SERS) (Michaels et al., 1999; Nie and Emory, 1997; Wang et al., 2003; Moskovits et al., 2002) is determined by the interaction of single molecules with M-modes. On the other hand, the excitation of plasmon molecules is possible with the help of scanning microscopes.

Illustrations of plasmon oscillations in the cluster of two spherical nanoparticles considered above were obtained with the help of the numerical solution of the systems of Eqs. (10.36) and (10.37). To obtain the analytical description of plasmon oscillations, the system of equations in Eq. (10.37) (for antisymmetric oscillations, i.e., for L-modes) and the system of equations in Eq. (10.36) (for M- and T-modes) may be written in the form of matrix equations:

$$[(\varepsilon - 1)\sinh\eta + (B + \cosh\eta A)(C + \varepsilon)]\,\gamma = 0 \qquad (10.38)$$

and

$$[(\varepsilon - 1)\sinh\eta + (B + \cosh\eta A)(T + \varepsilon)]\alpha = 0, \tag{10.39}$$

correspondingly.

In these expressions, the relatively simple infinite matrices are introduced:

$$A = -\begin{bmatrix} 2m+1 & 0 & 0 & 0 \\ 0 & 2m+3 & 0 & 0 \\ 0 & 0 & 2m+5 & 0 \\ 0 & 0 & 0 & \ldots\ldots \end{bmatrix} \tag{10.40}$$

$$B = \begin{bmatrix} 0 & 2m+1 & 0 & 0 \\ 1 & 0 & 2m+2 & 0 \\ 0 & 2 & 0 & \ldots \\ 0 & 0 & \ldots & 0 \end{bmatrix} \tag{10.41}$$

$$C = \begin{bmatrix} \coth\left(\frac{2m+1}{2}\right)\eta_0 & 0 & 0 & 0 & 0 \\ 0 & \coth\left(\frac{2m+3}{2}\right)\eta_0 & 0 & 0 & 0 \\ 0 & 0 & \coth\left(\frac{2m+5}{2}\right)\eta_0 & 0 & 0 \\ 0 & 0 & 0 & \coth\left(\frac{2m+7}{2}\right)\eta_0 & 0 \\ 0 & 0 & 0 & 0 & \ldots \end{bmatrix} \tag{10.42}$$

$$T = \begin{bmatrix} \tanh\left(\frac{2m+1}{2}\right)\eta_0 & 0 & 0 & 0 & 0 \\ 0 & \tanh\left(\frac{2m+3}{2}\right)\eta_0 & 0 & 0 & 0 \\ 0 & 0 & \tanh\left(\frac{2m+5}{2}\right)\eta_0 & 0 & 0 \\ 0 & 0 & 0 & \tanh\left(\frac{2m+7}{2}\right)\eta_0 & 0 \\ 0 & 0 & 0 & 0 & \ldots \end{bmatrix} \tag{10.43}$$

The matrix formulations in Eqs. (10.38) and (10.39) are especially convenient for analytical calculations.

Using Eq. (10.36), in the case of $\eta_0 \to 0$ (approaching nanospheres) asymptotic expressions for resonance parameters $\varepsilon = \varepsilon_1/\varepsilon_2$ can be found. For M-modes we have

$$\begin{aligned} \varepsilon_m^{(M)} &= \frac{\varepsilon_1}{\varepsilon_2} = -\left(\left(m + M + \delta_m\right)\eta_0\right) \\ m &= 0,\ 1,\ 2, \ldots ;\ M = 1,\ 2,\ 3, \ldots \end{aligned} \tag{10.44}$$

The corrections δ_m are independent on the mode number and equal to [1/2, −0.08578, −0.2639, −0.33, −0.3769, −0.4, −0.4172, −0.4289, −0.4377, −0.4446] for $m = 0 \ldots 9$ correspondingly. For $m > 1$, these corrections can be described by the series:

$$\delta_m = -\frac{1}{2} - \frac{1}{2m} + \frac{1}{8m^3} - \frac{1}{16m^5} + \frac{5}{128m^5} + \ldots \tag{10.45}$$

For high values of m, these corrections tend to $-1/2$.

These expressions show that when approaching nanospheres the relative resonance electric permittivity of a dimer tends to zero in the case of M-modes.

Vanishing of the relative electric permittivity corresponds to two physically different cases, two metal nanospheres ($\varepsilon_1 = -\infty$, $\varepsilon_2 = 1$) and two cavities in metal ($\varepsilon_1 = 1$, $\varepsilon_2 \to -\infty$).

Analogously, the following analytical expression can be obtained for relative resonance electric permittivities of L-modes:

$$\begin{aligned} \varepsilon_{m,L} &= \frac{\varepsilon_1}{\varepsilon_2} = -\left(m + L - \frac{1}{2}\right)^{-1} \eta_0^{-1} + \ldots \\ &(m = 0,\ 1,\ 2,\ 3,\ \ldots; \quad L = 1,\ 2,\ 3,\ \ldots), \end{aligned} \tag{10.46}$$

that is, when the nanospheres approach each other, the resonance values of ε tend to $-\infty$.

Eigenvectors of L- and M-modes, which correspond to the resonance electric permittivity, Eqs. (10.46) and (10.44), correspondingly, are also calculated analytically:

$$\alpha_{0,M} = \frac{1}{M}\begin{pmatrix} \left.\begin{matrix} -1 \\ -1 \\ \vdots \\ -1 \end{matrix}\right\} M \\ M \\ 0 \\ \vdots \\ 0 \end{pmatrix}, \qquad \gamma_{m,L} = \begin{pmatrix} \left.\begin{matrix} 0 \\ 0 \\ \vdots \\ 0 \end{matrix}\right\} (L-1) \\ 1 \\ 0 \\ \vdots \\ 0 \end{pmatrix}. \tag{10.47}$$

Let us note that the analogous expression for $\delta_{m,L}$ coincides with the expression in Eq. (10.47) for $\gamma_{m,L}$.

Knowing the eigenvectors α_{mn}, γ_{mn}, expressions for the spatial potentials (wavefunctions) of plasmons can be found. For the potentials of M-modes ($m = 0$) we obtain ($M = 1, 2, 3, \ldots$)

$$\Phi_{0,M}\Big|_{\eta_0\to 0} \approx -\frac{1}{aM}\sqrt{\cosh\,\eta - \cos\xi} \left(\begin{array}{l} \sum_{n=1}^{M} \frac{\cosh\,((n-\frac{1}{2})\eta)}{\cosh\,((n-\frac{1}{2})\eta_0)} P_{n-1}(\cos\xi) + \\ \quad - M\frac{\cosh\,((M+\frac{1}{2})\eta)}{\cosh\,((M+\frac{1}{2})\eta_0)} P_M(\cos\xi) \end{array} \right) \tag{10.48}$$

and

$$\begin{array}{l} \Phi^{(1)}_{0,M}\Big|_{\eta_0\to 0} \approx -\frac{1}{aM}\sqrt{\cosh\,\eta - \cos\xi} \\ \left(\begin{array}{l} \sum_{n=1}^{M} e^{(n+\frac{1}{2})(\eta+\eta_0)} P_{n-1}(\cos\xi) - \\ \quad - Me^{(M+\frac{1}{2})(\eta+\eta_0)} P_M(\cos\xi) \end{array} \right), \\ \Phi^{(2)}_{0,M}\Big|_{\eta_0\to 0} \approx -\frac{1}{aM}\sqrt{\cosh\,\eta - \cos\xi} \\ \left(\begin{array}{l} \sum_{n=1}^{M} e^{-(n-\frac{1}{2})(\eta-\eta_0)} P_{n-1}(\cos\xi) - \\ \quad - Me^{-(M+\frac{1}{2})(\eta-\eta_0)} P_M(\cos\xi) \end{array} \right). \end{array} \tag{10.49}$$

For the potentials of L-modes we have ($L = 1, 2, 3, \ldots$)

$$\Phi_{m,L}\Big|_{\eta_0\to 0} \approx \frac{1}{a}\sqrt{2\,(\cosh\,\eta - \cos\xi)}\,\sinh \left(\left(L - \frac{1}{2}\right)\eta\right) P^m_{L-1}(\cos\xi)\cos(m\varphi) \tag{10.50}$$

and

$$\begin{array}{l} \Phi^{(1)}_{m,L}\Big|_{\eta_0\to 0} \approx -\frac{\sqrt{2(\cosh\eta - \cos\xi)}}{a}\sinh\left(\left(L - \frac{1}{2}\right)\eta_0\right) e^{\left(L-\frac{1}{2}\right)(\eta+\eta_0)} \\ \quad P^m_{L-1}(\cos\xi)\cos(m\varphi), \\ \Phi^{(2)}_{m,L}\Big|_{\eta_0\to 0} \approx \frac{\sqrt{2\,(\cosh\,\eta - \cos\xi)}}{a}\sinh\left(\left(L - \frac{1}{2}\right)\eta_0\right) e^{-\left(L-\frac{1}{2}\right)(\eta-\eta_0)} \\ \quad P^m_{L-1}(\cos\xi)\cos(m\varphi). \end{array} \tag{10.51}$$

Knowing these potentials, the surface charge accumulating on each nanosphere can be calculated:

$$4\pi\sigma^{(i)} = \frac{1}{a}\left(\cosh\eta_0 - \cos\xi\right)\frac{\eta_i}{|\eta_i|}\left(\varepsilon_1\frac{\partial\Phi^{(i)}}{\partial\eta} - \varepsilon_2\frac{\partial\Phi}{\partial\eta}\right)\Bigg|_{\eta=\eta_i} \quad (i = 1, 2). \tag{10.52}$$

Using the asymptotic expressions for the potentials, Eqs. (10.48)–(10.51), we obtain the following asymptotic dependency for the M-mode ($m = 0$):

$$\sigma^{(1)}_{0,M} \approx -\frac{\varepsilon_2\eta_0\left(1-\cos\xi\right)^{1/2}}{8\pi a^2}\left(-\frac{1}{M}\sum_{n=0}^{M-1}\left(1-(2n+1)\left(1-\cos\xi\right)\right).\right.$$
$$\left. P_n(\cos\xi) + \left(1-(2M+1)\left(1-\cos\xi\right)\right)P_M\left(\cos\xi\right)\right) \approx \sigma^{(2)}_{0,M}, \tag{10.53}$$

and for the L-mode:

$$\sigma^{(1)}_{m,L} \approx -\frac{\varepsilon_2(1-\cos\xi)^{3/2}}{4\pi a^2\eta_0}\left(\frac{2L-1}{2(L+m)-1}\right)P^m_{L-1}(\cos\xi)\cos m\varphi \approx -\sigma^{(2)}_{m,L}. \tag{10.54}$$

In the case of $m \to \infty$, the leading terms in Eqs. (10.36) and (10.37) (which are proportional to m) become uncoupled, and the eigenvalues of electric permittivity and the corresponding eigenvectors can be found again:

$$\begin{array}{l}\varepsilon_{m,L} = -\coth\left((m+L-1/2)\eta_0\right) \\ m \gg 1;\ L = 1, 2, 3, \ldots\end{array}. \tag{10.55}$$

$$\begin{array}{l}\varepsilon_{m,M} = -\tanh\left((m+M-1/2)\eta_0\right) \\ m \gg 1;\ M = 1, 2, 3, \ldots\end{array}. \tag{10.56}$$

$$\alpha_{m,M} = \begin{pmatrix} \left.\begin{matrix}0\\0\\\vdots\\0\end{matrix}\right\}(M-1) \\ 1 \\ 0 \\ \vdots \\ 0 \end{pmatrix}, \quad \gamma_{m,L} = \begin{pmatrix} \left.\begin{matrix}0\\0\\\vdots\\0\end{matrix}\right\}(L-1) \\ 1 \\ 0 \\ \vdots \\ 0 \end{pmatrix}. \tag{10.57}$$

It is interesting to note that the eigenvalues, Eqs. (10.55) and (10.56), look very similar to the eigenvalues of permittivity in the case of two nanowires (see Eqs. (10.22) and (10.24)).

In the intermediate case $\eta_0 < 1, m > 1$, good asymptotics are obtained when the nondiagonal matrix B, Eq. (10.41), is simply neglected in Eqs. (10.38) and (10.39). As a result, the approximated expressions for the resonance values of electric permittivity take the form

$$\varepsilon_{mL} = -\frac{(L+m)\coth\left((L+m-1/2)\eta_0\right)+\tanh\,\eta_0}{(L+m)-\tanh\,\eta_0},$$
$$m = 0, 1, 2, \ldots; L = 1, 2, 3, \ldots \qquad (10.58)$$
$$\varepsilon_{mM} = -\frac{(M+m)\tanh\left((M+m-1/2)\eta_0\right)+\tanh\,\eta_0}{(M+m)-\tanh\,\eta_0},$$
$$m = 0, 1, 2, \ldots; M = 1, 2, 3. \qquad (10.59)$$

The plasmon oscillations of T-modes have no clearly defined peculiarities when the nanospheres are approaching each other, and thus their characteristics can be found with the help of numerical methods, if it is necessary.

Eqs. (10.44), (10.46), (10.58), and (10.59) are only the first terms of the asymptotics for $\eta_0 \to 0$ and $m \to \infty$, correspondingly. In principle, following terms can be found in context of perturbation theory as it is done in quantum mechanics (Davydov, 1973).

Above, we have considered the case of nonoverlapping spheres. When the spheres come in contact, their plasmonic spectrum changes drastically (Romero et al., 2006).

10.4 Local Fields' Enhancement in a Cluster of Two Nanospheres

One of the most significant problems of nanoplasmonics is focusing of optical fields into nanometric regions, that is, their substantial enhancement. A cluster of two nanoparticles is the most important system, which allows one to arrange this enhancement. A homogeneous external field is enhanced by the mode $m = 0, L = 1$ especially effectively. To estimate this enhancement, the help of the asymptotic expressions found in the previous section can be invaluable.

To perform these estimations, it is natural to use the general extension for the fields in the presence of plasmon nanoparticles (Chapter 5) and to keep the resonance term only:

$$\mathbf{E} \approx \mathbf{E}_0 + \frac{(1-\varepsilon(\omega))\mathbf{e}_L \int \mathbf{e}_L \cdot \mathbf{E}_0 dV}{(\varepsilon(\omega)-\varepsilon_L)\int \mathbf{e}_L^2 dV} \approx -\frac{\varepsilon_L \mathbf{e}_L \int \mathbf{e}_L \cdot \mathbf{E}_0 dV}{(\varepsilon(\omega)-\varepsilon_L)\int \mathbf{e}_L^2 dV}, \tag{10.60}$$

where ε_L, $\mathbf{e}_L = -\nabla\varphi_L$ is the eigenvalue of permittivity and the eigenfunction of the resonance plasmon electric field, and the integral is taken over the volume of both particles. Substituting asymptotics found in the previous section, Eqs. (10.44) to (10.60), in the case of the most bright longitudinal mode ($m = 0$, $L = 1$) we obtain the following expression for the enhancement coefficient:

$$G = \left|\frac{E_{\max}}{E_0}\right| \approx \frac{8}{3}\frac{R_0}{\Delta}\frac{1}{\varepsilon''(\omega)}, \tag{10.61}$$

where R_0 is the radius of the spheres, Δ is the width of the gap between the nanoparticles, and $E_{\max}$ is the maximal field on the external surface of the particle. It is important to note that Eq. (10.61) is very similar to the enhancement factor in the presence of two nanowires, Eq. (10.30). This expression allows to estimate that the field amplitude enhancement in a cluster of two identical spheres will be of the order of 10–1,000. A more substantial enhancement can be achieved by using more complicated clusters of nanoparticles (see, e.g., Sections 10.6.3 and 10.7).

10.5 Plasmons in a Cluster of Two Different Nanospheres and in Nonsymmetric Nanoshells

A wide range of practically interesting geometries having the boundaries coinciding with the coordinate surfaces of the bispherical system coordinates, can be investigated complementary to clusters consisting of two identical spheres.

The perfectly analogous approach can also be applied to a case of two spheres of different radii and compositions. In Fig. 10.9, practically interesting nanosystems are shown, which can be

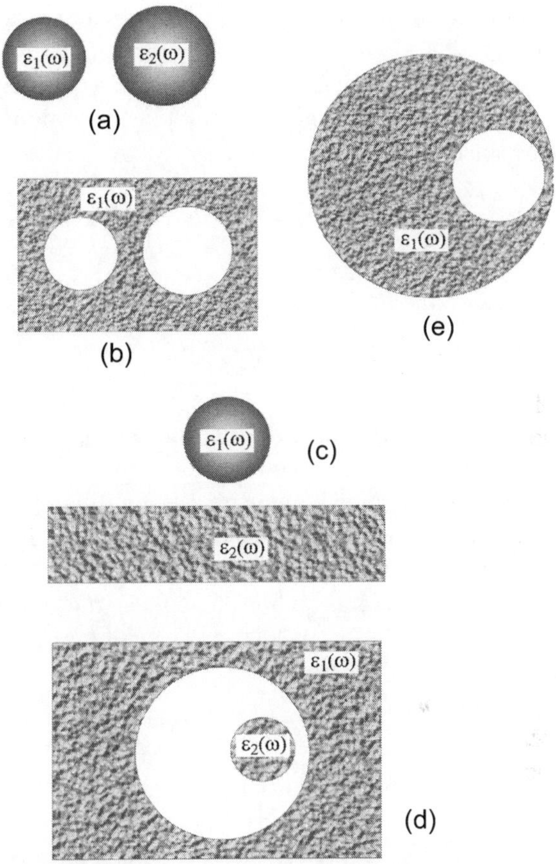

Figure 10.9 The variety of geometries that can be described in the context of the bispherical system of coordinates. (a) Two spherical nanoparticles of arbitrary sizes and composition; (b) two bubbles in metal; (c) a plasmon nanoparticle near the surface; (d) a nanoparticle in the cavity in metal (nanoegg); and (e) a nonsymmetric cavity in a plasmon nanoparticle (nanoegg).

described in the bispherical system of coordinates successfully. The system of recurrent correlations for the geometries shown in Fig. 10.9 has a slightly more complex form (Klimov and Guzatov, 2007a, 2007b, 2007c) in comparison with Eqs. (10.36) and (10.37). The main peculiarity of this system of equations is that now there are no symmetric or antisymmetric plasmon oscillations. Nevertheless, the solution of this system can be found both by numerical and by asymptotic methods.

In the case of a small distance between the spherical surfaces relative to their radii, the spectrum of plasmon modes can be described by the expression

$$\left(\frac{\varepsilon_1 - \varepsilon_3}{\varepsilon_1 + \varepsilon_3}\right)\left(\frac{\varepsilon_2 - \varepsilon_3}{\varepsilon_2 + \varepsilon_3}\right) \approx \exp\left[(2N + 2m - 1)(\eta_2 - \eta_1)\right], \quad (10.62)$$
$$N = 1, 2, 3, \ldots \quad m = 0, 1, 2, \ldots$$

where the parameter $\eta_2 - \eta_1$ is determined by the expression

$$\cosh\,(\eta_2 - \eta_1) = \left|\frac{R_{12}^2 - R_1^2 - R_2^2}{2R_1R_2}\right|, \quad (10.63)$$

and ε_1, ε_2, ε_3 are electric permittivities of the first sphere, the second sphere, and the interspace between them, correspondingly.

In the case of two metal nanospheres of different radii and compositions with a small interspace between them (Fig. 10.9a), of nanospheres in the vicinity of the metal surface (Fig. 10.9c), and of nanoparticles in the metal cavity (Fig. 10.9d), the solution of the dispersion equation Eq. (10.60) can be presented in the form

$$(\omega_m)^2 = 1/4\left(\omega_{\text{pl1}}^2 + \omega_{\text{pl2}}^2 \pm\sqrt{\left(\omega_{\text{pl1}}^2 - \omega_{\text{pl2}}^2\right)^2 + 4\omega_{\text{pl1}}^2\omega_{\text{pl2}}^2 \exp\left(-(2n + 2m + 1)(\eta_2 - \eta_1)\right)}\right),$$
$$n = 1, 2, 3, \ldots; m \gg 1$$

where ω_{pl1}, ω_{pl2} are the bulk plasmon frequencies of the substance in the regions 1 and 2 and it is assumed that medium 3 has an electric permittivity equal to 1.

In the case of two nanobubbles in metal (Fig. 10.9b) or of a nonsymmetric bubble in a nanoparticle (Fig. 10.9e), the expressions for the plasmon oscillation frequencies can be written as

$$\omega_{\pm} = \frac{\omega_{\text{pl1}}}{\sqrt{2}}\sqrt{1 \pm \exp(-(n + m + 1/2)(\eta_2 - \eta_1))}, \quad (10.64)$$

where it is assumed that the rest regions of the space have the electric permittivity equal to 1, for simplicity.

Methods of synthesis of nonsymmetric nanoshells (nanoeggs) were developed by Wang et al. (2006b). In this paper, numerical

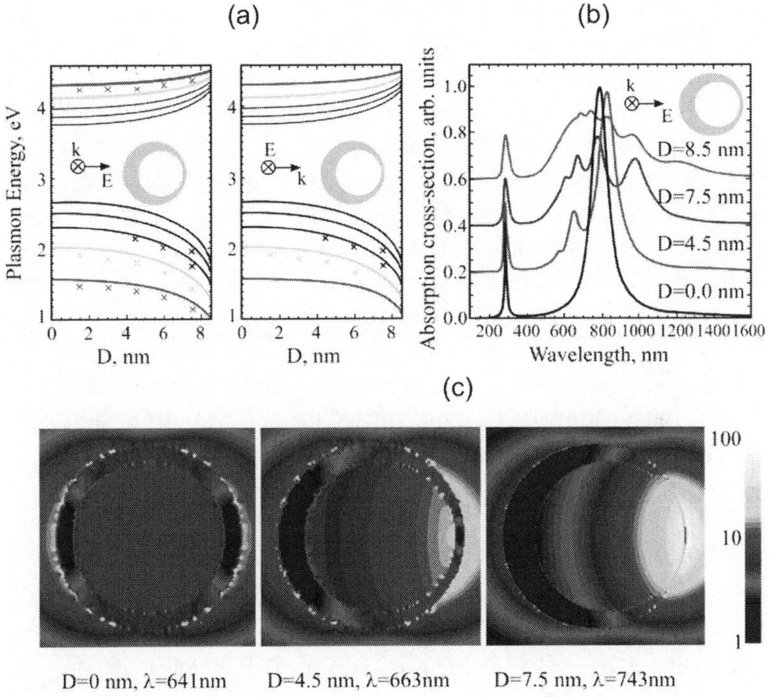

Figure 10.10 Results of numerical calculations of plasmon properties of nonsymmetric gold nanoshells (nanoeggs) with the internal radius of 39 nm and the external radius of 48 nm. (a) Plasmon spectra for different polarizations of the external field, (b) the absorption cross section depending on the nanoegg asymmetry degree, and (c) the distribution of the electric field intensity for different shifts of the core of SiO_2 and for different wavelengths of the horizontally polarized exciting field; (left): the shift is 0 (the symmetric spherical nanoshells), and the wavelength is 641 nm; (middle): the shift is 4.5 nm, and the wavelength is 663 nm; (right): the shift is 7.5 nm, and the wavelength is 743 nm (Wang et al., 2006b).

calculations of properties of such particles were also stated. In Fig. 10.10, results of these calculations are shown are shown. From these pictures, it is seen that there is a significant field enhancement as in the case of a cluster of two nanoparticles; however, in the case of a nanoegg this enhancement occurs outside the particle but not in the gap between the nanoparticles.

10.6 Plasmons in a Cluster of Two Nonspherical Nanoparticles of Finite Volume

This problem is even more complex than the problems considered above. However, from a general point of view, it is possible to say that for rather long distances between the nanoparticles, properties of the cluster are determined mainly by properties of the particles composing it with dipole–dipole interactions between the particles taken into account (see Section 10.1.2). At short distances, the existence of highly localized plasmon molecules (M-modes) with parameters determined by the local radii of the nanoparticles' curvature is possible. Investigations of plasmon oscillations in a cluster of nonspherical nanoparticles are not numerous, and here we consider only three geometries.

10.6.1 *Plasmon Oscillations in a Cluster of Two Nanocubes*

This geometry (Fig. 10.11) was investigated with the help of the numerical discrete dipole approximation (DDA) method (Ruppin, 1997). In Fig. 10.12a, results of calculations of the extinction cross section for the incident light polarization perpendicular and parallel to the lines joining the cubes are shown. From this picture, it is seen that these spectra differ substantially both from each other and from the spectrum of a single nanocube (Fig. 10.12b; see also Chapter 9).

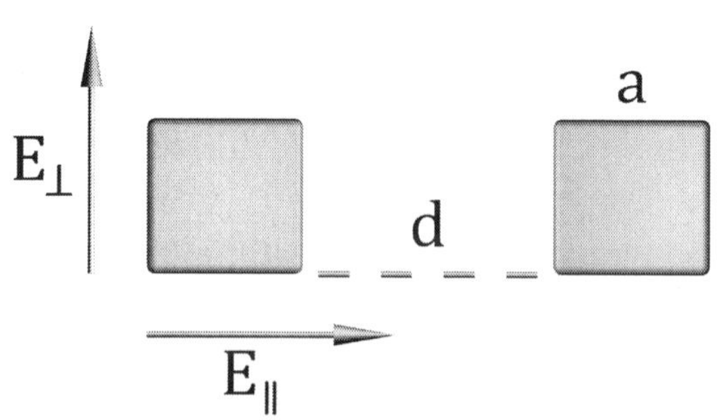

Figure 10.11 The geometry of the problem of a cluster of two nanocubes.

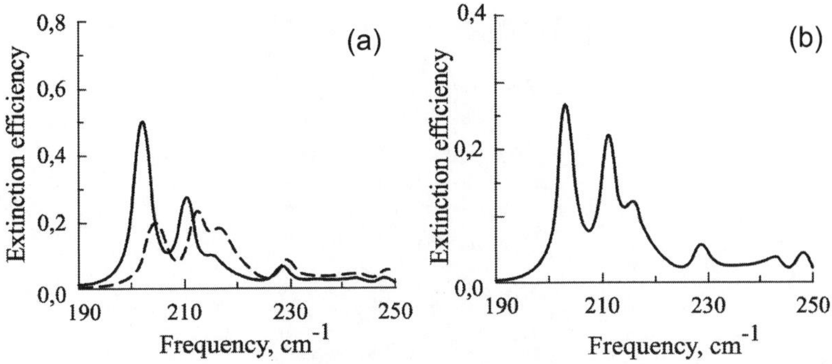

Figure 10.12 (a) The normalized extinction cross section of a cluster of two NaCl nanocubes with the side $a = 0.1$ μm and the distance between them $d = a/2$. The solid curve is the parallel polarization, and the dashed curve is the perpendicular polarization. (b) The extinction cross section of a single cube (Ruppin, 1997).

10.6.2 *Experimental Study of Plasmon Oscillations in a Cluster of Two Gold Nanodiscs*

The cluster of two gold nanodiscs of the diameter of 150 nm and of the height of 17 nm with different distances between their centers (450, 399, and 159 nm) was investigated by Rechberger et al. (2003). These systems were fabricated with the help of electron beam lithography, and their scanning electron microscopy (SEM) images are shown in Fig. 10.13.

Corresponding experimentally measured extinction cross sections (including the specimen with the interparticle distance of 200 nm) are shown in Fig. 10.14. In Fig. 10.14a, the polarization of incident light is parallel to the line joining the nanodiscs, while in Fig. 10.14b the polarization is normal to this line. From these pictures, the red shift of plasmon resonance for the parallel polarization due to a decrease of the distance between the nanodiscs is seen clearly. For the perpendicular polarization, the smaller shift of the peak to the blue region occurs. For the maximal distance between the nanodiscs (450 nm), the lattice is square, and the extinction spectra coincide for the both polarizations, naturally.

The position of the plasmon resonances peaks as a function of the interparticle distance is shown in Fig. 10.15 for both polarizations.

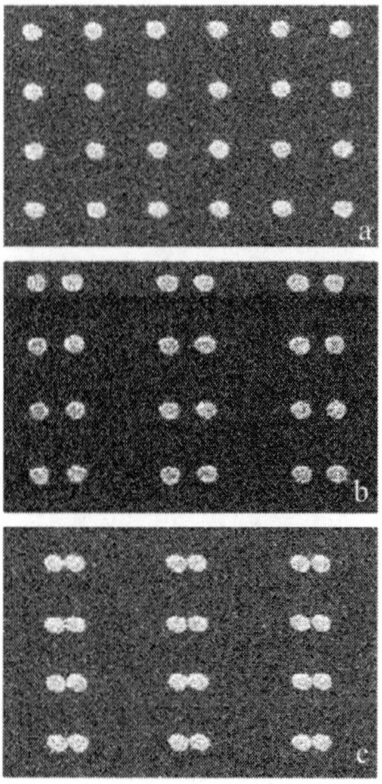

Figure 10.13 SEM images of clusters of nanodiscs with different distances between their centers: (a) 450 nm, (b) 300 nm, and (c) 150 nm. The nanodiscs' diameter is 150 nm, and the height is 17 nm (Rechberger et al., 2003).

Results of plasmon oscillations' interaction modeling within the two-dipole approximation are shown in this picture as well (see Section 10.1.2). At that, polarizabilities of a single disc were considered equal to the polarizability of the gold oblate spheroid approximating it. As it is seen from the picture, even this simple model allows to describe the experimental data satisfactorily.

10.6.3 *Plasmon Oscillations in a Cluster of Two Nanospheroids*

In previous sections, we have presented results on plasmonic properties of clusters of two cubes or two discs. These results were obtained numerically (cubes) or experimentally (discs). However, to understand plasmonic properties of a cluster made of nonspherical particles, one needs more detailed studies. As far as we know, there is only one analytical investigation of optical properties of two-

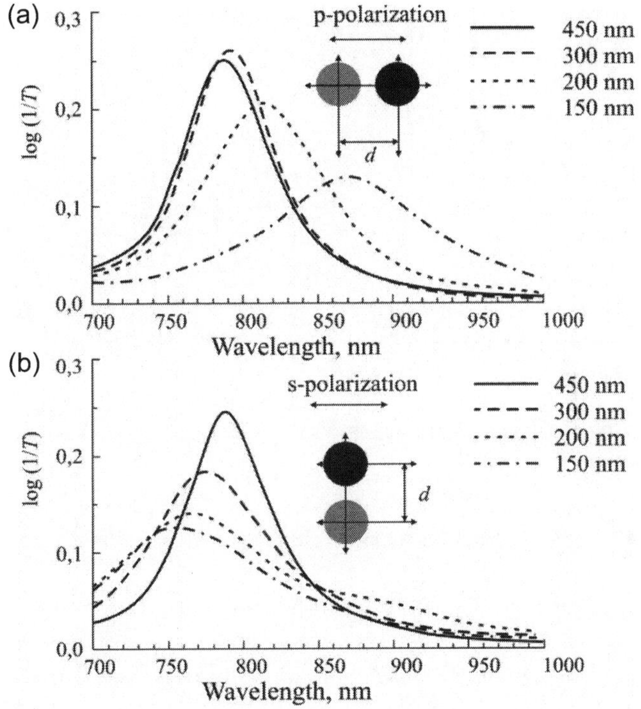

Figure 10.14 Extinction spectra of a two-dimensional lattice of pairs of gold nanoparticles for different distances between their centers (see Fig. 10.13). The polarization of the incident light is parallel (a) and perpendicular (b) to the line joining the nanodiscs (Rechberger et al., 2003).

nanospheroid clusters carried out with the help of the plasmon hybridization method (Willingham et al., 2008). In this paper, the interaction between unperturbed plasmonic modes of prolate spheroids was calculated by numerical integration. Fortunately, it is possible to find a fully analytical solution for plasmonic properties of a cluster of two nanospheroids and to find optical coefficients and decay rates of an atom placed near such a cluster (Guzatov and Klimov, 2011). Here, we will present this new analytical solution, which is obtained with the help of the new translational addition theorem (Guzatov and Klimov, 2010). It allows us to find an analytical solution for a cluster of two spheroidal nanoparticles

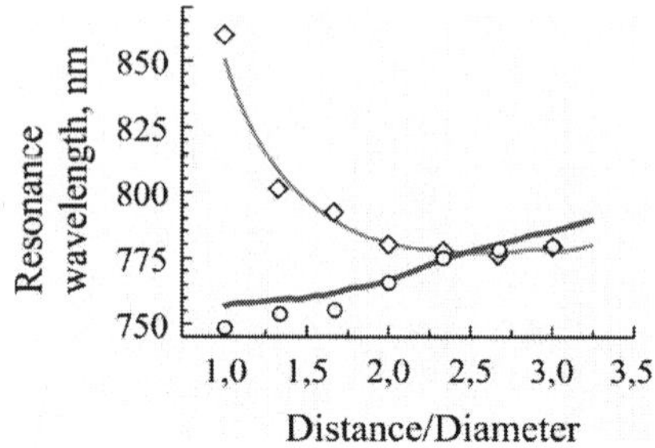

Figure 10.15 The position of spectra maxima, depending on the distance between the nanoparticles (four corresponding spectra are shown in Fig. 10.14) for the incident emission polarization parallel to the line joining the nanodiscs (the diamonds) and perpendicular to it (the circles). The solid lines show results of the modeling with the help of the point dipole model (Rechberger et al., 2003).

placed in an arbitrary external field. The geometry of the problem is shown in Fig. 10.16.

Investigations of prolate and oblate spheroids are similar, and here for brevity we outline only the solution for prolate spheroids. For derivation of formulas for oblate spheroids, see (Guzatov and Klimov, 2011).

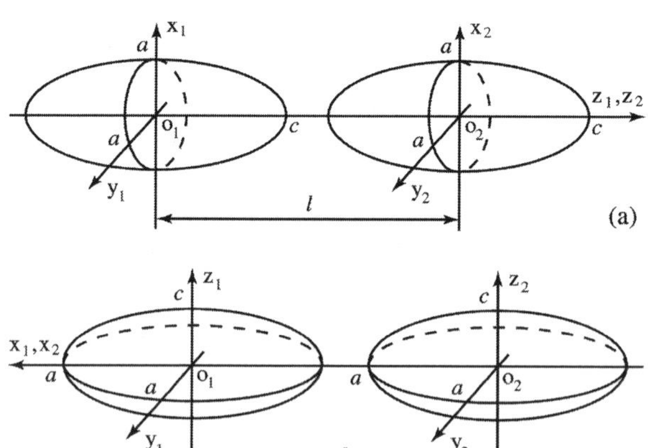

Figure 10.16 The geometry of a two-nanospheroid cluster. (a) The case of prolate nanospheroids and (b) the case of oblate nanospheroids.

For studying plasmon oscillations and other optical properties of clusters of two nanospheroids, it is enough as usual to find the solution of the Laplace equation:

$$\begin{aligned}
&\Delta\varphi_\nu^{\text{in}} = 0, \quad \text{inside the nanoparticle,}\\
&\Delta\varphi_\nu^{\text{out}} = 0, \quad \text{outside the nanoparticle,}\\
&\left.\begin{aligned}
&\varphi_\nu^{\text{in}}\big|_S = \varphi_\nu^{\text{in}}\big|_S\\
&\varepsilon_\nu \frac{\partial\varphi_\nu^{\text{in}}}{\partial\mathbf{n}}\bigg|_S = \frac{\partial\varphi_\nu^{\text{out}}}{\partial\mathbf{n}}\bigg|_S
\end{aligned}\right\} \text{at the surface of the nanoparticles}
\end{aligned} \tag{10.65}$$

In Eq. (10.65), φ_ν^{in}, φ_ν^{out} are the potentials of plasmon eigenfunctions inside and outside the nanoparticle, correspondingly, and $\partial\varphi_\nu/\partial\mathbf{n}|_S$ denotes the normal derivative at the nanoparticles' surface S. The last equation in Eq. (10.65) provides continuity of the normal components of electrical induction.

The systems of equations obtained in such a way have nontrivial solutions only for some negative values of permittivity ε_ν defining frequency of plasmon oscillations. In the case of Drude theory, $\varepsilon(\omega) = 1 - \omega_{\text{pl}}^2/\omega^2$, the frequency of plasmon oscillations ω_ν can be found from the expression

$$\omega_\nu = \frac{\omega_{\text{pl}}}{\sqrt{1-\varepsilon_\nu}} < \omega_{\text{pl}}, \tag{10.66}$$

where ω_{pl} is the bulk plasmon frequency of the metal composing the nanoparticles. Our approach allows us to investigate arbitrary spheroids, but for simplicity in the present section, we examine equations for plasmon oscillations in a cluster of two identical metal nanospheroids.

To find plasmonic spectra of a two-nanospheroid cluster, it is natural to use two local systems of spheroidal coordinates (ξ_j, η_j, φ_j), $j = 1, 2$ with the origins o_j placed in the centers of corresponding nanospheroids (see Fig. 10.16a) (see Chapter 7 for definitions of spheroidal coordinates). Coordinates (and all other values) related to the first or second nanospheroid will be denoted by the index "1" or "2," respectively. The potential inside the jth nanospheroid can be presented in the form (Smythe, 1950) ($j = 1$ and 2):

$$\varphi_j^{\text{in}} = \sum_{n=0}^{\infty}\sum_{m=0}^{n} P_n^m(\xi_j)P_n^m(\eta_j)\left(A_{mn}^{(j)}\cos(m\varphi_j) + B_{mn}^{(j)}\sin(m\varphi_j)\right), \tag{10.67}$$

where $P_n^m(\eta)$ is an associated Legendre function defined in the region $-1 \le \eta \le 1$ and $P_n^m(\xi)$ is an associated Legendre function defined in a complex plane with the branch cut from $-\infty$ to $+1$ (Abramowitz Stegun, 1965). The partial potential outside the jth nanospheroid can be presented as ($j = 1$ and 2):

$$\varphi_j^{\text{out}} = \sum_{n=0}^{\infty}\sum_{m=0}^{n} Q_n^m(\xi_j)P_n^m(\eta_j)\left(C_{mn}^{(j)}\cos(m\varphi_j) + D_{mn}^{(j)}\sin(m\varphi_j)\right), \tag{10.68}$$

where $Q_n^m(\xi)$ is an associated Legendre function of the second kind defined in a complex plane with the branch cut from $-\infty$ to $+1$ (Abramowitz and Stegun, 1965). The total potential outside the particles is a sum of partial potentials, Eq. (10.68):

$$\varphi^{\text{out}} = \varphi_1^{\text{out}} + \varphi_2^{\text{out}}. \tag{10.69}$$

By the construction, the potentials, Eqs. (10.67) and (10.68), are solutions of the Laplace equation. Therefore, to find a solution of Eq. (10.65), one should use only the boundary conditions

$$\begin{aligned} \varphi_1^{\text{in}}\Big|_{\xi_1=\xi_0} &= \varphi^{\text{out}}\Big|_{\xi_1=\xi_0}, \quad \varepsilon\frac{\partial\varphi_1^{\text{in}}}{\partial\xi_1}\Big|_{\xi_1=\xi_0} = \frac{\partial\varphi^{\text{out}}}{\partial\xi_1}\Big|_{\xi_1=\xi_0}, \\ \varphi_2^{\text{in}}\Big|_{\xi_2=\xi_0} &= \varphi^{\text{out}}\Big|_{\xi_2=\xi_0}, \quad \varepsilon\frac{\partial\varphi_2^{\text{in}}}{\partial\xi_2}\Big|_{\xi_2=\xi_0} = \frac{\partial\varphi^{\text{out}}}{\partial\xi_2}\Big|_{\xi_2=\xi_0}, \end{aligned} \tag{10.70}$$

where $\xi_0 = c/\sqrt{c^2 - a^2} = c/f$ is the local radial coordinate defining surfaces of the nanospheroids, and ε is permittivity of the materials of the nanoparticles. To reduce the boundary conditions Eq. (10.70) to a system of linear equations, we apply the translational addition theorem for wavefunctions of the prolate nanospheroid (Guzatov and Klimov, 2010). In the case of two identical coaxial nanospheroids, this theorem gives (j, $s = 1$ and 2, $j \neq s, \varphi_1 = \varphi_2$)

$$Q_n^m(\xi_j)P_n^m(\eta_j) = \sum_{q=m}^{\infty} S_{mqmn}^{(j)} P_q^m(\xi_s)P_q^m(\eta_s), \tag{10.71}$$

where the expansion coefficients are

$$\begin{aligned} S_{mqmn}^{(1)} &= \frac{(-1)^q(2q+1)(q-m)!(n+m)!}{(q+m)!(n-m)!} \\ &\times \sum_{r=0}^{\infty}\sum_{k=0}^{\infty} \frac{(q+n+2r+2k)!}{(2q+2r+1)!!(2n+2k+1)!!(2r)!!(2k)!!}\left(\frac{f}{l}\right)^{2r+2k+q+n+1}, \\ S_{mqmn}^{(2)} &= (-1)^{q+n} S_{mqmn}^{(1)} \end{aligned} \tag{10.72}$$

where l and f stand for distance between the centers of spheroids and half of the focal distance, respectively.

Applying the boundary conditions, Eq. (10.70), and making use of the cluster symmetry, one can obtain the following system of equations ($n = 0, 1, 2, \ldots; m = 0, 1, 2, \ldots, n$):

$$\begin{aligned}&\left(\varepsilon\frac{dP_n^m(\xi_0)}{d\xi_0}Q_n^m(\xi_0) - P_n^m(\xi_0)\frac{dQ_n^m(\xi_0)}{d\xi_0}\right)C_{mn}^{(1)}\\ &\pm(\varepsilon-1)P_n^m(\xi_0)\frac{dP_n^m(\xi_0)}{d\xi_0}\sum_{q=m}^{\infty}(-1)^{m+n}S_{mnmq}^{(1)}C_{mq}^{(1)} = 0\end{aligned}, \qquad (10.73)$$

where the "+" and "−" signs correspond to modes that are symmetric (+) or antisymmetric (−) relative to the $z \to -z$ transformation. It is important to note that separation of the spectra into symmetric and antisymmetric plasmon modes is possible only in the case when there is a plane of symmetry. When m is even, the antisymmetric modes have a nonzero dipole moment and they are "bright" modes. In contrast, the symmetric modes have a zero dipole moment and are "dark" modes when m is even. In the case of odd m, the "bright" and "dark" modes correspond to the symmetric and antisymmetric modes, respectively. One can expect that the antisymmetric mode with $m = 0$ will provide the largest polarizability and thus will be the "brightest" one for the excitation of our cluster with a longitudinally (along the z axis) polarized plane wave. The system of equations for $D_{mn}^{(j)}$ is identical to Eq. (10.73) and gives no additional information for plasmonic spectra of coaxial spheroids. Therefore, we shall not consider it further.

To study plasmon oscillations in clusters of two prolate spheroidal nanoparticles, one should solve the eigenvalue problems in Eq. (10.73) numerically. In Fig. 10.17, the normalized plasmon frequency $\omega/\omega_{\text{pl}}$ of a cluster of two prolate nanospheroids (see Fig. 10.16a for the geometry), corresponding to the first four plasmon modes, is shown as a function of normalized distances $l/2c$ between the nanoparticles' centers. The eigenvalues ε have been obtained as a nontrivial solution of the equations systems in Eq. (10.73) in the case of an axis-symmetric problem ($m = 0$). Then, the found solutions have been substituted in Eq. (10.66) to obtain plasmon oscillations' frequencies.

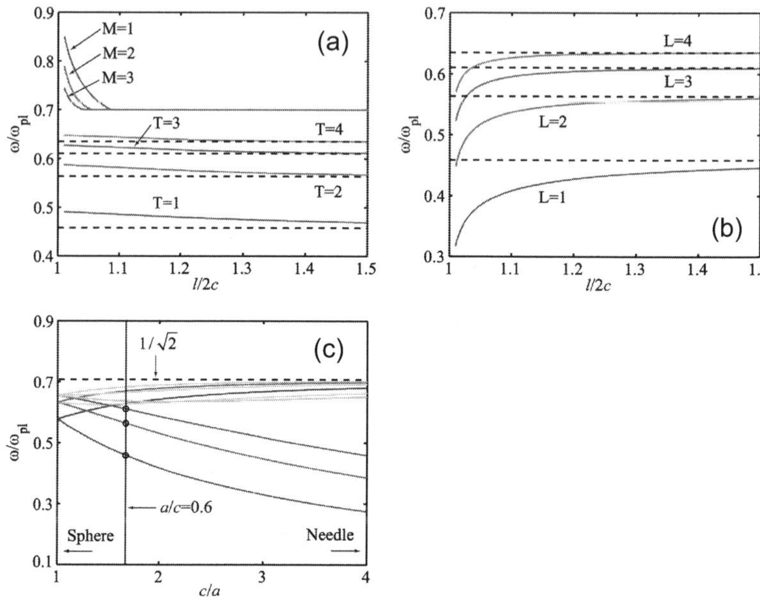

Figure 10.17 Normalized frequencies of the first few plasmonic oscillations in a cluster of two identical prolate spheroidal nanoparticles as a function of the distance between the nanospheroids' centers. The axis-symmetrical case ($m = 0$) is considered, and the aspect ratio of a single spheroid is $a/c = 0.6$. (a) Symmetric modes (the eigenvalues of Eq. (10.73) with a $+$ sign), and (b) antisymmetric modes (the eigenvalues of Eq. (10.73) with a $-$ sign). Dashed lines show plasmon frequencies of a single prolate nanospheroid:

$$\omega/\omega_{\text{pl}} = \{P_n^{m'}(\xi_0)\, Q_n^m(\xi_0)/(P_n^{m'}(\xi_0)\, Q_n^m(\xi_0) - P_n^m(\xi_0)\, Q_n^{m'}(\xi_0))\}^{1/2}.$$

(c) Plasmon frequencies of a single prolate nanospheroid as a function of the inverse aspect ratio c/a. The vertical line corresponds to $a/c = 0.6$ and allows us to select asymptotic values for panels (a) and (b).

In Fig. 10.17, one can observe that plasmon frequencies of a cluster of two prolate nanospheroids tend to plasmon frequencies of a single nanospheroid (see Fig. 10.17c) if distance between the nanospheroids is large enough. When the gap width between the nanospheroids tends to zero, the symmetric and antisymmetric solutions of Eq. (10.73) behave very differently. For symmetric modes (Fig. 10.17a), there are two branches, T-modes and M-modes. Modes of the "T" type can be obtained by the method of hybridization of plasmon modes of a single prolate nanospheroid, analogously to the case of a two spherical nanoparticle cluster (see Section 10.2). When the gap width between the nanoparticles is decreasing to zero, normalized plasmonic frequencies of T-modes tend to various values in the range from 0 to $1/\sqrt{2}$, on the analogy with a two-sphere cluster. T-modes with higher indices (not shown for clarity) will concentrate near $\omega/\omega_{\text{pl}} = 1/\sqrt{2}$. In Fig. 10.17a, one can also see that at very short distances between the nanospheroids ($l/2c < 1.1$), a new type of plasmonic modes (M-modes) appears. M-modes are characterized by strong spatial localization in the gap between the nanoparticles. As a result, they can be effectively excited only by a strongly nonuniform electric field of a molecule or a quantum dot or a nanoscope tip. Values of plasmonic frequencies of these modes lie in the range $\omega_{\text{pl}}/\sqrt{2} < \omega < \omega_{\text{pl}}$. As the gap width decreases to zero, the plasmon frequency of M-modes tends to the bulk plasmon frequency ω_{pl}.

In Fig. 10.17b, nontrivial antisymmetric solutions of the equations system in Eq. (10.73) in an axial symmetric case ($m = 0$) are shown. By analogy with a two-sphere cluster, we shall call these modes "L-modes" (longitudinal) because they are "bright" only for longitudinal excitation. These modes can be described by hybridization of plasmon oscillations of single nanospheroids forming the cluster under consideration. As the width of the gap between the prolate nanospheroids decreases to zero, normalized plasmon frequencies of these modes tend to zero as it also takes place in the case of spherical nanoparticles. Plasmonic frequencies of L-modes of higher orders (not shown) tend to $\omega_{\text{pl}}/\sqrt{2}$, and concentration of an infinite number of L-modes occurs near this value.

In Fig. 10.18, distribution of the surface charge of plasmonic modes of the lowest order in clusters of two identical prolate nanospheroids is shown. It is well seen in this figure that the T- and M-modes have symmetric distribution of the surface charge in contrast to the antisymmetric $L = 1$ mode. This behavior, of course, is in agreement with symmetry of the equations in Eq. (10.73). Another interesting feature is that the surface charge of the T-mode is distributed over the whole surface of nanoparticles for any distances between them, while for M- and L-modes it is concentrated

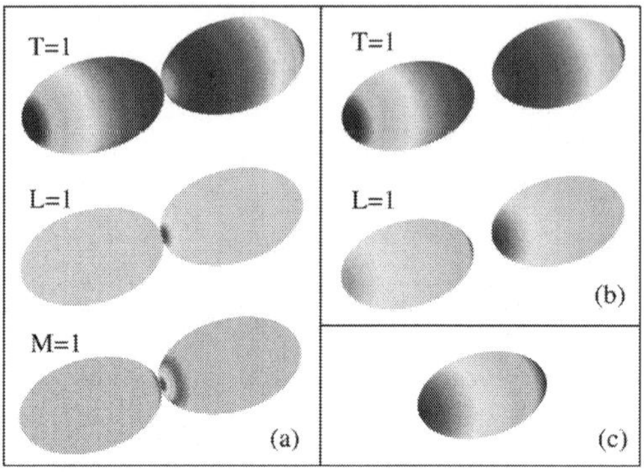

Figure 10.18 Distribution of a surface charge of the lowest plasmon modes in a cluster of two identical prolate spheroidal nanoparticles. The axial symmetrical case ($m = 0$) is considered, and the aspect ratio of a single spheroid is $a/c = 0.6$. The distance between the nanospheroids' centers $l/2c = 1.03$ (a) and $l/2c = 1.2$ (b). Distribution of a surface charge in a single spheroid is shown in panel (c). The red color corresponds to the positive charge and the blue one to the negative one.

near the gap between the nanospheroids if the distance between them is sufficiently small. It is interesting to note also that the surface charge of M-modes is more concentrated in comparison with that of L-modes. Indeed, due to the electroneutrality requirement, the total surface charge on each nanospheroid should be equal to zero. Here, both positive and negative charges of the M-mode are localized near the gap between the nanoparticles so that on the rest of the nanoparticles the charge is almost equal to zero, as it is well seen in Fig. 10.18a. At the same time, in the case of L-modes for each nanospheroid, near the gap a charge of only one sign is concentrated, and a charge of the opposite sign is distributed with a small magnitude over the remaining surface of the nanoparticles. Therefore, strictly speaking, the surface charge in the L-mode is distributed over the whole surface of the cluster nanoparticles, though it is not clearly seen at small distances between the nanoparticles (see Fig. 10.18a). As the distance increases, the charge distribution changes in the cluster: it spreads over the nanoparticles' surface, tending in the limit to a distribution corresponding to single prolate nanospheroids (see Fig. 10.18c).

Let us now consider plasmonic properties of a cluster made of oblate spheroids (Guzatov and Klimov, 2010, 2011). In this geometry, one should also use local systems of coordinates $(\xi_j, \eta_j, \varphi_j)$, $j = 1, 2$ that are connected with each spheroid and have origins in their centers (see Fig. 10.16b). Now, the electric potential inside the jth nanospheroid can be presented in the form ($j = 1$ and 2):

$$\varphi_j^{\text{in}} = \sum_{n=0}^{\infty}\sum_{m=0}^{n} P_n^m(i\xi_j)P_n^m(\eta_j)(A_{mn}^{(j)}\cos(m\varphi_j) + B_{mn}^{(j)}\sin(m\varphi_j)), \tag{10.74}$$

and the partial potential outside the jth oblate nanospheroid will look like ($j = 1$ and 2):

$$\varphi_j^{\text{out}} = \sum_{n=0}^{\infty}\sum_{m=0}^{n} Q_n^m(i\xi_j)P_n^m(\eta_j)(C_{mn}^{(j)}\cos(m\varphi_j) + D_{mn}^{(j)}\sin(m\varphi_j)). \tag{10.75}$$

The total potential outside the nanospheroids will be expressed by Eq. (10.69). As the boundary conditions for the potential, one should use Eq. (10.70), where $\xi_0 = c/\sqrt{a^2 - c^2} = c/f$.

For the case of two identical oblate nanospheroids, there are again two types of plasmon oscillations, symmetric and antisymmetric, relative to the symmetry plane ($x \to -x$ transformation). As a result, we shall obtain the following system of equations (Guzatov and Klimov, 2010, 2011):

$$\begin{aligned}&\left(\varepsilon\frac{dP_n^m(i\xi_0)}{d\xi_0}Q_n^m(i\xi_0) - P_n^m(i\xi_0)\frac{dQ_n^m(i\xi_0)}{d\xi_0}\right)C_{mn}^{(1)}\\ &\pm(\varepsilon - 1)\,P_n^m(i\xi_0)\frac{dP_n^m(i\xi_0)}{d\xi_0}\sum_{q=0}^{\infty}\sum_{p=0}^{q} M_{mnpq}^{(0)}C_{pq}^{(1)} = 0,\end{aligned} \tag{10.76}$$

where the $+$ and $-$ signs correspond to modes that are symmetric ($+$) or antisymmetric ($-$) relative to the $x \to -x$ transformation (see Fig. 10.16b). The explicit expression for matrix $M_{mnpq}^{(0)}$ can be found in (Guzatov and Klimov, 2010, 2011).

In Fig. 10.19, the dependence of normalized plasmon frequencies $\omega/\omega_{\text{pl}} = 1/\sqrt{1-\varepsilon}$ of a cluster of two identical oblate nanospheroids on normalized distances $l/2a$ between the nanoparticles' centers is shown for the first four plasmon modes. Eigenvalues ε were obtained as a solution of the equation systems in Eq. (10.76).

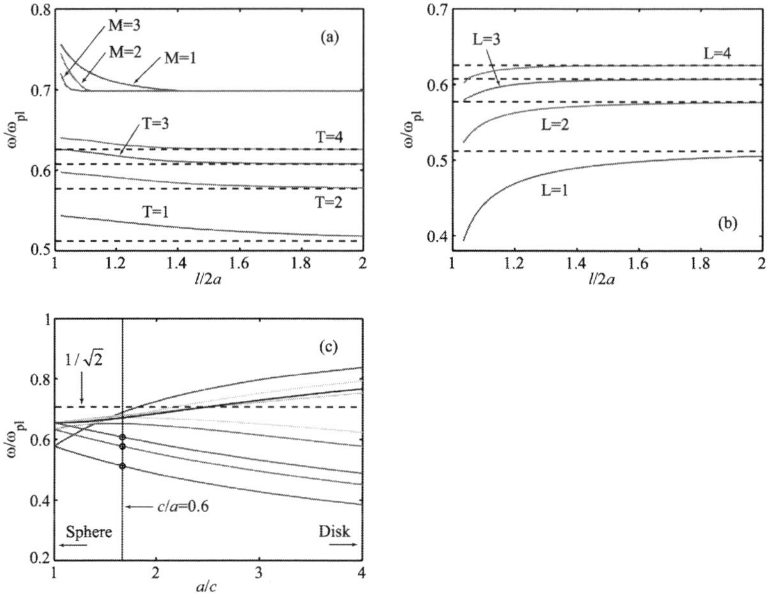

Figure 10.19 Normalized plasmon frequencies of the first few modes in a cluster of two identical oblate nanospheroids as a function of the distance between their centers. (a) Symmetric modes (the eigenvalues of Eq. (10.76) with a + sign) and (b) antisymmetric modes (the eigenvalues of Eq. (10.76) with a − sign). By the dashed lines, the plasmon frequencies of a single oblate nanospheroid

$$\omega/\omega_{\text{pl}} = \{P_n^{m'}(i\xi_0)Q_n^m(i\xi_0)/(P_n^{m'}(i\xi_0)Q_n^m(i\xi_0) - P_n^m(i\xi_0)Q_n^{m'}(i\xi_0))\}^{1/2}$$

are shown. The aspect ratio of a single oblate nanospheroid is $c/a = 0.6$. (c) Plasmon frequencies of a single oblate nanospheroid as a function of the inverse aspect ratio a/c. The vertical line corresponds to $c/a = 0.6$ and allows us to select asymptotic values for panels (a) and (b).

One can see in Fig. 10.19 that in clusters of two oblate nanospheroids, modes of T-, M-, and L-types, which are analogous to the T-, M-, and L-modes of a cluster made of two prolate spheroids (see Fig. 10.17), can exist. T- and M-modes are solutions of the system Eq. (10.76) with a "+" sign, while L-modes are a solution of the system Eq. (10.76) with a "−" sign. T- and L-modes can be derived by the method of hybridization of plasmonic modes of two oblate nanospheroids, and their plasmonic frequencies lie in the range $0 < \omega < \omega_{\text{pl}}/\sqrt{2}$. An infinite number of plasmonic frequencies of higher L- and T-modes lie near $\omega_{\text{pl}}/\sqrt{2}$. When the width of the gap decreases to 0, the ratio $\omega/\omega_{\text{pl}}$ for T-modes tends to various values in the range from 0 to $1/\sqrt{2}$, while plasmonic frequencies of L-modes approach 0 by analogy with L-modes in a cluster of two spherical nanoparticles (see Section 10.3). Plasmonic frequencies of strongly localized M-modes (Fig. 10.19a) lie in the range $\omega_{\text{pl}}/\sqrt{2} < \omega < \omega_{\text{pl}}$, as it happens in a cluster of two spherical nanoparticles. As the width of the gap between oblate nanospheroids decreases to zero, plasmon frequencies of M-modes tend to bulk plasmon frequency ω_{pl} analogously to the case of a two-sphere cluster. For large distances between the spheroids, M-modes disappear, and plasmon frequencies of L- and T-modes of a cluster of two oblate nanospheroids tend to plasmonic frequencies of a single spheroid (see. Fig. 10.19c) and can be found by means of a self-consistent model with approximation of spheroids by anisotropic point dipoles.

In Fig. 10.20, the distribution of a surface charge of plasmon modes of lower orders in a cluster of two identical oblate nanospheroids is shown. One can see in this figure that the charge distribution is symmetric in T- and M-modes, while in the L-mode it is antisymmetric in agreement with the definition of these modes. For small distances between nanospheroids, charges in M- and L-modes are strongly localized near the gap. On the contrary, when the distance between the spheroids increases, the charge distribution tends to a symmetric or antisymmetric combination of surface charge in a single oblate nanospheroid (see Fig. 10.19c).

Thus, in a cluster of two oblate or prolate spheroidal nanoparticles, fundamental symmetric and antisymmetric plasmon modes of T-, M-, and L-types can be excited, and it is these modes that define all optical properties of a two-nanospheroid cluster.

The approach used to find plasmonic modes in a cluster of two nanospheroids allows us to find also optical coefficients, local field enhancement, and decay rates of an atom or molecule placed near such a cluster.

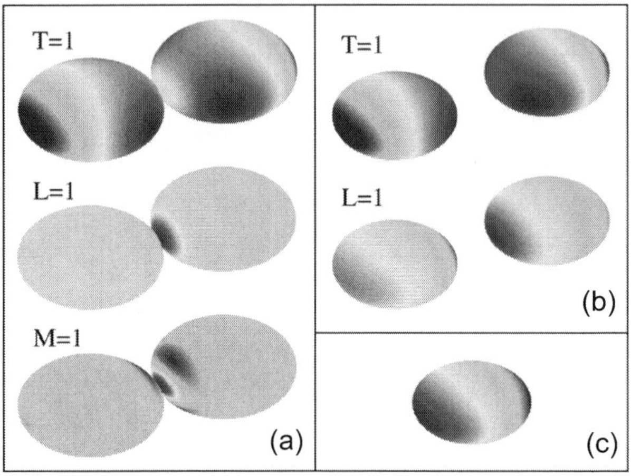

Figure 10.20 Distribution of a surface charge of the lowest plasmon modes in a cluster of two identical oblate spheroidal nanoparticles according to the solution of the equations in Eq. (10.76). The aspect ratio is $c/a = 0.6$. The distance between the centers of the nanospheroids $l/2a = 1.05$ (a) and $l/2a = 1.4$ (b). Distribution of a surface charge in a single spheroid is shown in panel (c). The red color corresponds to the positive charge and the blue one to the negative one.

In Fig. 10.21, the absorption cross section of a cluster of two identical prolate nanospheroids made of silver is shown as a function of wavelength. For longitudinal (z) polarization, the cross section has two peaks, which correspond to longitudinal plasmonic oscillations with $L = 1$ and 2 (see Fig. 10.17b). It is very important that both peaks are split substantially relative to the case of a single spheroid (the "z" dashed curve) due to a strong interaction between the nanospheroids.

On the contrary, for transverse (x or y) polarization one can see only one peak due to excitation of the symmetrical $T = 1$ mode, and this peak is shifted just slightly relative to the single spheroid resonance (the "y" dashed line). It means that transverse (x or y polarization) excitation of a two-spheroid cluster induces only a weak interaction between the nanospheroids (see the dispersion curves for T-modes in Fig. 10.17a). Due to this weak interaction, absorption cross sections are approximately equal to doubled cross-sections of a single spheroid, respectively.

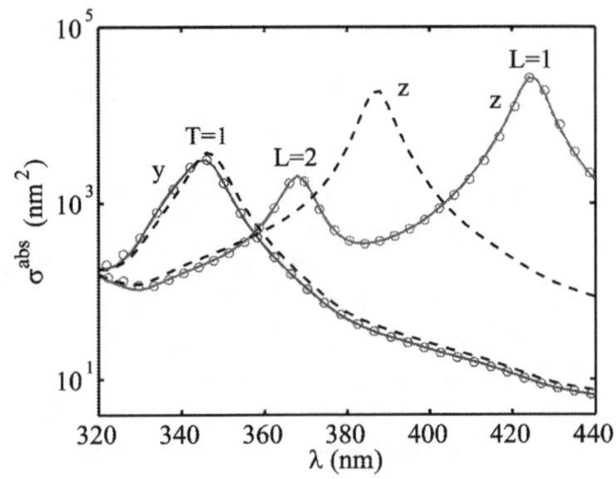

Figure 10.21 Absorption cross section of a cluster of two identical prolate nanospheroids made of silver as a function of wavelength. The large semiaxes of the nanospheroids are $c = 15$ nm, the aspect ratios are $a/c = 0.6$, and the distance between the nanospheroids' centers is $l/2c = 1.05$. The labels y and z correspond to polarizations of an incident wave along the y and z axes. The circles correspond to finite element simulations with Comsol Multiphysics® software. The doubled absorption cross section of a single nanospheroid is shown by the dashed curve.

It should be noticed that in Fig. 10.21 the maxima of absorption, which correspond to plasmon oscillations of M-type that should lie in the interval $\omega_{\text{pl}}/\sqrt{2} < \omega < \omega_{\text{pl}}$, which corresponds to $326 < \lambda < 337$ nm (Johnson and Christy, 1972) for silver, are not visible. It is related to the fact that M-modes interact with a homogeneous electric field weakly and can be effectively excited only by a source of radiation that is nonuniform at the scale, which is small in comparison with a size of the gap between the nanoparticles (Klimov and Guzatov, 2007a, 2007b, 2007c) (see Fig. 10.17).

In Fig. 10.22, the absorption cross section of a cluster of two identical oblate nanospheroids made of silver is shown as a function of wavelength. For longitudinal (x) polarization, the cross section has two peaks, which correspond to antisymmetric plasmonic oscillations with $L = 1$ and 2 (see Fig. 10.19b). It is very important that now only one peak ($L = 1$) is shifted substantially relative to the case of a single spheroid (the "x, y" dashed curves) due to a strong interaction between the nanospheroids. The $L = 2$ mode suffers only a small shift in agreement with Fig. 10.19b.

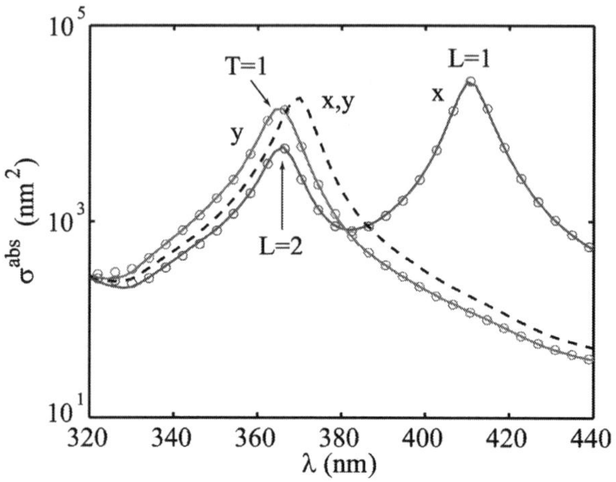

Figure 10.22 The absorption cross section of a cluster of two identical oblate nanospheroids made of silver as a function of wavelength. The large semiaxes of the nanospheroids are $a = 15$ nm, the aspect ratios are $c/a = 0.6$, and the distance between the nanospheroids' centers is $l/2a = 1.05$. The labels x and y correspond to polarizations of an incident wave along the x and y axes. The circles correspond to finite element simulations with Comsol Multiphysics®. The doubled absorption cross section of a single nanospheroid is shown by the dashed curve.

For transverse (y) polarization, one can see only one peak due to excitation of the symmetrical $T = 1$ mode, and this peak is only slightly shifted relative to the single spheroid resonance (the dashed line). It means that transverse (y polarization) excitation of a two-spheroid cluster results only in a weak interaction between the nanospheroids (see dispersion curves for T-modes in Fig. 10.19a). Due to this weak interaction, absorption cross sections for this polarization are approximately equal to doubled cross sections of a single spheroid, respectively. It is also interesting that the plasmon frequency of the $L = 2$ mode is very close to the plasmonic frequency of the $T = 1$ mode. This fact can be easily understood from analysis of Fig. 10.19. Indeed, when the width of the gap tends to zero, the plasmonic frequency of $L = 2$ modes also decreases to zero, while the plasmonic frequency of the $T = 1$ mode increases slightly. So, at some point these modes will intersect and have the same frequencies, and we observe this situation in Fig. 10.22.

It should be noticed that in Fig. 10.22 maxima of absorption corresponding to plasmon oscillations of M-type again are not visible. It is related to the fact that M-modes interact with a homogeneous electric field weakly and can be effectively excited only by a nanolocalized source of radiation (Klimov and Guzatov, 2007a, 2007b, 2007c) (see Fig. 10.20).

The most important characteristic of nanoparticle clusters is the factor of incident field enhancement in the gap between the nanoparticles. It is this characteristic that determines the excitation rate of molecules near the nanoparticles or the intensity of surface-enhanced Raman scattering (SERS) (Kneipp et al., 2006). Moreover, high values of this factor are the main goal of optical nanoantenna development.

Distribution of the squared electric field for the $L = 1$ resonance in a cluster of two silver prolate spheroids is shown in Fig. 10.23, which shows that, indeed, the maximal field enhancement takes place in the gap between the nanoparticles on their surfaces. The field maxima are also present on the outer side of the cluster; however, field amplitude is essentially less there. According to general theorems for harmonic functions, the field maximum can be reached only on the boundaries. In our case, the field maxima are reached in those points of the spheroids' surface where the distance between the spheroids is minimal.

For clusters of two identical prolate spheroids in the considered configuration (Fig. 10.16a), one can obtain the following expression for the field maximum in the case of an incident field polarized along the z axis:

$$G = \frac{|\mathbf{E}|^2}{|E_{0z}|^2} = \left| 1 - \frac{1}{E_{0z} f} \sum_{n=1}^{\infty} C_{0n}^{(1)} \left(\frac{dQ_n(\xi_1)}{d\xi_1} + \frac{dQ_n(\xi_2)}{d\xi_2} \right) \right|^2, \tag{10.77}$$

where $\xi_1 = c/f$ and $\xi_2 = (l - c)/f$. In the most interesting case of a small gap and strongly prolate spheroids, ξ_1, $\xi_2 \approx 1$, and one may

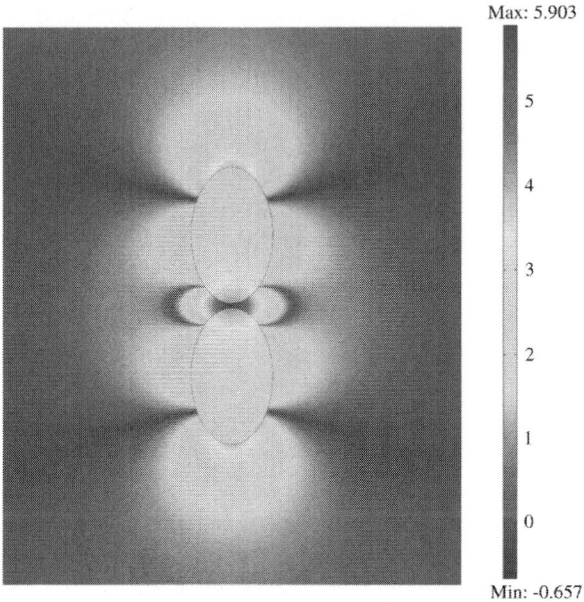

Figure 10.23 Spatial distribution of $\log_{10}(|\mathbf{E}|^2/|\mathbf{E}_0|^2)$ in a cluster of two prolate silver nanospheroids for $\lambda = 425$ nm at longitudinal polarization of the excitation field ($L = 1$ plasmon resonance, the common logarithmic scale).

use the asymptotic form $dQ_n(\xi)/d\xi|_{\xi\approx 1} \approx -1/(2(\xi-1))$. As a result, the field enhancement factor takes the form

$$G = \frac{|\mathbf{E}|^2}{|E_{0z}|^2} \approx \left| \frac{1}{2E_{0z}f} \left(\frac{1}{\xi_1 - 1} + \frac{1}{\xi_2 - 1} \right) \sum_{n=1}^{\infty} C_{0n} \right|^2 . \quad (10.78)$$

In Fig. 10.24, the dependence of squared electric field enhancement for clusters of two identical silver nanospheroids on the wavelength is shown. Comparing the peaks' positions with the dispersion curves in Fig. 10.17 and Fig. 10.19, one can come to a conclusion that only L-types of plasmon modes are excited in the clusters for the considered configurations of nanospheroids and incident electromagnetic wave polarizations (along the line joining the nanoparticles' centers). In particular, excitation of the $L = 1$ and $L = 2$ modes is noticeable. At that, the position of squared field enhancement peaks agrees with the maxima of the absorption cross sections shown in Figs. 10.21 and 10.22 by the solid lines z

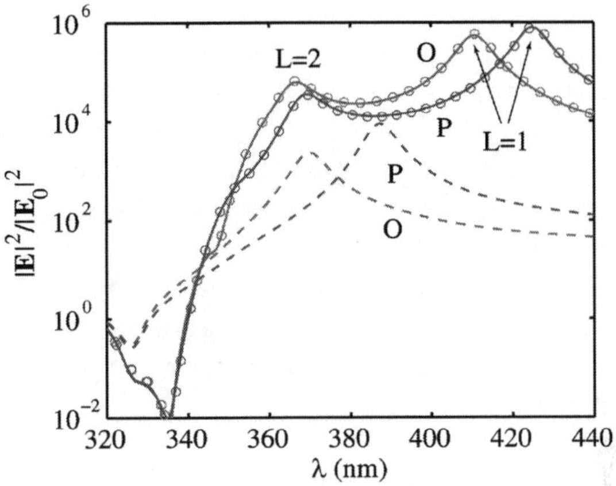

Figure 10.24 Enhancement of $|\mathbf{E}|^2$ in the gap between two identical silver nanospheroids as a function of wavelength. The P and O labels correspond to prolate and oblate spheroids, correspondingly. The large semiaxes of the nanospheroids are 15 nm, the aspect ratios are 0.6, and $l/2c = 1.05$ for a cluster of prolate spheroids and $l/2a = 1.05$ for a cluster of oblate spheroids. Enhancement factors for single nanospheroids are shown by the dashed curve. Polarization of incident light is along the line connecting spheroids' centers.

and x, correspondingly. It should be mentioned that the value of squared field enhancement near a cluster of two nanospheroids can reach up to 10^6. In the case of single nanoparticles, this value is almost two orders less than that of clusters (see the solid and dashed curves in Fig. 10.24). This fact determines a greater attractiveness of metal nanoparticle clusters in comparison to single nanoparticles for investigation of SERS and surface-enhanced fluorescence (SEF). Let us note that great field enhancement factors obtained can be slightly less in practice since for small particles and for small gaps between them, nonlocal and other effects not considered here become essential.

10.7 Plasmons in the Region of the Nanocontact of Two Plasmon Bodies of Infinite Volume

In the case of closely adjacent semi-infinite bodies, the situation turns out to be more complex since in this case the part of the charges can move arbitrarily far from the region of the contact. Strictly speaking, in systems of this kind the plasmon spectrum becomes continuous and one should speak about surface plasmons rather than localized plasmons. However, also in this case there are stable plasmon oscillations analogous to antisymmetric (L-modes) and to symmetric (M-modes) plasmonic oscillations in a two-sphere or two-spheroid cluster.

If the region of the nanocontact is approximated by two semi-infinite hyperboloids (Fig. 10.25), then the variables in the Laplace equation are separated in the coordinates of a prolate ellipsoid of rotation (Morse and Feshbach, 1953):

$$\begin{aligned} x &= f\sqrt{\eta^2-1}\sqrt{1-\mu^2}\cos\varphi, \\ y &= f\sqrt{\eta^2-1}\sqrt{1-\mu^2}\sin\varphi, \\ z &= f\eta\mu \end{aligned} \tag{10.79}$$

where $1<\eta<\infty$, $-1<\mu<1$, $0<\varphi<2\pi$ and $2f$ is the distance between the foci of the ellipse forming the system of coordinates. The following relations between the coordinates take place as well:

$$\begin{aligned} (\eta-\mu)f &= \sqrt{R^2+f^2-2fz}, \\ (\eta+\mu)f &= \sqrt{R^2+f^2+2fz}, \\ R^2 &= x^2+y^2+z^2. \end{aligned} \tag{10.80}$$

In these coordinates, the arbitrary solution of the Laplace equation has the form

$$\Phi = P^m_{-1/2+iq}(\eta)\left(A(q)P^m_{-1/2+iq}(\mu)+B(q)P^m_{-1/2+iq}(-\mu)\right)e^{im\varphi}, \tag{10.81}$$

where $m=0,\pm1,\pm2,\ldots;0<q<\infty$ are dimensionless constants of separation characterizing particular solutions and $P^m_{-1/2+iq}(\mu)$ are the associated Legendre functions. In the case of two symmetrically situated metal hyperboloids with the surfaces determined by the equation $\mu = \pm\mu_0$, there are symmetric and antisymmetric plasmon eigenoscillations in the system. In the space between the

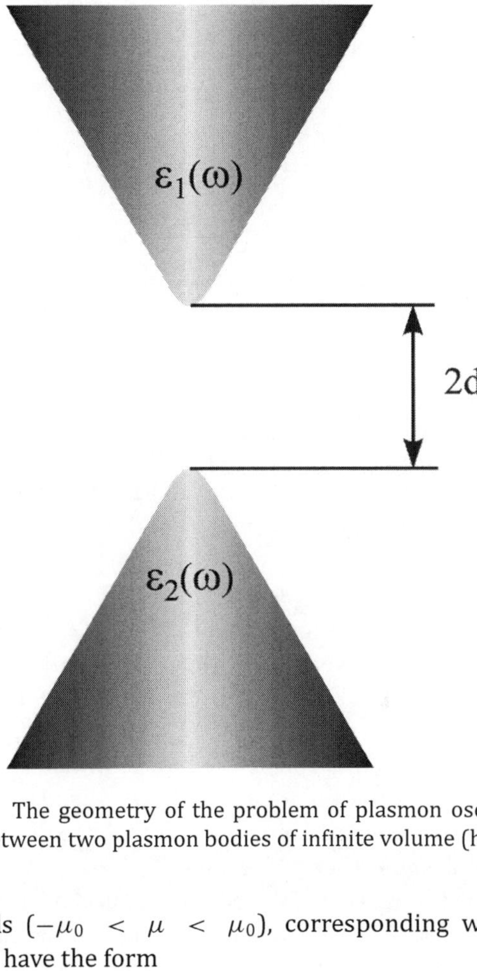

Figure 10.25 The geometry of the problem of plasmon oscillation in the interspace between two plasmon bodies of infinite volume (hyperboloids).

hyperboloids ($-\mu_0 < \mu < \mu_0$), corresponding wavefunctions (potentials) have the form

$$\begin{aligned}\Phi_m^{(L)} &= P^m_{-1/2+iL}(\eta)\left(P^m_{-1/2+iL}(\mu) - P^m_{-1/2+iL}(-\mu)\right),\\ \Phi_m^{(M)} &= P^m_{-1/2+iM}(\eta)\left(P^m_{-1/2+iM}(\mu) + P^m_{-1/2+iM}(-\mu)\right).\end{aligned} \quad (10.82)$$

Inside the hyperboloids, expressions analogous to Eq. (10.82) can be written.

Corresponding eigenvalues of electric permittivity and of frequency (if the Drude model is used) have the form (Klimov and Guzatov, 2007b):

$$\omega_m^{(L,M)} = \frac{\omega_{\text{pl}}}{\sqrt{1-\varepsilon_m^{(L,M)}}},$$

$$\varepsilon_m^{(L)} = \frac{P^m_{-1/2+iL}(\mu_0)}{P'^m_{-1/2+iL}(\mu_0)}\left(\frac{P'^m_{-1/2+iL}(\mu_0)+P'^m_{-1/2+iL}(-\mu_0)}{P^m_{-1/2+iL}(\mu_0)-P^m_{-1/2+iL}(-\mu_0)}\right),$$

$$\varepsilon_m^{(M)} = \frac{P^m_{-1/2+iM}(\mu_0)}{P'^m_{-1/2+iM}(\mu_0)}\left(\frac{P'^m_{-1/2+iM}(\mu_0)-P'^m_{-1/2+iM}(-\mu_0)}{P^m_{-1/2+iM}(\mu_0)+P^m_{-1/2+iM}(-\mu_0)}\right). \tag{10.83}$$

In Eqs. (10.82) and (10.83), the indices L- and M-, characterizing the modes, are changing from 0 to infinity continuously, $\mu_0 = d/f$, where d is half of the smallest distance between the hyperboloids and f is the dimensional constant (half of the distance between the foci of hyperboloids).

The spectra of plasmon oscillations for $m = 0$ and for $L, M = 1, 3, 5$, and 10 are shown in Fig. 10.26. The interesting fact is that for small distances between the hyperboloids $\mu_0 = d/f < 0.5$ the spectra of these oscillations are quite similar to the spectra of the oscillations in a cluster of two nanospheres (see Fig. 10.7a and Fig. 10.49). At that, symmetric T-modes, which could exist in a cluster of two nanospheres, are absent in the spectrum. It allows us to consider the localized plasmon oscillations in semi-infinite bodies as the virtual plasmon L- and M-modes.

The spatial distribution of the potential, Eq. (10.82), for the cases of $m = 0$ and $L, M = 1$ is shown in Fig. 10.27. From this picture, it is seen that in cases of both symmetric and antisymmetric plasmon oscillations a strong modes localization takes place, providing an effective interaction with ordinary atoms, molecules, and quantum dots.

With the help of the coordinates of a prolate ellipsoid of rotation, even more complex configurations can be considered. In particular, Passian et al. (2005) considered plasmon oscillations in the system shown in Fig. 10.28. Obviously, this geometry is very useful for a description of operation of different types of scanning microscopes.

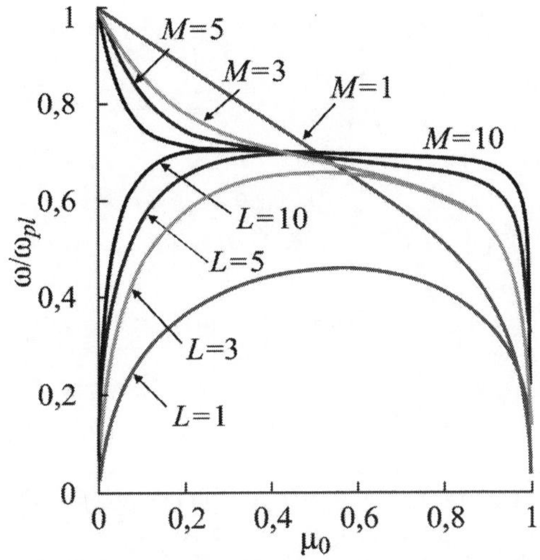

Figure 10.26 The plasmon oscillations spectra, Eq. (10.83), in the interspace between two infinite nanobodies (hyperboloids), depending on the gap width.

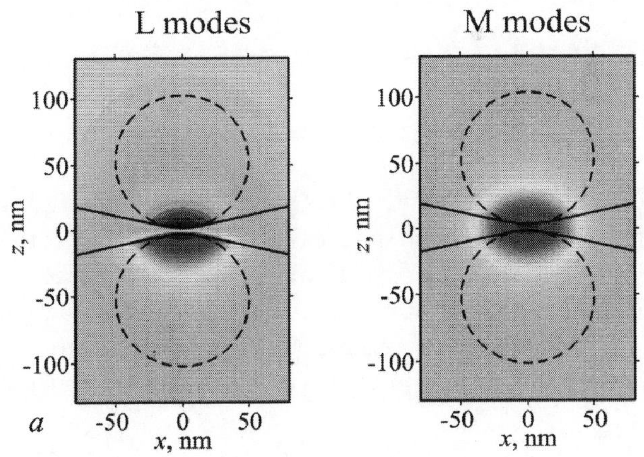

Figure 10.27 The spatial structure (potential) of the plasmon oscillations near the contact region of two infinite nanobodies (hyperboloids) with the curvature radii 50 nm and the gap width $2d = 5$ nm. The hyperboloids' surfaces are shown by the solid lines, and the inscribed spheres surfaces are shown by the dotted lines. These distributions are similar to the potential distributions in a cluster of two nanowires or two nanospheres (see Fig. 10.5 and Fig. 10.8, the top row).

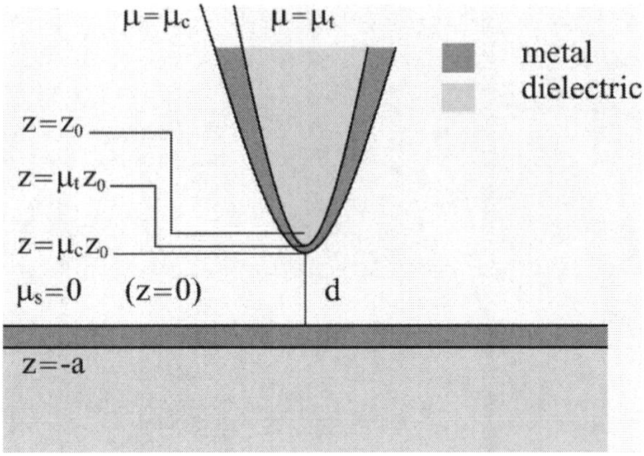

Figure 10.28 A more complex geometry where plasmon oscillations can be found with the help of prolate spheroidal coordinates (Passian et al., 2005). Evidently, this geometry is very useful for a description of scanning microscope's operation.

10.8 Plasmons Oscillations in a Cluster of More Than Two Particles

In clusters of three and more nanoparticles, plasmon oscillations are even more complicated in structure. At that, in practice the plasmon oscillations strongly localized in interspaces between particles (*M*-modes) do not sense the presence of other particles and therefore they can exist in any configurations almost independently. All of the rest of the plasmon oscillations are related to excitation of the whole cluster. When initially the oscillations of one particle are excited, its neighbor particles are affected due to electrostatic interactions, then their neighbour particles are, and the energy propagates over the whole cluster, that is, the cluster becomes a plasmon waveguide.

10.8.1 *Plasmon Properties of Linear Clusters: Quasi-Static Approximation*

A linear cluster of identical (e.g., spherical) nanoparticles is one of the most interesting nanostructures; its structure is shown in Fig. 10.29.

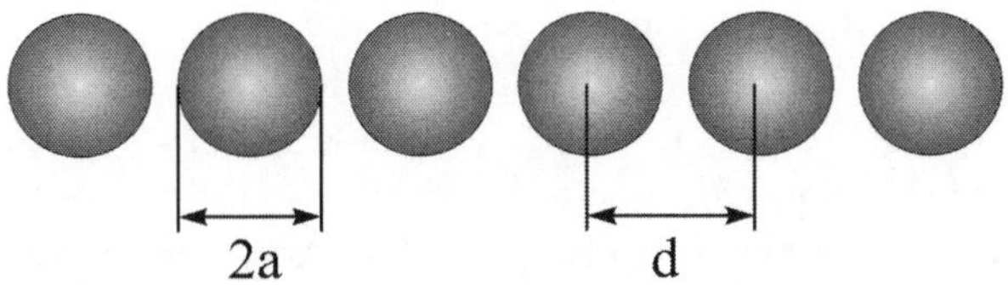

Figure 10.29 The geometry of a linear cluster of identical spheres.

Owing to the nanoparticles' interaction, a mode with longitudinal orientation of dipole moments and two modes with transverse orientation of dipole moments can coexist in a linear cluster. In the systems of this kind, the excitation propagation was investigated within different approaches.

Satisfactory comprehension of physical processes taking place in linear chains can be achieved by using the model that relates only one dipole moment to each particle and neglects quadrupole and higher moments (see Section 10.1.2 and (Brongersma et al., 2000)).

In an infinite chain of interacting nanoparticles, the spectrum of plasmon oscillations is not discreet and can be described by the dispersion relation, that is, by dependency of the oscillations' frequency on the longitudinal wavevector $\mathbf{k}$ conveniently. At that, the mode is determined by a set of amplitudes of all dipoles, depending on time, $m(t)$, $m = -\infty \ldots \infty$. If the distance between the nanoparticles is smaller than the wavelength, it is possible to describe the interaction between the dipoles within the quasi-static approximation where the electric field of the mth dipole at the point $\mathbf{r}$ is (see Section 10.1.2):

$$\mathbf{E}_m(\mathbf{r}, t) = -\frac{\mathbf{d}_m(t)}{r^3} + 3\frac{\mathbf{r}(\mathbf{d}_m(t)\mathbf{r})}{r^5}. \qquad (10.84)$$

The field of the mth dipole in the chain nodes $m+1$, $m-1$, is described by the expression

$$E_{i,m}(\mathbf{d}, t) = \frac{\gamma_i d_{i,m}(t)}{\varepsilon d^3}, \; i = x, y, z, \qquad (10.85)$$

where γ_i is the constant depending on the mode polarization, that is, on the dipoles' orientation:

$$\gamma_x = \gamma_y = \gamma_{\mathrm{T}} = 1, \; \gamma_z = \gamma_L = -2, \qquad (10.86)$$

and ε is permittivity of the space where the cluster is situated. Due to rapid dipole field decay ($1/r^3$), interactions of neighbour dipoles dominate, and so we can restrict ourselves to consideration of the interaction of neighbour dipoles only.

For a chain of dipoles with $\mathbf{d}_m = q\mathbf{x}_m$, where q is the effective (negative) charge of the dipoles (the full charge of conduction electrons) and $\mathbf{x}_m$ is the shift of the effective charge from the state of equilibrium of the mth particle, the motion equation can be written in the form (Brongersma et al., 2000):

$$\ddot{d}_{i,m} = -\omega_0^2 d_{i,m} - \Gamma_1 \dot{d}_{i,m} + \frac{\Gamma_R}{\omega_0^2} \dddot{d}_{i,m} - \gamma_i \omega_1^2 (d_{i,m-1} + d_{i,m+1}). \quad (10.87)$$

In this equation, the first term on the right side describes the plasmon oscillation with a frequency ω_0 of a single particle, which is assumed to be spherical, and thus the plasmon frequency does not depend on the polarization. The second ($\sim \Gamma_1$) and third ($\sim \Gamma_R$) terms in the right part describe the internal and radiative decay of plasmon oscillations in an isolated nanoparticle, and, finally, the fourth term describes the electrostatic interaction with the nearest neighbours, that is, with the dipoles situated in the $m+1$ and $m-1$ nodes. This term is specifically responsible for the origin of propagating plasmon modes. The interaction strength is determined by the frequency

$$\omega_1^2 = \frac{qe}{\varepsilon m^* d^3} = \frac{1}{3\varepsilon}\left(\frac{a}{d}\right)^3 \omega_{\text{pl}}^2, \quad (10.88)$$

where q is the value of the charge oscillating in the nanoparticle, m^* is the effective mass of the outer electron, and e is the electron charge. For a cluster of silver nanoparticles of the radius of 25 nm spaced at intervals of 75 nm in vacuum, one can take $q = 6.1 \times 10^{-13} C$ ($n_e = 5.85 \times 10^{22}$ cm^{-3}) (Kittel, 1986) and $m^* = 8.7 \times 10^{-28}$ g for Ag (Johnson and Christy, 1972). As a result, typical interaction frequency, Eq. (10.88), can be estimated as $\omega_1 = 1.4 \times 10^{15}$c^{-1}. Let us note that in the considered case the order of the ω_1 value coincides with the plasmon frequency of the single particle dipole oscillations $\omega_0 = 5 \times 10^{15}$c^{-1}, and thus the interaction between nanoparticles in the cluster turns out to be rather strong.

The solution of Eq. (10.87) in the form of propagating waves may be sought in the form

$$\mathbf{d}_m = \mathbf{d}_0 \exp(-\alpha m d + i(\omega t \pm k m d)), \quad (10.89)$$

where $m = 0, 1, 2, 3$ and $\omega t - kmd$ correspond to the case when the phase and group rates are parallel, $\omega t + kmd$ correspond to the case when the phase and group rates of the wave are antiparallel, $\mathbf{d}_0$ determines the oscillation amplitude of the dipole situated in the node $m = 0$, and, finally, α is the wave attenuation coefficient. Substituting Eq. (10.89) in Eq. (10.87) and separating the real and imaginary parts, we obtain two equations:

$$\omega^2 = \omega_0^2 + 2\gamma_i\omega_1^2 \cos(kd)\cosh(\alpha d), \tag{10.90}$$

$$0 = \omega\Gamma_1 + \frac{\omega^3\Gamma_R}{\omega_0^2} + 2\gamma_i\omega_1^2\sin(kd)\sinh(\alpha d). \tag{10.91}$$

These equations determine the dispersion law and the plasmon oscillations' spectrum in a linear chain. In the case of weak attenuation, $\alpha d \ll 1$, Eq. (10.90) is reduced to

$$\omega^2 = \omega_0^2 + 2\gamma_i\omega_1^2 \cos(kd). \tag{10.92}$$

In Fig. 10.30 (Maier et al., 2001), the spectrum of plasmon oscillations determined by Eq. (10.92) for longitudinal ($\gamma_L = -2$) and for transverse ($\gamma_T = 1$) oscillations is shown (solid lines). The dashed curves show the dispersion laws that are obtained by taking into account the interaction with five nearest neighbours. The comparison of these curves shows that the interaction between the nearest neighbours is dominant, indeed.

From this picture it is seen that the bandwidth of the longitudinal oscillations $\Delta\omega_L$ is larger than the bandwidth of the transverse oscillations $\Delta\omega_T$. The group velocity of both longitudinal and transverse waves can be found easily from the dispersion equations:

$$v_{g,i} = \frac{d\omega}{dk} = -\frac{\gamma_i\omega_1^2 \sin(kd)d}{\omega}. \tag{10.93}$$

From Eq. (10.93), it is seen that absolute values of the group velocity of longitudinal oscillations are always higher than absolute values of the group velocity of transverse oscillations since $|\gamma_L| > \gamma_T$, and the phase and group velocities of transverse modes are directed oppositely. This circumstance makes the transverse modes very interesting for realization of metamaterials, that is, materials with unusual optical properties (for more on unusual properties of plasmon metamaterials see Chapter 11 and Chapter 13).

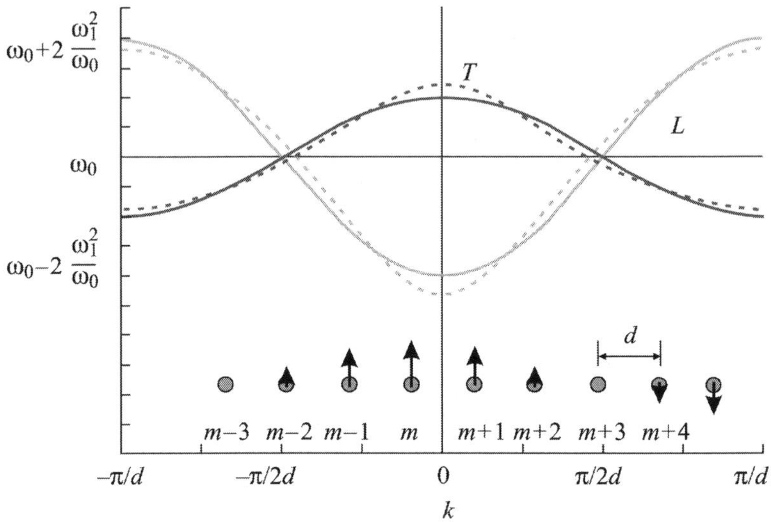

Figure 10.30 The dispersion curves of the plasmon modes in a linear cluster calculated in the context of the point dipole model and showing the doubly degenerate transverse (T-) and longitudinal (L-) modes. The solid curve is calculations considering the nearest neighbours only, and the dotted line is calculations considering the five nearest neighbours. The small differences of these curves testify that the plasmon mode propagation is determined mainly by the near fields (Maier et al., 2001).

In the case of the resonance between transverse and longitudinal waves, that is, for $\omega = \omega_0$, the group velocities of both modes reach the maximum, and $|v_{g,L}| = 2|v_{g,T}|$. This happens because of a stronger interaction of longitudinal dipoles.

In the low attenuation case, it can be found from Eq. (10.91) that the attenuation coefficient is determined by the relation

$$\alpha = \frac{\Gamma_1 + (\omega^2/\omega_0^2)\Gamma_R}{2v_g}, \tag{10.94}$$

which shows that by propagating along a linear cluster the wave attenuation is determined by the relation between the plasmon oscillations' total decay rate in a single nanoparticle and the duplicated group velocity.

The analysis of the plasmon oscillations in a chain of metal nanoparticles stated above is very demonstrative, though rather simplified. Fortunately, the complete problem solution considering all types of multipolarity can be found in the quasi-static approximation. This solution can be built by expansion of the electric potential in spherical harmonics with the centers in each sphere in an infinite chain (Bergman, 1979). The details of the calculations can be found in (Zabkov, 2011a), and here we will only outline the calculations' course.

The field potential inside each particle of a cluster is sought in the form of a linear combination of normalized potentials of a single sphere:

$$\Phi_{lm}^{0(\alpha)} = \left(\frac{r^{(\alpha)}}{a}\right)^{l} \frac{Y_l^m\left(\theta^{(\alpha)}, \varphi^{(\alpha)}\right)}{\sqrt{al}}, \tag{10.95}$$

(see Eq. (6.43)), that is, in the form

$$\Phi^{(\alpha)} = \sum_{l,m} A_{lm}^{(\alpha)} \Phi_{lm}^{0(\alpha)}, \tag{10.96}$$

where α denotes one of the particles and the local spherical coordinates are $r^{(\alpha)}$, $\theta^{(\alpha)}$, $\varphi^{(\alpha)}$. Substituting Eq. (10.96) in the volume integral equation of the ε-method, Eq. (5.36), and using the properties of Eq. (5.7), this integral equation is reduced to the form

$$\begin{aligned} sA_{lm}^{(\alpha)} &= s_l A_{lm}^{(\alpha)} + \sum_{\beta \neq \alpha} Q_{lm,l'm'}^{(\alpha,\beta)} A_{l'm'}^{(\beta)}, \\ Q_{lm,l'm'}^{(\alpha,\beta)} &= \frac{1}{4\pi} \int_{V_\alpha} d\mathbf{r} \int_{V_\beta} d\mathbf{r}' \frac{\partial \Phi_{lm}^{0(\alpha)}(\mathbf{r})}{\partial r_\mu} \frac{\partial \Phi_{l'm'}^{0(\alpha)}(\mathbf{r}')}{\partial r'_\nu} \frac{\partial^2}{\partial r_\mu \partial r_\nu} \frac{1}{|\mathbf{r} - \mathbf{r}'|}, \end{aligned} \tag{10.97}$$

where $s = (1 - \varepsilon/\varepsilon_H)^{-1}$, $s_l = l/(2l+1)$.

The system, Eq. (10.97) is applicable for any cluster when corresponding potentials are used for isolated particles. In the case of periodically situated nanoparticles, the Bloch theorem is true, and the expansion coefficients $A_{nm}^{(\alpha)}$ can be presented as

$$A_{nm}^{(\alpha)} = A_{nm}(\mathbf{k}) e^{i\mathbf{k}\mathbf{r}^{(\alpha)}}, \tag{10.98}$$

where $\mathbf{k}$ is the vector of the reciprocal lattice and $\mathbf{r}^{(\alpha)}$ is the vector characterizing the position of the center of the αth particle. Substituting Eq. (10.98) in Eq. (10.97), we obtain the following equation:

$$sA_{lm}(\mathbf{k}) = s_l A_{lm}(\mathbf{k}) + \sum_{n'm'} Q_{lm,l'm'}(\mathbf{k}) A_{l'm'}(\mathbf{k}), \tag{10.99}$$

where

$$Q_{lm,l'm'}(\mathbf{k}) = \sum_{\beta \neq \alpha} Q^{(\alpha,\beta)}_{lm,l'm'} e^{-i\mathbf{k}(\mathbf{r}^{(\alpha)}-\mathbf{r}^{(\beta)})}. \tag{10.100}$$

The infinite system of linear interlocking equations in Eq. (10.99) will have nonzero solutions only for certain values of the longitudinal wavevector $\mathbf{k}$ and electric permittivity $\varepsilon(k)$. In the case of spherical particles, the coefficients $Q_{nm,n'm'}(\mathbf{k})$ can be calculated with the help of the spherical harmonics addition theorem (Bergman, 1979), and it appears to be that the matrix $Q_{nm,n'm'}(\mathbf{k})$ is diagonal with respect to the azimuthal quantum number m, precisely as it was in the case of two spheres. As a result, the resonance values of electric permittivity are determined by the vanishing determinant of this system of equations:

$$\det\left[s\mathbf{I} - \mathbf{H}\right] = 0,$$

$$H_{lm,l'm}(\mathbf{k}) = \frac{l}{2l+1}\delta_{l,l'} + Q_{lm,l'm}(\mathbf{k}), \tag{10.101}$$

$$s = \frac{\varepsilon_{\mathrm{H}}}{\varepsilon_{\mathrm{H}} - \varepsilon_m},$$

where the matrix $Q_{lm,l'm}(\mathbf{k})$ is determined by the correlations

$$Q_{lm;l'm}(\mathbf{k}) = \left(\frac{a}{d}\right)^{l+l'+1} \sum_{n=1}^{\infty} \frac{\cos(nkd)}{n^{l+l'+1}} K_{ll'm},$$

$$Q_{lm;l'm}(\mathbf{k}) = i\left(\frac{a}{d}\right)^{l+l'+1} \sum_{n=1}^{\infty} \frac{\sin(nkd)}{n^{l+l'+1}} K_{ll'm}, \tag{10.102}$$

for even and odd $l + l'$ correspondingly, where

$$K_{ll'm} = 2(-1)^{l'+m}\left(\frac{ll'}{(2l+1)(2l'+1)}\right)^{1/2} \times \frac{(l+l')!}{\left[(l+m)(l'+m)(l-m)(l'-m)\right]^{1/2}}. \tag{10.103}$$

In Fig. 10.31, dispersion curves obtained by the numerical diagonalization, Eq. (10.101), for $m = 0, \pm 1$ are shown. The solution, Eq. (10.92), is also shown in this picture for comparison, obtained by keeping the first term only in the infinite matrix, Eq. (10.101):

$$H_{1m;1m}(\mathbf{k}) = \frac{1}{3} + \left(\frac{a}{d}\right)^{3} \sum_{n=1}^{\infty} \frac{\cos(nkd)}{n^{3}} K_{11m}, \tag{10.104}$$

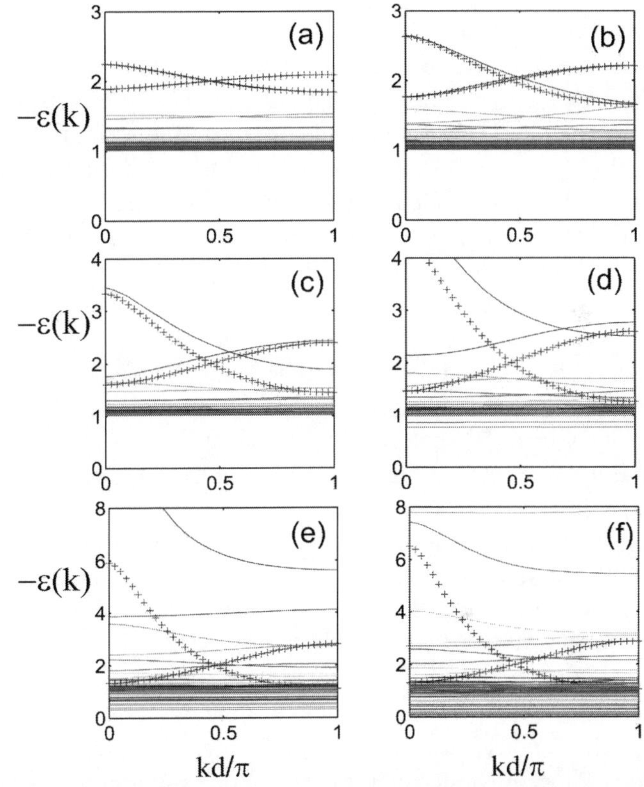

Figure 10.31 The dispersion curves $\varepsilon_m(\mathbf{k})$ of the plasmons propagating in a linear cluster of spherical nanoparticles as a function of the wavevector **k**. (a) $a/d = 0.25$, (b) $a/d = 0.33$, (c) $a/d = 0.4$, (d) $a/d = 0.45$, (e) $a/d = 0.49$, and (f) $a/d = 0.4999$ (the spheres contact). The solid lines correspond to the case of $m = 0$ and $m = \pm 1$, taking into account the $l = 80$ harmonics in the Hamiltonian, Eq. (10.98). The crosses correspond to calculations in the dipole approximation, that is, with the consideration of the first harmonics $l = 1$ only. Attention should be paid to the fact that for small distances between the particles, the phase transition takes place in the spectrum and new M-modes appear in the region of $0 < -\varepsilon < 1$ (Zabkov, 2011a).

where $K_{110} = -4/3$ and $K_{11\pm1} = 2/3$. Corresponding curves Eq. (10.92) are denoted by crosses on the picture.

As the picture implies, the plasmon dispersion curves differ from simple curves, Eq. (10.92), substantially when higher multipoles are considered. For rather long distances between nanoparticles ($d > 3a$) (Fig. 10.31a,b) the precise dispersion curves are quite similar to the dispersion curves obtained in the simple approximation (Eq. (10.104) or Eq. (10.92)). It is not astonishing since in this case only the dipole–dipole interactions between nanoparticles are significant. When the nanoparticles approach each other, the quadrupole, octupole, etc., interactions take greater importance, and as a result the exact quasi-static solution turns out to be completely distinct from the approximate ones, and the band structure changes radically. There is another important feature. Beginning from the distance $a/d > 0.4$, the branches with $0 > \varepsilon > -1$ or $\omega > \omega_{\text{pl}}/\sqrt{2}$ appear in the spectrum. These branches arise as a result of the interaction of modes highly localized in the gap (symmetric M-modes), which were discussed thoroughly above in Section 10.3.

The plasmonic frequencies ω_{pl} of a linear cluster are plotted in Fig. 10.32 as functions of the distance $d/2a$ between the particles. By analogy with a two-sphere cluster (see Section 10.3), in this figure it is possible to recognize three types of dispersion curves. The modes with normalized frequencies going to zero or to one for touching spheres correspond to antisymmetric L-modes or symmetric M-modes. The modes with normalized frequencies lying between 0 and $\omega_{\text{pl}}/\sqrt{2}$ for touching spheres correspond to symmetric T-modes. When speaking about symmetry of potential, we have in mind the symmetry plane situated in the gap between the spheres. It is convenient to enumerate these modes with indices that increase with increasing plasmonic frequencies ω_{pl} (L- and T-modes) or with decreasing plasmonic frequencies ω_{pl} (M-modes).

When the distance between the particles $d/2a$ goes to infinity, the plasmonic frequencies tend to plasmonic frequencies of a single sphere $\omega/\omega_{\text{pl}} = \sqrt{l/(2l+1)}$, where $l = 1, 2, 3, \ldots$ for $m = 0$, $l = m, m+1, m+2, \ldots$ for $m \neq 0$. At that, in the case of $k = 0$, the frequencies of antisymmetric modes tend to $\omega/\omega_{\text{pl}} = \sqrt{l/(2l+1)}$, with odd l for even m and even l for odd m. On the contrary, the frequencies of symmetric modes tend to $\omega/\omega_{\text{pl}} = \sqrt{l/(2l+1)}$, with

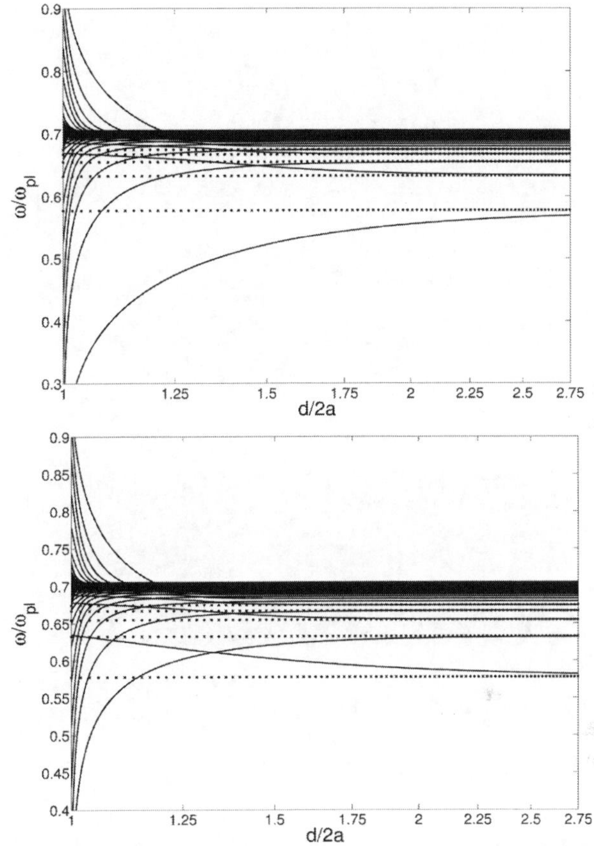

Figure 10.32 Frequencies ω_{pl} of plasmonic oscillations in a linear cluster made of spherical particles as a function of the distance between the particles $d/(2a)$ for $k=0$ and $m=0$ and $m=1$ (bottom). The solid lines correspond to numerical simulations, taking into account the $l=85$ harmonics. The horizontal dotted lines correspond to the first five plasmonic frequencies of a single sphere, $\omega/\omega_{\text{pl}} = \sqrt{l/(2l+1)}$, $l = 1\ldots5$.

odd l for odd m and even l for even m. In the case $k = \pi/d$, we have the opposite situation.

Let us now consider the spatial structure of the electric potential corresponding to different modes (Figs. 10.33 and 10.34). In the axial symmetric case ($m = 0$, Fig. 10.33a,b), the spatial structure of antisymmetric $L1$ and symmetric $T1$ modes correspond to modes

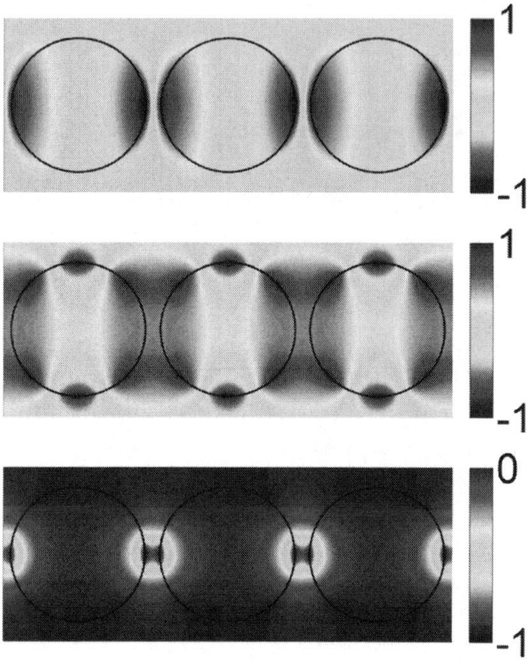

Figure 10.33 Spatial distribution of potential for $L1(\omega/\omega_{\rm pl} = 0.408)$, $T1(\omega/\omega_{\rm pl} = 0.665)$, and $M1$ $(\omega/\omega_{\rm pl} = 0.751)$ modes in the axial symmetric case $m = 0$, $kd = 0$, $d/2a = 1.11$.

of a single sphere $\varphi \approx r^l P_l(\cos(\theta))$, $l = 1, 2$, correspondingly. The interaction between the spheres results only in some redistribution of potential. At that, the L-mode is a bright one, that is, this mode has a nonzero (in fact, infinite) electric dipole momentum and can be excited by a plane electromagnetic wave. On the contrary, the T-mode is a dark one and cannot be excited by far fields.

In the case of symmetric M-mode ($m = 0$, Fig. 10.33c), the situation is different, and we have a strong concentration of the negative charge near the gap between the spheres, while the surface charge of the opposite sign is distributed over the rest of the particle surface. In fact, far from the gap the potential is nearly equal to zero. It is worth to note that all M-modes are characterized by a strong localization of charge and potential near the gap, but for $M2$ and

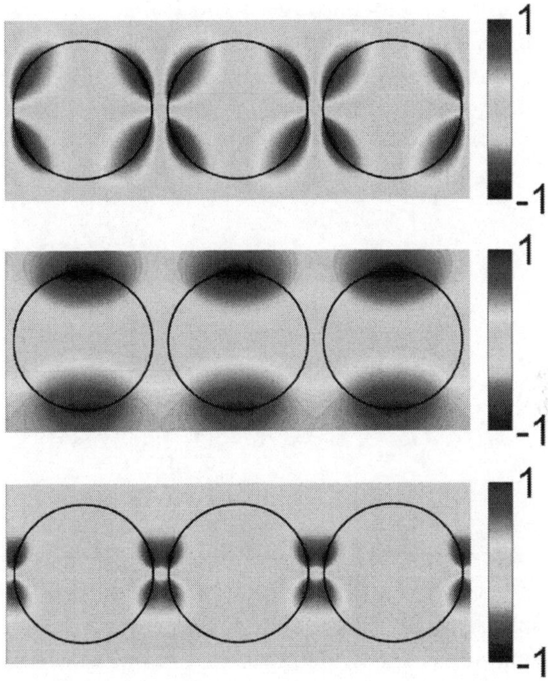

Figure 10.34 Spatial distribution of potential for $L1(\omega/\omega_{\text{pl}} = 0.564)$, $T1(\omega/\omega_{\text{pl}} = 0.626)$, and $M1(\omega/\omega_{\text{pl}} = 0.73)$ modes for $m = 1$, $kd = 0$, $d/2a = 1.11$.

higher modes, in contrast to $M1$ mode shown in Fig. 10.33c, the charges of both signs are concentrated near the gap.

In the case $m = 1$ (Fig. 10.34), the situation is analogous, and for L- and T-modes the interaction results again in some redistribution of potential. The only difference is that now L-modes are dark, while T-modes become bright. M-modes in this case are also bright, but their dipole moments are small in comparison with T-modes.

When the distance between the spheres increases, the localization of M-modes becomes weaker, and for the critical distance $d/2a \approx 1.2$ these modes disappear, while L- and T-modes suffer no substantial changes at this point and their frequencies go to frequencies of a single sphere.

After we have found plasmonic spectra of a linear cluster, it is important to estimate propagation lengths of different modes. The propagation length is determined by the imaginary part of the longitudinal wavenumber $k = k' + ik''$, which can be found from the relation

$$\varepsilon_n(k) = \varepsilon(\omega), \tag{10.105}$$

or

$$\varepsilon_n'(k' + ik'') - i\varepsilon_n''(k' + ik'') = \varepsilon'(\omega) + i\varepsilon''(\omega). \tag{10.106}$$

If we assume that $k'' \ll k'$, $\varepsilon' \gg \varepsilon''$, the solution of Eq. (10.106) can be easily found:

$$k'' = \frac{2\pi}{L} = \frac{\varepsilon''(\omega) + \varepsilon_n''(k')}{d\varepsilon_n'(k')/dk'}, \tag{10.107}$$

where $\varepsilon''(\omega)$, $\varepsilon_n''(k')$ describe Joule's and radiative losses, respectively. For nanoparticles, the radiative losses $\varepsilon_n''(k')$ tend to zero and one can neglect them in the first approximation. In Fig. 10.35, the propagation lengths for the $L1$ mode are shown for some values of the wavenumber. These values were obtained with the help of Eq. (10.107) for $\varepsilon_n''(k') = 0$. From this figure, one can see that the propagation length for the $L1$ mode can reach several microns, and it is enough to apply this type of a waveguide in optical nanocircuits. The propagation length can be further increased by placing the cluster inside a host matrix with a high refractive index.

10.8.2 *Plasmon Properties of Linear Clusters: Retardation Effects*

In the previous section, the plasmonic properties of a linear cluster were investigated within the quasi-static approximation. However, the linear cluster is an infinite system, and smallness of nanoparticles forming it does not guaranty that the quasi-static approximation will be good enough. To check this point, one should investigate plasmonic properties of a linear cluster, taking retardation effects into account. Within this approach, it is very difficult to find an analytical solution, and one should either simplify the system or use direct numeric simulation.

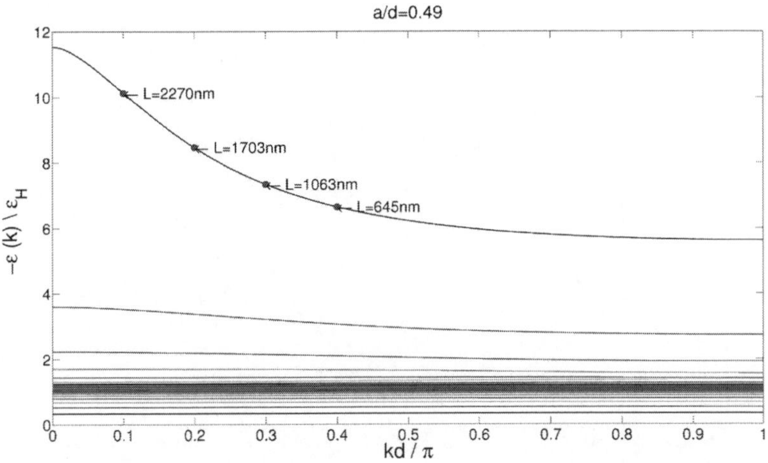

Figure 10.35 Resonant values of ε in a linear cluster as a function of the longitudinal wavevector k for $m = 1$, $a = 5$ nm, and $d = 10.2$ nm. The propagation lengths were calculated for $kd/\pi = 0.1$, 0.2, 0.3, and 0.4. The imaginary part is taken as $\varepsilon''(\omega) = 0.5$, and it is assumed that the cluster is placed in a host matrix with permittivity $\varepsilon_{\mathrm{H}} = 3$.

The first approach was used by Weber and Ford (2004), who investigated the plasmon oscillations' spectrum in a linear chain of spherical nanoparticles within the point dipole approximation but taking retardation effects into account, that is, in the context of the full system of Maxwell's equations. They have shown that, indeed, for transverse plasmon oscillations there is a substantial deviation from the quasi-static solution in the vicinity of the light line $\omega = ck$ (Fig. 10.36). Physically, the explanation of this anomaly is related to the fact that at this point the phase synchronism arises between the quasi-static plasmons with transverse polarization and the usual transverse photons. There are no corresponding photons for the plasmon waves with longitudinal polarization (Fig. 10.37), and therefore the retardation effect results only in quantitative change (about 10%) of the dispersion curves.

To take into account higher multipoles, the numerical simulation of plasmonic oscillations in a linear cluster was also carried out within Comsol Multiphysics® (Zabkov, 2011b). Two polarizations of incident radiation were considered, S- and P-polarizations. The

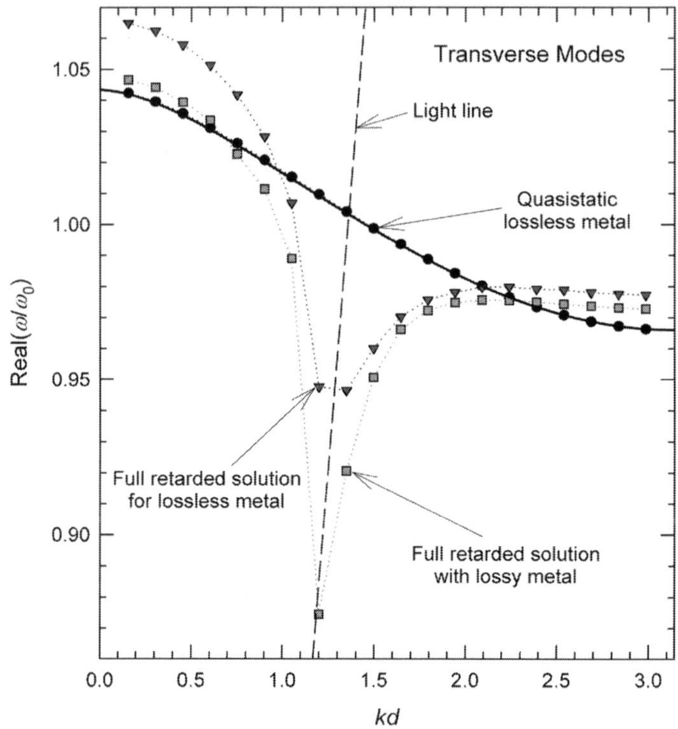

Figure 10.36 The dispersion curves for the transverse excitation of a linear cluster with retardation taken into account. The solid curve is the same as in Fig. 10.30 for the infinite cluster. The results of the calculations for the cluster of 20 nanoparticles are shown by these points: the black circles (•) denote the quasi-static approximation, the green squares (■) denote the solution for metal with retardation taken into account, and the red triangles (▼) denote the solution for the lossless metal with retardation. The dashed line denotes the dispersion law of free photons, $\omega = ck$ (Weber Ford, 2004).

first of them corresponds to the polarization perpendicular to the cluster axis, while the second corresponds to the polarization along the axis for normal incidence. In Figs. 10.38, 10.39, and 10.40, the absorption cross sections are shown as a function of permittivity of spheres. Figure 10.40 corresponds to an enlarged part of Fig. 10.39. The vertical dotted lines correspond to the position of plasmon resonances in the quasi-static approximation.

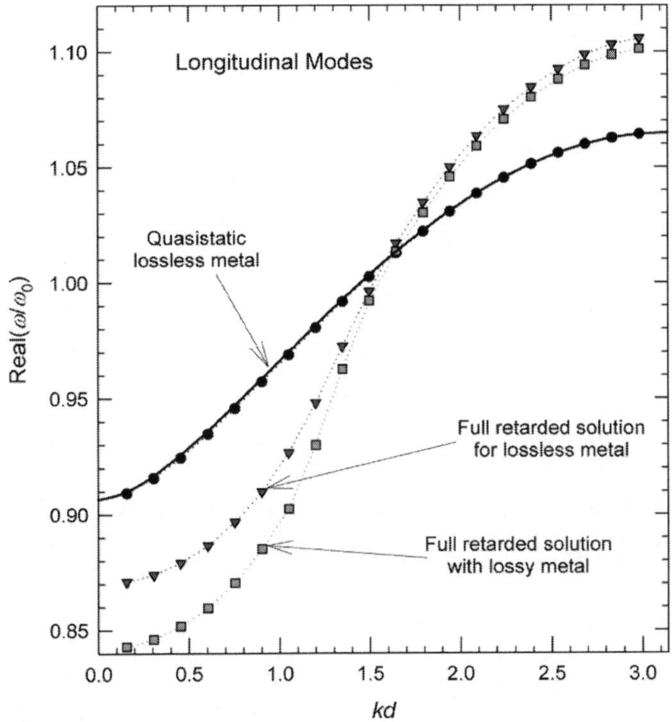

Figure 10.37 The dependencies same as shown in Fig. 10.36 but for the longitudinal polarization of the exciting field (Weber Ford, 2004).

Corresponding resonances were identified by comparing values of permittivities and also spatial distributions of mode eigenpotentials.

For S-polarization ($m = 1$, Fig. 10.38), one can see four pronounced resonances: three symmetrical ($T1$, $T2$, $T3$) and one antisymmetric ($L2$) (See Fig. 10.32, bottom). The $T1$ resonance is the strongest one and corresponds to a bright mode with dipole orientation across the axis that is along the excitation electric field. $T2$ and $T3$ resonances correspond to symmetric oscillations of higher order multipoles. It is important to note that $T1$, $T2$, and $T3$ modes can be excited by a corresponding electric field in the quasi-static approximation, while $L2$ mode can be excited only when retardation is taken into account.

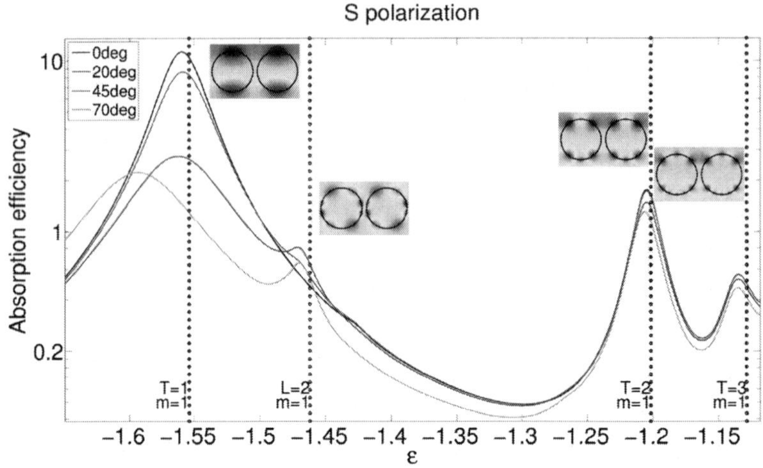

Figure 10.38 The absorption efficiency for a sphere as a function of the incidence angle and the real part of the spheres' permittivity for S-polarization of incident radiation obtained by simulation with Comsol MultiPhysics®. The vertical dotted lines correspond to the position of plasmon resonances obtained analytically in the quasi-static approximation for $k = 0$. On the insets, the corresponding distribution of potential is shown. Simulation parameters: the sphere radius $a = 5$ nm, the distance between the spheres $d/2a = 0.45$, the wavelength $\lambda = 500$ nm, the imaginary part of permittivity $\varepsilon'' = 0.01$, and the permittivity of host medium is $\varepsilon_H = 1$ (vacuum).

In the case of P-polarization (Figs. 10.39 and 10.40), one can observe five resonances: $L1$, $L2$, $L3$ ($m = 0$), and $L1$ and $T1$ ($m = 1$). Now, the $L1$ resonance is the strongest one and corresponds to a bright mode with dipole orientation along the chain axis that is along the excitation electric field. The $T1$ ($m = 1$) mode appears only for oblique incidence and is a result of transverse electric field component appearance. When the angle of incidence increases, the amplitude of this resonance increases correspondingly. It is important to note that the $T1$ ($m = 1$) resonance cannot be excited in the quasi-static approximation; that is why it has a pure retardation nature.

Thus, the quasi-static approximation gives a reasonable description of plasmonic properties of a linear cluster and can be used for fast estimations. Of course, for a more precise result one can

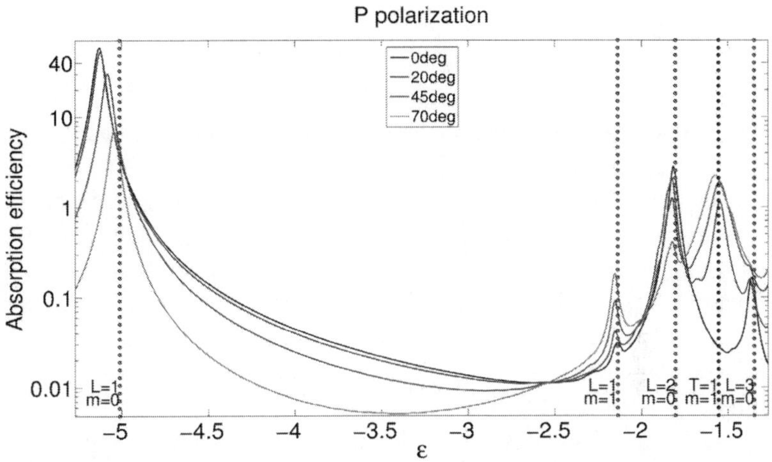

Figure 10.39 The same as in Fig. 10.38 but for P-polarization and $\varepsilon'' = 0.02$.

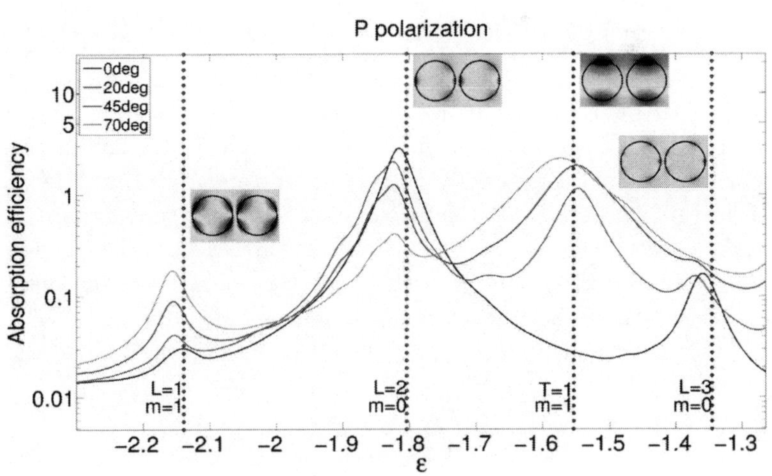

Figure 10.40 The same as in Fig. 10.39 but with higher resolution of resonances.

use numerical methods, especially for transverse polarizations near $\omega = ck$.

10.8.3 *Plasmon Properties of Self-Similar Clusters*

The construction of devices able to focus optical fields into nanoscale regions is also an important problem of nanoplasmonics. Usual lenses can focus fields only into the region of the order of the wavelength. In the gaps between the nanoparticles and in nonsymmetric plasmon nanoshells, the local fields $10 \sim 100$-fold strengthening is possible (see Eqs. (10.30) and (10.61)). For higher concentration of optical fields, Li et al. (2003) proposed to use a self-similar cluster with decreasing diameters. The geometry of this cluster is shown in Fig. 10.41.

Self-similarity implies that the radii i and $i+1$ of the spheres and the distances between them are related by the ratios

$$R_{i+1} = \kappa R_i; \quad d_{i+1,i+2} = \kappa d_{i,i+1}, \tag{10.108}$$

where $\kappa < 1$. It is assumed that the latter condition provides only a weak inverse influence of the smaller sphere on the preceding, larger one. The self-similarity is not a necessary condition of focusing into nanoscale regions; however, it results in an elegant formulation and in a solution of the problem. At that, all particles and the whole system are considered in the quasi-static approximation. Now, if each particle enhances the incident field by factor α, the cumulative effect will be of the order of α^n, where n is the number of particles in a cluster. In other words, the field enhanced by a large nanoparticle is the exciting field for the next smaller nanoparticle, and so on. For the system of three nanoparticles with a moderate enhancement $\alpha = 10$, the resulting effect will be of the order of 1,000.

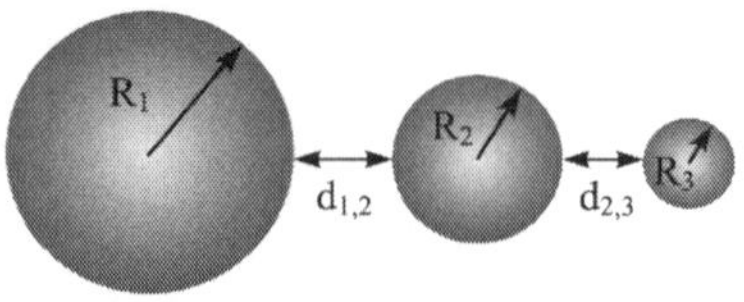

Figure 10.41 The geometry of a self-similar cluster with decreasing sizes of nanoparticles and distances between them.

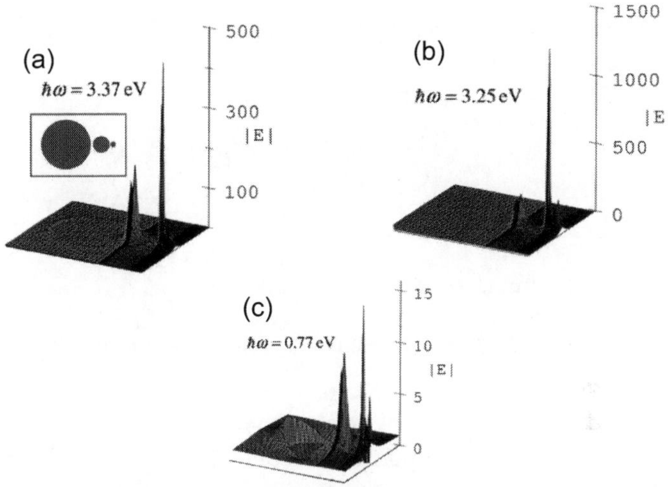

Figure 10.42 The values distribution of local electric fields normalized to incident fields' amplitudes in the equatorial symmetry plane of a self-similar cluster of three silver nanospheres. The ratio between the radii of neighbour spheres is determined by the relation $\kappa = R_{i+1}/R_i = 1/3$; the distance between neighbour nanospheres is determined by the relation $d_{i,i+1} = 0.6R_{i+1}$ (Li et al., 2003).

In Fig. 10.42, the electric field distribution is shown, which was obtained as a result of the solution of the quasi-static problem in the geometry of Fig. 10.41 by the method of spherical harmonics expansion of the fields and applying of the continuity equations of the potential and normal component of electric displacement on the spheres' boundaries. In this picture, it is seen well that near the smaller sphere, more than 1,000-fold enhancement of the field's absolute value occurs. Apparently, this enhancement allows explaining of the giant combinational light scattering on a single molecule situated near metal clusters.

10.8.4 *Plasmon Properties of Starlike Clusters*

An interesting class of clusters was investigated by Poponin and Ignatov (2005) (see Fig. 10.43). The objective of the investigation was an attempt to find the plasmon oscillation with the intensity maximum situated outside the nanoparticles' surface. By analyzing

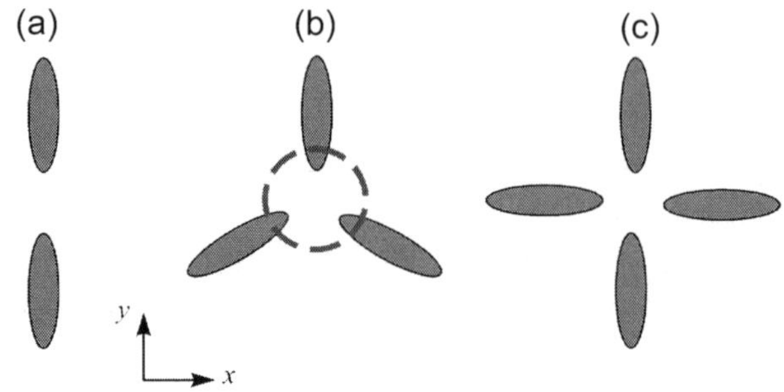

Figure 10.43 Starlike clusters of prolate spheroids.

the general relations of electrostatics, however, it was shown that the intensity maxima can be reached only on the interfaces in the points of the maximal curvature. Only minimum and saddle points of intensity are possible between the nanoparticles. An example of plasmon oscillations calculation in systems of this kind is shown in Figs. 10.44 and 10.45, where it is seen that, indeed, the maxima are situated on the nanoparticles' surfaces only. This circumstance is extremely important for control of the atom motion and for construction of nanotraps and nanopincers for atoms and molecules since it allows only the field minima as effective trap configurations for atoms. For more on atom nano-optics see (Balykin et al., 2006).

10.9 Influence of Plasmon Resonances in a Cluster of Nanoparticles on the Radiation of Atoms and Molecules

As it was repeatedly stated above, plasmon resonances in nanoparticles influence the atoms' and molecules' emission substantially. From this point of view, the influence of clusters of nanoparticles is especially interesting since an essential enhancement of local fields takes place in clusters, resulting in a substantial increase of spontaneous decay rates and in a radiation frequency shift. Since strongly localized plasmon oscillations interact especially

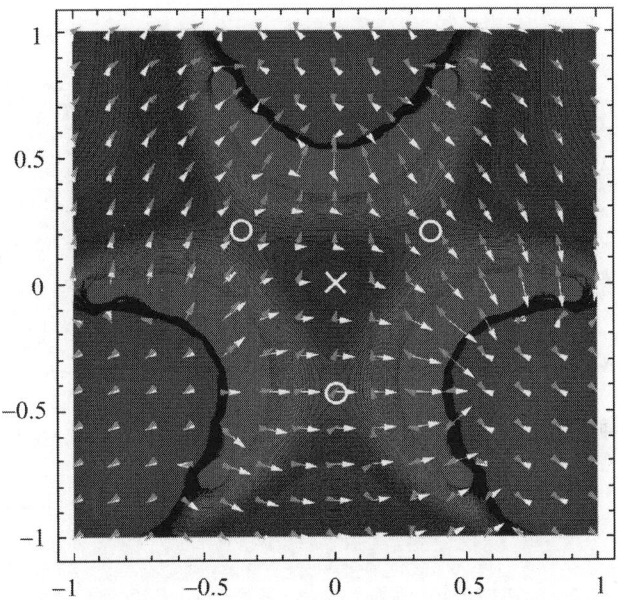

Figure 10.44 The distribution of the electric field absolute value in the plane $z = 0$ in the vicinity of the center of a three-nanospheroid cluster (Fig. 10.43b) for the circular polarization of the incident wave. The green arrows show Re**E**, and the white arrows show Im**E**. Different saddle points of intensity are shown by the signs O and X (Poponin and Ignatov, 2005).

effectively with strongly nonhomogeneous electric fields of atoms and molecules, the construction of different detectors of single molecules or quantum dots is possible.

As it is shown in Appendix A1, to describe both the radiative decay rate and the total decay rate, it is sufficient to solve the quasi-static problem and to find the electric field reflected from the cluster and the electric dipole moment induced in the cluster. In the case of two spherical nanoparticles, these characteristics can be found by solving the Laplace equation in the bispherical system of coordinates, as it was done to find the plasmon resonance in a cluster of two nanoparticles (Section 10.3). This solution is rather tedious, and we omit its details, which can be found in the original publication (Klimov and Guzatov, 2007c).

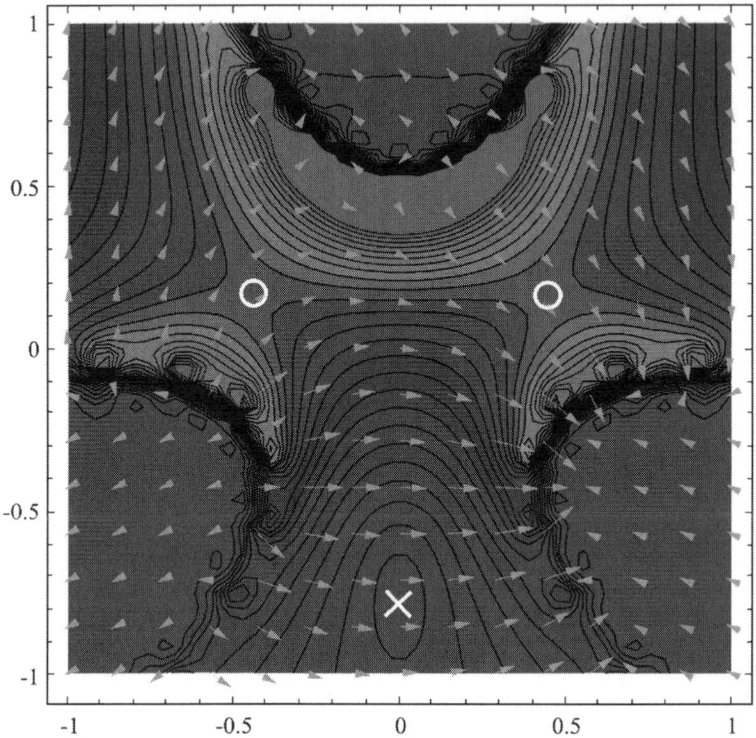

Figure 10.45 The same as in Fig. 10.44 but for the linear (vertical) polarization of the incident wave (Poponin and Ignatov, 2005).

In particular, the expression for the radiative rate $\gamma^{\text{radiative}}$ of spontaneous emission of an atom (molecule) in the vicinity of any nanobody has the form (Klimov and Ducloy, 2004):

$$\frac{\gamma^{\text{radiative}}}{\gamma_0} = \frac{|\mathbf{d}_{\text{total}}|^2}{|\mathbf{d}_0|^2}, \tag{10.109}$$

where γ_0 is the rate of the atoms' spontaneous decay in vacuum in the absence of a nanobody and $\mathbf{d}_0$, $\mathbf{d}_{\text{total}} = \mathbf{d}_0 + \delta\mathbf{d}$ are the dipole moment of the transition and the total dipole moment of the system "emitting atom + nanobody," correspondingly. In the case of a cluster of two nanospheres, the expression for the induced dipole moment $\delta\mathbf{d}$ has the form (Guzatov and Klimov, 2005):

$$\begin{aligned}
\delta d_x &= -\sqrt{2}a\sum_{n=1}^{\infty} n(n+1)(\mathbf{d}_0\cdot\nabla')\alpha_{1n},\\
\delta d_y &= -\sqrt{2}a\sum_{n=1}^{\infty} n(n+1)(\mathbf{d}_0\cdot\nabla')\beta_{1n},\\
\delta d_z &= \sqrt{2}a\sum_{n=0}^{\infty} (2n+1)(\mathbf{d}_0\cdot\nabla')\gamma_{0n},
\end{aligned} \tag{10.110}$$

where α_{1n}, β_{1n}, and γ_{0n} are solutions of nonhomogenous analogues of Eqs. (10.36) and (10.37), taking into account the electric field of a single point charge situated in the place of the atom or molecule location $\mathbf{r}'$, and differentiation is over $\mathbf{r}'$.

Since the point charge has a substantially nonhomogeneous electric field, in the spontaneous emission spectrum one should observe the peaks related to the excitation of plasmon modes described in Section 10.3 and corresponding to the indices $m = 0$ and 1 (see Eq. (10.110)).

In Fig. 10.46, the dependence of the radiative part of the spontaneous decay rate of different molecules with the vertical orientation of the dipole moment on the wavelength of this molecules emission is shown. It is assumed that the nanospheres are made of either of Na with the plasmon resonances in the visible band (Fig. 10.46a) or of SiC with phonon-polariton resonances in the infrared band (Fig. 10.46b). Let us remember that the photon-polariton resonances are determined by the lattice oscillations in the frequency region where the real part of the electric permittivity becomes negative. The small internal attenuation of the photon-polariton resonances is their important peculiarity, which allows observing of the fine structure of localized plasmon spectra in nanoparticles.

The peaks at the right part of the pictures correspond to the interaction between the molecule and the plasmon T-mode, while the peaks at the left correspond to the interaction with M-modes (see Section 10.3 and Fig. 10.7b).

The analysis of these pictures shows that for the small distances between the nanospheres the interaction between the radiating molecule and M-modes (plasmon molecules) is more important than the interaction with symmetric T-modes (hybridized plasmon

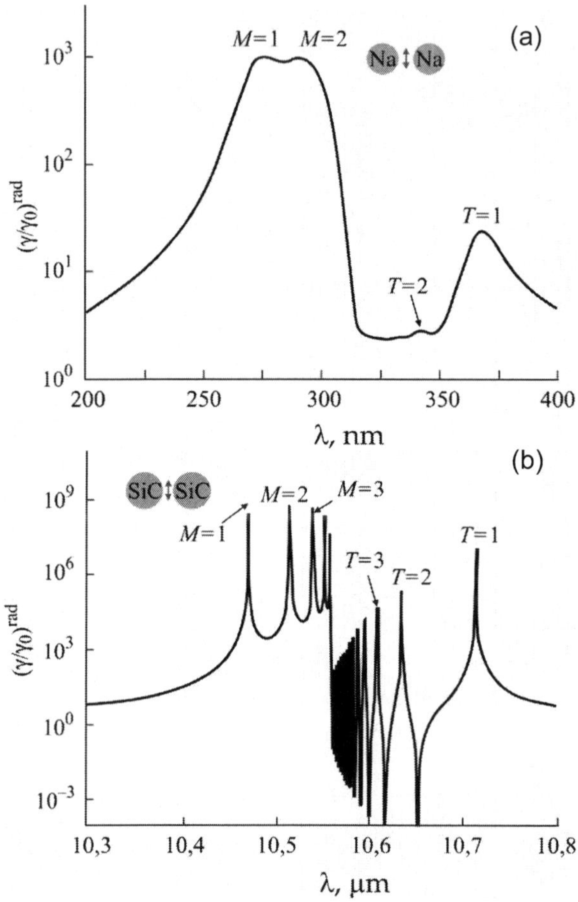

Figure 10.46 The dependencies of the spontaneous decay rate of a molecule situated in the gap between nanoparticles on the wavelength of its emission. (a) Nanospheres of Na and (b) nanospheres of SiC (Klimov and Guzatov, 2007b).

atoms) (see Fig. 10.7b). It is also extremely important that in the case of a homogeneous external field, M-modes are excited ineffectively. The result of these circumstances is that the clusters of two and more nanospheres with the geometry allowing the excitation of highly localized plasmon M-modes with a certain frequency can be used for development of effective nanodetectors of

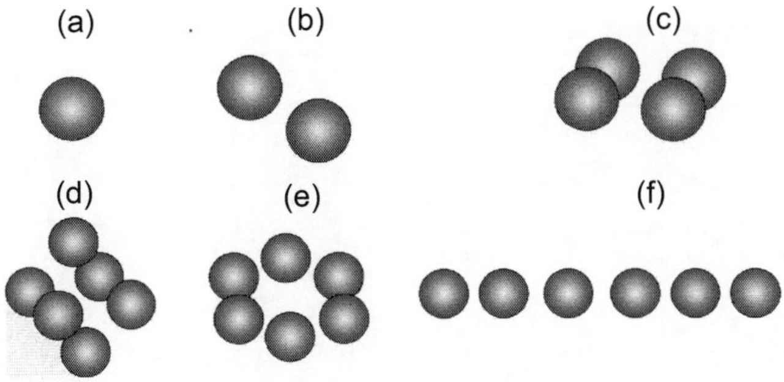

Figure 10.47 The geometry of different clusters of nanoparticles of radius $R = 0.2c/\omega_{\text{pl}}$. (a) A single nanosphere, (b) two spheres, (c) a square cluster, (d) a cluster of double pyramids, (e) a hexagonal cluster, and (f) a linear cluster of six nanospheres.

single molecules with a high signal-to-noise ratio since the external noise fields cannot excite M-mode at its oscillations' frequency. For large distances between the nanospheres ($R_{12} > 2.4R_0$), M-modes do not exist, and only the resonances related to T-modes (the right sides in Fig. 10.46) remain in the spectra.

On the other hand, the effective interaction of M-modes and usual molecules can be used for the development of both spasers (Bergman and Stockman, 2003; Vinogradov et al., 2012; Lisyansky et al., 2012) and nanolasers (Bergman and Stockman, 2004; Protsenko et al., 2005; Noginov et al., 2009) and of a whole series of these devices. Let us remember that spasers are systems where the gain of the coherent plasmon oscillations' amplitude is provided by optical pumping.

The influence of the clusters' structure consisting of two and more nanoparticles on the atoms' spontaneous emission was investigated by Yannopapas and Vitanov (2007), where the clusters shown in Fig. 10.47 were considered. The calculations were performed in the context of the full system of Maxwell's equations by field expansion over the spherical harmonics related to single nanoparticles and providing usual boundary conditions on the nanospheres' surfaces.

The results of calculations of the spontaneous emissions' total rate (i.e., the sum of the radiative and nonradiative rates) on its frequency and the cluster geometry are shown in Fig. 10.48a for

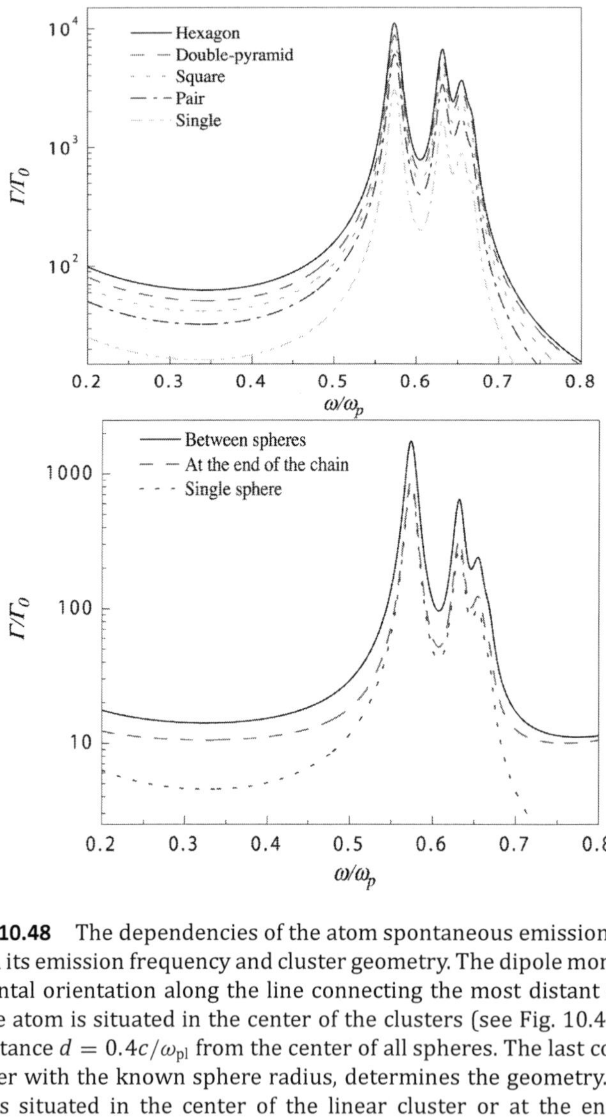

Figure 10.48 The dependencies of the atom spontaneous emission relative rate on its emission frequency and cluster geometry. The dipole moment has horizontal orientation along the line connecting the most distant spheres. (a) The atom is situated in the center of the clusters (see Fig. 10.47b–e) at the distance $d = 0.4c/\omega_{\mathrm{pl}}$ from the center of all spheres. The last condition, together with the known sphere radius, determines the geometry. (b) The atom is situated in the center of the linear cluster or at the end of the chain. In this case the distance between the nanospheres' centers is c/ω_{pl}, and the distance between the atom and the center of the nearest sphere is $d = 0.5c/\omega_{\mathrm{pl}}$ (Yannopapas and Vitanov, 2007).

clusters a–e and in Fig. 10.48b for a linear chain of six nanospheres. The analysis of Fig. 10.48 shows that the spontaneous decay rate increases with an increase in the number of nanoparticles in the cluster. This tendency is especially perceptible at frequencies different from resonance frequencies. Physically, this phenomenon is understandable since the polarizability and the induced dipole moment of the cluster $\delta\mathbf{d}$ are proportional to its volume. Certainly, it is not true for an arbitrary case. So, for example, in the case of dielectric periodic photon crystals, the spontaneous emission of an atom can be suppressed absolutely in some frequency regions due to interference effects (Li et al. 2000; Li and Xia, 2001; Wang et al., 2002; Wang et al., 2003).

Another interesting peculiarity in Fig. 10.48a is the dependency of the atom decay rate on the three-dimensional topological structure of the cluster. In particular, the atom radiation inside a double pyramid is always weaker than inside a hexagon, though the total number of nanospheres and the distances between them are the same in both cases. This fact also has a simple physical explanation since in the considered case of the horizontal (x axis) orientation of the atom dipole moment, in the case of a double pyramid the atom dipole moment is oriented radially relative to two spheres and tangentially relative to four spheres. In the case of a hexagon, the dipole moment is oriented radially relative to two spheres and partially radially relative to four. Taking into account the fundamental fact that the atom radial orientation results in acceleration of the decay rate, and the tangential orientation results in its slowing down (see Chapter 6), one can conclude that in the case of a hexagonal cluster the spontaneous emission rate is stronger.

The results of the investigation of the atom spontaneous emission in the presence of a linear cluster of six nanoparticles are shown in Fig. 10.48b for two positions of an atom, in the center of the cluster and at the end of the chain. In both cases, the distance from the atom to the center of the nearest sphere is $d = 0.5c/\omega_{\text{pl}}$. From this picture, it is seen that the decay rate of the excited state of an atom near the cluster edge is essentially smaller since it interacts effectively only with one sphere. Generally, from Fig. 10.48 it is seen that in resonance regions the decay rates depend mostly on the nearest nanospheres. Let us note that the modeling performed by

(Yannopapas and Vitanov, 2007) did not show the presence of highly localized plasmon modes (M-modes), which are to appear above the surface plasmon frequency $\omega_{sp} = \omega_{pl}/\sqrt{2} \approx 0.707\omega_{pl}$ (see Fig. 10.46). Apparently, it is related to the selected atom orientation and to a relatively large distance between the atom and the nearest spheres, resulting in the absence of M-modes' excitation.

10.10 Plasmon Nanoparticles Influence on the van der Waals Forces between Nanoparticles

As it is known, van der Waals forces are related to the density spatial dependency on the vacuum fluctuations' energy (Barash and Ginsburg, 1984; Sernelius, 2001). In the case of closely spaced plasmon nanoparticles, the main contribution to van der Waals energy is due to "zero" oscillations of all plasmon modes presented in the system (Klimov and Lambrecht, 2009).

In the case of two identical plasmon nanospheres, a contribution to van der Waals energy is due to "zero" oscillations of antisymmetric and symmetric plasmon modes (L-, T-, and M-modes):

$$\begin{aligned} U_{vdW} &= U_{vdW}^{M} + U_{vdW}^{L} + U_{vdW}^{T} \\ &= \frac{\hbar}{2}\left[\sum_{M=1}^{\infty}\omega_{M0} + \sum_{L=1}^{\infty}\omega_{L0} + \sum_{T=1}^{\infty}\omega_{T0}\right] \\ &\quad + \hbar\left[\sum_{M,m=1}^{\infty}\omega_{Mm} + \sum_{L,m=1}^{\infty}\omega_{Lm} + \sum_{T,m=1}^{\infty}\omega_{Tm}\right], \end{aligned} \tag{10.111}$$

where we have taken into account that eigenfrequencies with $m \neq 0$ are twice degenerated. In the case of the Drude dispersion law

$$\varepsilon(\omega) = 1 - \frac{\omega_{pl}^2}{\omega^2}, \tag{10.112}$$

the plasmon frequencies can be easily expressed through the electric permittivity eigenvalue:

$$\omega_n = \frac{\omega_{pl}}{\sqrt{1-\varepsilon_n}}. \tag{10.113}$$

As a result, the components of van der Waals energy can be written in the form

$$U_{vdW}^{L} = \sum_{m,L=0}^{\infty} \frac{\hbar\omega_m^{(L)}}{2} = \frac{\hbar\omega_{\text{pl}}}{2}\left[\sum_{L=0}^{\infty} \frac{1}{\sqrt{1-\varepsilon_0^{(L)}}} + 2\sum_{m=1,L=0}^{\infty} \frac{1}{\sqrt{1-\varepsilon_m^{(L)}}}\right], \quad (10.114)$$

$$U_{vdW}^{M} = \sum_{m=0,M=1}^{\infty} \frac{\hbar\omega_m^{(M)}}{2} = \frac{\hbar\omega_{\text{pl}}}{2}\left[\sum_{M=1}^{\infty} \frac{1}{\sqrt{1-\varepsilon_m^{(M)}}} + 2\sum_{m=1,M=1}^{\infty} \frac{1}{\sqrt{1-\varepsilon_m^{(M)}}}\right], \quad (10.115)$$

$$U_{vdW}^{T} = \sum_{m=0,T=1}^{\infty} \frac{\hbar\omega_m^{(T)}}{2} = \frac{\hbar\omega_{\text{pl}}}{2}\left[\sum_{T=1}^{\infty} \frac{1}{\sqrt{1-\varepsilon_m^{(T)}}} + 2\sum_{m=1,T=1}^{\infty} \frac{1}{\sqrt{1-\varepsilon_m^{(T)}}}\right]. \quad (10.116)$$

In spite of the formal identity of expressions for van der Waals energy for different types of modes, these contributions result in entirely different physical consequences. From Fig. 10.49, it is seen that the energies of antisymmetric L-modes grow as the distance increases, resulting in attraction between particles. Analogously, the energy of symmetric M-modes decreases as the distance does, resulting in the repulsion of nanospheres. Symmetric T-modes also result in very weak repulsion.

The attraction and repulsion determined by different plasmon oscillations have a simple physical nature. Let us first consider the case of the axially symmetric oscillations ($m = 0$). In this case, longitudinal antisymmetric L-modes consist of two parallel dipoles, which are attracted according to the laws of electrostatics (see Fig. 10.3 and Fig. 10.50). In the case of symmetric oscillations, dipoles have opposite directions, resulting in their repulsion (see Fig. 10.3 and Fig. 10.50).

In the case of $m = 1$, that is, when wavefunctions are described by the law $\cos\varphi$ or $\sin\varphi$ (where φ is the angle describing

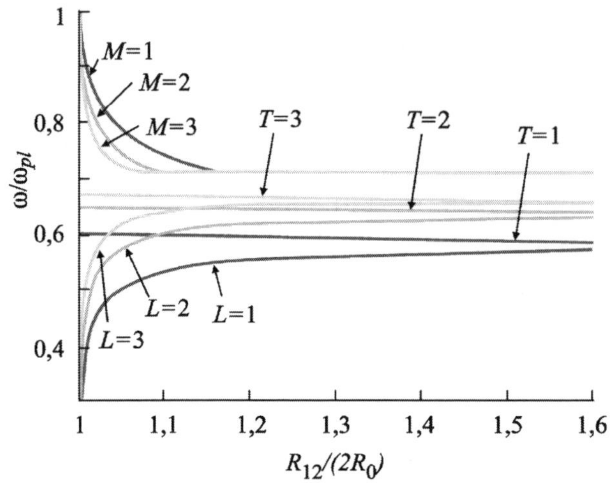

Figure 10.49 The spectrum of axially symmetric ($m = 0$) plasmon oscillations in a cluster of two identical spherical nanoparticles. In the upper part of the graph ($\omega > \omega_{\mathrm{pl}}/\sqrt{2}$), only the symmetric M-modes exist. The energy of the antisymmetric L-modes decreases as the distance between the nanospheres does; this corresponds to attraction. The energy of the symmetric (T-, M-) modes increases as the distance between the nanospheres decreases, and this corresponds to repulsion (Klimov and Guzatov, 2007b).

the rotation relative to the symmetry axis), directions of dipoles corresponding to different plasmon states change. In the case of antisymmetric L-modes, dipole moments in different spheres are to be directed perpendicular to the symmetry axis and to have opposite orientations (Fig. 10.3 and Fig. 10.50). It is obvious that again it results in the repulsion effect for these modes. For symmetric (M-modes), the dipole moments in different spheres are also to be oriented across the symmetry axis, though due to the symmetry relative to the plane perpendicular to the symmetry axis, they are to have the same dipole moments (see Fig. 10.50). Again, it results in the repulsion of nanospheres.

In the region of small distances, the contribution of symmetric T-modes may be neglected, and the asymptotic value of the electric permittivity eigenvalues, Eqs. (10.58) and (10.59), may be used for estimation of contributions of the rest of the L- and M-modes. Using

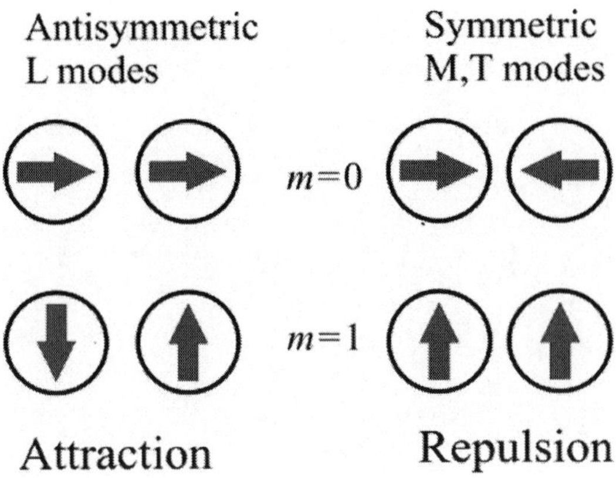

Figure 10.50 The illustration of the spheres' attraction and repulsion determined by different plasmon modes.

of the Euler–MacLaurin summation formula allows one to estimate different plasmon modes' contributions to van der Waals energy (Klimov and Lambrecht, 2009):

$$U_{vdW}^{M} = \frac{\hbar\omega_{\rm pl}}{\eta^2}\int_{\eta/2}^{\infty} dQQ\left[\frac{1}{\sqrt{1+\tanh Q}} - \frac{1}{\sqrt{2}}\right] \approx \frac{0.0838\hbar\omega_{\rm pl}}{\eta^2}, \tag{10.117}$$

$$U_{vdW}^{L} = \frac{\hbar\omega_{\rm pl}}{\eta^2}\int_{\eta/2}^{\infty} dQQ\left[\frac{1}{\sqrt{1+\coth Q}} - \frac{1}{\sqrt{2}}\right] \approx -\frac{0.096\hbar\omega_{\rm pl}}{\eta^2}. \tag{10.118}$$

In Fig. 10.51, the dependence of contributions of different plasmon modes to van der Waals energy on the distance between the spheres is shown (Klimov and Lambrecht, 2009). These contributions were obtained by direct summation of all plasmon modes in Eq. (10.111). As it was clear in advance, symmetric M-modes result in repulsion, while antisymmetric L-modes result in attraction of the nanoparticles. It is unexpected that the repulsive contribution of M-modes is nearly equal to the attractive contribution of L-modes. As a result, van der Waals total energy increases as the distance does, that

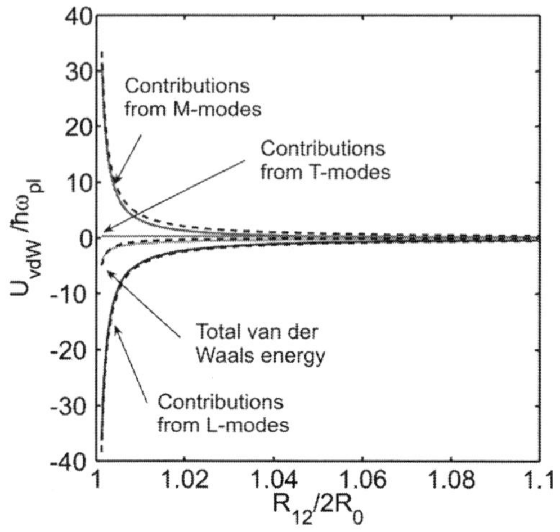

Figure 10.51 The dependence of van de Waals energy between two spherical nanoparticles on the distance between them and its components determined by different plasmon modes. The dashed lines are the asymptotic expressions Eqs. (10.117) and (10.118) (Klimov and Lambrecht, 2009).

is, it is the energy of attractions, but the magnitude of this energy is by an order less than that obtained without the consideration of the *M*-modes' contribution.

The direct measurement of van der Waals forces between plasmonic nanoparticles may form a basis for the experimental validation of the *M*-modes' existence.

10.11 Plasmon Resonance Excitation in a Cluster of Nanoparticles

As we have seen in previous sections, plasmon oscillations' spectra become more and more complex as the number of particles in a cluster increases and its geometry becomes more complicated. Thus, the question of these plasmon oscillations' excitation and observation arises. Homogeneous and weakly nonhomogeneous at nanoscale fields can excite effectively only a small number of so-called "bright" plasmon modes, that is, modes characterized by a large dipole momentum. This kind of modes includes, for example, antisymmetric ($m = 0$) longitudinal L-modes and symmetric ($m = 1$) T-modes in a cluster of two nanoparticles (see Fig. 10.7) and longitudinal or transverse oscillations with zero wavevector $\mathbf{k} = 0$ in an infinite linear cluster of spherical nanoparticles. However, the majority of the modes are "dark" and have zero or a very small dipole momentum and therefore they interact weakly with homogeneous on the nanoscale optical fields. For example, symmetric ($m = 0$) M- and T-modes, antisymmetric ($m = 1$) L-modes, and all modes in the infinite linear chain with $k \neq 0$ may be referred to as "dark" modes.

The most effective method of excitation of certain modes in clusters of nanoparticles is strongly nonhomogeneous in nanometer scale optical fields, which are effectively produced by the emission of atoms, molecules, and nanocrystal quantum dots. In Section 10.9, the influence of plasmon nanoparticle clusters on atoms' emission was analyzed; however, excited plasmon modes were not considered there.

Effectiveness of excitation of plasmon modes by a nanolocalized light source was used by Maier et al. (2003) for finite-difference time domain (FDTD) simulation of the plasmon pulse propagation through a plasmon waveguide made of nine spherical nanoparticles of 50 nm diameters with a 75 nm interparticle distance. The plasmon pulse was excited by the impulse of the point dipole (atom) emission with the central frequency $\hbar\omega_0 = 2.4$ eV, which corresponds to $kd = \pi/2$ and to the maximal propagation group velocity (see Fig. 10.30). The width of the excitation impulse was chosen to be equal to 95% of the bandwidth of the dispersion curves shown in Fig. 10.30. This allowed providing of the pulse width of 30 fs for the total modeling time of 125 fs. In the right upper corner of Fig. 10.52, the distribution of the x component of the electric field in the xy plane for longitudinal polarization is shown by pseudocolor. The periodicity of the field distribution over the chain confirms that the plasmon impulse is concentrated near $kd = \pi/2$, that is, its wavelength is equal to four distances between the particles. Let us emphasize that this dark mode pulse cannot be excited by a spatially uniform optical field. The analogous distribution for the y components of the electric field for transverse polarization is

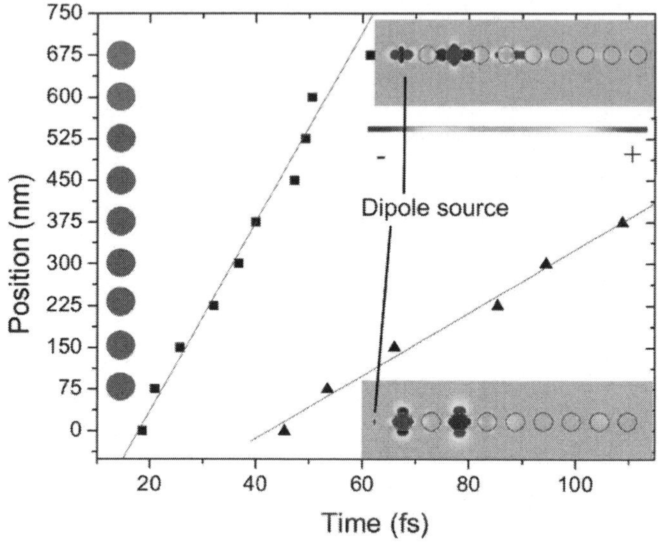

Figure 10.52 The position of the pulse peak, depending on time in a plasmon waveguide consisting of spherical particles for longitudinal (the black squares) and two transverse (the black triangles) polarizations. The red spheres show the position of Au nanoparticles in the y direction. The spatial distribution of the $x(y)$ components of the electric field in the xy plane for the longitudinal (transverse) polarization is shown in the upper (lower) part of the picture (Maier et al., 2003a).

shown in the right lower part of Fig. 10.52. In the main part of Fig. 10.52, the dependence of the plasmon pulse position on time for both longitudinal (black squares) and transverse (black triangles) polarizations of excitations is shown. Linear data approximations allow determination of the group velocities $v^*_{\mathrm{gL}} = 1.7 \times 10^7$ m/s, $v^*_{\mathrm{gT}} = 5.7 \times 10^6$ m/s, which are in good agreement with estimations of the group velocities in the context of the dipole model, Eq. (10.93), $v_{\mathrm{gL}} = 0.94 v^*_{\mathrm{gL}}$, $v_{\mathrm{gT}} = 1.15 v^*_{\mathrm{gT}}$. The attenuation length obtained as a result of pulse propagation simulation is $a_L = 3dB/140$ nm, $a_T = 3dB/43$ nm, correspondingly. Both values are two to three times higher than the attenuation length value calculated in the context of the dipole model where the time of plasmon decay is 4 fs. This discrepancy may be conditioned by a finite length of the considered waveguide, resulting in a discreet spectrum instead of a continuous one. Nevertheless, these calculations in the context of the FDTD method confirm numerically the possibility of the plasmon waveguide use for information transmission with the group velocity $v_g > 0.01c$. Calculations in the context of the FDTD method also show distinctively the negative phase velocity appearance in these nanowaveguides for transverse polarization of plasmon oscillations. For a transverse pulse, the group velocity and thus the energy propagation direction and the phase velocity of the plasmon pulse individual components are antiparallel according to Eq. (10.93). Therefore, the plasmon waveguides could be used as a relatively simple model for investigation of systems with negative phase velocity and for speculation about unique properties of media with a negative refraction index ((Veselago, 1967; Smith and Kroll, 2000; Pendry, 2000), see also Chapters 11 and 13).

In principle, excitations of highly localized modes (plasmon molecules) in clusters and lattices can be performed with the help of homogeneous optical fields of corresponding frequency. However, this method of excitation is ineffective due to small polarizability of M-modes. The virtue of this ineffectiveness is the M-modes' weak sensitivity to external noise fields. The M-modes' excitation with the help of spontaneous or stimulated emission of usual molecules is more effective. In this case, depending on usual molecules' orientation and position relative to the excitation field, highly localized M-modes can be excited in preassigned places of a cluster or a lattice, and therefore the assignment of different initial conditions for the information procession and transmission can be realized. In Fig. 10.53, the principal scheme of excitation of plasmon molecules of predefined configuration with the help of usual molecules' emission is shown. In the initial stage (the left picture), both usual molecules and M-modes (plasmon molecules) are in an unexcited state. In the next stage, a homogeneous external field excites usual molecules at the resonance frequency of their absorption. In the final stage, a strongly nonhomogeneous emission of usual molecules excites the sought M-modes (plasmon molecules).

However, excitation of arbitrary plasmon oscillations in clusters with the help of the molecules' emission is effective, but it is a rather complex procedure at the modern stage of nanotechnology develop-

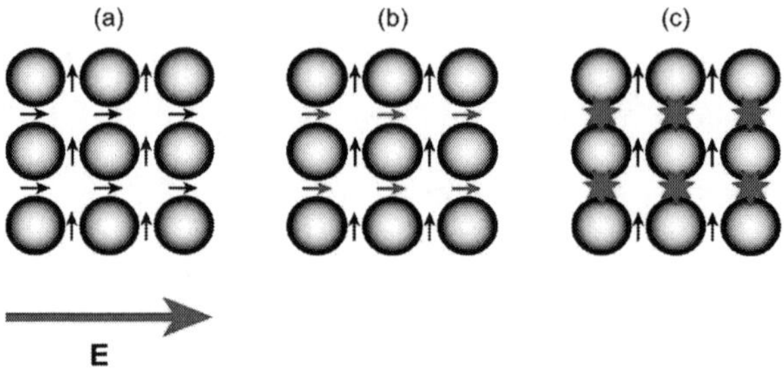

Figure 10.53 The scheme of excitation of plasmon nanoparticle clusters with the help of usual molecules in resonance with plasmon molecules. (a) The nonexcited system of nanospheres with usual molecules between them. (b) The homogeneous longitudinal field excites usual molecules but cannot excite highly localized plasmon modes. (c) The strongly nonhomogeneous field of the spontaneous emission of usual molecules excites plasmon molecules (M-modes).

ment. Cluster excitation with the help of near-field microscopes is simpler. Near the tip of an optical scanning microscope, a strongly nonhomogeneous optical field arises, and this field can be used for excitation of predetermined plasmon oscillations in a cluster of nanoparticles. Thus, Maier et al. (2003b) have used the near field of an aperture scanning microscope to excite the pulse in a waveguide made of silver nanoparticles. Dye molecules embedded in standard plastic nanospheres were used for detection of the excited plasmon pulse (see Fig. 10.54).

In Fig. 10.55, the SEM microphotograph of $100 \times 100\,\mu\text{m}$ of a lattice of rod-shaped Ag nanoparticles is shown. The lattices themselves were constructed with the help of electron beam lithography. In the figure, the vertical series of prolate nanoparticles are seen, which are forming plasmon waveguides. The prolate form of nanoparticles of the size of $90 \times 30 \times 30$ nm spaced at intervals of 50 nm provides their strong interaction.

This lattice was coated with a layer of arbitrarily situated fluorescent nanospheres, some of which happen to be outside or on the waveguides. A scanning optical microscope with an aperture of 100 nm diameter was used as the inhomogeneous field source necessary for waveguide excitation. For each position of the

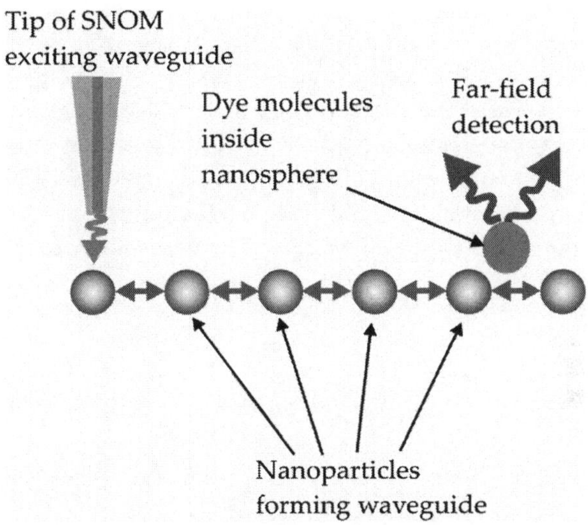

Figure 10.54 The scheme of excitation of a linear cluster with the help of a near field of a scanning microscope and measurements of the plasmon propagation along it with the help of dye molecules embedded in a plastic nanosphere.

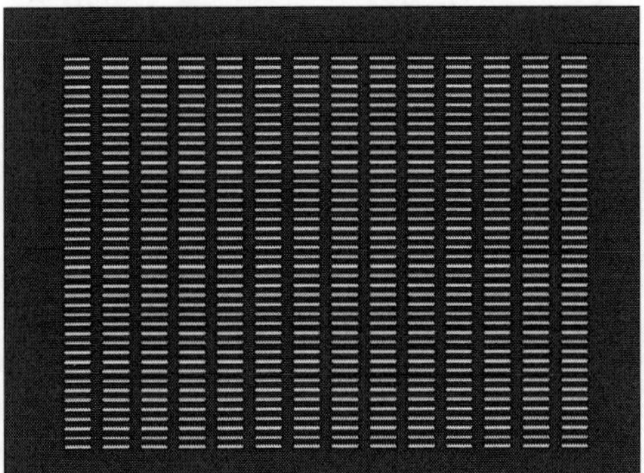

Figure 10.55 The scheme of the 100 μm × 100 μm lattice of rod-shaped Ag nanoparticles. Vertical series of rod-shaped particles form the plasmon nanowaveguides. The prolate form of nanoparticles provides a strong interaction between them.

needle of the scanning microscope, the fluorescence intensity was measured in the far region, and thus scan images of fluorescent nanospheres were obtained. The width of the scan images of the nanospheres situated outside the waveguides is determined by their size and the size of the aperture, while the width of the scan images of the nanosphere situated on the waveguide will be also determined by the length of plasmon propagation through the waveguide.

Measurement results of the width of the fluorescent nanospheres images along the waveguides' direction are shown in Fig. 10.56. The curves A and B correspond to the intensity distribution of

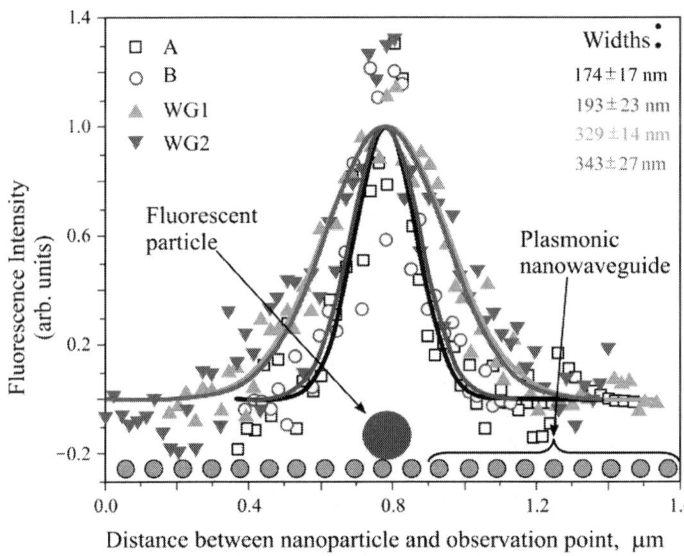

Figure 10.56 Results of experiments on investigation of plasmon nanowaveguides through measurement of the width of the scan image of fluorescent nanospheres along the nanowaveguide. The curves A and B correspond to intensity distribution of nanospheres outside the waveguides. The curves WG1 and WG2 correspond to nanospheres situated on the nanowaveguide. The curves WG1 and WG2 are nearly two times wider than the controls A and B, and this testifies that a plasmon propagates along the waveguide with an attenuation of 6 dB at a distance of 195 ± 28 nm (Maier et al., 2003b). This data is in good agreement with results of theoretical simulation, which predicts attenuation of 6 dB at a distance of 200 nm.

the nanospheres that happen to be outside the waveguides. The curves WG1 and WG2 correspond to nanospheres situated on the waveguide. The curves WG1 and WG2 are nearly two times wider than the control curves A and B. It testifies the plasmon propagation through the waveguide with an attenuation of 6 dB at a distance of 195±28 nm, and it is in good agreement with the results of the theoretical modeling predicting attenuation of 6 dB at a distance of 200 nm (Maier et al., 2003b).

References

Abramowitz, M., Stegun, I.A. (1965) *Handbook of Mathematical Functions* (Dover, New York).

Balykin, V.I., Klimov, V.V., Letokhov, V.S. (2006) Atom nanooptics, in *Handbook of Theoretical and Computational Nanotechnology*, ed. Michael Rieth, Wolfram Schommers, **7**, 1–78.

Barash, Y.S., Ginsburg, V.L. (1984) Some questions of the Van-der-Waals theory, *Uspekhi Fizicheskih Nauk*, **143**, 345–389.

Bergman, D.J. (1979) Dielectric constant of a two-component granular composite: a practical scheme for calculating the pole spectrum, *Phys. Rev. B*, **19**, 2359–2368.

Bergman, D.J., Stockman, M.I. (2003) Surface plasmon amplification by stimulated emission of radiation: quantum generation of coherent surface plasmons in nanosystems, *Phys. Rev. Lett.*, **90**, 027402.

Bergman, D.J., Stockman, M.I. (2004) Can we make a nanoscopic laser? *Laser Phys.*, **14**, 3, 409–411.

Blanco, L.A., Garcia de Abajo, F.J. (2004) Spontaneous emission enhancement near nanoparticles, *J. Quant. Spectrsc. Radiat. Transfer*, **89**, 37–42.

Brongersma, M.L., Hartman, J.W., Atwater, H.A. (2000) Electromagnetic energy transfer and switching in nanoparticles chain arrays below the diffraction limit, *Phys. Rev. B*, **62**, 16356–16359.

Buchholz, G. (1961) *Calculation of Electric and Magnetic Fields* (Foreign literature, Moscow).

Chaumet, P.C., Dufour, J.P. (1998) Electric potential and field between two different spheres, *J. Electrostat.*, **43**, 145–159.

Claro F. (1982) Absorption spectrum of neighbouring dielectric grains, *Phys. Rev. B*, **25**, 7875–7876.

Davydov, A.S. (1965) Quantum Mechanics (Pergamon).

Genov, D.A., Sarychev, A.K., Shalaev, V.M., Wei, A. (2004) Resonant field enhancement from metal nanoparticle arrays, *Nano Lett.*, **4**, 153–158.

Guzatov, D.V., Klimov, V.V. (2005) Properties of spontaneous radiation of an atom located near a cluster of two spherical nanoparticles, *Quantum Electron.*, **35**, 891–900.

Guzatov, D.V., Klimov, V.V. (2010) Optical properties of a two-nanospheroid cluster: analytical approach, arXiv:1010.5760v1.

Guzatov, D.V., Klimov, V.V. (2011) Optical properties of a plasmonic nano-antenna: analytical approach, *New J. Phys.*, **13**, 053034.

Jackson, J.D. (1975) *Classical Electrodynamics* (Wiley, New York).

Johnson, P.B., Christy, R.W. (1972) Optical constants of the noble metals, *Phys. Rev. B*, **6**, 4370–4379.

Kittel, C. (1986) *Introduction to Solid State Physics* (Wiley, New York), p. 134.

Klimov, V.V., Ducloy, M. (2004) Spontaneous emission rate of an excited atom placed near a nanofiber, *Phys. Rev. A*, **69**(1), 013812.

Klimov, V.V., Guzatov, D.V. (2007a) Strongly localized plasmon oscillations in a cluster of two metallic nanospheres and their influence on spontaneous emission of an atom, *Phys. Rev. B*, **75**, 024303.

Klimov, V.V., Guzatov, D.V. (2007b) Plasmonic atoms and plasmonic molecules, *Appl. Phys. A*, **89**, 305–314.

Klimov, V.V., Guzatov, D.V. (2007c) Optical properties of an atom in the presence of a cluster of two nanospheres, *Quantum Electron.*, **37**(3), 209–230.

Klimov, V.V., Lambrecht, A. (2009) Van der Waals forces between plasmonic nanoparticles, *Plasmonics*, **4**, 31–36.

Kneipp, K., Moskovits, M., Kneipp, H., eds. (2006) *Surface-Enhanced Raman Scattering* (Springer, Berlin).

Li, K., Stockman, M.I., Bergman, D.J. (2003) Self-similar chain of metal nanospheres as an efficient nanolens, *Phys. Rev. Lett.*, **91**, 227402.

Li, Z.-Y., Lin, L.-L., Zhang, Z.-Q. (2000) Spontaneous emission from photonic crystals: full vectorial calculations, *Phys. Rev. Lett.*, **84**, 4341–4344.

Li, Z.-Y., Xia, Y. (2001) Full vectorial model for quantum optics in three-dimensional photonic crystals, *Phys. Rev. A*, **63**, 043817.

Lisyansky, A.A., Andrianov, E.S., Dorofeenko, A.V., Pukhov, A.A., Vinogradov, A.P. (2012) Forced Spaser Oscillations, *Proc. of SPIE*, **16**, 8457, 84570X.

Maier, S.A., Brongersma, M.L., Kik, P.G., Meltzer, S., Requicha, A.A.G., Koel, B.E., Atwater, H.A. (2001) Plasmonics: route to nanoscale optical devices, *Adv. Mater.*, **13**, 1501–1505.

Maier, S.A., Kik, P.G., Atwater, H.A. (2003a) Optical pulse propagation in metal nanoparticle chain waveguides, *Phys. Rev. B*, **67**, 205402.

Maier, S.A., Kik, P.G., Atwater, H.A., Meltzer, S., Harel, E., Koel, B.E., Requicha, A.A.G. (2003b) Local detection of electromagnetic energy transport below the diffraction limit in metal nanoparticle plasmon waveguides, *Nat. Mater.*, **2**, 229–232.

Michaels, A.M., Nirmal, M., Brus, L.E. (1999) Surface enhanced Raman spectroscopy of individual rhodamine 6G molecules on large Ag nanocrystals, *J. Am. Chem. Soc.*, **121**, 9932–9939.

Morse, P.M., Feshbach, H. (1953) *Methods of Theoretical Physics*, vol. 2 (McGraw-Hill, New York).

Moskovits, M., Tay, L.-L., Yang, J., Haslett, T. (2002) SERS and the single molecule, *Top. Appl. Phys.*, **82**, 215–227.

Nie, S., Emory, S.R. (1997) Probing single molecules and single nanoparticles by surface-enhanced Raman scattering, *Science*, **275**, 1102–1106.

Noginov, M.A., Zhu, G., Belgrave A.M., Bakker R., Shalaev V.M., Narimanov E.E., Stout S., Herz E., Suteewong T., Wiesner U. (2009) Demonstration of a spaser-based nanolaser, *Nature*, 460, 1110–1112.

Nordlander, P., Oubre, C., Prodan, E., Li, K., Stockman, M.I., (2004) Plasmon hybridization in nanoparticle dimers, *Nano Lett.*, **4**, 899–903.

Olivares, I., Rojas, R., Claro, F. (1987) Surface modes of a pair of unequal spheres, *Phys. Rev. B*, **35**, 2453–2455.

Paley, A.V., Radchik, A.V., Smith, G.B. (1993) Quasistatic optical response of a pair of touching spheres of arbitrary dielectric permeability, *J. Appl. Phys.*, **73**, 3446–3453.

Passian, A., Ritchie, R.H., Lereu, A.L., Thundat, T., Ferrell, T.L. (2005) Curvature effects in surface plasmon dispersion and coupling, *Phys. Rev. B*, **71**, 115425.

Pendry, J.B. (2000) Negative refraction makes a perfect lens, *Phys. Rev. Lett.*, **85**, 3966–3969.

Poponin, V., Ignatov, A., (2005) Local field enhancement in star-like sets of plasmon nanoparticles, *J. Korean Phys. Soc.*, **47**, 222–228.

Prikulis, J., Svedberg, F., Kall, M., Enger, J., Ramser, K., Goksor, M., Hanstorp, D. (2004) Optical spectroscopy of single trapped metallic nanoparticle in solution, *Nano Lett.*, **4**, 115–118.

Protsenko, I.E., Uskov, A.V., Zaimidoroga, O.A., Samoilov, V.N., O'Reilly, E.P. (2005) Dipole nanolaser, *Phys. Rev. A*, **71**, 063812.

Rechberger, W., Hohenau, A., Leitner, A., Krenn, J.R., Lamprecht, B., Aussenegg, F.R. (2003) Optical properties of two interacting gold nanoparticles, *Optics Commun.*, **220**, 137–141.

Romero, I., Aizpurua, J., Bryant, G.W., Garcia de Abajo, F.J. (2006) Plasmons in nearly touching metallic nanoparticles: singular response in the limit of touching dimers, *Opt. Express*, **14**, 9988–9999.

Ruppin, R. (1982) Surface modes of two spheres, *Phys. Rev. B*, **26**, 3440–3444.

Ruppin, R. (1989) Optical absorption of two spheres, *J. Phys. Soc. Japan*, **58**, 1446–1451.

Ruppin, R. (1997) Infrared absorption of cube pair, *Opt. Commun.*, **136**, 395–398.

Sernelius, B.E. (2001) *Surface Modes in Physics* (VCH, Wiley, New York).

Smith, D.R., Kroll, N. (2000) Negative refractive index in left-handed materials, *Phys. Rev. Lett.*, **85**, 2933.

Smythe, W.R. (1950) *Static and Dynamic Electricity* (McGraw-Hill, New York).

Su, K., Wei, Q., Zhang, X., Mock, J., Smith, D., Schultz, S. (2003) Interparticle coupling effects on plasmon resonances of nanogold particles, *Nano. Lett.*, **3**, 1087–1090.

Tamaru, H., Kuwata, H., Miyazaki, H.T., Miyano, K. (2002) Resonant light scattering from individual Ag nanoparticles and particle pairs, *Appl. Phys. Lett.*, **80**, 1826–1828.

Vinogradov, A.P., Andrianov, E.S., Pukhov, A.A., Dorofeenko, A.V., Lisyansky, A.A. (2012) Quantum plasmonics of metamaterials: loss compensation using spasers, *Phys. Usp.*, **55**, 1046.

Veselago, V.G. (1967) The electrodynamics of substances with simultaneously negative values ε and μ, *Uspekhi Fizicheskih Nauk*, **92**, 3, 517–526.

Wang, H., Brandl, D.W., Nordlander, P., Halas, N.J. (2006a) Plasmonic nanostructures: artificial molecules, *Acc. Chem. Res.*, **40**, 53–62.

Wang, H., Wu, Y., Lassiter, B., Nehl, C.L., Hafner, J.H., Nordlander, P., Halas, N.J. (2006b) Symmetry breaking in individual plasmonic nanoparticles, *PNAS*, **103**, 10856–10860.

Wang, X.-H., Gu, B.-Y., Wang, R., Xu, H.-Q. (2003) Decay kinetic properties of atoms in photonic crystals with absolute gaps, *Phys. Rev. Lett.*, **91**, 113904.

Wang, X.-H., Wang, R., Gu, B.-Y., Yang, G.-Z. (2002) Decay distribution of spontaneous emission from an assembly of atoms in photonic crystals with pseudogaps, *Phys. Rev. Lett.*, **88**, 093902.

Wang, Z., Pan, S., Krauss, T.D., Du, H., Rothberg, L.J. (2003) The structural basis for giant enhancement enabling single molecule Raman spectroscopy, *Proc. Natl. Acad. Sci. USA*, **100**, 8638–8643.

Weber, W.H., Ford, G.W. (2004) Propagation of optical excitations by dipolar interactions in metal nanoparticle chains, *Phys. Rev. B*, **70**, 125429.

Willingham, B., Brandl, D.W., Nordlander, P. (2008) Plasmon hybridization in nanorod dimers, *Appl. Phys. B*, **93**, 209–216.

Xu, H., Kall, M. (2002) Surface-plasmon-enhanced optical forces in silver nano-aggregates, *Phys. Rev. Lett.*, **89**, 246802.

Yannopapas, V., Vitanov, N.V. (2007) Spontaneous emission of a two-level atom placed within clusters of metallic nanoparticles, *J. Phys.: Condens. Matter*, **19**, 096210.

Zabkov, I.V., Klimov, V.V., Treshin, I.V., Glazov, O.A. (2011a) Plasmon oscillations in a linear cluster of spherical nanoparticles, *Quantum Electron.*, **41**(8), 742–747.

Zabkov, I.V. (2011b) *Optical Properties of Cluster of Plasmonic Nanoparticles and Its Possible Applications*, Master thesis. MIPT (in Russian).

Chapter 11

Optical Properties of Metamaterials and Nanoparticles Made from Them

In the past 20 years, the possibility of creation of artificial materials with predefined nontrivial electromagnetic and optical properties has been investigated actively. Here, photonic crystals, that is, materials with refractive index changing periodically, are widely known. This periodicity is achieved by usage of closely packed dielectric spheres of the size of 200–500 nm (artificial opals) or dielectric wires of different cross sections. In photonic crystals of this kind, one can select such refractive indices of nanospheres and their environment and such a geometry of their disposition, which provide the band gaps' occurrence at some frequencies, where radiation with frequencies falling into the band gaps cannot propagate (Bykov and Shepelev, 1986; Yablonovitch, 1987; John, 1987; Yablonovitch et al., 1991; Soukoulis, 2001). In particular, an excited atom situated in a photonic crystal and having an emission frequency within the band gap will have an infinitely long lifetime. It is important that for the origin of the band gap in a photonic crystal, the sizes of spheres (and other inclusions) are to be comparable with the radiation wavelength.

Nanoplasmonics
Vasily Klimov

ISBN 978-981-4267-16-8 (Hardcover), 978-981-4267-42-7 (eBook)
www.panstanford.com

So-called metamaterials are even more interesting and intriguing (see, e.g., (Cai and Shalaev, 2010)). In contrast to photonic crystals, in metamaterials sizes of scattering elements and the distances between them are much less than the wavelength under consideration. Thus, artificial materials of this kind may be considered by analogy with usual materials as consisting of some nanoparticles (meta-atoms) instead of atoms and molecules. Using the same reasoning as was used for derivation of Maxwell's macroscopic equations, macroscopic Maxwell's equations can be derived for metamaterials. As in usual Maxwell's equations, metamaterials properties can be described by the effective permittivity $\varepsilon(\omega, k)$, the permeability $\mu(\omega, k)$, and, possibly, other characteristics. A crucially important circumstance is that at the nanoscale electric and magnetic fields can be considered independently, and consequently, one can control values of the effective electric permittivity $\varepsilon(\omega)$ and the magnetic permeability $\mu(\omega)$ independently by use of nanoparticles of different shapes.

In the first part of this chapter, we will consider optical properties of metamaterials where both electric permittivity and magnetic permeability can take negative values. In contrast to metals, in materials of this kind, propagating waves with unusual properties exist. These effects are related to the fact that in these media the refractive index has a negative value. The physics of the media with a negative refractive index is very rich, and thus in this chapter we will only touch upon the properties of these media, the ways of their fabrication, and unusual plasmon properties. A closer look at the properties of these unusual media is taken in more special literature, for example, in (Cai and Shalaev, 2010; Cui et al., 2010; Jakšić et al., 2006), where many useful references can also be found. Some of the most representative applications of these media will be considered in Chapter 13.

In the second part of this chapter we will consider another example of metamaterials—so-called chiral media. Interesting properties of media of this kind are related to the fact that here nanoparticles act as scattering elements, which scatter right-circularly and left-circularly polarized light differently. A special importance of chiral metamaterials is that their optical properties are to some extent similar to optical properties of live (organic) matter, where the principal difference between left and right is presented. Proceeding from this analogy, many devices for medical and biological diagnostics can be elaborated on the base of chiral media.

11.1 Optics of Particles with a Negative Refractive Index

11.1.1 *Main Properties of Media with a Negative Refractive Index*

The main properties of media with a negative refractive index were discovered by Veselago (1968), and in the subsequent text we will follow this fundamental article. However, one can trace the history of negative refraction down to 1904 (Lamb, 1904; Schuster, 1904; Pocklington, 1905).

Tensors of electric permittivity ε and magnetic permeability μ are basic characteristics determining electromagnetic waves propagation in a medium. It is related to the fact that they are the only parameters of the medium entering in the dispersion equation:

$$\left|\frac{\omega^2}{c^2}\sum_{l=1}^{3}\varepsilon_{il}\mu_{lj} - \mathbf{k}^2\delta_{ij} + \mathbf{k}_i\mathbf{k}_j\right| = 0;\ i,\ j = 1,\ 2,\ 3, \tag{11.1}$$

which determines the relation between the frequency ω of a plane monochromatic wave and its wavevector $\mathbf{k}$. In the case when a substance is isotropic and spatial dispersion is absent, Eq. (11.1) is reduced to the known dispersion equation:

$$\mathbf{k}^2 = \frac{\omega^2}{c^2}n^2, \tag{11.2}$$

where n^2 is the square of the medium refractive index and is equal to

$$n^2 = \varepsilon\mu. \tag{11.3}$$

If losses are not taken into account and n, ε, and μ are assumed to be real numbers, from Eqs. (11.2) and (11.3) it is seen that the simultaneous change of the signs of ε and μ does not influence these relations. This fact can have different consequences. First, it can be admitted that the substance properties do not depend, actually, on the simultaneous sign change of ε and μ. Second, it can turn out to be that simultaneous negativity of ε and μ contradicts some fundamental laws of nature, and thus a substance with $\varepsilon < 0$ and $\mu < 0$ cannot exist. Finally, it should be admitted that a medium with negative ε and μ has some properties different from a medium with positive ε and μ. As we shall see further, the third case exists in actuality.

To discover electrodynamic peculiarities related to signs of ε and μ, one should refer to expressions where ε and μ enter separately but not as the product as in Eqs. (11.1)–(11.3).

First of all, ε and μ enter separately into Maxwell's and constitutive equations:

$$\text{rot}\,\mathbf{E} = -\frac{1}{c}\frac{\partial \mathbf{B}}{\partial t}; \quad \text{rot}\,\mathbf{H} = \frac{1}{c}\frac{\partial \mathbf{D}}{\partial t};$$
$$\mathbf{B} = \mu\mathbf{H}; \quad \mathbf{D} = \varepsilon\mathbf{E}. \tag{11.4}$$

For a plane monochromatic wave with all field components proportional to $e^{i\mathbf{kr}-i\omega t}$, Eq. (11.4) is reduced to

$$[\mathbf{kE}] = k_0\mu\mathbf{H}; \quad [\mathbf{kH}] = -k_0\varepsilon\mathbf{E}, \tag{11.5}$$

where $k_0 = \omega/c$.

From these expressions it is seen immediately that if $\varepsilon > 0$ and $\mu > 0$, then **E**, **H**, and **k** form a right-handed triplet of vectors, and if $\varepsilon < 0$ and $\mu < 0$, then they form a left-handed set. If one introduces direction cosines for **E**, **H**, and **k** and denotes them as α, β, γ, correspondingly, a wave propagating in given media will be characterized by the matrix

$$G = \begin{pmatrix} \alpha_E & \alpha_H & \alpha_k \\ \beta_E & \beta_H & \beta_k \\ \gamma_E & \gamma_H & \gamma_k \end{pmatrix}, \tag{11.6}$$

The determinant of this matrix $\Xi = \det G$ is equal to $+1$ when the triplet of the vectors **E**, **H**, and **k** is right-handed, and it is equal to -1 when the triplet of these vectors is left-handed. One may say that Ξ characterizes the "rightness" of the medium.

The energy flow transferred by the wave is determined by the Umov–Poynting vector **S**:

$$\mathbf{S} = \frac{c}{4\pi}[\mathbf{EH}]. \tag{11.7}$$

According to Eq. (11.7), the vector **S** forms always a right triplet with the vectors **E** and **H**. Thus, for right-handed substances **S** and **k** are in the same direction, and for left-handed ones they are in opposite directions. Since the vector **k** direction coincides with the phase velocity and the vector **S** direction coincides with the group velocity, it is clear that left-handed substances are substances with a so-called negative group (or phase) velocity. Hereafter, we will use the term "left-handed" substance for brevity, meaning exceptionally that this term is equivalent to the term "a substance with a negative group velocity" or with a negative refractive index. Now such media are also referred to as double-negative media, or DNG media.

Let us now consider consequences of the circumstance that in left-handed substances the phase velocity is directed oppositely to the energy flow. First of all, in left-handed substances the reversed Doppler effect can be observed. Indeed, let a radiation receiver situated in a left-handed medium, for example, move with the velocity v to a motionless source of radiation emitting the frequency ω_0, as it is shown in Fig. 11.1b. During the motion, the receiver will overtake the wave points corresponding to a certain phase. The frequency perceived by the receiver will be smaller than ω_0, in contrast to a usual (right-handed) medium (Fig. 11.1a). Using the "rightness" parameter Ξ introduced above, the formula for the Doppler shift for an arbitrary medium may be written as

$$\omega - \omega_0 = \omega_0 \Xi \frac{v}{u}. \tag{11.8}$$

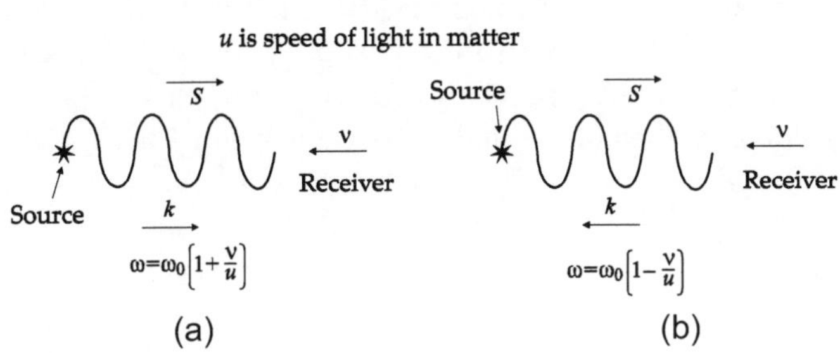

Figure 11.1 The illustration of the Doppler effect in usual media (a) and in media with a negative refractive index (b).

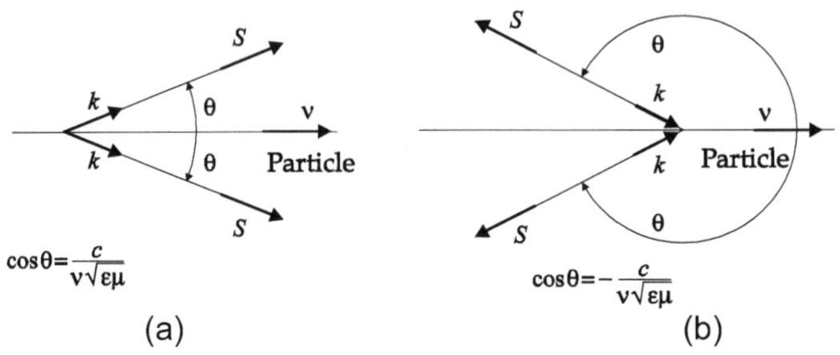

Figure 11.2 The illustration of the Vavilov–Cherenkov effect in a usual medium (a) and in a medium with a negative refractive index (b).

At that, the receiver velocity v is considered to be positive by approaching the source. The velocity of the energy flow u is considered to be positive in both cases.

The Vavilov–Cherenkov effect will be also reversed as the Doppler effect is. When a particle moves in a medium at the velocity v (Fig. 11.2) along the z axis, its radiation can be described by the cylindrical wave $e^{i(k_z z+k_r r-\omega t)}$, and the radiation wavevector along the z axis is equal to $\mathbf{k}_z = |\mathbf{k}\cos\theta| = |n(\omega)(\omega/c)\cos\theta|$ and directed along $\boldsymbol{v}$. The direction of the radial wavevector $\mathbf{k}_r$ will be different in different media according to the expression

$$\mathbf{k}_r = \Xi\left|\sqrt{\mathbf{k}^2 - \mathbf{k}_z^2}\right|, \tag{11.9}$$

where Ξ is the "rightness" parameter Ξ introduced above. This choice of the sign of the root in Eq. (11.9) will always provide the energy outgoing from the radiating particle to infinity in the radial direction. At that, it is obvious that for a left-handed medium the vector $\mathbf{k}_r$ will be directed to the particle path, and the radiation cone will be directed backward relative to the particle motion. It corresponds to the obtuse angle θ between v and $\mathbf{S}$. This angle may be determined for media of any "rightness" from the expression

$$\cos\theta = \Xi\sqrt{\frac{c^2}{v^2 n(\omega)^2}}. \tag{11.10}$$

When a light beam transits from medium 1 to medium 2, the continuity conditions of the tangential components of $\mathbf{E}$ and $\mathbf{H}$

$$E_{t1} = E_{t2};\ H_{t1} = H_{t2} \tag{11.11}$$

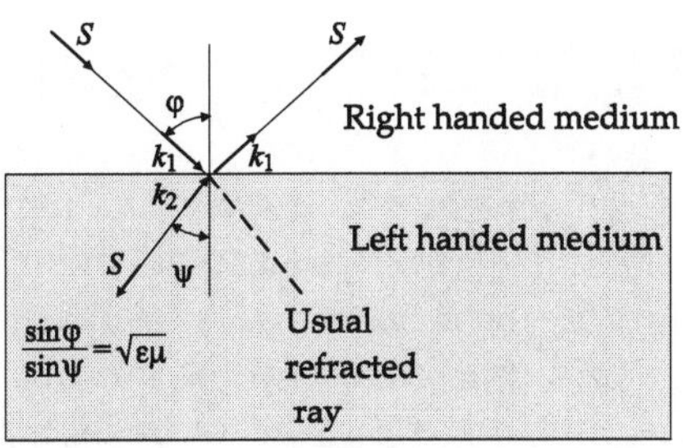

Figure 11.3 The illustration of refraction laws between a usual medium and a medium with a negative refractive index.

are to be fulfilled independently of the same or different "rightness" of the media. Eq. (11.11) implies that in the refracted beam the tangential to interface components of the wavevector retain their direction independently of the "rightness" of both media. As to the wavevector component normal to the surface, it retains its direction only in the case when the "rightnesses" of both media are equal. When the "rightnesses" differ, the normal component of the wavevector changes its sign. The path of the refracted beam obtained as a result of such reasoning is shown in Fig. 11.3.

As we see, the refracted beam in the second left-handed medium will be symmetric relatively to the z axis in contrast to the case when the second medium is the usual right-handed one (the dashed line). It should be noted that the reflected beam has always a unique direction independent of the "rightness" of both media. From Fig. 11.3, it is seen that the regular Snell's law:

$$\frac{\sin\varphi}{\sin\psi} = n_{1,2} = \sqrt{\frac{\varepsilon_2\mu_2}{\varepsilon_1\mu_1}}, \tag{11.12}$$

requires attention when media 1 and 2 have different "rightnesses." The correct formulation is:

$$\frac{\sin\varphi}{\sin\psi} = n_{1,2} = \frac{\Xi_2}{\Xi_1}\sqrt{\frac{\varepsilon_2\mu_2}{\varepsilon_1\mu_1}}, \tag{11.13}$$

where Ξ_1 and Ξ_2 are the "rightnesses" of the first and second media, correspondingly. From the expression it is clear that the relative refractive index of two media can be negative when their "rightnesses" are different in sign. In particular, the refractive index of left-handed media is negative relative to vacuum.

Another interesting feature of left-handed metamaterials is related to the quantum nature of light. In a left-handed medium, a monochromatic wave can be regarded as a flow of photons, each of which has the momentum $\mathbf{p} = \hbar\mathbf{k}$, and the vector $\mathbf{k}$ is directed to the radiation source, but not from it, as it is in a right-handed medium. Therefore, by the incidence on the reflecting body the light beam propagating in left-handed medium transfers the momentum $\mathbf{p} = 2N\hbar\mathbf{k}$ (N is the number of incident photons) directed to the radiation source, to it. In this connection, in left-handed substances the light pressure typical for usual (right-handed) substances is replaced by the light attraction.

Thus, indeed, left-handed media or media with a negative refractive index have a range of really interesting and intriguing properties, and a question of the existence of natural or artificial materials having a negative refractive index arises. This question will be considered in the next section.

11.1.2 *Experimental Realization of Media with a Negative Refractive Index*

A regular three-dimensional lattice of metal wires with micron diameters (Pendry et al., 1996, 1998) is an important example of metamaterials where low frequency plasmons can exist (in contrast to optical ones considered above). It can be shown that electric properties of such a system can be described effectively by means of the free electrons model of a density determined by the space portion occupied by wires. Since the electron density determines the plasma frequency, in this system plasmon oscillations will have low frequency in comparison with optical frequencies of plasmon oscillations in a solid metal. Generally, properties of a wire lattice in relation to microwave radiation will be analogous to metal properties in relation to optical radiation. According to it, one of the arguments requiring metamaterials development is the necessity to shift the plasmon resonance in natural materials to the microwave region (e.g., for development of invisibility coatings for *stealth* aircraft).

Another, exactly the opposite, argument in favour of metamaterials is the necessity of development of materials with nontrivial

Figure 11.4 The schematic image of an SRR made for development of metamaterials with negative magnetic permeability $\mu(\omega)$.

magnetic properties in the region of higher frequencies than those in natural materials. In particular, development of metamaterials with magnetic properties and resonances in the optical field is of considerable interest. These magnetic resonances in the optical range allow us to talk about development of metamaterials with a negative refractive index.

While the magnetism of natural materials is related to spins of unpaired electrons (Kittel, 1996), the magnetism of metamaterials is determined entirely by resonances depending on the geometry of the particles composing the metamaterials. The geometry of double split-ring resonators (SRRs), shown in Fig. 11.4 is of special interest in this context. This nanoelement consists of two concentric nanorings with a split situated in one plane. Pendry et al. (1999) showed that regular lattices of such elements will have resonance magnetic properties under the condition that the elements' sizes and the distances between them are much less than the considered wavelength.

To be simpler, in split rings the time-varying magnetic field induces a magnetic moment owing to induced currents' origin in the main part of the ring. This, generally speaking, small magnetic moment can be amplified through the resonance. Thus, the structure shown in Fig. 11.4 is a nanosized LC resonator with inductance L and capacity C, and so the magnetic moment and the magnetic permeability μ experience resonance at the frequency

$$\omega_{LC} = \frac{c}{\sqrt{LC}}. \qquad (11.14)$$

It is important that the magnetic permeability becomes less than zero, $\mu < 0$, for $\omega > \omega_{LC}$, as it happens in all resonance processes for a frequency exceeding the resonance frequency. As it will be shown below, the combination of a lattice of split-ring nanoresonators and a lattice of nanowires results in development of metamaterials with negative permittivity and permeability and thus in creation of a medium with a negative refractive index.

Initially, a possibility of the existence of metamaterials with negative magnetic permeability was demonstrated for microwave frequencies (see, e.g., the review (Smith et al., 2004)) and then in the THz range (Yen et al., 2004). Measured experimentally, the magnetic permeability of materials of this kind with split-ring microresonators can be described in the form of Lorentzian

$$\mu(\omega) = 1 - \frac{F\omega^2}{\omega^2 - \omega_{LC}^2 + i\Gamma\omega}, \tag{11.15}$$

where ω_{LC} is the resonance frequency, F is the factor depending on the geometry, and, finally, Γ describes the Ohmic losses in the ring resonator. For $\omega \ll \omega_{LC}$, the induced magnetic moment oscillates in phase with the exciting field, as it happens in other resonance processes. In this region, the metamaterial demonstrates paramagnetic properties, that is, $\mu > 1$. As the exciting field frequency increases, the current begins to lag in phase behind the exciting field, and for $\omega \gg \omega_{LC}$ the magnetic dipole oscillates in phase opposite to the exciting field. In this region, the material becomes diamagnetic ($\mu < 1$). In the region of frequencies slightly exceeding the resonance frequency ω_{LC}, the magnetic permeability becomes negative ($\mu < 0$). Let us emphasize that only induced magnetic moments are discussed here, and there is no permanent magnetic moment in nanoparticles.

Thus, in a lattice consisting of SRRs and wire dipoles parameters can be chosen in such a manner that both electric permittivity and magnetic permeability would be negative, $\varepsilon < 0$ and $\mu < 0$, in a certain frequency range. In turn, this results in a negative refractive index, $n < 0$, of the metamaterial in this frequency range.

Shelby et al. (2001) demonstrated the existence of metamaterials with a negative refractive index in the microwave region. A direct proof of negative refraction (see Fig. 11.3) was demonstrated with the help of a prism made of a metamaterial (Smith et al., 2004).

In the THz region, existence of a negative refractive index was demonstrated by means of planar metamaterials (Moser et al., 2005).

In microwave and THz ranges, resonance properties of metamaterials consisting of SRRs and wire dipoles include an important scaling property. That is, the metamaterial resonance frequency is inversely proportional to the typical spatial scale a:

$$\omega_{LC} \approx \frac{c}{a}. \tag{11.16}$$

The size of the SRR may be taken as such a spatial scale. However, at further increasing of frequency the scaling law, Eq. (11.16), is violated since at higher frequencies the frequency dispersion of the electric permittivity in metals is to be taken into account.

Theoretical investigations have shown that imperfections of this kind result in saturation in the resonance frequency increase at frequencies $f > 100$ THz ($\lambda_0 < 3$ μm) (Zhou et al., 2005). Using gold SRRs of the size of 35 nm, Klein et al. (2006) showed that the resonance in the magnetic permittivity can be shifted to the near-infrared region, to the wavelength $\lambda = 900$ nm. Today, it is not clear if the magnetic permittivity resonance can be shifted to the region of the visible spectrum with the help of the SRR concept.

There are other nanostructures that can be used for development of metamaterials with a negative refractive index. In particular, Shalaev et al. (2005) demonstrated that in metal with a lattice formed by pairs of gold nanorods at the wavelength $\lambda = 1.5$ μm the refractive index is negative, $n = -0.3$.

In Fig. 11.5, a scheme and a scanning electron microscopy (SEM) image of a metamaterial of this kind are shown. Each element consists of two gold nanorods of the thickness of 50 nm separated by a dielectric layer (SiO_2) of the thickness of 50 nm (Fig. 11.5a). Apparently, in this structure the magnetic permeability resonance is related to the resonance in the LC resonator formed by the upper and lower gold rods, as it is shown in Fig. 11.5a, as it happens in the case of the SRR. Metal rods provide inductance, and the dielectric layer between them provides capacity. Pairs of such nanorods form a planar structure with parameters shown in Fig. 11.5c.

The dependency of the refractive index of such a metamaterial on the wavelength in the near-infrared range is shown in Fig. 11.6. From this picture it is seen that in the vicinity of frequencies

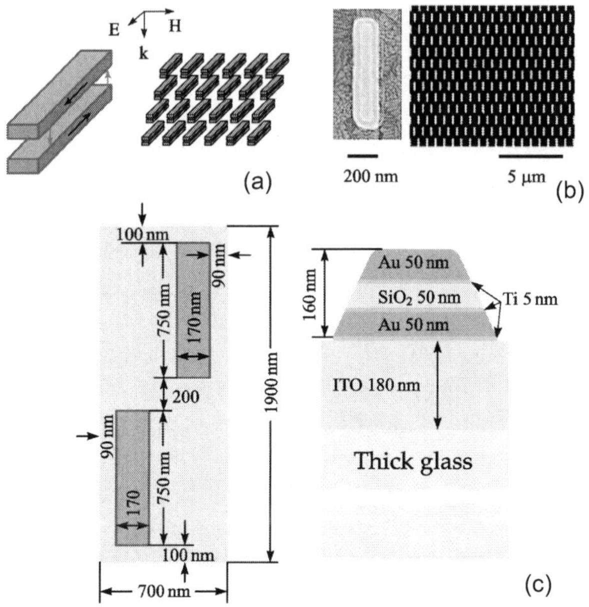

Figure 11.5 (a) The scheme and (b) the SEM image of planar metamaterials consisting of parallel gold nanorods and (c) the parameters of an elementary cell of planar metamaterials consisting of nanorod pairs (Shalaev et al., 2005).

corresponding to the region of telecom frequencies ($\lambda = 1500$ nm), the refractive index becomes negative, $n < 0$. Let us note, however, in this experiment an essential attenuation (see Fig. 11.6a) takes place, in addition to the negative refractive index, and this reduces its conclusiveness.

It is important to note that in this case an element's typical size is of the wavelength order in contrast to the case of a metamaterial with SRRs. Moreover, it should be kept in mind that in the case of nanorods the similarity law, Eq. (11.16), is broken as the elements' sizes decrease due to increase both of heat losses and significance of plasmon effects in nanorods, as it happens in the case of a metamaterial with SRRs. For development of metamaterials with a negative refractive index, U-shaped nanoparticles (Sarychev et al., 2006) were also used. In this metamaterial, the magnetic permeability negative value is provided by plasmon effects rather than by *LC* resonance.

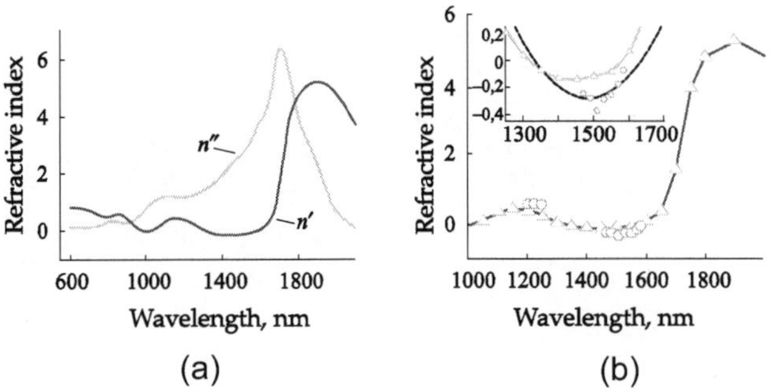

Figure 11.6 (a) The real and imaginary parts of the refractive index n', n'' of the metamaterials shown in Fig. 11.5 and obtained as a result of numerical modeling. (b) The comparison of the results of modeling (the triangles) and the experimental results (the circles) for the real part of the refractive index. In the inset, the scaled-up region of the negative refractive index is shown (Shalaev et al., 2005).

The existence of a metamaterial with a negative refractive index in the visible spectrum was demonstrated by Grigorenko et al. (2005). In this case, the metamaterial consisted of pairs of domelike gold nanoparticles. The antisymmetric state of the localized plasmon resonances in such pairs results in reduction of external field magnetic components and thus in the negative value of the magnetic permittivity $\mu < 0$. Unfortunately, recently these important results have been disputed (Kildishev et al., 2007), and development of metamaterials with a negative refractive index in the visible spectrum remains the most significant problem of modern nanoplasmonics.

In Fig. 11.7, a progress in development of metamaterials with a negative refractive index or with negative magnetic permittivity is shown. For recent achievements in search for the negative refractive index of metamaterials in the visible range, see (Garcia-Meca et al., 2009, 2011; Xiao et al., 2009).

All the metamaterials considered above are planar structures. For further development of this area, three-dimensional nanostructures are required. In spite of technical difficulties in this direction, certain results are achieved (Dolling et al., 2007; Liu et al., 2008; Valentine et al., 2008; Liu et al., 2009; Gansel et al., 2009). In

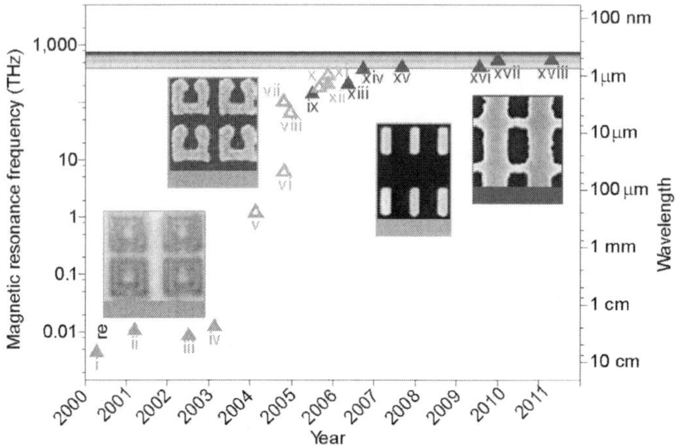

Figure 11.7 Progress in metamaterial operating frequency over the past decade. The operating frequency of metamaterials with negative magnetic permeability μ (empty triangles) and negative index of refraction n (solid triangles) is shown on a logarithmic scale ranging from microwave to visible wavelengths. Orange: structures based on double SRRs; green: Ushaped SRRs; blue: metallic cut-wire pairs; red: negative-index double-fishnet structures. The four insets show optical or electron micrographs of the four types of structure. Data points are from: i) Smith et al. (2000); ii) Shelby et al. (2001); iii) Bayindir et al. (2002); iv Greegor et al. (2003); v) Yen, T. J. et al. (2004); vi) Katsarakis, N. et al. (2005); vii) Linden, S. et al. (2004); viii) Zhang, S. et al. (2005a); ix) Zhang, S. et al. (2005b); x) Enkrich, C. et al. (2005); xi) Dolling et al. (2005); xii) Shalaev, V. M. et al. (2005); xiii Dolling et al. (2006); xiv) Dolling et al. (2006; xv) Chettiar, U. K. et al. (2007); xvi) Garcia-Meca et al. (2009); xvii) Xiao et al. (2009); and xviii) Garcia-Meca et al. (2011). (Soukoulis, Wegener, 2011 with permission).

particular, Liu et al. (2008) developed a technology for development of a three-dimensional lattice of SRRs, which is shown in Fig. 11.8.

11.1.3 *Focusing Properties of a Slab Made of a Metamaterial with a Negative Refractive Index*

In the previous section, it was shown that fabrication of negative refraction metamaterials is in progress. It allows us to think about unusual optical devices based on wonderful properties of left-handed metamaterials. An example of a system of this kind is an

Figure 11.8 The artist view of three-dimensional nanostructure of gold SRRs.

infinite slab of thickness d made of a left-handed substance with the refractive index $n = -1$ situated in vacuum (Veselago, 1968; Pendry, 2000; Klimov, 2009a, 2009b; Klimov et al., 2011a).

At first sight, the ray picture of this system has a remarkable view (Veselago, 1968), and all rays outgoing from the point source A will be focused after the slab at the point B without reflection at the interfaces (see Fig. 11.9).

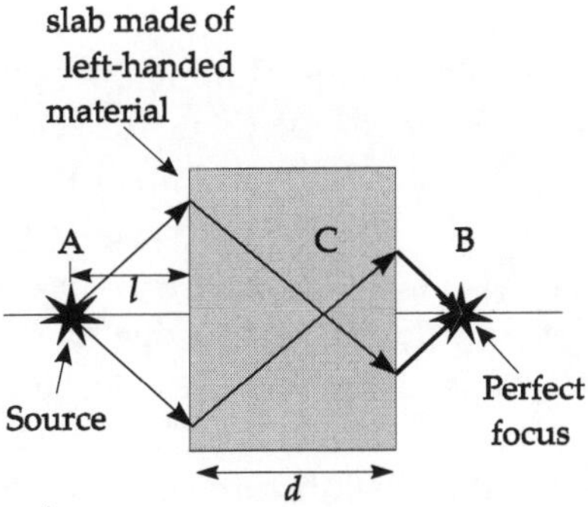

Figure 11.9 The ray picture in a negative refraction slab according to Veselago (Veselago's lens, 1968). The gray region in the middle of the figure denotes the infinite negative refraction slab with the refractive index $n = -1$.

However, despite its simple geometry this system becomes more complicated if one considers a description of it within the full Maxwell theory. Here, one should distinguish clearly between a realistic case of small losses in a negative refraction slab and a hypothetical case of a lossless negative refraction material. In the first case, any point source excites surface plasmon waves, and one cannot speak about a true lens at all. In the second case, there is no stationary solution of the problem, and one should think of how to reformulate it.

Let us consider the first case in more detail. Due to the presence of Ohmic losses, the problem is well defined, and the spatial distribution of fields from the dipole source near an arbitrary slab can be described by the well-known Sommerfeld integrals (see, e.g., (Novotny and Hecht, 2006), Appendix C).

When the thickness of the slab is greater than the wavelength, one can see three-dimensional focusing of energy in agreement with the ray picture shown in Fig. 11.9. However, in this case the focal spot size is diffraction-limited and the negative refraction slab has no advantages in comparison with conventional lenses (see, e.g., (Tran et al., 2011)).

When the thickness of the slab becomes smaller than the wavelength, the situation changes drastically. The energy flows found by calculating Sommerfeld's integrals are shown in Fig. 11.10 (the upper row) for different distances l between the source and the left-handed slab. If the source is far enough ($l > d$, Fig. 11.10a), energy flows are straight lines suffering negative refraction at interfaces, and one cannot see anything special for this geometry. However, when the distance becomes equal to the slab thickness ($l = d$, Fig. 11.10b), the streamlines become more complicated, and by further approaching of the source to the slab one can see a drastic change of energy flows, which is due to excitation of surface plasmons waves propagating along the interfaces between right-handed and left-handed media ($l = d/2$, Fig. 11.10c). This image is definitely far from the ideal picture shown in Fig. 11.9, and this setup cannot be used for focusing light from point A to point B, as shown in Fig. 11.9. Nevertheless, it can be used for the focusing of light from a source placed at $l = d$ on the right interface (see Fig. 11.10b and Section 13.7).

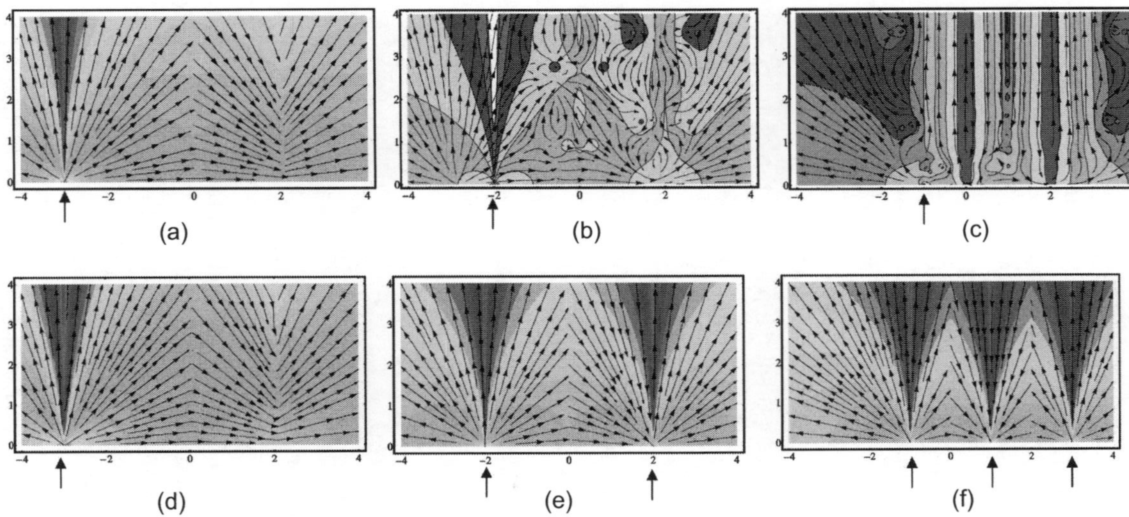

Figure 11.10 Energy flows in the presence of a negative refraction slab for different positions of the initial source. The upper row corresponds to a single source near a left-handed slab with small losses. The bottom row corresponds to the solution of Eqs. (11.17) and (11.18) with two sources and one sink. All distances are measured in $\lambda/(2\pi)$ units. The thickness of the slab is taken to be $2\pi d/\lambda = 2$. Pseudocolor shows fields intensities. The arrows below the figures show positions of the sources and sinks.

Let us now consider a hypothetical case when a slab material with $n = -1$ has no losses at all. It means that imaginary parts of permittivity and permeability are exactly equal to zero. The first attempt to treat this problem within wave theory was carried out by Pendry (2000), who argued that perfect focusing could be achieved in this case. However, it can be shown mathematically that in the lossless case there is no solution of the monochromatic problem, and one cannot again speak about a perfect lens in the sense of (Veselago, 1968; Pendry, 2000). Nevertheless, the ray picture shown in Fig. 11.4 still has a luring character, and below we will show that it is possible to change the formulation of the problem and to elaborate a new type of focusing devices based on negative refraction domains (Klimov, 2009a, 2009b; Klimov et al., 2011a).

Indeed, let us consider a source placed at the point $\mathbf{r}_A$ far ($l > d$) from a lossless negative refraction slab ($\varepsilon = \mu - 1$, $n = -1$) (see Fig. 11.11). In this case, it is possible to find an exact solution of Maxwell's equations:

$$\mathbf{E} = \nabla(\nabla\mathbf{\Pi}) + \varepsilon\mu\left(\frac{\omega}{c}\right)^2\mathbf{\Pi},$$
$$\mathbf{H} = -i\varepsilon\left(\frac{\omega}{c}\right)\nabla\times\mathbf{\Pi}, \qquad (11.17)$$

where the Hertz $\mathbf{\Pi}$ potential is described by

$$\mathbf{\Pi}^{(1)} = \mathbf{d}_0\frac{\exp(ik_0|\mathbf{r}-\mathbf{r}_A|)}{|\mathbf{r}-\mathbf{r}_A|} \quad \text{(in the region of source A)},$$

$$\mathbf{\Pi}^{(2)} = \mathbf{d}_0\frac{\exp(ik_0|\mathbf{r}-\mathbf{r}_C|)}{|\mathbf{r}-\mathbf{r}_C|} \quad \text{(inside the slab)}, \qquad (11.18)$$

$$\mathbf{\Pi}^{(3)} = \mathbf{d}_0\frac{\exp(ik_0|\mathbf{r}-\mathbf{r}_B|)}{|\mathbf{r}-\mathbf{r}_B|} \quad \text{(in the region of image B)},$$

and $\mathbf{r}_B$, $\mathbf{r}_C$ should be found by ray tracing, taking into account the negative refraction at the interfaces.

For a distant position of the source, the points $\mathbf{r}_B$, $\mathbf{r}_C$ are "imaginary" in the sense that they are not situated in the region of the image or inside the slab (see Fig. 11.11).

Despite the fact that Eqs. (11.17) and (11.18) are found for a distant source, this solution can be analytically continued to any other position of the source. If the distance l between the source and the slab is smaller than the slab thickness d, the points $\mathbf{r}_B$, $\mathbf{r}_C$

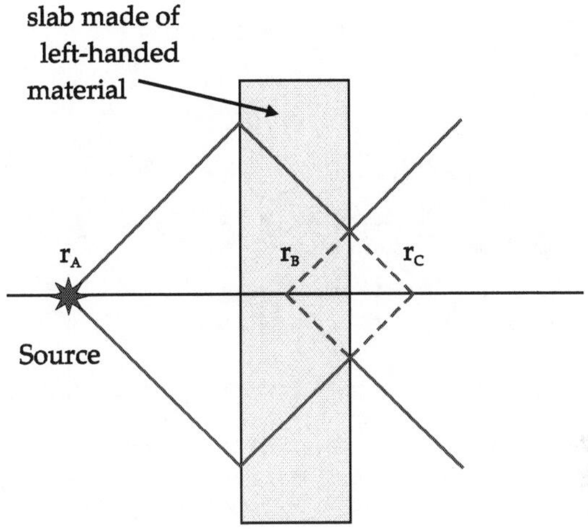

Figure 11.11 The ray picture for the case when the source is far ($l > d$) from the slab.

become real and a new source and a sink appear. This effect is similar to pair creation in the Klein paradox (Klimov et al., 2011). So, one can consider Eqs. (11.17) and (11.18) as a general solution of Maxwell's equation for a slab with negative refraction.

Energy flows found with the help of Eqs. (11.17) and (11.18) are shown in Fig. 11.10 (the bottom row). For a distant position of the source ($l > d$, Fig. 11.10d), energy flows are similar to energy flows in the case of a slab with small losses (Fig. 11.10a). However, when the distance becomes equal to the slab thickness ($l = d$, Fig. 11.10e) and for further approaching of the source to the slab ($l < d$, Fig. 11.10f), straight rays that strongly remind one of the ray pictures in Fig. 11.9 can be seen again. The only difference is that energy flows directions are different between points B and C (compare Figs. 11.9 and 11.12).

Despite this difference, a system with two sources (or emitters) and one sink (or receiver) can have interesting applications ((Klimov, 2009a, 2009b; Klimov et al., 2011a), see also Chapter 13). In fact, the solution of Eqs. (11.17) and (11.18) and Fig. 11.12 put forward a new paradigm for subwavelength wave guiding and focusing. The main feature of this paradigm is that both sources and receivers are needed to provide effective operation of nanodevices.

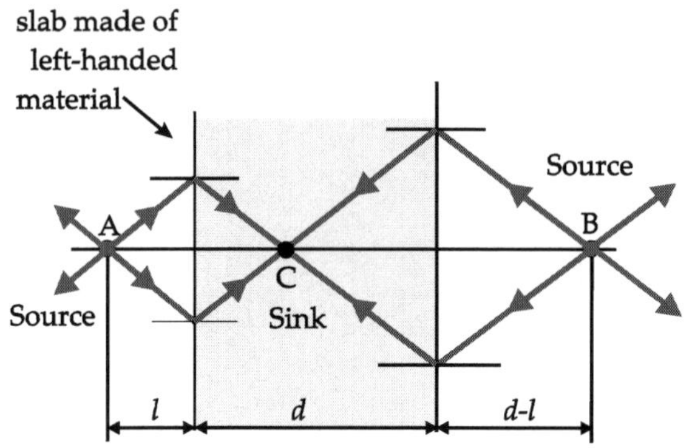

Figure 11.12 The rays' paths in a perfect lens according to (Klimov, 2009a, 2009b; Klimov et al., 2011). Note the presence of two sources and one energy sink.

The new paradigm can be applied for different geometries (sphere, cylinder, wedge, etc.) and wave phenomenon (optics, acoustics, high-energy physics, solid-state physics) where negative refraction is present.

11.1.4 *Plasmon Resonances in a Sphere of a Material with a Negative Refractive Index and Their Influence on the Radiation of Atoms and Molecules*

A more or less proven existence of metamaterials with a negative refractive index allows raising of a question of resonance oscillations in particles made of such a substance. For the first time, a case of a spherical particle made of a metamaterial with a negative refractive index was considered by Klimov (2002). The investigation of spherical particles with a negative refractive index is made easier by the fact that all the formulas obtained for usual substances (Chapter 6) can be applied also to the case of negative permittivities and permeabilities. In particular, expressions for the reflection coefficients of spherical transverse magnetic (TM) and transverse electric (TE) waves from a spherical interface (see Fig. 6.1) have the following form:

$$R_n^{\mathrm{TM}} = 1 - 2q_n;\ R_n^{\mathrm{TE}} = 1 - 2p_n, \tag{11.19}$$

where the coefficients p_n and q_n are determined by the relations from Eqs. (6.21) and (6.22):

$$q_n = \frac{\left[\varepsilon_1 \dfrac{d}{dz_2}\left[z_2 j_n(z_2)\right] j_n(z_1) - \varepsilon_2 \dfrac{d}{dz_1}\left[z_1 j_n(z_1)\right] j_n(z_2)\right]}{\left[\varepsilon_1 \dfrac{d}{dz_2}\left[z_2 h_n^{(1)}(z_2)\right] j_n(z_1) - \varepsilon_2 \dfrac{d}{dz_1}\left[z_1 j_n(z_1)\right] h_n^{(1)}(z_2)\right]}, \tag{11.20}$$

$$p_n = \frac{\left[\mu_1 \dfrac{d}{dz_2}\left[z_2 j_n(z_2)\right] j_n(z_1) - \mu_2 \dfrac{d}{dz_1}\left[z_1 j_n(z_1)\right] j_n(z_2)\right]}{\left[\mu_1 \dfrac{d}{dz_2}\left[z_2 h_n^{(1)}(z_2)\right] j_n(z_1) - \mu_2 \dfrac{d}{dz_1}\left[z_1 j_n(z_1)\right] h_n^{(1)}(z_2)\right]}, \tag{11.21}$$

where $z_{1,2} = \sqrt{\varepsilon_{1,2}\mu_{1,2}}k_0 a = k_{1,2}a$. In what follows, we will consider the case of a particle placed in vacuum, $\varepsilon_2 = \mu_2 = 1$, $\varepsilon_1 = \varepsilon$, $\mu_1 = \mu$.

Equations (11.20) and (11.21) are analytical functions of ε and μ, and therefore they are true for any complex values of these variables, including negative and even complex values of both ε and μ.

As usual, to investigate resonance properties of spheres made of an arbitrary metamaterial, it is sufficient to find the poles of the reflection coefficients or zeros of the denominators q_n or p_n. Existence of resonance is related to a very complex dependence of Eqs. (11.20) and (11.21) on the sphere size a, the wavevector $\mathbf{k}_0$, the permittivity ε, and the permeability μ. To illustrate these dependencies, let us consider properties of the coefficient $q_4(ka, \varepsilon, \mu)$ in more detail (Figs. 11.13 and 11.14).

In Fig. 11.13, the coefficients q_4 (TM modes) are shown as a function of the sphere size and electric permittivity for some values of magnetic permeability. For positive values of magnetic permeability, the resonances' behavior is generally similar to the resonances' behavior in a nonmagnetic sphere (compare Fig. 6.2 and Fig. 11.13a,b,c). Namely, there is a range of high-quality whispering gallery modes for positive values of electric permittivity, while for negative values of electric permittivity (for $n = 4$, $\varepsilon_4 = -(4+1)/4 = -1.25$) high-quality plasmon resonances take place only for very small spheres (nanospheres).

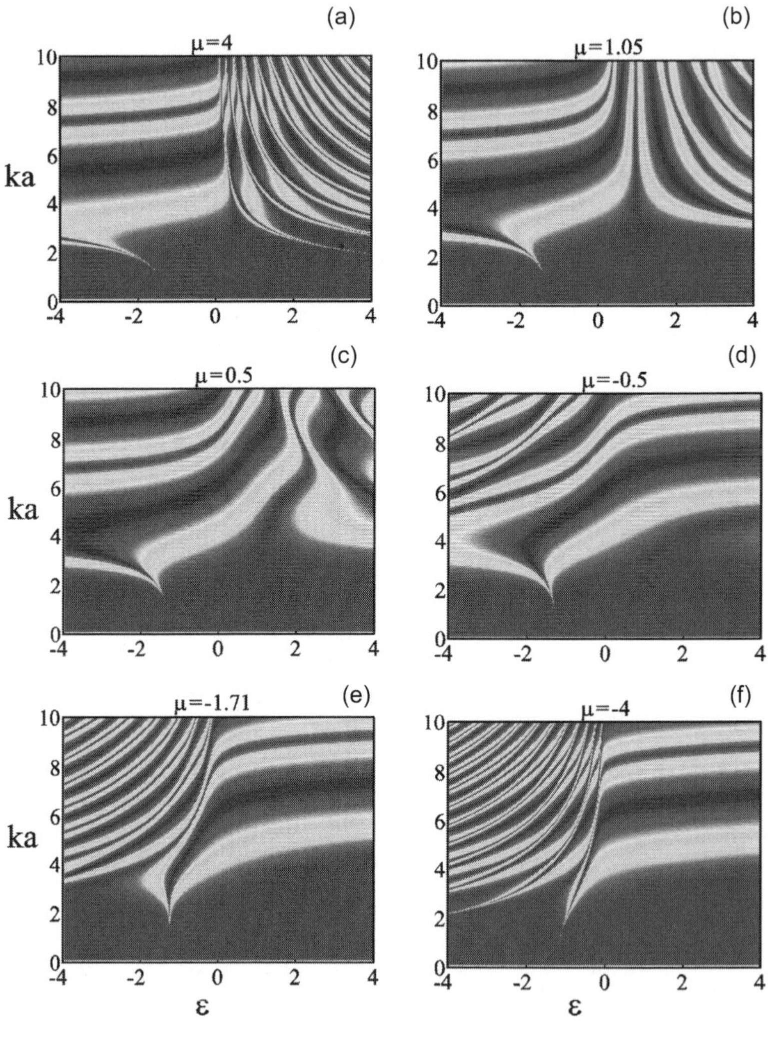

Figure 11.13 The dependencies of the Mie coefficients q_4 (TM modes) on the permittivity and on the sphere radius for different values of permeability. The red color corresponds to the resonant reflection ($R^{TM} = 1$), while the blue color corresponds to the absence of reflection ($R^{TM} = 0$).

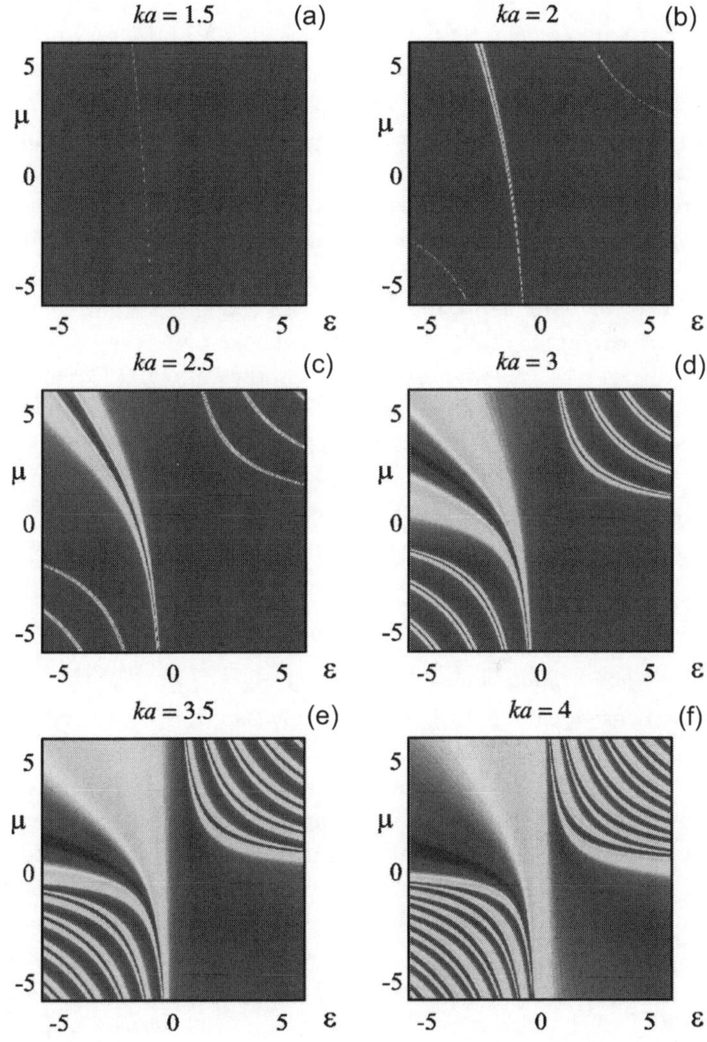

Figure 11.14 The dependencies of the Mie coefficients q_4 (TM modes) on electric permittivities and magnetic permeabilities for different values of the sphere radius ka. The red color corresponds to the resonant reflection ($R^{\mathrm{TM}} = 1$), while the blue color corresponds to the absence of reflection ($R^{\mathrm{TM}} = 0$).

For negative values of permeability, the situation changes substantially. More specifically, now the whispering gallery modes can exist only for *negative* values of electric permittivity, while the plasmon resonance remains in the region of negative permittivities (see Fig. 11.13d,e,f).

In turn, for negative values of electric permittivity the properties of plasmon resonances depend on magnetic permeability essentially. While magnetic permittivity is not quite negative, the plasmon properties remain analogous to the plasmon properties in a nonmagnetic sphere. However, as the magnetic permittivity decreases further, the plasmon properties change substantially, and it is more correct to talk about left-handed plasmons.

In Fig. 11.14, resonances of the coefficients q_4 (TM modes) are shown as functions of electric permittivity and magnetic permeability for different values of the sphere size *ka*. From Fig. 11.14a, it is seen that for small spheres the plasmon resonances' position is almost independent of the magnetic permeability value (the vertical line). As the sphere size increases in the transparency regions ($\varepsilon > 0$, $\mu > 0$ or $\varepsilon < 0$, $\mu < 0$), the whispering gallery modes appear (in the right upper and left lower parts of the pictures), in addition to plasmon resonances.

A perfectly analogous picture takes place also for TE modes (resonances in the coefficients *p*4), only electric permittivity and magnetic permeability are to be interchanged.

In Fig. 11.15, plasmon spectra in spherical nanoparticles are shown, that is, dependencies on the sphere size *ka*, for which the plasmon TM resonance arises in it, on the orbital number *n*. These dependencies are determined by vanishing of denominators in Eq. (11.20). For usual (right-handed) plasmons, the resonant sphere radius increases with an increase of the orbital quantum number $n > n_{\min}$ (i.e., of the multipolarity degree) (Fig. 11.15a), while for left-handed plasmons the situation is directly opposite, and as the orbital number *n* increases, the resonance arises at smaller values of the sphere radius, and for $n > n_{\max}$ there are no resonances at all (Fig. 11.15b). The left-handed plasmons also have other unusual properties (Klimov, 2002) that are related to negativity of the refractive index.

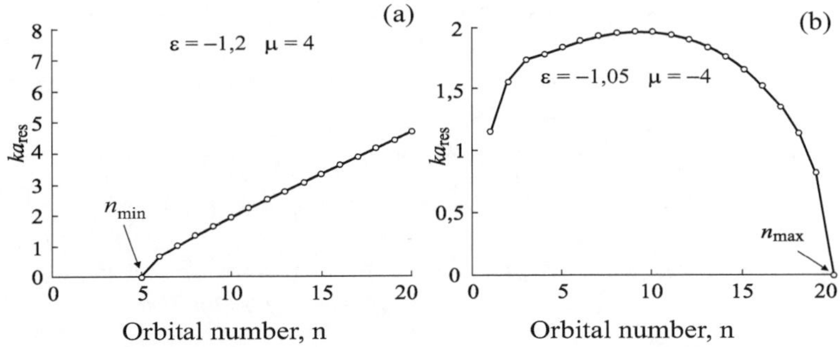

Figure 11.15 The spectra of usual plasmons ($\varepsilon = -1.2$, $\mu = 4$) (a) and of plasmons in a medium with a negative refractive index ($\varepsilon = -1.05$, $\mu = -4$) (b).

A generalized picture of plasmon resonances of a sphere with arbitrary permittivity and permeability is shown schematically in Fig. 11.16 (see also Fig. 11.14). Positions of plasmon resonances in the plane $\varepsilon - \mu$ are shown by the red arrows. On this scheme,

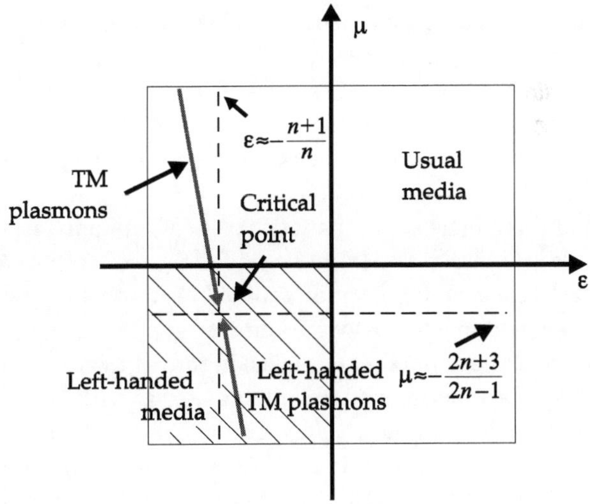

Figure 11.16 The regions of existence of different plasmon oscillations on the plane $\varepsilon - \mu$. The positions of the plasmon resonances in the plane $\varepsilon - \mu$ are shown by the red arrows.

Table 11.1 Oscillations classification in particles with arbitrary dielectric and magnetic properties

Medium properties	Material	Resonance oscillations
$\varepsilon > 0, \mu > 0$	Conventional dielectrics	Propagating waves and whispering gallery modes exist in rather big (in comparison with the wavelength) particles.
$\varepsilon < 0, \mu > 0$	Electric plasma, metals	There are no propagating waves, but electrostatic plasmons exist in small particles.
$\varepsilon < 0, \mu < 0$	Metamaterials with a negative refractive index	Propagating waves and whispering gallery modes exist in rather big (in comparison with the wavelength) particles. Plasmon oscillations of different types exist in small particles.
$\varepsilon > 0, \mu < 0$	Magnetic plasma, metamaterials	There are no propagating waves; only magnetostatic plasmon oscillations exist in small particles.

the critical point corresponds to the transition from regular right-handed plasmons to left-handed ones. This transition is distinctively perceptible in Fig. 11.13, where the plasmon branch (the "tail" at the bottom) changes its orientation as the permeability changes.

The classification of resonances in a sphere of an arbitrary size with arbitrary values of electric permittivity and magnetic permeability is presented in Table 11.1.

Plasmon resonances in nanospheres of metamaterials with a negative refractive index influence light scattering and the emission of atoms and molecules situated near substantially, as any resonance in nanoparticles does. As an example of this influence, the dependency of the spontaneous emission rate of an atom situated in the vicinity of a sphere made of a metamaterial with $\varepsilon = -4$, $\mu = -1.05$ on its radius (Klimov, 2002) is shown in Fig. 11.17. From the picture, it is seen that, indeed, both whispering gallery modes (on the right side of the picture) and left-handed plasmon modes (on the left side of the picture) exist simultaneously in nanoparticles of this kind. Moreover, for some parameters the simultaneous existence of both whispering gallery modes and right-handed and left-handed localized plasmons is possible.

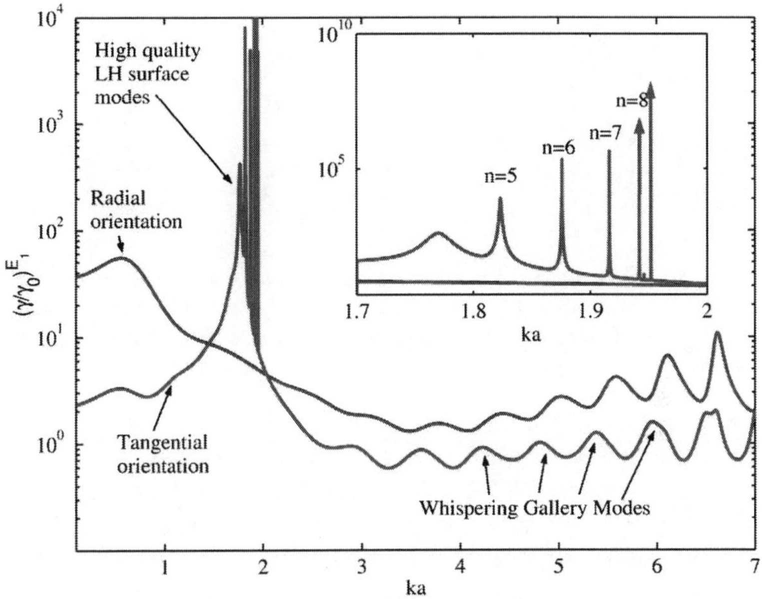

Figure 11.17 The dependence of the spontaneous emission rate of an atom situated near the surface of a sphere made of a metamaterial with $\varepsilon = -4$, $\mu = -1.05$ on the sphere radius *ka*.

11.2 Optical Properties of Chiral Particles

11.2.1 *Main Properties and Methods of Chiral Media Implementation*

The word "chirality" is derived from the Greek root *kheir* meaning "hand." Thus, the term "chirality" denotes such a property of an object that is also a property of the human hand. This term was introduced by Lord Kelvin in his famous "Baltimore Lectures": "I call any geometrical figure, or any group of points, chiral, and say it has chirality, if its image in a plane mirror, ideally realized, cannot be brought to coincide with itself" (Kelvin, 1904). This definition implies that, first, chirality is a geometric property of an object; second, only spatial, that is, three-dimensional, objects possess this property. Plane (two-dimensional) or linear (one-dimensional) objects do not possess this property in three-dimensional space.

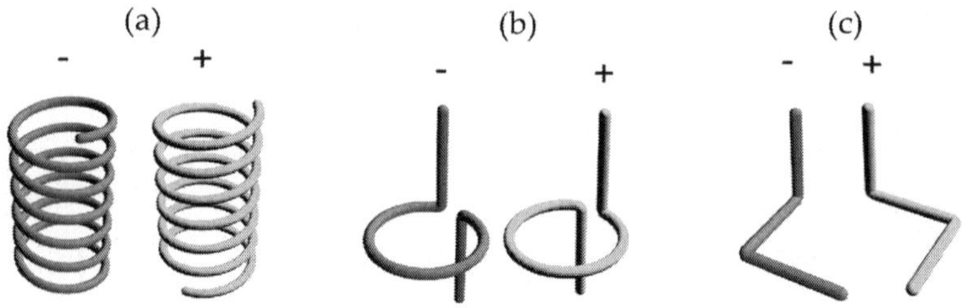

Figure 11.18 A schematic image of chiral meta-atoms used to fabricate chiral metamaterials. (a) Cylindrical spirals, (b) rings with orthogonal linear ends (Ω particles), and (c) broken wires with linear parts along the coordinate axes. The signs "+" and "−" denote right and left chiral elements, correspondingly.

However, let us note that planar nanostructures can also bring about the effects similar to the Faraday effect (Fedotov et al., 2006).

Chiral objects exist in two types, an object and its twin having the form of its mirror reflection, for example, left and right hands, right-hand and left-hand screw, right-hand and left-hand spiral, etc. Lord Kelvin has investigated the chiral properties of crystals and molecules of different substances and their twins.

Examples of chiral meta-atoms for metamaterials and their twins are shown in Fig. 11.18. Metal spirals conducting electric current play an important role in development of chiral metamaterials (Fig. 11.18a). The second pair of chiral elements (Fig. 11.18b) is a simplified version of a spiral and is formed by split rings with linear horns directed normally to the ring plane but in opposite directions. The elements differ by the horns bend direction at the split of the ring. When moving along the wire upward, in the first element right rotation takes place by passing the ring and in the second element left rotation does. Linear parts of the third pair of chiral particles (Fig. 11.18c) are directed along the axes of the left and right Cartesian coordinate system, correspondingly. These elements are a further simplification of previous ones and can have a simpler technological realization.

To understand the physics of chiral particles, let us now consider (see, e.g., (Feynman et al., 2005)) what happens if an electromagnetic wave is incident on a small conducting spiral (Fig. 11.18a). Let the field in the area of the spiral have the components

E_z and H_z along the spiral axis. Then the longitudinal (along the spiral axis) component of the electric dipole moment d_z is formed both by the component E_z of the electric field producing current along the spiral axis and by the component H_z of the magnetic field penetrating the spiral rings and inducing the ring current with a longitudinal component. Thus,

$$d_z = \alpha_{EE} E_z + \alpha_{EH} H_z, \qquad (11.22)$$

where α_{EE} is regular electric polarizability of the spiral and α_{EH} is so-called electromagnetic cross-polarizability of the particle. Analogously, the longitudinal component of the spiral magnetic dipole moment m_z is formed by the ring electric current produced by the magnetic field penetrating the spiral and the ring current produced by the electric field since the current runs along the spiral axis only by passing through the spiral rings. As a result we obtain

$$m_z = \alpha_{HH} H_z + \alpha_{HE} E_z, \qquad (11.23)$$

where α_{HH} is regular magnetic polarizability of the spiral and α_{HE} is magnetoelectric cross-polarizability of the particle. Besides, the spiral is characterized by some other (transverse relatively to the spiral axis) components of electric and magnetic dipole moments, which also have a chiral origin. The above arguments become even more illustrative when applied to the second pair of elements shown in Fig. 11.18b.

It is very important that electromagnetic cross-polarizabilities are of inductive nature and are related to time dependence of the incident fields. Thus, for the harmonic time dependence of the electromagnetic fields, $\exp(-i\omega t)$, the electromagnetic cross-polarizabilities can be presented in the form $\alpha_{EH} = i\beta$; $\alpha_{HE} = -i\beta$; where β is a real number, positive for left-handed spirals and negative for right-handed spirals. So, the longitudinal structure of the spiral dipole moments can be rewritten as

$$\begin{aligned} d_z &= \alpha_{EE} E_z + i\beta H_z, \\ m_z &= -i\beta E_z + \alpha_{HH} H_z. \end{aligned} \qquad (11.24)$$

Properties of spirals (helices) were considered in more detail in (Jaggard et al., 1979; Arnaut and Davis, 1993; Zouhdi et al., 1992; Lindell et al., 1994; Vinogradov, 2001). In particular, (Jaggard et al., 1979) have demonstrated that for a chiral meta-atom shown in Fig. 11.18b, the expressions for cross-polarizabilities can be presented in the form

$$\beta = \alpha_{HH}\frac{L}{k(\pi R^2)} = \alpha_{EE}\frac{k(\pi R^2)}{L}, \tag{11.25}$$

where L is the total height of the element in Fig. 11.18b and R is the radius of its ring part.

After polarization properties of single chiral nanoparticles have been determined, a question of optical properties of metamaterials made of chiral meta-atoms may be considered. A chiral medium can be either an ordered structure in the form of a spatial lattice of chiral nanoparticles or a chaotic mixture of chiral elements. As an exciting example of a regular chiral structure, a planar chiral metamaterial made from three-dimensional gold nanohelices is shown in Fig. 11.19 (Gansel et al., 2009). This structure was recently fabricated by laser "writing" of helical cavities in a photoresist with subsequent deposition of gold inside these holes.

In any case, known mixing rules (see, e.g., (Vinogradov, 2001; Lindell et al., 1994)) can be used to elaborate the macroscopic description from the microscopic one. The application of this procedure to chiral metamaterials with chaotically situated nanoparticles, taking Eq. (11.24) into account, results in the following constitutive equations:

$$\mathbf{D} = \varepsilon\mathbf{E} + i\kappa\mathbf{H}; \qquad \mathbf{B} = \mu\mathbf{H} - i\kappa\mathbf{E}, \tag{11.26}$$

where $\kappa \sim \beta$ is the chirality parameter. Often, the constitutive equations in Eq. (11.26) are written as

$$\mathbf{D} = \varepsilon(\mathbf{E} + \tilde{\beta}\text{rot}\mathbf{E}); \qquad \mathbf{B} = \mu(\mathbf{H} + \tilde{\beta}\text{rot}\mathbf{H}), \tag{11.27}$$

where $\tilde{\beta} = \beta/(\varepsilon\mu\omega/c)$. Using Maxwell's equations, the equivalence of these representations can be shown easily. Further, we will use the constitutive equations in the form of Eq. (11.27). This symmetrical form of the notation of a constitutive equation in chiral media is called the Born–Fedorov form (Bokut' et al., 1970).

It is important that the chiral parameter κ or β is proportional to the wavevector $\mathbf{k}$ or to the ratio a/λ, where a is the size of the nanoparticle and λ is the wavelength. For $a/\lambda \to 0$, that is, when the particle sizes decrease, the chiral properties of the media disappear. Thus, consideration of chiral properties

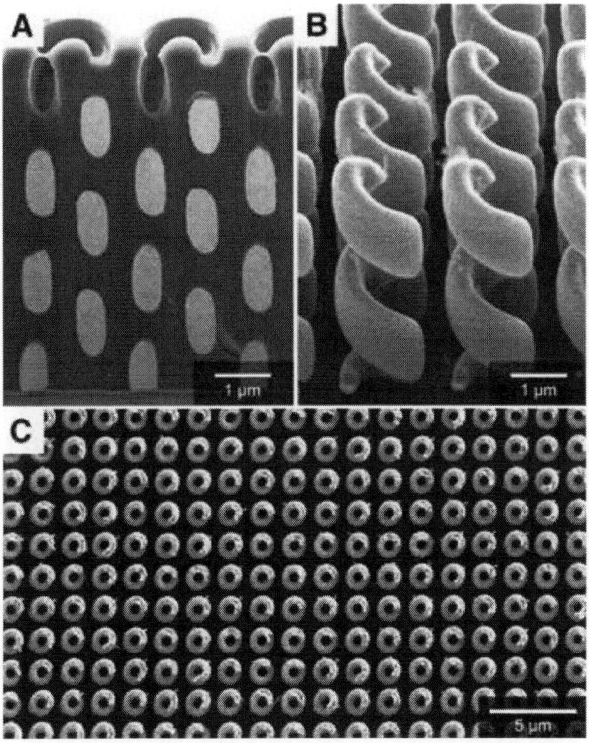

Figure 11.19 Planar chiral metamaterial made of three-dimensional gold nanohelices. (A) FIB cut of a polymer structure partially filled with gold by electroplating. (B) Oblique view of a left-handed helix structure after removal of the polymer by plasma etching. (C) Top view image revealing the circular cross section of the helices and homogeneity on a larger scale. The lattice constant of the square lattice is $a = 2\mu\text{m}$ (Gansel et al., 2009). *Abbreviation*: FIB, focused ion beam.

means consideration of nanoparticles' (or molecules) nonvanishing sizes, that is, spatial dispersion. We have already ensured the importance of spatial dispersion influence on optical properties of spherical nanoparticles of sizes less than 10 nm (Chapter 6); however, there the effects were proportional to the square of the wavevector **k**.

In the optics of natural media, the value of the ratio a/λ turns out to be of the order of 10^{-4}. Indeed, the size of a molecule or an atom is several angstroms, that is, of the order of 10^{-8} cm, and the wavelength in the optical region is about 1 micron, that is, of the order of 10^{-4} cm. In organic substances, for example, polymers, molecules' sizes are much greater than simple molecules, and therefore chiral properties are stronger here. Unfortunately, many polymers are hardly transparent for optical waves.

In spite of their smallness, chiral effects (or effects of optical activity) are widely used in biochemical investigation for:

1. detection of small quantities of organic and biological molecules;
2. determination of whether a protein is folded or not and thus for the determination of its secondary and tertiary structure;
3. comparison of the structure of the proteins obtained from different sources and/or the structures of different mutations of the initial protein;
4. examination of the protein conformational stability relative to different changes of environment (temperature, acidity); and
5. determination of the influence of interaction between proteins on their structure.

Quite a different situation can happen for chiral metamaterials where the chirality parameter $\kappa = Ca/\lambda$ can be increased substantially by an increase of the ratio a/λ, and, above all, by an increase of the coefficient C by use of resonance properties of chiral particles forming the medium. For example, one can use the resonance of the current along the spiral element. The external linear spiral sizes can be small relative to the wavelength, but the length of the wire forming the spiral can be of the order of the wavelength providing the resonance condition. In this case, chirality is not a small effect, and the properties of chiral media will differ substantially from the properties of nonchiral media even without making use of small effect accumulation by means of a large length of the optical device. Moreover, chiral nanoparticles usage can increase the effectiveness of chiral molecules' (proteins and DNA) properties investigation substantially (see the section on spontaneous emission of chiral molecules near chiral nanoparticles at the end of this chapter).

11.2.2 *Optical Properties of a Spherical Chiral Particle*

After the material properties of chiral media have been stated (see the expressions in Eq. (11.27)), the optical properties both of these media and particles composing them can be investigated. The simplest example of a particle made of a chiral substance is a spherical drop of sugar solution of an arbitrary small size.

All particles considered in previous chapters were assumed to consist of linear isotropic substances with the constitutive equations

$$\mathbf{D} = \varepsilon \mathbf{E}; \mathbf{B} = \mu \mathbf{H}. \tag{11.28}$$

Unfortunately, any change of these constitutive equations results in a substantial complication of their optical properties' description. Nevertheless, for spherical nanoparticles made of a chiral substance, the analytical solution of Maxwell's equations can be found (Bohren and Huffman, 1983; Lindel et al., 1994). As we will see further, in these media only right-polarized and left-polarized waves can propagate without a change in their polarization. This is an essential physical difference between chiral and usual media, where waves with any polarization can propagate without changing it.

11.2.3 *Waves in an Infinite Uniform Chiral Medium*

Before starting the investigation of chiral particles' properties, let us first consider waves in an infinite chiral medium. Since in standard textbooks on optics and electrodynamics such media are not considered, we will consider this question in more detail here. During the presentation, we will follow (Bohren and Huffman, 1983; Lindel et al., 1994).

Substituting constitutive equations of Eq. (11.27) in Maxwell's equations, the following system of equations for electromagnetic fields can be derived:

$$\begin{aligned}
&\operatorname{rot} \mathbf{E} - i k_0 \mu \beta \operatorname{rot} \mathbf{H} = i k_0 \mu \mathbf{H}, \\
&\operatorname{rot} \mathbf{H} + i k_0 \varepsilon \beta \operatorname{rot} \mathbf{E} = -i k_0 \varepsilon \mathbf{E}, \\
&\operatorname{div} \mathbf{D} = \operatorname{div} \mathbf{E} = 0, \\
&\operatorname{div} \mathbf{B} = \operatorname{div} \mathbf{H} = 0,
\end{aligned} \tag{11.29}$$

or in the matrix form

$$\begin{bmatrix} 1 & -ik_0\mu\beta \\ ik_0\varepsilon\beta & 1 \end{bmatrix} \operatorname{rot} \begin{bmatrix} \mathbf{E} \\ \mathbf{H} \end{bmatrix} = \begin{bmatrix} 0 & ik_0\mu \\ -ik_0\varepsilon & 0 \end{bmatrix} \begin{bmatrix} \mathbf{E} \\ \mathbf{H} \end{bmatrix},$$
$$\operatorname{div} \begin{bmatrix} \mathbf{E} \\ \mathbf{H} \end{bmatrix} = \begin{bmatrix} 0 \\ 0 \end{bmatrix}, \tag{11.30}$$

where $k_0 = \omega/c$ is the wavenumber of vacuum, and we have to omit the tilde sign over β.

At first sight, these equations differ substantially from Maxwell's equations, which were considered in previous chapters. However, these equations may be quite easily reduced to a usual vector Helmholtz equation.

Indeed, by inverting the matrix in the left part of Eq. (11.30) it can be rewritten as

$$\operatorname{rot} \begin{bmatrix} \mathbf{E} \\ \mathbf{H} \end{bmatrix} = \hat{K} \begin{bmatrix} \mathbf{E} \\ \mathbf{H} \end{bmatrix},$$
$$\operatorname{div} \begin{bmatrix} \mathbf{E} \\ \mathbf{H} \end{bmatrix} = \begin{bmatrix} 0 \\ 0 \end{bmatrix}, \tag{11.31}$$

where the matrix $\hat{K}$ has the form

$$\hat{K} = \frac{ik_0}{1 - k_0^2\beta^2\varepsilon\mu} \begin{bmatrix} -ik_0\varepsilon\mu\beta & \mu \\ -\varepsilon & -ik_0\varepsilon\mu\beta \end{bmatrix}. \tag{11.32}$$

After that, applying the operation rot to the first equation, we obtain the equation similar to the Helmholtz equation:

$$\nabla^2 \begin{bmatrix} \mathbf{E} \\ \mathbf{H} \end{bmatrix} = \hat{K}^2 \begin{bmatrix} \mathbf{E} \\ \mathbf{H} \end{bmatrix}. \tag{11.33}$$

The only, but not crucial, difference between Eq. (11.33) and the Helmholtz equations is that $\hat{K}^2$ is rather a matrix than a square of the scalar wavenumber.

By the linear transformation of the electromagnetic field

$$\begin{bmatrix} \mathbf{E} \\ \mathbf{H} \end{bmatrix} = \hat{A} \begin{bmatrix} \mathbf{\Omega}_L \\ \mathbf{\Omega}_R \end{bmatrix}, \tag{11.34}$$

where

$$\hat{A} = -i \begin{bmatrix} i & \sqrt{\mu/\varepsilon} \\ \sqrt{\varepsilon/\mu} & i \end{bmatrix}, \tag{11.35}$$

the matrix $\hat{K}$ can be diagonalized:

$$\hat{\Lambda} = \hat{A}^{-1}\hat{K}\hat{A} = \begin{bmatrix} k_L & 0 \\ 0 & -k_R \end{bmatrix}. \tag{11.36}$$

The transformation Eq. (11.34) is called "Bohren transformation." In Eq. (11.36)

$$k_L = \frac{k_0\sqrt{\varepsilon\mu}}{1 - k_0\sqrt{\varepsilon\mu}\beta} = k_0 N_L,\ k_R = \frac{k_0\sqrt{\varepsilon\mu}}{1 + k_0\sqrt{\varepsilon\mu}\beta} = k_0 N_R, \tag{11.37}$$

are wavevectors and refractive indices of left- and right-polarized waves, which can be determined experimentally. In turn, knowing these wavevectors, the chirality parameter β can be found:

$$\beta = \frac{1}{2}\left(\frac{1}{k_R} - \frac{1}{k_L}\right). \tag{11.38}$$

After the transformation Eq. (11.34), each component of the transformed field $\mathbf{\Omega}_\mathrm{L}$, $\mathbf{\Omega}_\mathrm{R}$ satisfies the Helmholtz equation independently:

$$\nabla^2\mathbf{\Omega} + \tilde{k}^2\mathbf{\Omega} = 0. \tag{11.39}$$

They also will satisfy the following equations:

$$\mathrm{rot}\,\mathbf{\Omega} = \tilde{k}\mathbf{\Omega} \tag{11.40}$$

$$\mathrm{div}\,\mathbf{\Omega} = 0 \tag{11.41}$$

In Eqs. (11.39) and (11.40), $k = k_L$ for $\mathbf{\Omega} = \mathbf{\Omega}_L$, and $k = -k_R$ for $\mathbf{\Omega} = \mathbf{\Omega}_R$. Thus, two waves, $\mathbf{\Omega}_L$, $\mathbf{\Omega}_R$ with different wavevectors, Eq. (11.37), are fundamental solutions (eigenmodes) in a chiral medium.

To understand the physical meaning of these modes, let us consider a plane wave, $\sim\exp(i\mathbf{kr})$, where $\mathbf{k} = k_{L,R}\hat{\mathbf{e}}$ and $\hat{\mathbf{e}}$ are unit vectors in the direction of propagation. Eq. (11.40) implies that

$$i\hat{\mathbf{e}} \times \mathbf{\Omega} = \pm\mathbf{\Omega}, \tag{11.42}$$

where the "+" sign corresponds to $\mathbf{\Omega}_L$ and the "−" sign corresponds to $\mathbf{\Omega}_R$. In turn, Eq. (11.41) implies that the vector $\mathbf{\Omega}$ is transverse relatively the direction of wave propagation, that is, $\hat{\mathbf{e}} \cdot \mathbf{\Omega} = 0$, and therefore it can be written as

$$\mathbf{\Omega} = \mathbf{\Omega}_1\hat{\mathbf{e}}_1 + \mathbf{\Omega}_2\hat{\mathbf{e}}_2, \tag{11.43}$$

where $\hat{\mathbf{e}} \times \hat{\mathbf{e}}_2 = \hat{\mathbf{e}}_1, \hat{\mathbf{e}} \times \hat{\mathbf{e}}_1 = -\hat{\mathbf{e}}_2$, that is, $\mathbf{e}_1$ and $\mathbf{e}_2$ are unit vectors perpendicular to the propagation direction and between themselves. Now substituting Eq. (11.43) in Eq. (11.42) and making the coefficients of $\hat{\mathbf{e}}_1$, $\hat{\mathbf{e}}_2$ equal, we obtain

$$i\mathbf{\Omega}_2 = \pm\mathbf{\Omega}_1, \tag{11.44}$$

and, consequently

$$\mathbf{\Omega} = \mathbf{\Omega}_1(\hat{\mathbf{e}}_1 \mp i\hat{\mathbf{e}}_2), \tag{11.45}$$

where the "+" sign corresponds to $\mathbf{\Omega}_L$ and the "−" sign corresponds to $\mathbf{\Omega}_R$. Eq. (11.45) is the definition of the left and right circular polarizations, and thus it reveals us that solutions of Eq. (11.39) in the form of plane waves, $\mathbf{\Omega}_L$ or $\mathbf{\Omega}_R$, are to have left or right polarizations, correspondingly. Corresponding electric and magnetic fields can be found with the help of Eq. (11.34) and also will have either right or left polarizations if only $\mathbf{\Omega}_R$ or $\mathbf{\Omega}_L$ is different from zero.

However, polarization states of waves with nonzero both $\mathbf{\Omega}_R$ and $\mathbf{\Omega}_L$ will change during propagation in chiral media. In particular, if at the point $z = 0$ (at the entry to the chiral medium) the light polarization is linear, the light can be polarized elliptically at the point $z = L$ (at the exit from the chiral medium). This effect is explained by the fact that when entering chiral media, the linearly polarized light can be presented as a superposition of right- and left-polarized waves propagating at different velocities determined by the complex refractive index for the right- and left-polarized waves:

$$N_L = n_L + i\kappa_L;\ N_R = n_R + i\kappa_R, \tag{11.46}$$

where N_L, N_R are determined in Eq. (11.37).

If the extinction coefficients κ_L, κ_R are not equal, the medium is said to have circular dichroism, and different circular polarizations attenuate differently during propagation. If the refractive indices n_L, n_R are different, as the light passes through the chiral medium, the rotation of the polarization plane of the initially linearly polarized wave takes place. Both effects are called effects of optical activity and used for investigation of biomolecules widely.

Let us stress again that in a chiral medium only right- and left-polarized waves can propagate without a change of their polarization.

11.2.4 *Spherical Waves in Chiral Media*

To investigate optical properties of a chiral sphere, it is natural to use expansions in spherical rather than plane waves. It is widely known that the Helmholtz vector equation Eq. (11.39) has a solution in the form of spherical harmonics (Stratton, 1941) **M** and **N**:

$$\begin{aligned} \mathbf{M}_{emn} &= \nabla \times (\mathbf{r}\psi_{emn}); \mathbf{M}_{omn} = \nabla \times (\mathbf{r}\psi_{omn}), \\ \mathbf{N}_{emn} &= \tfrac{1}{\tilde{k}}\nabla \times \mathbf{M}_{emn}; \mathbf{N}_{omn} = \tfrac{1}{\tilde{k}}\nabla \times \mathbf{M}_{omn}, \end{aligned} \tag{11.47}$$

where ψ_{emn} and ψ_{omn} are even and odd solutions of the Helmholtz scalar equation:

$$\begin{aligned} \psi_{emn} &= \cos m\phi P_n^m(\cos\theta) z_n(\tilde{k}r), \\ \psi_{omn} &= \sin m\phi P_n^m(\cos\theta) z_n(\tilde{k}r). \end{aligned} \tag{11.48}$$

In Eq. (11.48), $z_n(\tilde{k}r)$ is any one of the Bessel spherical functions $j_n(\tilde{k}r)$, $y_n(\tilde{k}r)$, $h_n^{(1)}(\tilde{k}r)$, $h_n^{(2)}(\tilde{k}r)$, and $P_n^m(\cos\theta)$ are associated Legendre polynomials (Abramowitz and Stegun, 1965). Hereafter, we will indicate only the spherical harmonics dependence on the radius since this dependence contains the information on the chirality.

In the case of chiral media, an arbitrary solution of equations in Eq. (11.39) can be expressed through the vector spherical harmonics in spherical coordinates, as well as in the case of the regular media. At that, owing to Eq. (11.40), which mixes **M** and **N** spherical harmonics, the linear combinations of **M** and **N** spherical harmonics are to be used in a general solution of Eq. (11.39):

$$\begin{aligned} \mathbf{\Omega}_L &= \sum_{n=1}^{\infty} \left\{ \begin{array}{l} f_{onm}\left[\mathbf{M}_{omn}(k_{1L}r) + \mathbf{N}_{omn}^{(k_{1L}r)}\right] \\ + f_{enm}\left[\mathbf{M}_{emn}(k_{1L}r) + \mathbf{N}_{emn}(k_{1L}r)\right] \end{array} \right\}, \\ \mathbf{\Omega}_R &= \sum_{n=1}^{\infty} \left\{ \begin{array}{l} g_{onm}\left[\mathbf{M}_{omn}(k_{1R}r) - \mathbf{N}_{omn}(k_{1R}r)\right] \\ + g_{enm}\left[\mathbf{M}_{emn}(k_{1R}r) - \mathbf{N}_{emn}(k_{1R}r)\right] \end{array} \right\}. \end{aligned} \tag{11.49}$$

The expansion, Eq. (11.49), can be verified by substituting it in Eq. (11.40) and by making use of the definitions, Eq. (11.47). When deriving Eq. (11.49), it is assumed that the considered medium has the electric permittivity and the magnetic permeability ε_1, μ_1 and the chirality parameter β, and the wavevectors $\mathbf{k}_{1L}$ and $\mathbf{k}_{1R}$ are determined by formulas from Eq. (11.37), where ε, μ should be replaced by ε_1, μ_1. Electric and magnetic fields can be found from the expansion Eq. (11.49) with the help of the matrix equation Eq. (11.34). Let us emphasize that different wavenumbers enter as arguments of the Bessel spherical functions in Eq. (11.49). This is an essential difference from the spherical harmonics expansion in regular isotropic media. An analogous situation takes place also in the case of usual spatial dispersion (see Chapter 6, Section 6.5) when transverse and longitudinal spherical waves have different wavevectors.

Thus, expressions in Eq. (11.49) together with Eq. (11.34) provide the expansion of the arbitrary solution of Maxwell's equations in vector spherical harmonics in a chiral medium with the constitutive equations, Eq. (11.27).

11.2.5 *Optical Properties of a Spherical Chiral Particle Placed in the Field of a Plane Wave*

After the vector spherical harmonics expansion of an arbitrary solution of Maxwell's equations have been performed in chiral media, investigation of optical properties of a chiral spherical particle can be carried out in full analogy with the case of a usual medium (Bohren, 1974). Namely, the exciting field, the scattered field, and the field inside the particle are sought in the form of vector spherical harmonics expansion, and unknown coefficients are found from the continuity of tangential components of **E** and **H** at the sphere surface.

In this section, we will consider the case of scattering of a plane wave $\mathbf{E} = \{E_0, 0, 0\}e^{ikz}$ by a chiral sphere of radius a placed at the origin of the coordinate system. We also assume that this wave propagates in a nonchiral medium with parameters ε_2, μ_2.

The expansion of this plane wave in vector spherical harmonics, which is finite at $r = 0$, has the form (Stratton, 1941):

$$\mathbf{E}_{in} = \{E_0, 0, 0\}e^{ik_2 z} = \sum_{n=1}^{\infty} E_n(\mathbf{M}^{(1)}_{o1n}(k_2 r) - i\mathbf{N}^{(1)}_{o1n}(k_2 r)). \quad (11.50)$$

In Eq. (11.50) and further, $\mathbf{M}^{(1)}$, $\mathbf{N}^{(1)}$ are vector spherical harmonics with $j_n(x)$ as radial functions, and $E_n = i^n E_0(2n+1)/$

$[n(n+1)]$. For this incident wave, fields inside the chiral sphere can be written as

$$\begin{aligned}\boldsymbol{\Omega}_L &= \sum_{n=1}^{\infty} E_n \Big\{ f_{on} \left[\mathbf{M}^{(1)}_{o1n}(k_{1L}r) + \mathbf{N}^{(1)}_{o1n}(k_{1L}r)\right] \\ &\quad + f_{en} \left[\mathbf{M}^{(1)}_{e1n}(k_{1L}r) + \mathbf{N}^{(1)}_{e1n}(k_{1L}r)\right] \Big\}, \\ \boldsymbol{\Omega}_R &= \sum_{n=1}^{\infty} E_n \Big\{ g_{on} \left[\mathbf{M}^{(1)}_{o1n}(k_{1R}r) - \mathbf{N}^{(1)}_{o1n}(k_{1R}r)\right] \\ &\quad + g_{en} \left[\mathbf{M}^{(1)}_{e1n}(k_{1R}r) - \mathbf{N}^{(1)}_{e1n}(k_{1R}r)\right] \Big\},\end{aligned} \tag{11.51}$$

and here again the Bessel function $j_n(k_{L,R}r)$ is to be taken as the spherical function $z_n(x)$ since the field inside the spheres is to be finite.

Correspondingly, for the scattered field we have

$$\begin{aligned}\mathbf{E}_{\text{sca}} &= \sum_{n=1}^{\infty} E_n \Big\{ i a_n \mathbf{N}^{(3)}_{e1n}(k_2 r) - b_n \mathbf{M}^{(3)}_{o1n}(k_2 r) \\ &\quad + c_n \mathbf{M}^{(3)}_{e1n}(k_2 r) - i d_n \mathbf{N}^{(3)}_{o1n}(k_2 r) \Big\}, \\ \mathbf{H}_{\text{sca}} &= \frac{k_2}{\omega\mu_2} \sum_{n=1}^{\infty} E_n \Big\{ a_n \mathbf{M}^{(3)}_{e1n}(k_2 r) \\ &\quad + i b_n \mathbf{N}^{(3)}_{o1n}(k_2 r) - i c_n \mathbf{N}^{(3)}_{e1n}(k_2 r) - d_n \mathbf{M}^{(3)}_{o1n}(k_2 r) \Big\},\end{aligned} \tag{11.52}$$

where the spherical Hankel function of the first kind $h_n^{(1)}(kr)$ is to be taken as the spherical function $z_n(x)$ for description of outgoing waves $\mathbf{M}^{(3)}(k_2 r)$, $\mathbf{N}^{(3)}(k_2 r)$.

From the requirement of continuity of the tangential components **E** and **H** at the sphere surface, coefficients in the expression for a scattered field can be found (Bohren, 1974; Bohren and Huffman, 1983):

$$\begin{aligned}a_n &= \frac{V_n(R)A_n(L) + V_n(L)A_n(R)}{V_n(R)W_n(L) + V_n(L)W_n(R)}, \\ b_n &= \frac{B_n(R)W_n(L) + B_n(L)W_n(R)}{V_n(R)W_n(L) + V_n(L)W_n(R)}, \\ c_n &= i\frac{A_n(L)W_n(R) - A_n(R)W_n(L)}{V_n(R)W_n(L) + V_n(L)W_n(R)} = -d_n.\end{aligned} \tag{11.53}$$

The coefficients A_n, B_n, V_n, and W_n are expressed through the Rikatti–Bessel spherical functions ($\psi_n(x) = xj_n(x)$, $\xi_n(x) = xh_n^{(1)}(x)$):

$$\begin{aligned}W_n(J) &= m\psi_n(z_{1J})\xi_n'(z_2) - \xi_n(z_2)\psi_n'(z_{1J}),\\ V_n(J) &= \psi_n(z_{1J})\xi_n'(z_2) - m\xi_n(z_2)\psi_n'(z_{1J}),\\ A_n(J) &= m\psi_n(z_{1J})\psi_n'(z_2) - \psi_n(z_2)\psi_n'(z_{1J}),\\ B_n(J) &= \psi_n(z_{1J})\psi_n'(z_2) - m\psi_n(z_2)\psi_n'(z_{1J}),\end{aligned} \quad (11.54)$$

where $z_2 = k_2a$, $z_{1J} = k_{1J}a$ and the index J takes the values L and R. The relative refractive indices m_L, m_R and the average refractive index m are determined by the relations

$$m_L = \frac{k_{1L}}{k_2};\, m_R = \frac{k_{1R}}{k_2};\, \frac{1}{m} = \frac{1}{2}\left(\frac{1}{m_L} + \frac{1}{m_R}\right)\frac{\mu_1}{\mu_2}. \quad (11.55)$$

The difference Δm between the refraction indices of the right- and left-polarized waves m_R and m_L is small usually, and it can be considered that

$$m \approx \frac{1}{2}(m_L + m_R). \quad (11.56)$$

If the nanoparticles' material is not chiral, then $m_L = m_R$, and the coefficients a_n and b_n go over into scattering coefficients of a nonchiral sphere q_n and p_n (Eqs. (11.20) and (11.21)), correspondingly, and the coefficients c_n and d_n vanish. The scattering of a plane wave of different polarizations by a chiral sphere can be investigated analogously.

After the scattering coefficients a_n, b_n, c_n, and d_n have been calculated, all optical characteristics of chiral particles can be found. It is important that the scattering cross section of chiral particles depend substantially on the incident radiation polarization in contrast to usual spherical particles.

For scattering and extinction cross sections of a chiral sphere illuminated by plane waves of right and left polarization (the indices

R and L), the following expressions can be obtained (Bohren and Huffman, 1983):

$$S_{\text{sca},L} = \frac{2\pi}{k^2}\sum_{n=1}^{\infty}(2n+1)[|a_n|^2 + |b_n|^2 + 2|c_n|^2 - 2\text{Im}((a_n+b_n)c_n^*)],$$

$$S_{\text{sca},R} = \frac{2\pi}{k^2}\sum_{n=1}^{\infty}(2n+1)[|a_n|^2 + |b_n|^2 + 2|c_n|^2 + 2\text{Im}((a_n+b_n)c_n^*)], \tag{11.57}$$

$$S_{\text{ext},L} = \frac{2\pi}{k^2}\sum_{n=1}^{\infty}(2n+1)\text{Re}(a_n + b_n - 2ic_n),$$

$$S_{\text{ext},R} = \frac{2\pi}{k^2}\sum_{n=1}^{\infty}(2n+1)\text{Re}(a_n + b_n + 2ic_n). \tag{11.58}$$

Let us emphasize again that the scattering and extinction cross sections of chiral particles depend substantially on the incident radiation polarization in contrast to usual spherical particles. This dependency becomes apparent in different signs before the scattering coefficients c_n.

As usual, the resonance properties of chiral spherical particles are determined by poles of the coefficients a_n, b_n, and $c_n = -d_n$. As we have already seen (Section 11.1.4), for nonchiral spherical particles made of metamaterials of arbitrary electric permittivity and magnetic permeability, resonances have a very complex and interesting structure (Section 11.1.2, Figs. 11.13 and 11.14). The resonances in chiral nanoparticles have an even more interesting structure and will be discussed later in this chapter.

Spherically layered chiral particles can be investigated by perfectly analogous methods. However, expressions obtained are rather tedious, and we refer to the original literature (Bohren, 1975; Li et al., 2000). Optical properties of chiral particles in the form of a cylinder can be also found in the analytical form (Bohren, 1978).

11.2.6 *Spontaneous Emission of Optically Active Molecules Induced by the Presence of Nearby Chiral Nanoparticles*

In Section 11.2.5, the scattering of a plane wave by a chiral sphere was considered. However, in nano-optics highly nonuniform

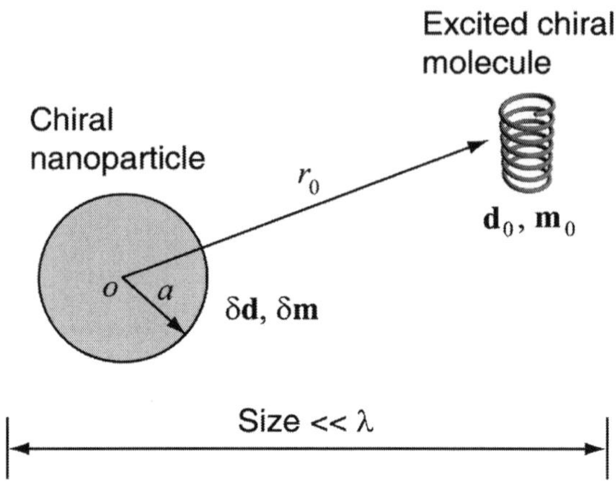

Figure 11.20 The geometry of the problem of radiation of an optically active (chiral) molecule placed near an arbitrary chiral nanoparticle.

near fields have even greater importance. In this section, we will consider the interaction of chiral nanoparticles with the radiation of an optically active molecule and show that very interesting and important plasmonic effects arise here (Klimov, Guzatov, Ducloy, 2011; Guzatov and Klimov, 2012). The geometry of the problem is shown in Fig. 11.20.

In the model of an optically active molecule, we assume that its radiation is both of electric and of magnetic dipole nature, that is, that the Hamiltonian of the molecule interaction with the field has the form (Schellman, 1975):

$$\hat{H}_{\text{int}} = -\hat{\mathbf{d}}\mathbf{E} - \hat{\mathbf{m}}\mathbf{H}. \tag{11.59}$$

Matrix elements of dipole operators will be written as $\mathbf{d}_0$ and $-i\mathbf{m}_0 = -i\xi\mathbf{d}_0$ due to the fact that the magnetic moment operator is purely imaginary. The parameter ξ characterizes the relative value of the magnetic dipole moment. Within this model, we will refer to a molecule for which $\mathbf{d}_0$ and $\mathbf{m}_0$ are parallel as a "right" molecule, whereas a molecule in which the magnetic moment is antiparallel to the electric dipole moment will be referred to as a "left" molecule. Thus, we are considering a molecule where electrons are constrained to move along a helical path. This picture also explains the choice of the phase difference between electric and magnetic dipole moments.

To control the radiation of a molecule with particular chirality, it is necessary to create such a nanoenvironment that forces interference between electric and magnetic dipole radiation of the molecule. Such fields can be created with the help of chiral nanoparticles, that is, nanoparticles that can be described with the Drude–Born–Fedorov constitutive relations, Eq. (11.27). The dimensionless chirality parameter χ is defined by the relation $\chi = \beta k_0$, where $k_0 = \omega/c$ is the wavenumber in vacuum.

Here, for simplicity we will consider a homogeneous spherical nanoparticle, where the substance is described by the constitutive equations in Eq. (11.27). In Fig. 11.21, the structure of TM plasmonic resonances for a spherical nanoparticle of radius a without chirality and with the chirality factor $\chi = \beta k_0 = 0.1$, found from Eq. (11.53) with $n = 1$, is shown.

From this figure (see also Fig. 11.14), it is seen that TM modes in a nonchiral nanoparticle have electric dipole plasmon resonance at $\varepsilon \approx -2$, nearly independent of magnetic permeability. Any chirality—even small—of the nanoparticle (see Fig. 11.21b) leads to a radical change in the shape of the resonance line, which becomes dependent in a nontrivial way on both permittivity and permeability of the nanoparticle material. Below, we will see that these hybrid chiral-plasmon modes in the nanoparticle provide effective interference between radiation of electric and magnetic dipole moments of the chiral molecule.

To estimate the interaction of a radiating optically active molecule with a chiral *nanoparticle*, we can neglect retardation effects (see Fig. 11.20). In this case, in the field of radiating chiral molecules the near fields are dominating:

$$\mathbf{E}_0 = \frac{3\mathbf{n}(\mathbf{n}\mathbf{d}_0) - \mathbf{d}_0}{r^3}, \quad \mathbf{H}_0 = -i\frac{3\mathbf{n}(\mathbf{n}\mathbf{m}_0) - \mathbf{m}_0}{r^5}, \tag{11.60}$$

where r is the distance between the molecule and the nanoparticle, $\mathbf{n}$ is the unit vector in the direction from the molecule to the observation point, and the monochromatic time dependence of fields ($e^{-i\omega t}$) is omitted.

The fields, Eq. (11.60), induce the electric $\delta\mathbf{d}$ and magnetic $\delta\mathbf{m}$ dipole moments in the nanoparticle:

$$\begin{aligned} \delta\mathbf{d} &= \alpha_{EE}\mathbf{E}_0(\mathbf{r}_0) + \alpha_{EH}\mathbf{H}_0(\mathbf{r}_0), \\ \delta\mathbf{m} &= \alpha_{HE}\mathbf{E}_0(\mathbf{r}_0) + \alpha_{HH}\mathbf{H}_0(\mathbf{r}_0), \end{aligned} \tag{11.61}$$

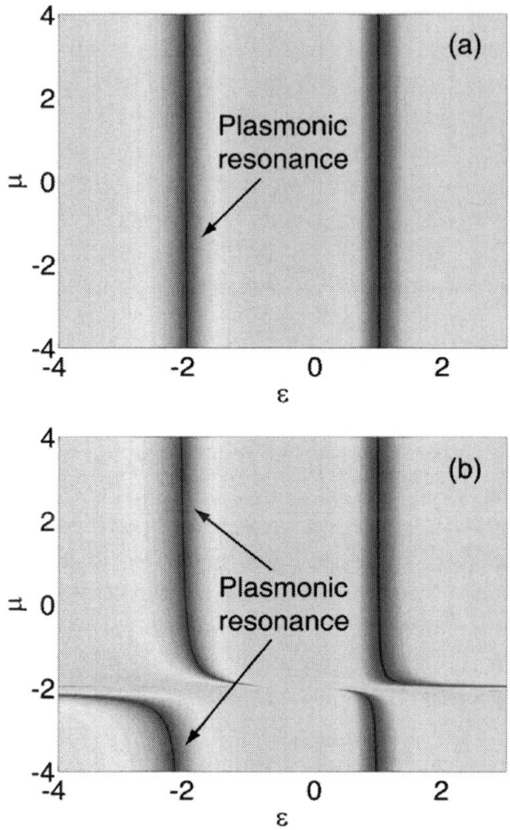

Figure 11.21 Structure of TM ($n = 1$) resonances of a nanoparticle without chirality (a) and with the chirality factor $\chi = \beta k_0 = 0.1$ (b). The size parameter of the nanosphere is $k_0 a = 0.1$.

where α_{EE}, α_{EH}, α_{HE}, and α_{HH} are the tensors of electromagnetic polarizability of nanoparticle in *uniform* external fields. Possibility to apply the formula Eq. (11.61) to *strongly nonuniform* fields Eq. (11.60) is associated with the unique property of ellipsoids and spheres, that is, the tensor of polarizability of an ellipsoid in a nonuniform field of a dipole source is defined by usual polarizabilities and by the dipole field, which is averaged over the volume of the particle (Guzatov and Klimov, 2005). In the case of spheres, such averaging gives the value of the dipole field in the center of the sphere, that is, $\mathbf{E}_0(\mathbf{r}_0)$ and $\mathbf{H}_0(\mathbf{r}_0)$, respectively.

For a spherical particle with constitutive relations, Eq. (11.27), it is easy to find that quasi-static electromagnetic polarizabilities are

$$\begin{aligned}\alpha_{EE} &= a^3\frac{(\varepsilon-1)(\mu+2)+2\varepsilon\mu\chi^2}{(\varepsilon+2)(\mu+2)-4\varepsilon\mu\chi^2},\\ \alpha_{HH} &= a^3\frac{(\mu-1)(\varepsilon+2)+2\varepsilon\mu\chi^2}{(\varepsilon+2)(\mu+2)-4\varepsilon\mu\chi^2},\\ \alpha_{EH} &= a^3\frac{3\chi\varepsilon\mu i}{(\varepsilon+2)(\mu+2)-4\varepsilon\mu\chi^2},\\ \alpha_{HE} &= a^3\frac{-3\chi\varepsilon\mu i}{(\varepsilon+2)(\mu+2)-4\varepsilon\mu\chi^2},\end{aligned} \tag{11.62}$$

where a is the nanosphere radius and $\chi = \beta k_0$, ε, μ are the dimensionless chirality parameter, electric permittivity, and magnetic permeability of the nanoparticle, respectively, and it is assumed that the particle is located in vacuum.

Since the total size of the "molecule + nanoparticle" system is much less than the wavelength, the phases of dipole moments of the molecule and the particle are equal, and the intensity of radiation from this composite system is described by the decay rate:

$$\begin{aligned}\Gamma_{\text{rad}} &= \frac{4k_0^3}{3\hbar}\left\{|\mathbf{d}_0+\delta\mathbf{d}|^2+|\mathbf{m}_0+i\delta\mathbf{m}|^2\right\},\\ &= \frac{4k_0^3}{3\hbar}\left\{\begin{array}{l}\left|\mathbf{d}_0+\alpha_{EE}/r_0^3(3\mathbf{n}(\mathbf{n}\mathbf{d}_0)-\mathbf{d}_0)-i\alpha_{EH}/r_0^3(3\mathbf{n}(\mathbf{n}\mathbf{m}_0)-\mathbf{m}_0)\right|^2+\\ \left|\mathbf{m}_0+i\alpha_{HE}/r_0^3(3\mathbf{n}(\mathbf{n}\mathbf{d}_0)-\mathbf{d}_0)+\alpha_{HH}/r_0^3(3\mathbf{n}(\mathbf{n}\mathbf{m}_0)-\mathbf{m}_0)\right|^2\end{array}\right\},\end{aligned} \tag{11.63}$$

where $\hbar$ is Planck's constant.

Since the orientation of the molecule can be arbitrary relative to the particle, to obtain an effective decay rate one should average Eq. (11.63) over orientations of the molecule or, equivalently, over the unit vector $\mathbf{n}$. As a result, we obtain

$$\begin{aligned}\Gamma_{\text{rad,eff}} = \frac{4k_0^3|\mathbf{d}_0|^2}{3\hbar}&\left\{1+\frac{2}{r_0^6}|\alpha_{EE}-i\xi\alpha_{EH}|^2+|\xi|^2\right.\\ &\left.+\frac{2}{r_0^6}|i\alpha_{HE}+\xi\alpha_{HH}|^2\right\},\end{aligned} \tag{11.64}$$

where ξ is defined by the relation $\mathbf{m}_0 = \xi\mathbf{d}_0$.

Usually, the electric dipole moment of a molecule is much larger than the magnetic dipole moment $|\mathbf{d}_0| \gg |\mathbf{m}_0|$ or $\xi \ll 1$. The chirality parameter, even in hypothetical metamaterials, is also small, $\chi \ll 1$. Because of this, the first (electric) term in Eq. (11.63) is usually greater than the second (magnetic) one. Therefore, the interference between electric and magnetic fields is possible only when two nontrivial conditions take place:

1. The system must have chiral-plasmon resonance, that is, the following condition should be satisfied:

$$(\varepsilon + 2)(\mu + 2) - 4\varepsilon\mu\chi^2 = 0, \qquad (11.65)$$

(see Fig. 11.21b). Under this condition, contribution of magnetic radiation increases.
2. The electric dipole moment induced in the nanoparticle must be zero, that is, the following condition should take place:

$$\alpha_{EE} - i\xi\alpha_{EH} = 0. \qquad (11.66)$$

The simultaneous solution of Eqs. (11.65) and (11.66) determines values of electric permittivity and magnetic permeability of the nanoparticle where the radiative decay rate of chiral molecules is minimal and, therefore, the interference between electric and magnetic dipoles radiations is maximal:

$$\mu^* \to -\frac{2d_0}{d_0 + 2m_0\chi}; \quad \varepsilon^* \to -\frac{2m_0}{m_0 + 2d_0\chi}. \qquad (11.67)$$

The equations in Eq. (11.67) are the key equations for chiral molecule discrimination and enantiomer selectivity.

It is very important to note that when the sign of m_0 is changed, that is, when molecule's chirality is changed, the "resonance" magnetic permeability varies slightly and remains approximately equal to -2. On the other hand, the "resonance" permittivity has different signs for molecules of different chirality. This means that both nanoparticles with simultaneously negative ε and μ (left-handed or DNG metamaterials) and nanoparticles with negative μ and positive ε (μ-negative [MNG] metamaterials or magnetic plasma) are suitable for effective control of chiral molecules radiation. For practical implementation of such a control, MNG nanoparticles seem to be the most suitable because they can be realized within a well developed technology of SRRss (Smith et al., 2000, Pendry et al., 1999).

Figure 11.22 shows decay rates of right-handed (top) and left-handed (bottom) molecules on electric permittivity and magnetic permeability of the material of a nanoparticle.

This figure shows that, indeed, for values of electric permittivity and magnetic permeability, which are determined by the equations in Eq. (11.67), the decay rates of right-handed (Fig. 11.22a) or left-handed (Fig. 11.22b) molecules are close to zero. It is this fact that determines the possibility of their separation.

To determine a quantitative measure of discrimination, the ratio of left-handed molecules decay rate to that of right-handed ones and vice versa is shown in Fig. 11.23.

This figure shows that, indeed, for the chosen parameters, there is a difference in the radiative decay rates of right- and left-handed molecules up to 15 or 60 or more times, depending on chirality of the molecule considered as a reference one. In other words, nanoparticles with the parameters, Eq. (11.67), will enhance the radiation of right-handed molecules and slow down the radiation of left-handed molecules, and vice versa. Let us stress that left-handedness of a chiral sphere (DNG material) or its negative μ (MNG material) is of crucial importance for such discrimination.

Until now, we have considered a usual case when the electric dipole moment of a molecule is greater than the magnetic dipole moment, that is, the case when $\xi \ll 1$. In a hypothetical case when the magnetic dipole moment is dominating, that is, in the case where $\xi \gg 1$, it follows from Eq. (11.67) that one needs to use DNG or epsilon-negative (ENG) materials to obtain an enhancement of the radiation of a right-handed or left-handed molecule.

The proposed theory allows one to estimate parameters of nanoparticles required for discrimination of the radiation of left-handed or right-handed molecules. It is very important that despite the simple character of the proposed theory, it is accurate enough. The accuracy of the present approach directly follows from the comparison with results of full quantum electrodynamics calculations performed by us without any approximations (Guzatov and Klimov, 2011).

For applications of this theory for drug enantiomers' discrimination, see Chapter 13.

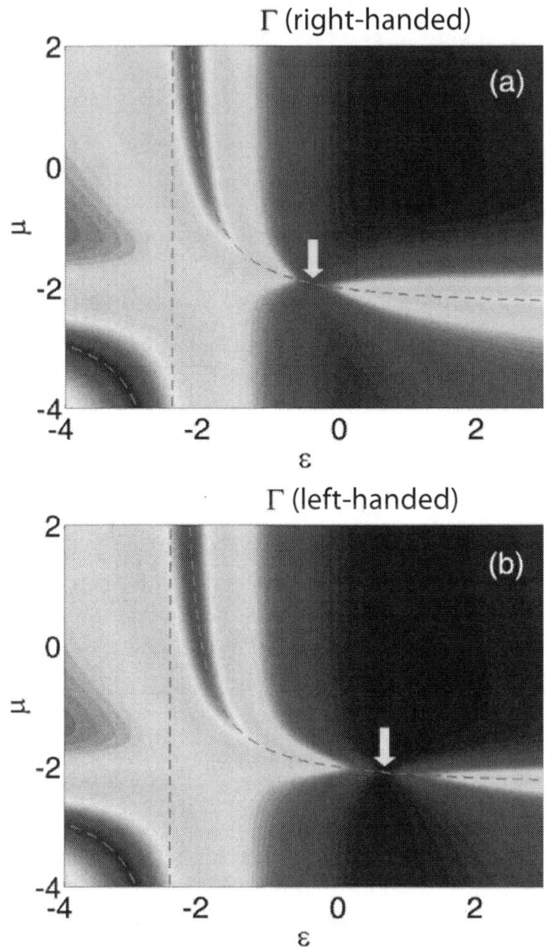

Figure 11.22 Dependence of the effective radiative decay rate, Eq. (11.64), of right-handed ($m_0/d_0 = +0.1$, top) and left-handed ($m_0/d_0 = -0.1$, bottom) molecules on permittivity and permeability of the material of a nanoparticle ($\chi = 0.2$, $\varepsilon'' = \mu'' = 0.1$). The dashed line corresponds to the position of the chiral-plasmon resonance, Eq. (11.65). The yellow arrow indicates the minimum rates on the resonance curve. The molecule is placed close to the sphere surface.

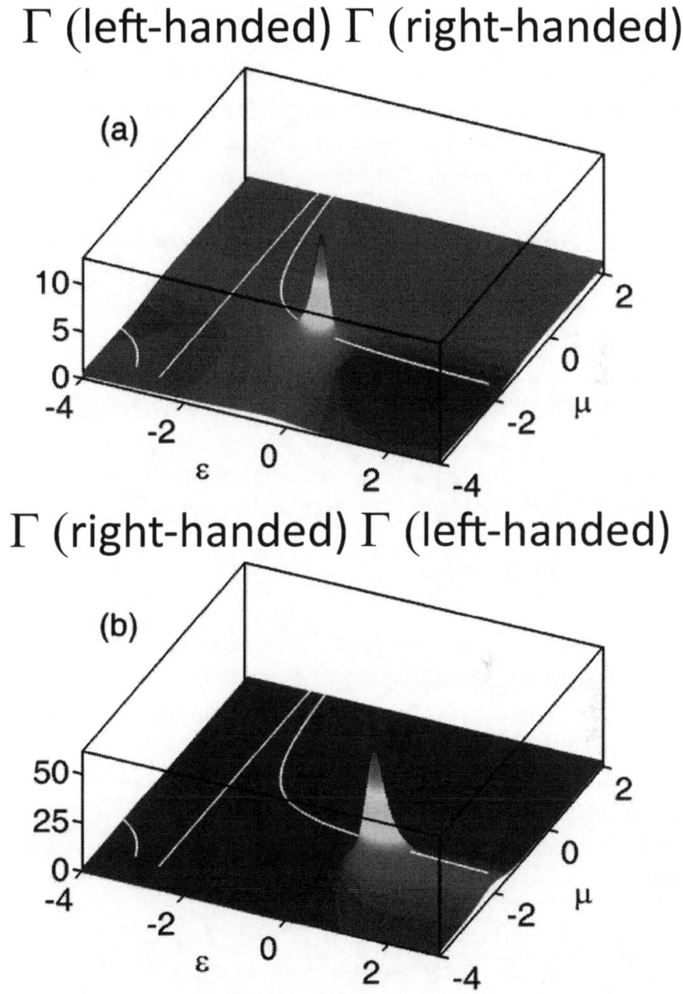

Figure 11.23 The ratio of the radiative decay rate of left-handed molecules to the radiative decay rate of right-handed (top), and vice versa (bottom) as a function of permittivity and permeability of the material of a nanoparticle. $\chi = 0.2$, $m_0/d_0 = \pm 0.1$; $\varepsilon'' = \mu'' = 0.1$. The white line shows the position of chiral-plasmon resonance of nanoparticles, Eq. (11.65) (cf. Fig. 11.21). The molecule is placed close to the sphere surface.

References

Abramowitz, M., Stegun, I., ed. (1965) *Handbook of Mathematical Functions with Formulas, Graphs, and Mathematical Tables* (Dover, New York).

Arnaut, L.R., Davis, L.E. (1993) Chiral properties of lossy n-turn helices in the quasi-stationary approximation using a transmission-line model, *23rd European Microwave Conference*, Sept. 6–10, 176–178 (in Madrid, Spain).

Bayindir, M., Aydin, K., Ozbay, E., Markos, P. Soukoulis, C.M. (2002) Transmission properties of composite metamaterials in free space, *Appl. Phys. Lett.*, **81**, 120–122.

Bohren, C.F. (1974) Light scattering by an optically active sphere, *Chem. Phys. Lett.*, **29**, 458–462.

Bohren, C.F. (1975) Scattering of electromagnetic waves by an optically active spherical shell, *Chem. Phys.*, **62**, 1566–1571.

Bohren, C.F. (1978) Scattering of electromagnetic waves by an optically active cylinder, *Colloid Interface Sci.*, **66**, 105–109.

Bohren, C.F., Huffman, D.R. (1983) *Absorption and Scattering of Light by Small Particles* (John Wiley-Interscience, New York).

Bokut', V., Serdyukov, A.N., Fedorov, F.I. (1970) On the phenomenological theory of optically active crystals, *Kristallografiya*, **15**, 1002 [*Soviet Phys. Crystallogr.*, **15**, 871 (1971)].

Bykov, V.P., Shepelev, G.V. (1986) *Atoms Emission Near Material Bodies* (Nauka, Moscow).

Cai, M., Shalaev, V. (2010) *Optical Metamaterials: Fundamentals and Applications* (Springer, New York), 200 p.

Chettiar, U.K. et al. (2007) Dual-band negative index metamaterial: Double negative at 813 nm and single negative at 772 nm. *Opt. Lett.*, **32**, 1671–1673.

Cui, T.J., Smith, D., Liu, R., eds. (2010) *Metamaterials: Theory, Design, and Applications* (Springer, New York), 368 p.

Dolling, G., Wegener, M., Linden, S. (2007) Realization of a three-functional-layer negative-index photonic metamaterial, *Opt. Lett.*, **32**, 551–553.

Dolling, G., Enkrich, C., Wegener, M., Soukoulis, C.M. Linden, S. (2006) Simultaneous negative phase and group velocity of light in a metamaterial. *Science*, **312**, 892–894.

Dolling, G., Enkrich, C., Wegener, M., Zhou, J. F. Soukoulis, C. M. (2005) Cutwire pairs and plate pairs as magnetic atoms for optical metamaterials. *Opt. Lett.*, **30**, 3198–3200.

Dolling, G., Wegener, M., Soukoulis, C.M. Linden, S. (2007) Negative-index metamaterial at 780 nm wavelength. *Opt. Lett.*, **32**, 53–55.

Enkrich, C. et al. (2005) Magnetic metamaterials at telecommunication and visible frequencies. *Phys. Rev. Lett.*, **95**, 203901.

Fedotov, V.A., Mladyonov, P.L., Prosvirnin, S.L., Rogacheva, A.V., Chen, Y., Zheludev, N.I. (2006) Asymmetric propagation of electromagnetic waves through a planar chiral structure, *Phys. Rev. Lett.*, **97**, 167401.

Feynman R.P., Leighton R.B., Sands M. (2005) *The Feynman Lectures on Physics including Feynman's Tips on Physics: The Definitive and Extended Edition*, 2nd ed., Vol. 1 (Addison-Wesley), Chap. 33, §5.

Gansel, J.K., Thiel, M., Rill, M. S., Decker, M., Bade, K., Saile, V., Freymann, G. von, Linden, S., Wegener, M. (2009) Gold helix photonic metamaterial as broadband circular polarizer, *Science*, **18**(September), 1513–1515.

Garcia-Meca, C. et al. (2011) Low-loss multilayered metamaterial exhibiting a negative index of visible of refraction at visible wavelengths. *Phys. Rev. Lett.*, **106**, 067402.

Garcia-Meca, C., Ortuno, R., Rodriguez-Fortuno, F.J., Marti, J., Martinez, A. (2009) Double-negative polarization-independent fishnet metamaterial in the visible spectrum, *Opt. Lett.*, **34**, 1603–1605 .

Greegor, R.B., Parazzoli, C.G., Li, K. Tanielian, M.H. (2003) Origin of dissipative losses in negative index of refraction materials, *Appl. Phys. Lett.*, **82**, 2356–2358.

Grigorenko, A.N., Geim, A.K., Gleeson, H.F., Zhang, Y., Firsov, A.A., Khrushchev, I.Y., Petrovic, J. (2005) Nanofabricated media with negative permeability at visible frequencies, *Nature*, **438**, 335–338.

Guzatov, D.V., Klimov, V.V. (2012) The influence of chiral spherical particles on the radiation of optically active molecules, *New J. Phys.*, **14**, 123009.

Guzatov, D.V., Klimov, V.V. (2005) Radiative decay engineering by triaxial nanoellipsoid, *Chem. Phys. Lett.*, **412**, 341–346.

Jaggard, D.L., Michelson, A.R., Papas, C.H. (1979) On electromagnetic waves in chiral media, *Appl. Phys.*, **18**, 211–216.

Jakšić, Z., Dalarsson, N., Maksimović, M. (2006) Negative refractive index metamaterials: principles and applications, *Microwave Rev.*, **June**, 36–49.

John, S. (1987) Strong localization of photons in certain disordered dielectric superlattices, *Phys. Rev. Lett.*, **58**, 2486–2489.

Katsarakis, N. et al. (2005) Magnetic response of split-ring resonators in the far infrared frequency regime, *Opt. Lett.*, **30**, 1348–1350.

Kelvin, Lord (1904) *Baltimore Lectures on Molecular Dynamics and the Wave Theory of Light* (C.J. Clay and Sons, London).

Kildishev, A.V., Drachev, V.P., Chettiar, U.K., Shalaev, V.M., Werner, D.H., Kwon, D.-H. (2007) Comment on negative index in artificial metamaterials, *Opt. Lett.*, **32**, 1510–1511.

Kittel, C. (1996) *Introduction to Solid State Physics* (John Wiley & Sons, New York).

Klein, M.W., Enkrich, C., Wegener, M., Soukoulis, C.M., Linden, S. (2006) Single-slit split-ring resonators at optical frequencies: limits of size scaling, *Opt. Lett.*, **31**, 1259–1261.

Klimov, V.V. (2002) Spontaneous emission of an excited atom placed near a "left-handed" sphere, *Opt. Commun.*, **211**, 183–196.

Klimov, V.V. (2009a) Novel approach to a perfect lens, *JETP Lett.*, **89**, 270–273.

Klimov, V.V. (2009b) Energy flows in a "perfect" lens, http://demonstrations.wolfram.com/EnergyFlowInANegativeIndexMaterial

Klimov, V.V., Baudon, J., Ducloy, M. (2011) Comparative focusing of Maxwell and Dirac fields by negative refraction half-space, *Europhys. Lett.*, **94**, 20006.

Klimov, V.V., Guzatov, D.V., Ducloy, M. (2011) Spontaneous emission of chiral molecule placed near chiral sphere (*PRL* to be published).

Lamb, H. (1904) On group-velocity, *Proc. London Math. Soc.*, **1**, 473–479.

Li, L.-W., You, D., Leong, M.-S., Yeo, T.-S., Kong, J.A. (2000) Electromagnetic scattering by multilayered chiral-media structures: a scattering-to-radiation transform, *Prog. Electromagn. Res.*, **26**, 249–291.

Linden, S. et al. (2004) Magnetic response of metamaterials at 100 THz, *Science*, **306**, 1351–1353.

Lindell, I.V., Sihvola, A.H., Tretyakov, S.A., Viitanen, A.J. (1994) *Electromagnetic Waves in Chiral and Bi-Isotropic Media* (Artech House, London).

Liu, N., Guo, H., Fu, L., Kaiser, S., Schweizer, H., Giessen, H. (2008) Three-dimensional photonic metamaterials at optical frequencies, *Nat. Mater.*, **7**, 31.

Liu, N., Liu, H., Zhu, S., Giessen, H. (2009) Stereometamaterials, *Nat. Photonics*, **3**, 157.

Moser, H.O., Casse, B.D.F., Wilhelmi, O., Shaw, B.T. (2005) Terahertz response of a microfabricated rod-split-ring-resonator electromagnetic metamaterial, *Phys. Rev. Lett.*, **94**, 063901.

Novotny, L., Hecht, B. (2006) *Principles of Nano-Optics* (Cambridge University Press).

Pendry, J.B. (2000) Negative refraction makes a perfect lens, *Phys. Rev. Lett.*, **85**, 3966.

Pendry, J.B., Holden, A.J., Robbins, D.J., Stewart, W.J. (1998) Low frequency plasmons in thin-wire structures, *J. Phys.: Condens. Matter*, **10**(22), 4785–4809.

Pendry, J.B., Holden, A.J., Robbins, D.J., Stewart, W.J. (1999) Magnetism from conductors and enhanced nonlinear phenomena, *IEEE Trans. Microwave Theory Tech.*, **47** (11), 2075–2084.

Pendry, J.B., Holden, A.J., Stewart, W.J., Youngs, I. (1996) Extremely low frequency plasmons in metallic mesostructures, *Phys. Rev. Lett.*, **76** (25), 4773–4776.

Pocklington, H.C. (1905) Growth of a wave-group when the group velocity is negative, *Nature*, **71**, 607–608.

Sarychev, A.K., Shvets, G., Shalaev, V.M. (2006) Magnetic plasmon resonance, *Phys. Rev. E*, **73**, 036609.

Schellman, J.A. (1975) Circular dichroism and optical rotation, *Chem. Rev.*, **75**(3), 323.

Schuster (1904) *An Introduction to the Theory of Optics* (Edward Arnold, London).

Shalaev, V.M. et al. (2005) Negative index of refraction in optical metamaterials, *Opt. Lett.* **30**, 3356–3358.

Shelby, R.A., Smith, D.R., Schultz, S. (2001) Experimental verification of a negative index of refraction, *Science*, **292**, 77–79.

Smith, D.R., Padilla, W.J., Vier, D.C., Nemat-Nasser, S.C., Schultz, S. (2000) Composite medium with simultaneously negative permeability and permittivity, *Phys. Rev. Lett.*, **84**, 4184–4187.

Smith, D.R., Pendry, J.B., Wiltshire, M.C.K. (2004) Metamaterials and negative refractive index, *Science*, **305**, 788–792.

Soukoulis, C.M. (Ed.) (2001) *Photonic Crystals and Light Localization in the 21st Century* (Kluwer Academic, Dordrecht, the Netherlands).

Soukoulis, C.M., Linden, S., Wegener, M. (2007) Negative refractive index at optical wavelengths, *Science*, **315**, 47.

Stratton, J.A. (1941) *Electromagnetic Theory* (McGraw-Hill, New York).

Tran Minh Hien, Ho Trung Dung (2011) Image of an emitting dipole by a superlens, *Phys. Rev. A* (accepted Friday, Dec. 23, 2011).

Valentine, J., Zhang, S., Zentgraf, T., Ulin-Avila, E., Genov, D.A., Bartal, G., Zhang, X. (2008) Three-dimensional optical metamaterial with a negative refractive index, *Nature*, **455**, 376–379.

Veselago, V.G. (1968) The electrodynamics of substances with simultaneously negative ε and μ, *Soviet Phys. Uspekhi*, **10**(4), 509–514.

Vinogradov, A.P. (2001) *Electrodynamics of Composite Materials* (URSS, Moscow).

Xiao, S.M., Chettiar, U.K., Kildishev, A.V., Drachev, V.P., Shalaev, V.M. (2009) Yellow-light negative-index metamaterials, *Opt. Lett.*, **34**, 3478–3480.

Yablonovitch, E. (1987) Inhibited spontaneous emission in solid-state physics and electronics, *Phys. Rev. Lett.*, **58**, 2059–2062.

Yablonovitch, E., Gmitter, T.J., Leung, K.M. (1991) Photonic band-structure: the face-centered-cubic case employing nonspherical atoms, *Phys. Rev. Lett.*, **67**, 2295–2298.

Yen, T.J., Padilla, W.J., Fang, N., Vier, D.C., Smith, D.R., Pendry, J.B., Basov, D.N., Zhang, X. (2004) Terahertz magnetic response from artificial materials, *Science*, **303**, 1494–1496.

Zhang, S. et al. (2005a) Experimental demonstration of near-infrared negative-index metamaterials, *Phys. Rev. Lett.*, **95**, 137404.

Zhang, S. et al. (2005b) Mid-infrared resonant magnetic nanostructures exhibiting a negative permeability, *Phys. Rev. Lett.*, **94**, 037402.

Zhou, J., Koschny, Th., Kafesaki, M., Economou, E.N., Pendry, J.B., Soukoulis, C.M. (2005) Saturation of the magnetic response of split-ring resonators at optical frequencies, *Phys. Rev. Lett.*, **95**, 223902.

Zouhdi, S., Foumer-Lamer, A., Mariotte, F. (1992) On the relationships between constitutive parameters of chiral materials and dimensions of chiral objects (helices), *J. Phys. III (France)*, **2**, 337–342.

Chapter 12

Optical Properties of Nanoholes in Metal Films

Optical scanning microscopes of different types are one of the most important elements of nanoplasmonics, nano-optics, and nanotechnology in general. Scanning near-field optical microscopes (SNOMs) hold a significant place among these microscopes (nanoscopes) (Pohl and Courjon, 1992; Dunn, 1999). Recently, optical scanning microscopes were widely used even for the investigation of single molecules (Betzig and Chichester, 1993; Trautman et al., 1994; Xie and Dunn, 1994; Ambrose et al., 1994; Bian et al., 1995; Veerman et al., 1999; Gersen et al., 2000, 2001; Hosaka and Saiki, 2001; Butter and Hecht, 2006). Interpretation of these and other experiments (Shubeita et al., 2002, 2003; Aigouy et al., 2003; Wenger et al., 2005; Rigneault et al., 2005), in which a nanohole plays a crucial role, is related to necessity of understanding of properties of optical fields near nanoholes. On the other hand, in a case of an array of nanoholes in a metal film, sometimes anomalous light transmission through these holes arises (Genet and Ebbesen, 2007; Ebbesen et al., 1998; Grupp et al., 1999; Degiron et al., 2004; Degiron and Ebbesen, 2005; Wannemacher, 2001). Balykin et al., (2003, 2005) proposeduse unique optical properties of optical fields in nanohole

Nanoplasmonics
Vasily Klimov

ISBN 978-981-4267-16-8 (Hardcover), 978-981-4267-42-7 (eBook)
www.panstanford.com

arrays for control of atoms' and molecules' motion with nanoscale accuracy.

These and other properties of nanoholes make their investigation and description very significant, and in this chapter these and other important properties of one and many nanoholes will be analyzed. In Section 12.1, optical properties of a single nanohole in an infinitely thin, perfectly conducting screen will be described. In Section 12.2, the influence of localized surface plasmons on light transmission through a single nanohole in a metal film of finite thickness will be considered. In Section 12.3, extraordinary light transmission through nanohole arrays will be examined. In Section 12.4, highly directional emission from a nanostructured hole will be studied. Finally, in Section 12.5 nanoholes' influence on molecules' emission will be considered.

12.1 Optical Properties of a Circular Hole in an Infinitely Thin, Perfectly Conducting Screen (Bethe–Bouwkamp Theory)

A precise formal solution of the problem of a plane wave diffraction on a circular hole in an infinitely thin, perfectly conducting screen was obtained in the 1950s by a range of authors (Meixner and Andrejewski, 1950; Andrejewski, 1953; Nomura and Katsura, 1985). These solutions were expressed in series of spheroidal functions or hypergeometric polynomials, poorly converging in the short-wave limit. In the case of the long-wave limit, expressions for the wave zone field, accurate to the terms $(ka)^6$, were obtained by Bethe (1944), Bouwkamp (1950). Near-zone fields have a very complex structure. Their spatial distributions were found in (Bouwkamp, 1950) within the spheroidal coordinate system. Simpler analytical expressions for the near fields in the vicinity of a small hole in the Cartesian coordinates were obtained by Klimov and Letokhov (1994). The geometry of the problem is shown in Fig. 12.1.

For the *normal* incidence of an electromagnetic wave on a hole,

$$\mathbf{E}(\mathbf{r}, t) = \mathbf{E}_0 e^{i(kz-\omega t)}, \tag{12.1}$$

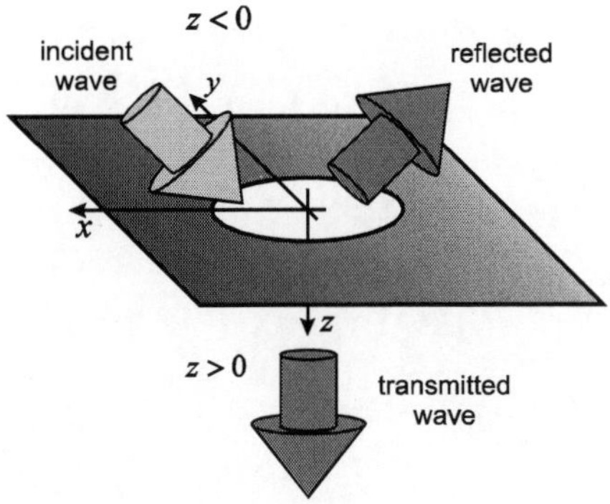

Figure 12.1 The geometry of the problem of a plane wave diffraction on a circular aperture in a perfectly conducting screen.

in the first approximation of long-wave perturbation theory, the expressions for the scattered electric fields behind the screen can be presented in the form ($z > 0$)

$$\mathbf{E}_{\tan} = \frac{ika}{3\pi}\left[[\mathbf{n}_z\mathbf{H}_0]A + \frac{(\mathbf{H}_0\mathbf{r})[\mathbf{n}_z\mathbf{r}]}{r^2}B\right], \tag{12.2}$$

$$E_z = -\frac{ika}{3\pi r}[\mathbf{H}_0\mathbf{r}]_z C, \tag{12.3}$$

where $\mathbf{H}_0$ is the magnetic field in the screen plane in the case of the hole's absence, $\mathbf{n}_z$ is the unit vector along the z axis, and E_z and $\mathbf{E}_{\tan}$ are components of the electric field along the z axis and in the screen plane, correspondingly.

In Eqs. (12.2) and (12.3), the following designations are also used:

$$A = R^-\left(\frac{2a^2}{R^*} + 2 - \frac{z^2}{r^2}\right) + za\left(\frac{R^+}{r^2} - \frac{3}{a^2}arctg\left(\frac{1}{R^+}\right)\right), \tag{12.4}$$

$$B = R^-\left(\frac{2z^2}{r^2} - \frac{2r^2 - z^2}{R^*}\right) + zR^+a\left(\frac{1}{R^*} - \frac{2}{r^2}\right) + \frac{3zr^2R^+}{aR^*\left(1 + R^{+2}\right)}, \tag{12.5}$$

$$C = \frac{2arR^+}{R^*\left(1 + R^{+2}\right)}, \tag{12.6}$$

$$R^* = ((R^2 - a^2)^2 + 4a^2z^2)^{1/2};\ R^{\pm} = \left(\frac{R^* \pm (R^2 - a^2)}{2a^2}\right)^{1/2}. \tag{12.7}$$

$$r^2 = x^2 + y^2;\ R^2 = r^2 + z^2$$

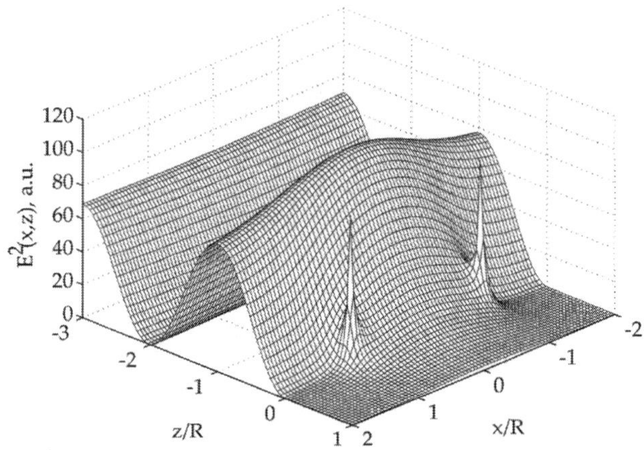

Figure 12.2 The distribution of the electric filed intensity near the aperture in the case of a circularly polarized, normally incident wave. The wave is incident from the left, where the standing wave is seen, with the maximum near the aperture. The sharp maxima arise at the aperture sharp rim.

In Fig. 12.2, the distribution of the electromagnetic field intensity in the vicinity of a hole in the case of circular polarization of a normally incident wave is shown. This picture illustrates a rather complex structure of the electric field: there are fields' maxima near the aperture rim, which disappear as moving away from it. The field's maximum appearing in the antinode of the standing wave (in front of the hole) is another interesting feature of the field distribution (Klimov and Letokhov, 1995). This field structure influences substantially both fluorescence of the molecule situated near the hole and atoms and molecules motion (atom traps and nanopincers).

At long distances from the screen ($a \ll R \ll \lambda$), the electric field, Eqs. (12.2) and (12.3), is determined by the formula

$$\mathbf{E} = \frac{2a^3 ik}{3\pi R^3}[\mathbf{RH}_0], \tag{12.8}$$

and is analogous to the magnetic dipole field with the moment

$$\mathbf{M} = -\frac{2a^3}{3\pi}\mathbf{H}_0. \tag{12.9}$$

In the far field (in the radiation zone, $R \gg \lambda$), electric and magnetic fields are also equal to magnetic dipole fields, Eq. (12.9).

Knowing far-zone fields, the total power of radiation transmission through the hole can be calculated. The ratio between this power and the power incident on the hole determines the hole's transmission coefficient T:

$$T = \frac{R^2 \int I_{sc}(\Omega) d\Omega}{S I_0}, \tag{12.10}$$

where $I_{sc}(\Omega)$ is the flow of the scattered energy in the direction Ω, I_0 is the flow of the energy incident on the film, $S = \pi a^2$ is the geometric area of the hole.

In the case of normal incidence of a plane wave on the hole, the transmission coefficient was calculated in (Bethe, 1944; Bouwkamp, 1950) and can be presented in the form

$$T = \frac{64}{27\pi^2}(ka)^4 \propto \left(\frac{a}{\lambda}\right)^4. \tag{12.11}$$

The factor $(a/\lambda)^4$ results in a very substantial weakening of the light transmission through the hole. Let us note that an analogous factor appears in the Rayleigh theory of light scattering by small particles (see Chapter 6, Eq. (6.48)). Let us emphasize that it is true for normal incidence of radiation of any polarization. For oblique incidence, the transmitting field in the far zone is a superposition of electric and magnetic dipoles fields. In this case, the transmission coefficient becomes dependent on the incident radiation polarization (Bethe, 1944).

12.2 A Circular Hole in a Screen of Finite Thickness

The Bethe–Bouwkamp theory of diffraction on a hole is based on two assumptions. The thickness of the conducting screen is considered to be infinitely small, and its conductivity is considered to be infinitely high providing full opacity of the screen. Consideration of screens of finite thickness is possible in the context of semianalytical and numerical methods. At that, two cases should be considered depending on whether propagating waves can exist in the hole. Bethe–Bouwkamp theory is more or less applicable in case when there are no propagating waves inside the hole. In the case of a circular hole of diameter d in a perfectly conducting screen, this condition is fulfilled for $d \leq 0.3\lambda$. In this case, the transmission coefficient decreases exponentially as the screen thickness h increases (Roberts, 1987, 1989). This behavior is typical for tunneling processes. These conclusions are not applicable to cases when propagating waves (of different origin) can exist in holes; here the transmission coefficient can be much higher. Known examples of waveguiding holes of this kind are holes of a radius greater than the cutoff radius (de Abajo, 2002), a one-dimensional slit where transverse electric magnetic (TEM) modes exist at any wavelength (Vainstein, 1988), ring-shaped holes (Baida and van Labeke, 2002), and C-shaped holes (Shi et al., 2003).

When considering optical properties of the hole, the finite conductivity of the film should be taken into account together with the finite screen thickness. For thin films, the screen is not perfectly opaque, and Bethe–Bouwkamp theory is again inapplicable. On the other hand, optically thick films become nontransparent when their thickness h is several times greater than the skin depth. In this case, localized plasmons can exist, resulting in a substantial increase of the transmission coefficient (Degiron et al., 2004).

12.2.1 *Localized Plasmons in a Nanohole*

Certainly, optical properties of a hole in a real metal screen are of the greatest interest. The special interest is to obtain a description of plasmon modes in this geometry since light transmission through a nanohole can be increased substantially with the help of plasmons. In this case, the problem is complicated in comparison with the case considered by Bethe and Bouwkamp. However, in the case of a small hole in a sufficiently thin metal screen, the quasi-static approximation (ε-method, Chapter 5) can be used, and all plasmon modes can be found. For this purpose, a screen of the shape of a hyperboloid of rotation as the model of a screen with a circular hole was considered (Becker et al., 1981; Klyuchnik et al., 2003). This geometry is shown in Fig. 12.3.

To find plasmon modes in this geometry, it is natural to use the system of coordinates $0 < \xi < \infty, -1 < \eta < 1, 0 < \varphi < 2\pi$ of an oblate ellipsoid of rotation (Smythe, 1969). Coordinates of

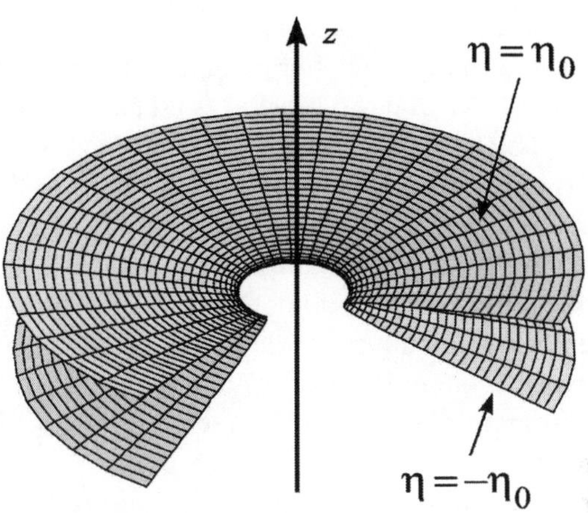

Figure 12.3 The geometry of a screen in the shape of a hyperboloid of rotation of one sheet. The upper and lower surfaces are described by the condition $\eta = \pm\eta_0$ in the system of coordinates of an oblate spheroid (see also Fig. 7.3).

this system are expressed through the Cartesian coordinates by the relations

$$z = f\xi\eta; \quad x = f\cos\varphi\sqrt{(\xi^2+1)(1-\eta^2)};$$
$$y = f\sin\varphi\sqrt{(\xi^2+1)(1-\eta^2)}, \tag{12.12}$$

where f is the focal distance of the ellipse. The radius a of the hole is determined by the relation

$$a = f\sqrt{1-\eta_0^2}, \tag{12.13}$$

where the parameter $\eta = \pm\eta_0$ characterizes the hyperbolic screen surface with the hole. A small η_0 corresponds to a thin screen, $\eta \approx 1$ corresponds to a thick screen.

In this system of coordinates, an arbitrary solution of the Laplace equation satisfying the boundary conditions on the hyperboloid surface may be written in the form of the superposition:

$$\Phi(\xi,\eta,\phi) = \sum_m \Phi^m(\xi,\eta)\exp(im\phi)$$
$$\Phi^m(\xi,\eta) = \oint d\nu A^m(\nu)\,\Phi^m_\pm(\eta,\nu)\left[P^m_\nu(i\xi) \pm P^m_\nu(-i\xi)\right]$$
$$\Phi^m_\pm(\eta,\nu) = \begin{cases} \left[P^m_\nu(\eta_0) \pm P^m_\nu(-\eta_0)\right] P^m_\nu(\eta),\ 1 > \eta > \eta_0 \\ P^m_\nu(\eta_0)\left[P^m_\nu(\eta) \pm P^m_\nu(-\eta)\right],\ \eta_0 > \eta > -\eta_0 \\ \pm\left[P^m_\nu(\eta_0) \pm P^m_\nu(-\eta_0)\right] P^m_\nu(-\eta),\ -\eta_0 > \eta > -1 \end{cases}, \tag{12.14}$$

where $A^m(\nu)$ is the amplitude of the plasmon mode, $P_\nu^m(x)$ is the associated Legendre functions hereafter, the upper index m will be omitted for simplicity, and the integral is taken over the contour in the complex plane ν, which provides necessary properties of the solution. It is convenient to take this contour in the form of $\nu = -1/2 + it$, $0 < t < \infty$, where the dimensionless parameter t is analogous to the radial wavenumber. The sign "$\pm$" corresponds to the evenness of the plasmon potential relative to the reflection in the *xy* plane: the upper and lower signs correspond to the symmetric and antisymmetric potentials, correspondingly.

The requirement of the potential and electric induction normal component continuity results in existance of the electric permittivity value providing the system's plasmon resonance for any $\nu = -1/2 + it, 0 < t < \infty$:

$$\varepsilon_{\text{plasmon}}(\omega) = \frac{P_\nu'(\eta_0)\,[P_\nu(\eta_0) \pm P_\nu(-\eta_0)]}{P_\nu(\eta_0)\left[P_\nu'(\eta_0) \mp P_\nu'(-\eta_0)\right]},\ \nu = -\frac{1}{2} + it. \quad (12.15)$$

The modes, Eq. (12.14), correspond to these resonant values of permittivity.

In Fig. 12.4, dispersion curves of symmetric and antisymmetric plasmon modes relative to the *xy* plane modes are shown for a screen of different thickness.

These plots were made under the assumption that the material of the screen follows the Drude dispersion law:

$$\varepsilon(\omega) = 1 - \frac{\omega_{\text{pl}}^2}{\omega^2}. \quad (12.16)$$

It is seen from this picture that for small values of the radial moment t, resonant plasmon frequencies occupy a band from 0 to ω_{pl}, while for large values of t values of all plasmon frequencies tend to the surface plasmon frequency ($\omega/\omega_{\text{pl}} = \sqrt{2}/2$).

In Fig. 12.4, dispersion curves of symmetric mode potential ($m = 0$) (a) and that of the symmetric mode ($m = 1$) (b) is shown for $t = 2$ and $\eta_0 = 0.05$. It is seen from these pictures that the electric field (which is normal to the equipotential lines) of the antisymmetric plasmon is mainly directed perpendicular to the surface (a), while the electric field of the symmetric plasmon is mainly directed along the screen (b). Apparently, it is symmetric plasmons that will play

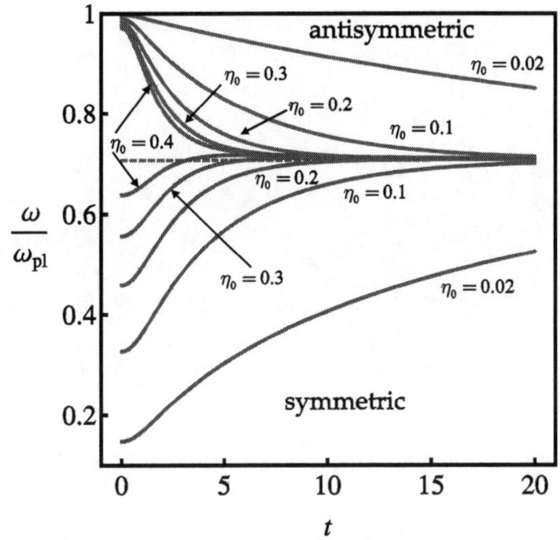

Figure 12.4 The dispersion curves $\omega/\omega_{\mathrm{pl}}$ of symmetric and antisymmetric plasmon modes with $m = 0$ for $\eta_0 = 0.02, 0.1, 0.2, 0.3$, and 0.4.

the main role by transmission of normally incident light, in which the electric field is parallel to the screen.

To investigate the process of light transmission through such a hole, one should know the value of the electric dipole moment of the hole's plasmon mode. In the case of the Bethe–Bouwkamp model, the hole can be characterized by the effective magnetic dipole moment, Eq. (12.9), while the electric dipole moment is equal to zero. For a screen of finite thickness, a nonzero dipole moment arises, and its value is described by the following expression:

$$\mathbf{d} = \int d^2 S \sigma(\mathbf{r}(S))\mathbf{r}(S), \tag{12.17}$$

where $\sigma(\mathbf{r}(S)) = \frac{\varepsilon_n - 1}{4\pi} \frac{\partial \Phi_n}{h_\eta \partial \eta}\Big|_{\eta=\eta_0}$ is the surface charges density, and the integral is taken over the screen surface. Here, $h_\eta = f(\xi^2 + \eta^2)^{1/2}/(1-\eta^2)^{1/2}$ is the Lame coefficient for the η coordinate.

Substituting Eq. (12.14) for the plasmon modes in the interval $-\eta_0 < \eta < \eta_0$ in Eq. (12.17), one can find the value of the dipole moments of different modes. In particular, for the bright ($m = 0$)

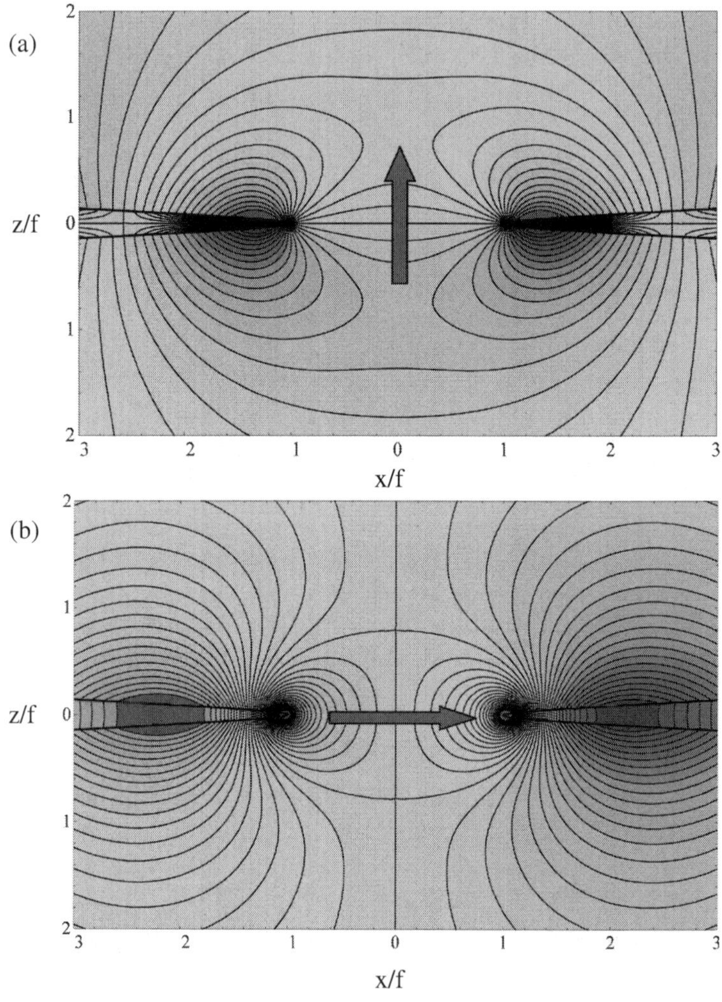

Figure 12.5 Equipotential lines of the antisymmetric mode ($m = 0$) (a) and the symmetric mode ($m = 1$) (b) for $t = 2$ and $\eta_0 = 0.05$. The red arrows show orientation of the dipole moment.

antisymmetric mode one can obtain the following estimation (Klyuchnik et al., 2003):

$$|\mathbf{d}| \sim \eta_0 \frac{(t^2 + 1/4)}{|\Gamma(5/4 + it)|^2} (\rho^*)^{3/2}. \tag{12.18}$$

Estimation of the dipole moment for the bright ($m = 1$) symmetric mode gives the analogous expression:

$$|\mathbf{d}| \sim \eta_0 \frac{(t^2 + 1/4)}{|\Gamma(5/4 + it)|^2} (\rho^*)^{3/2}. \qquad (12.19)$$

In Eqs. (12.18) and (12.19), $\rho^* = \lambda/a$ is the cutoff parameter in the quasi-static problem, which should be introduced due to a weak plasmon potential decreasing during integrating over the hole area (see below).

At that, the dipole moment of the antisymmetric mode ($m = 0$) is directed perpendicular to the screen surface, and the dipole moment of the antisymmetric mode with $m = 1$ is parallel to it.

Certainly, the quasi-static solution, Eqs. (12.14) and (12.15), has a rather restricted domain of applicability since at some distance from the axis effects related to retardation and propagating plasmon modes are to be considered necessarily. Nevertheless, the description of the fields' distribution near the hole with the help of Eq. (12.14) seems to be correct. Moreover, the results of Eqs. (12.18) and (12.19) divergent in the quasi-static approximation $\rho^* = \lambda/a \to \infty$ testify that the dipole moment is great and can provide effective light transmission from one side of the surface to another.

The numerical modelling (Popov et al., 2005; Bonod et al., 2005) has also confirmed possibility of excitation of plasmon resonances of this kind in the geometry analogous to the geometry of Fig. 12.3.

12.2.2 *Localized Surface Plasmons and Light Transmission through a Hole*

As it was noted above during the discussion of the applicability region of Bethe–Bouwkamp theory, the finite conductivity of a real metal, resulting in the origin of plasmons both localized near the hole and propagating along the surface, should be taken into account, even for an optically thick (and nontransparent) metal film. As the theory presented in the previous section implies, there is no sharp boundary between these plasmons ($t \to 0$ corresponds to localized plasmons, while $t \to \infty$ corresponds to a high frequency surface wave, that is, to surface plasmons with $\omega = \omega_{\mathrm{pl}}/\sqrt{2}$). In particular, from Fig. 12.5 it is seen that the surface plasmons are formed in the region of radius $r > 1.5a$, where a is the geometrical radius of the hole.

The localized plasmons' excitation in a nanohole in a metal film made of a real metal can influence the transmission coefficient $T(\lambda)$ in two ways. First of all, fields penetration in metal results in an effective increase of the hole's radius (see Fig. 12.5). In turn, this results in a substantial increase of the cutoff wavelength that determines the absence of waveguide modes in the hole in comparison with the cutoff length calculated on the basis of geometrical parameters.

Analytical and numerical investigations have shown (see Fig. 12.5) that existence of plasmons results in an increase of the cutoff wavelength λ_{max} by 40% (Gordon and Brolo, 2005), which is extremely important for consideration of holes of a diameter slightly smaller than the cutoff wavelength calculated on the basis of geometrical parameters of the hole. Moreover, the theoretical study of the problem of light transmission through a circular hole in a metal film with the dielectric function described by the Drude law, Eq. (12.16), has shown that propagating modes exist at frequencies lower than the plasmon frequency for any sizes of the hole (Shin et al., 2005; Webb and Li, 2006). The influence of these modes on the transmission of the hole has not been verified experimentally yet.

The second important consequence of the plasmon nature of light transmission through nanoholes is the transmission coefficient resonance's dependence on the sizes and geometry of the hole and on the wavelength of the incident radiation. An increase of the transmission coefficient is related to a substantial resonance enhancement of the electric field in the region of plasmon resonance.

Only in recent times, owing to achievements of nanotechnologies, a single nanohole has been formed successfully in a freely suspended metal film, that is, in a film having no dielectric substrate, and its plasmon properties have been investigated (Fig. 12.6, (Degiron et al., 2004)). In the case of perceptible light transmission by the hole, the maximum was found in the spectrum for a relatively thin, but opaque, film at its width of 200 nm (Fig. 12.6b). This maximum can be explained naturally with the help of a localized

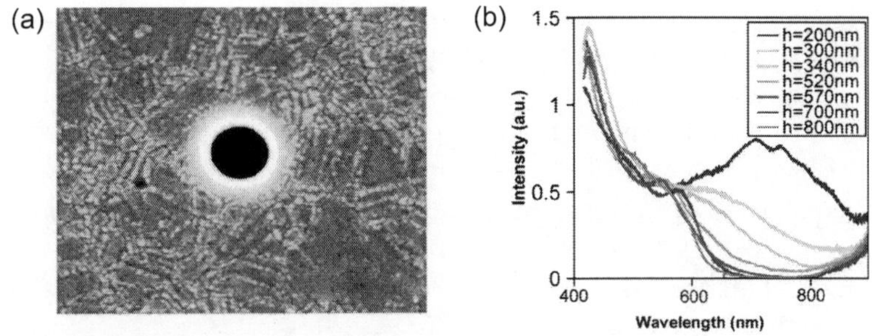

Figure 12.6 Light transmission through a circular aperture milled in an optically thick suspended silver film. (a) The SEM image of a single aperture in a silver film. (b) Spectra of transmission through an aperture of the diameter of 270 nm for different film widths. The transmission coefficient maximum at the wavelength of 700 nm for the film width of about 200 nm is determined by localized surface plasmons (Degiron et al., 2004a). *Abbreviation*: SEM, scanning electron microscopy.

plasmon. Moreover, in these experiments spatial structure and spectral peculiarities of such a plasmon were investigated with the help of excitation of this plasmon by an electron beam of high energy.

In Fig. 12.7a, spatial distribution of optical emission excited by an electron beam is shown. This picture was obtained by measurements of emission components of different polarizations. From this figure, it is seen that the emission maximum is always

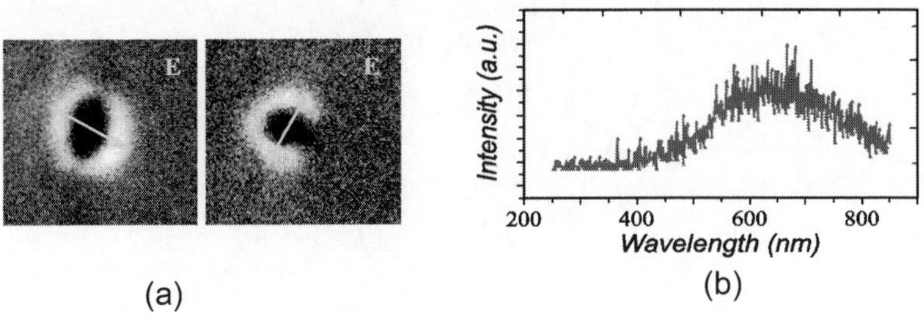

Figure 12.7 Emission of an aperture excited by an electron beam. (a) The image of cathodoluminiscence by observation of differently polarized emission components. (b) A corresponding emission spectrum (Degiron et al., 2004a).

situated at the hole points where its rim is perpendicular to the electric field vector. This fact agrees completely with structure of plasmon modes with $m = 1$, which were considered in Section 12.2.1. The emission spectrum shown in Fig. 12.7a is presented in Fig. 12.7b. This spectrum is in good agreement with the transmission coefficient's, $T(\lambda)$, dependence on the wavelength shown in Fig. 12.6. This circumstance gives arguments for the plasmon nature of the effect of extraordinary light transmission through a single hole.

12.3 Extraordinary Light Transmission through Arrays of Nanoholes

The coefficient of light transmission through a nanohole can be increased substantially when an array of periodically situated nanoholes is considered instead of a single hole (see Fig. 12.8). In Chapter 4, periodical lattices on a surface have been already considered as a method providing an effective interaction between light and surface plasmons. Likewise, in a periodical array of nanoholes surface plasmons can be excited, resulting in an increase of the field on the film surface. After the enhanced field has accomplished tunnel transmission through the holes, the inverse

Figure 12.8 SEM image of nanoapertures in a gold film on a silica substrate. The film width is 300 nm, the apertures' diameter is 200 nm, and the period of the square array is 600 nm (Krishnan et al., 2001).

transformation of surface plasmons into radiation fields occurs. As in the case of a single hole in a smooth surface, light transmission through an array of periodically situated holes takes place by means of the tunnel effect, resulting in an exponential decrease of the transmission coefficient with an increase of the film width. However, if the film width is of the order of the skin depth, the interaction between surface plasmons on the front and rear sides of the film becomes possible. This interaction becomes most effective (resonance) when environments of the film have the same dielectric properties on both sides.

For the given geometry of an array, excitation of surface plasmon waves is possible only for certain wavelengths of the incident radiation, and as a result the spectrum of the transmission $T(\lambda_0)$ has a well-defined structure, where peaks correspond to the case of the surface plasmon excitation. It is important that at these resonant wavelengths, the transmission coefficient (which is determined as the ratio between the energy incident on the hole and the energy transmitted through it) can be even more than one. This means that the intensity of the light incident on the hole is less than the intensity of the light transmitted by it. Certainly, it is not a paradox since the light is transmitted through the hole with the help of plasmons that are strongly localized near the hole and enhance the intensity of the light incident on the hole substantially. When the ratio between the energy transmitted through all apertures (nanoholes) and the energy incident on the whole film is considered, it is always less than one. According to it, it is natural to call this effect extraordinary light transmission through an array of nanoholes (Ebbesen et al., 1998). In Fig. 12.9, the transmission spectrum of the light incident normally on a silver film of the thickness $t = 200$ nm with nanoholes of the diameter $d = 150$ nm arranged in a square array with the period $a_0 = 900$ nm is shown as a typical example of extraordinary light transmission.

There is a range of rather wide peaks in the spectrum complementary to the sharp bulk plasmon peak in the ultraviolet region observed only in very thin films. Two of these peaks are situated at wavelengths greater than the period of the square array a_0. Apparently, the origin of these peaks and the fact that the coefficient of the transmission through the hole can be more than

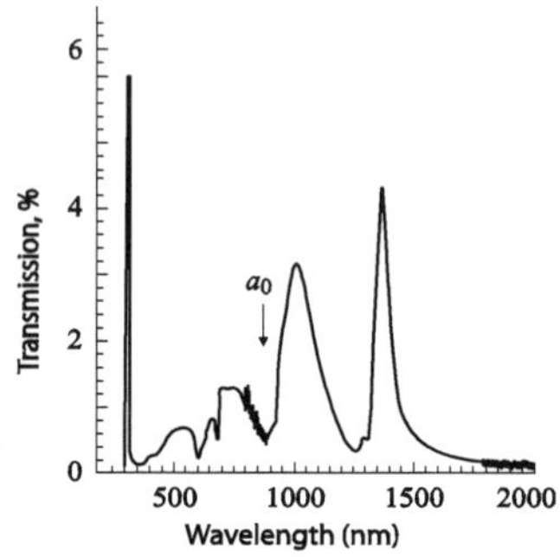

Figure 12.9 The transmission spectrum of the light incident normally on an array of nanoapertures of the diameter $d = 150$ nm and the period of the square array $a_0 = 900$ nm. The silver film thickness is 200 nm (Ebbesen et al., 1998).

1 cannot be explained without involving surface plasmons in the process of the light transmission. On the other hand, involving of the surface plasmons' conception allows one to explain the enormously high light transmission easily: the light incident on opaque screen areas is transferred to the aperture by the surface wave, and after that it can tunnel through the hole.

It should be kept in mind that precise experimental determination of the enormously high light transmission is complicated due to normalizing problems since according to Bethe–Bouwkamp theory the transmission coefficient is very sensitive to the radius of the hole ($\sim r^4$).

A study of dependence of spectral maxima position on the radiation incidence angle allows one to clarify the dispersion law of surface plasmons participating in the process of the light transmission. An example of the experimentally determined dispersion ratio is presented in Fig. 12.10. In this picture, reduced bands, that is, sections of the dispersion curve shifted by the vector of the reciprocal lattice $G = 2\pi/a_0$, are clearly seen. In Fig. 12.10, the intersection of the dispersion curves with the axis $k_x = 0$ determines energies and wavelengths at which excitation of surface plasmons takes place for normal incidence, and therefore it determines the peaks' position in Fig. 12.9.

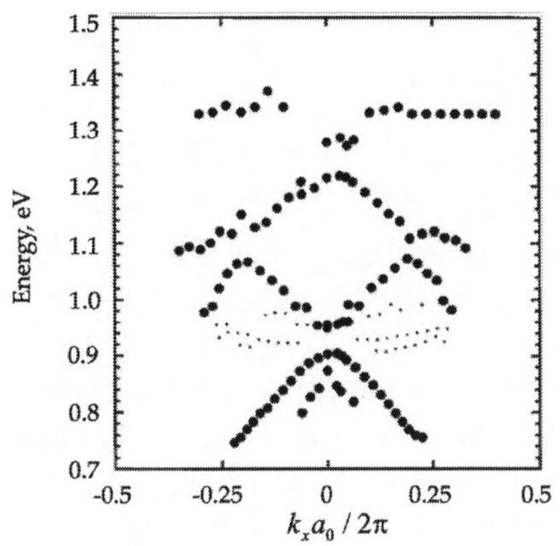

Figure 12.10 Surface plasmons' dispersion curves along the x axis, obtained from transmission spectra at different angles of light incidence (bold dots). For $k_x = 0$, resonances correspond to transmission coefficient maxima in Fig. 12.9 (Ebbesen et al., 1998).

Thus, the observed structure of the spectrum $T(\lambda)$ can be explained when it is assumed that the condition of the phase matching has the form

$$\mathbf{k}_{sp} = \mathbf{k} \pm \frac{2\pi n}{a_0}\mathbf{n}_x \pm \frac{2\pi m}{a_0}\mathbf{n}_y, \tag{12.20}$$

where $\mathbf{k}_{sp}$ is the longitudinal wavevector of the surface plasmon, $\mathbf{k}$ is the tangential component of the incident light wavevector, $\mathbf{n}_x$, $\mathbf{n}_y$ are unit vectors in the directions x and y, and n, m are integers.

Combining the spectrum of plasmons for a plane surface, Eq. (4.5):

$$\mathbf{k}_{sp} = k_0\sqrt{\frac{\varepsilon_1\varepsilon_2}{\varepsilon_1 + \varepsilon_2}}, \tag{12.21}$$

and Eq. (12.20), it is easy to show that for the normal incidence transmission spectra maxima are formed when (Ghaemi et al., 1998)

$$\lambda_{sp}(n, m) = \frac{n_{sp}a_0}{\sqrt{n^2 + m^2}}, \tag{12.22}$$

where $n_{sp} = \sqrt{\varepsilon_1\varepsilon_2/(\varepsilon_1 + \varepsilon_2)}$ is the effective refraction index for the surface plasmon, Eq. (12.21). Often, the simple expression in

Eq. (12.22) serves as a good first approximation for estimation of resonance conditions.

A more precise description of light transmission through an array of nanoapertures can be made with the help of numerical simulation (Krishnan et al., 2001). In Fig. 12.11, amplitude distributions of the electric field in the transverse cross section of an array of circular apertures on which a plane wave is incident from the left are shown. In this picture, three different cases are presented. In the case (b), there is no resonance between the light and plasmons and therefore the transmission is small (the transmission coefficient is $T = 0.005\%$). In the case (c), the light interacts resonantly with the surface plasmons on the rear side of the film (the transmission coefficient is $T = 0.5\%$), and, finally, in the symmetric case (d) the light interacts resonantly with the surface plasmons on both sides of the film (the transmission coefficient is $T = 20\%$). A comparison of Fig. 12.11a and Fig. 12.11b shows how the light interaction with surface plasmons results in electric field enhancement on the back (right) side of a film. When dielectric properties of the media adjacent to the film are equal (Fig. 12.11d), the surface plasmons are in resonance on both sides of the film, and a substantial growth of the field intensity occurs as compared to the nonsymmetric case (c). In turn, this field enhancement results in an essential increase of the transmission coefficient (up to 20%). It should be noted that the stored field energy is located mainly on the external film surfaces, that is, in the surface plasmons. Thus, an aperture may be regarded as an inverse Fabry-Perot resonator, where the energy is located on the external surface of the aperture volume, and the interaction between these surfaces is performed by means of exponentially decreasing near fields.

Since the phase matching of incident light and plasmons is of crucial importance for extraordinary transmission, an analogous effect is to take place also in other cases of phase matching of light and plasmons. In particular, anomalous transmission through a single aperture can occur when this aperture is surrounded by nanostructures that allow the phases to match. Indeed, this effect was discovered in the case of a single nanohole surrounded by a square array of pits on the surface of an opaque metal film (Grupp et al., 1999). A single aperture surrounded by concentric grooves

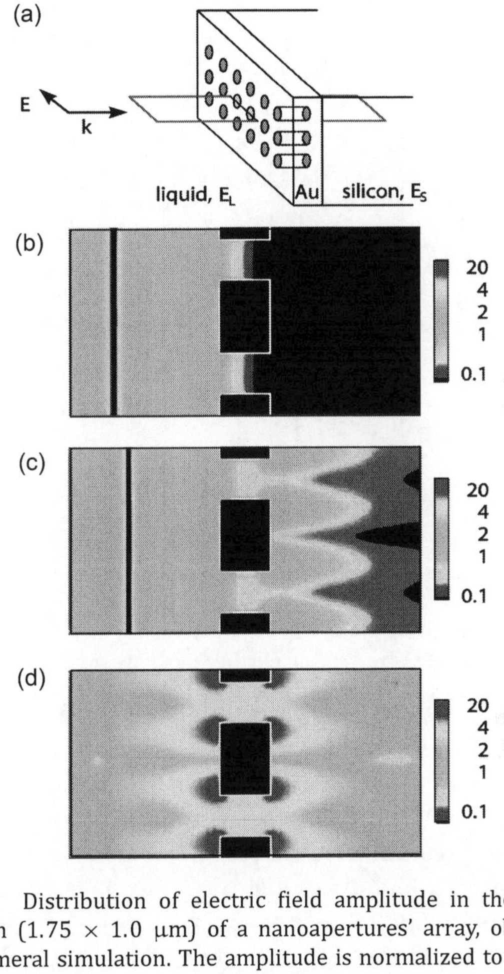

Figure 12.11 Distribution of electric field amplitude in the transverse cross section (1.75 × 1.0 μm) of a nanoapertures' array, obtained as a result of numeral simulation. The amplitude is normalized to the incident field amplitude. (a) The geometry of the array of nanoapetures and of the incident radiation. (b, c, d) The distribution of the electric field amplitudes for different conditions of the surface plasmon excitation on the array of holes: (b) a nonresonant asymmetric case ($\varepsilon_S = 2.31$, $\varepsilon_L = 1.0$, $\lambda = 1100$ nm, transmission coefficient $T = 0.005\%$); (c) the plasmon resonance on the boundary between a gold film and a silica substrate in the asymmetric environment ($\varepsilon_S = 2.31$, $\varepsilon_L = 1.0$, $\lambda = 957$ nm, transmission coefficient $T = 0.5\%$); (d) the plasmon resonance on both sides of the film for the symmetric environment ($\varepsilon_S = 2.31$, $\varepsilon_L = 2.31$, $\lambda = 966$ nm, transmission coefficient $T = 20\%$) (Krishnan et al., 2001).

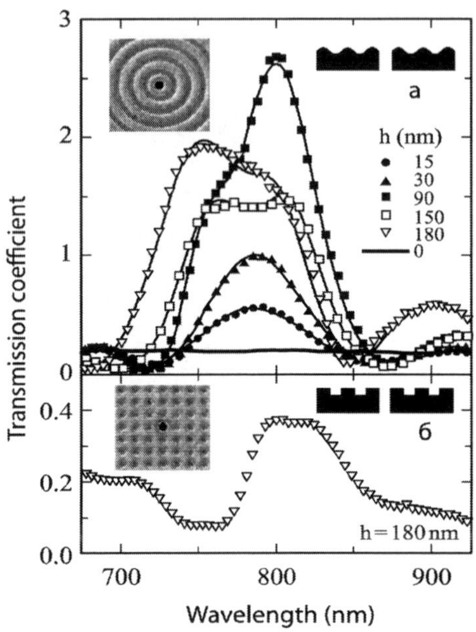

Figure 12.12 The transmission through a circular aperture of the diameter of 440 nm surrounded by (a) concentric rings with a sinusoidal profile and (b) a square array of nanopits of the height *h* on a film of Ag (100 nm)/Ni (300 nm)/Ag (30 nm) of the total thickness of 430 nm (Thio et al., 2001).

can be also used for matching the light and the surface plasmon wavevectors. In Fig. 12.12, transmission spectra of a nanohole surrounded by concentric grooves (a) or a square array of pits (b) are shown. In both cases, the transmission coefficient is more than predicted by Bethe theory, and for resonance conditions the transmission coefficient becomes more than one. It is also seen from Fig. 12.12 that the depth of the pits or grooves influences substantially the transmission coefficient since the effectiveness of the light and plasmons' interaction depends on this value.

A great number of research studies were devoted to experimental or numerical investigation of the influence of the film width (Shou et al., 2005), the aperture size (van der Molen et al., 2004), or the nanoholes' array symmetry (Wang et al., 2005) on transmission coefficient spectra. The final proof of the surface plasmons' role in anomalous light transmission by arrays of nanoapertures was presented with the help of a detailed measurement of angular and polarization characteristics of transmitted, reflected, and absorbed light (Ghaemi et al., 1998; Barnes et al., 2004).

An even more complex physical picture appears in the case when wave propagation is possible in the aperture, for example, as it happens for a primary TEM mode in a one-dimensional slit for which the cutoff wavelength is equal to infinity. In this case, the transmission coefficient can be modulated by resonances of waveguided modes owing to a change of the metal film width. Indeed, resonances of such kind were discovered experimentally for a set of parallel nanoslits (Porto et al., 1999). Furthermore, in a perfect analogy with the case discussed above of nanoholes transmission coefficient increase by means of modulation of the surrounding surface height, grooves situated periodically around the slit and parallel to it provoke a substantial increase of the coefficient of the single slit transmission; at that, here the transmission coefficient can exceed 1.

In the previous section, influence of plasmons localized on the rim of the aperture on light transmission through this aperture was considered. This effect also takes place in the case of a nanoapertures' array (Degiron and Ebbesen, 2005). However, its contribution is essentially less than effects related to surface plasmons (de Abajo et al., 2006; Chang et al., 2005).

12.4 A Pattern of Radiation Outgoing from a Nanoaperture

In the previous section, we have seen that periodical structures on the external side of a film with nanoapertures can increase the transmission coefficient substantially. Similarly, nanostructuring of the rear film side results in narrowing of the radiation pattern of the light transmitting through the aperture. This phenomenon is nontrivial since according to Bethe–Bouwkamp theory the nanohole field in the far zone is to have a dipole character and thus to be very weakly directional. (Let us remember that the dipole radiation pattern is proportional to $\sin^2\theta$, where θ is the angle measured from the dipole axis.) A range of investigations of scanning microscopes apertures was carried out to understand which dipole and quadrupole field components characterize the radiation of a nanoaperture in a film made of real metals (Obermüller and Karrai, 1995; Drezet et al., 2001, 2002; Shin et al., 2002; Shin et al., 2001; Decca et al., 1997; Sekatskii et al., 2003; Höppener et al., 2002; Naber et al., 2002; Aigouy et al., 2003).

Nevertheless, it was shown (Lezec et al., 2002) that nanostructuring of the reverse film side results in a substantial narrowing of the radiation pattern. Thus, nanostructuring of the front and rear film surfaces allows one to enhance the transmission coefficient and the pattern of the transmitting radiation simultaneously.

In Fig. 12.13, examples of this effect are shown, both for a circular aperture surrounded by concentric grooves and for a slit surrounded by parallel, equally spaced grooves. In nanostructures present in these pictures, the period of the grooves' disposition was $d = 600$ nm and $d = 500$ nm, correspondingly, and the grooves' depth was $h = 60$ nm. These structures were developed with the help of a focused ion beam (FIB) on a freely suspended metal film of the thickness of 300 nm.

In the considered case, both external and internal surfaces around the hole or the slit are nanostructured. For this geometry, the position of the transmission coefficient maximum and its value $T(\lambda_0)$ are determined by phase-matching conditions on the front film side, while the direction of the outgoing beam and its width are determined by the phase-matching conditions on the film's rear side. In these experiments, it was discovered that the outgoing beam has high directivity ($\Delta\theta \approx 6^\circ$, see Fig. 12.13). This effect can be explained by the fact that surface plasmons propagating on the back film side are scattered coherently by the system of grooves, whose overall size D is generally large in comparison with the wavelength. It is this size that determines the narrow beam width $\Delta\theta$, which can be easily estimated through the well-known relation $\Delta\theta \sim \lambda/D$, where $D \gg \lambda$ is a typical size of the whole system of pits or grooves.

It is interesting that different wavelengths form beams with different directions of propagation, and thus the nanostructuring of the surface near a nanohole has spatial filtering properties.

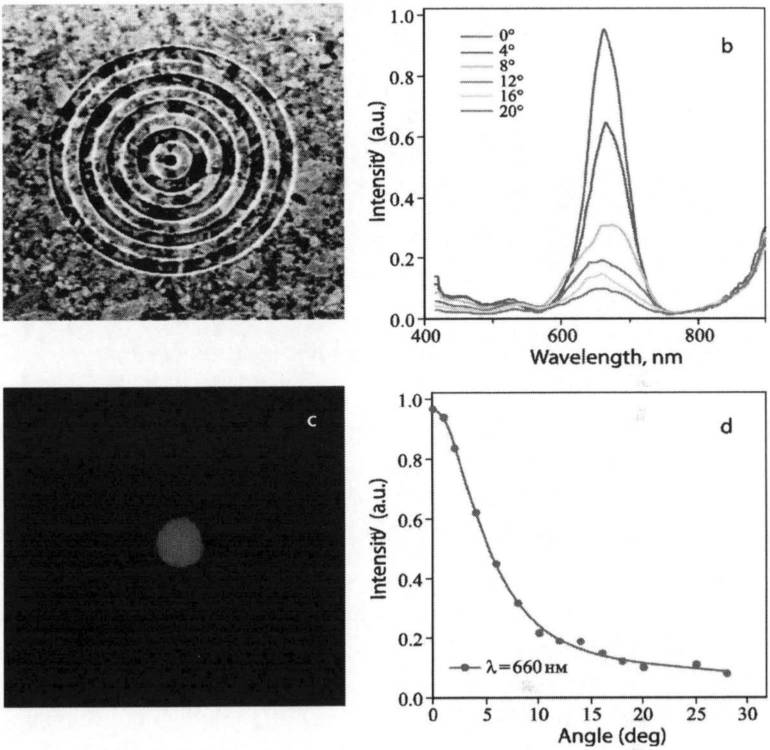

Figure 12.13 (a) The image of a nanostructure surrounding the aperture ("bull's-eye") in a silver film of the thickness of 300 nm formed with the help of FIBs (the grooves' period is 600 nm, the groves' depth is 60 nm, and the aperture diameter is 300 nm). (b) Transmission spectra measured in different directions of the outgoing light propagation (the angle θ is measured from the aperture axis). (c) The optical image of the directed emission for the aperture's maximal transmission. (d) Angular distribution of the transmitting light for the aperture's maximal transmission (Lezec et al., 2002). *Abbreviation*: FIB, focused ion beam.

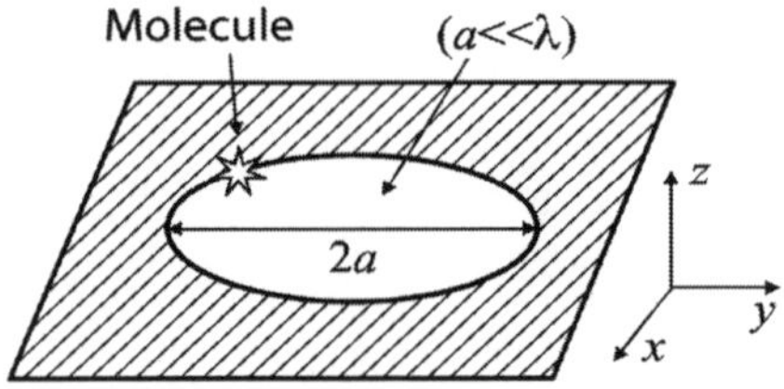

Figure 12.14 The geometry of the problem of spontaneous emission of a molecule near the aperture.

12.5 Fluorescence of Atoms and Molecules Near a Nanoaperture

12.5.1 *Influence of a Circular Nanoaperture in a Perfectly Conducting Screen on Spontaneous Emission of an Atom or a Molecule*

As it was said in the beginning of this chapter, a nanoaperture is often and effectively used for investigation of optical properties of single molecules. For interpretation of results of these investigations, it is necessary to know not only how the fields incident on a hole are modified but also how the nanohole influences the molecules' spontaneous emission rate. General approaches to calculation of this influence are given in Appendix A1. In this section, we will consider the circular aperture influence on the molecules' spontaneous emission rate. The geometry of the problem of molecule emission is shown in Fig. 12.14.

The case of an arbitrary aperture size. When a molecule is located on the aperture symmetry axis and its electric dipole moment is perpendicular to the axis, the spontaneous decay rate may be expressed by means of spheroidal functions for any relations between the hole's radius and the emission wavelength (Belkina, 1957). Nevertheless, when calculating the spontaneous emission rates with the help of formulas from (Belkina, 1957), some technical difficulties arise, which are related to calculation of spheroidal functions and their derivatives (Stratton et al., 1941; Abramowitz and Stegun, 1965), and a detailed analysis has not been conducted yet, although some applications for calculation of these functions have been developed for a well-known mathematical software package Mathematica®.

The only case where an analytical result can be obtained is the case of a molecule situated in the center of a small aperture and having a dipole moment parallel to the screen plane. According to the vectorial version of Babinet's principle (Vainstein, 1988), in this case electric and magnetic fields (**E**, **H**) of the electric dipole d in the vicinity of a circular *aperture* can be expressed through scattered electric and magnetic fields ($\delta\mathbf{E}'$, $\delta\mathbf{H}'$) of the magnetic dipole $m = d$ in the center of a perfectly conducting *disc*. For $z > 0$, we have

$$\begin{aligned} \mathbf{E} &= \delta\mathbf{H}', \\ \mathbf{H} &= -\delta\mathbf{E}', \end{aligned} \tag{12.23}$$

and for $z < 0$

$$\begin{aligned} \mathbf{E} &= -\delta\mathbf{H}', \\ \mathbf{H} &= \delta\mathbf{E}'. \end{aligned} \tag{12.24}$$

In the far field of the small aperture, the expression for the scattered fields $\delta\mathbf{E}'$, $\delta\mathbf{H}'$ from the magnetic dipole oriented along the x axis can be found in (Belkina, 1957):

$$\begin{aligned} -\delta H'_\theta = \delta E'_\phi &= k^2 m_x \frac{4ci}{3\pi}(2 - \sin^2\theta)\frac{e^{ikR}}{R}\cos\varphi, \\ \delta E'_\theta = \delta H'_\phi &= k^2 m_x \frac{8ci}{3\pi}\frac{e^{ikR}}{R}\cos\theta\sin\varphi, \end{aligned} \tag{12.25}$$

where c is the light velocity in vacuum and $m_x = d_x$ is the magnetic dipole moment and the standard notation for spherical coordinates is used. Let us note that expressions in Eq. (12.25) describe the radiation pattern of a more complex character than the dipole one. It is related to the fact that a plane divides the space in two almost independent parts.

Calculating the energy flow at the infinity with the help of Eq. (12.25), one can find a change of the spontaneous emission rate of a molecule situated in the center of a small aperture and having the dipole moment parallel to the screen plane in comparison with the case of a molecule in free space:

$$\left(\frac{\gamma}{\gamma_0}\right)_\rho = \frac{64}{15\pi^2}(ka)^2 + O(ka)^4. \tag{12.26}$$

Eq. (12.26) shows that in the hole center the dipole emission rate tends to zero as the diameter of the aperture decreases. Physically, it is rather clear since in this case the situation becomes similar to the situation when an electric dipole is placed on a perfectly conducting surface and when the dipole moment induced on the surface compensates the atom dipole moment.

Quasi-static approximation. The practical absence of an illustrative analytical solution to the problem of dipole emission in the vicinity of the aperture with the retardation effects taken into account makes the analytical solution of this problem in the quasi-static approximation especially important, because it allows one to estimate the spontaneous emission rates and radiation frequency shift (see Appendix A1) for an arbitrary position of an atom near the hole.

Within the context of the quasi-static approach, the spontaneous emission rate of a molecule is expressed through the effective dipole moments $d^{\pm}_{z,\mathrm{tot}}$ of the whole "molecule + aperture" system (d_0 is the dipole moment of the transition of a molecule in vacuum (Klimov, 2003):

$$\frac{\gamma}{\gamma_0} = \frac{1}{2}\left(\frac{d^{+\;2}_{z,\mathrm{tot}}}{d_0^2} + \frac{d^{-\;2}_{z,\mathrm{tot}}}{d_0^2}\right). \tag{12.27}$$

In Eq. (12.27), the first term describes the molecule emission into the upper half-space ($z > 0$), while the second term describes the emission into the lower half-space ($z < 0$). An important peculiarity of the considered approximation is that in the case of φ orientation of the dipole (in the cylindrical system of coordinates), the spontaneous emission is suppressed substantially for any position of the atom and described by terms of the order $(ka)^2$ (see also Eq. (12.26)), that is, by terms of a higher order of smallness in comparison with Eq. (12.27).

Analytical expressions for effective dipole moments in Eq. (12.27) for arbitrary positions and orientations of the molecule can be found in (Klimov, 2003). In typical cases, these expressions become quite simple. Thus, for example, in the case of a molecule situated on the system axis, the effective emission is possible only for the dipole moment oriented along the z axis. In this case, the dipole moments describing the emission into the upper and lower half-spaces have the form (Klimov, 2003):

$$\frac{d^{\pm}_{\text{tot},z}}{d_0} = 1 \pm \left(\frac{2}{\pi}\arctan\left(\frac{z}{a}\right) + \frac{2az}{\pi(a^2+z^2)}\right). \tag{12.28}$$

For $z = 0$ $\frac{d^{\pm}_{\text{tot},z}}{d_0} = \frac{d^{-}_{\text{tot},z}}{d_0} = 1$, that is, in the case of an atom placed in the aperture center, it emits just as if it emitted in free space. For $z \to \infty$, $\frac{d^{+}_{\text{tot},z}}{d_0} = 2$, $\frac{d^{-}_{\text{tot},z}}{d_0} = 0$, and the atom emits with the duplicated dipole moment into the upper half-space. Let us note that, certainly, in the context of the quasi-static approximation the limit $z \to \infty$ implies only $\lambda \gg z \gg a$. In the case when an atom is moving away from the nanohole on distances comparable with the radiation wavelength, the nanoaperture ceases to influence the emission, and one should use expressions for the spontaneous decay rates in the presence of the perfectly conducting plane without a hole:

$$\frac{\gamma}{\gamma_0} = \left(1 + 3\left(\frac{\sin(2kz)}{(2kz)^3} - \frac{\cos(2kz)}{(2kz)^2}\right)\right), \quad \frac{z}{a} \gg 1. \tag{12.29}$$

In Fig. 12.15, the total decay rate is shown as well as the rates of decays into the upper and lower half-spaces of the atom with z orientation of the dipole moment, depending on its position on the system axis.

Let us now consider decay rates of an atom with z orientation of the dipole moment and located in the plane $z = 0$. In this case, the atom decay rate does not change in the aperture in comparison with the case of free space for any position of it, that is, $d^{+}_{\text{tot},z}/d_0 = d^{-}_{\text{tot},z}/d_0 = 1$ for $z = 0$, $\rho < a$, where ρ is the radius in the cylindrical coordinates. In the case of an atom placed near the film surface, the dipole moments describing the emission into the upper and lower half-spaces take the form (Klimov, 2003):

$$\frac{d^{\pm}_{\text{tot},z}}{d_0} = 1 \pm \frac{2}{\pi}\left(\frac{a}{\sqrt{\rho^2 - a^2}} + \arcsin\frac{\sqrt{\rho^2 - a^2}}{a\rho}\right), \ \rho > a, \ z = 0. \tag{12.30}$$

In Fig. 12.16, the total decay rate of an atom with z orientation of the dipole moment is shown depending on its position on the plane $z = 0$. From Eq. (12.30) and from the picture, it is seen that near the rim of the hole a substantial (infinite) enhancement of the decay rate

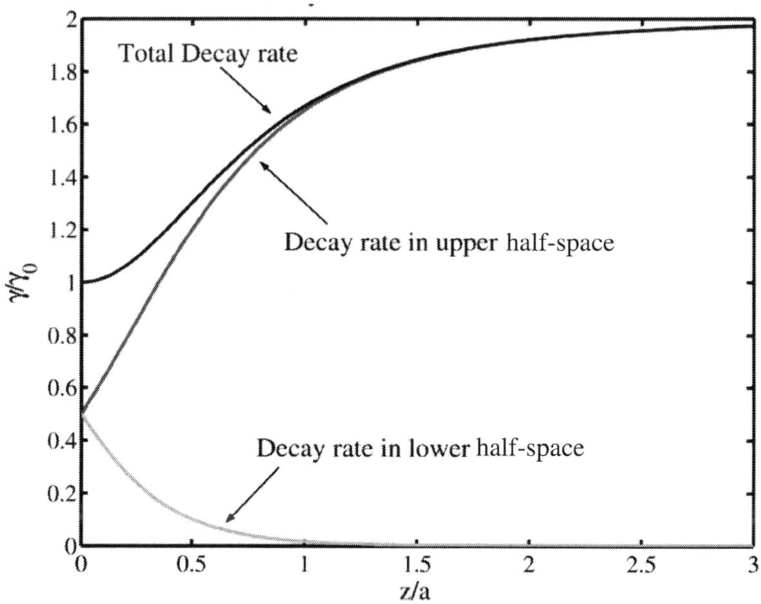

Figure 12.15 The total decay rate and the rate of decay into the upper and lower half-spaces of an atom with z orientation of the dipole moment, depending on its position on the system axis.

arises, and the radiation propagates in both upper and lower half-spaces. When the finiteness of the screen width and the finiteness of its conductivity are taken into account, the spontaneous decay rate of the atom becomes limited near the aperture rim.

Let us now consider decay rates for an atom with radial orientation of the dipole moment (in the cylindrical system of coordinates) and situated in the plane $z = 0$. (The decay rate of an atom with φ orientation of the dipole moment is identically equal to zero in the quasi-static approximation.) In the case of $\rho < a$, the dipole moments describing fields in the upper and lower half-spaces will have the form (Klimov, 2003):

$$\frac{d^{\pm}_{\text{tot},z}}{d_0} = \mp\frac{2}{\pi}\frac{\rho}{\sqrt{a^2-\rho^2}},\ \rho < a. \qquad (12.31)$$

Outside the aperture $\rho > a$, the total dipole moment will be equal to zero according to the boundary conditions on a perfectly

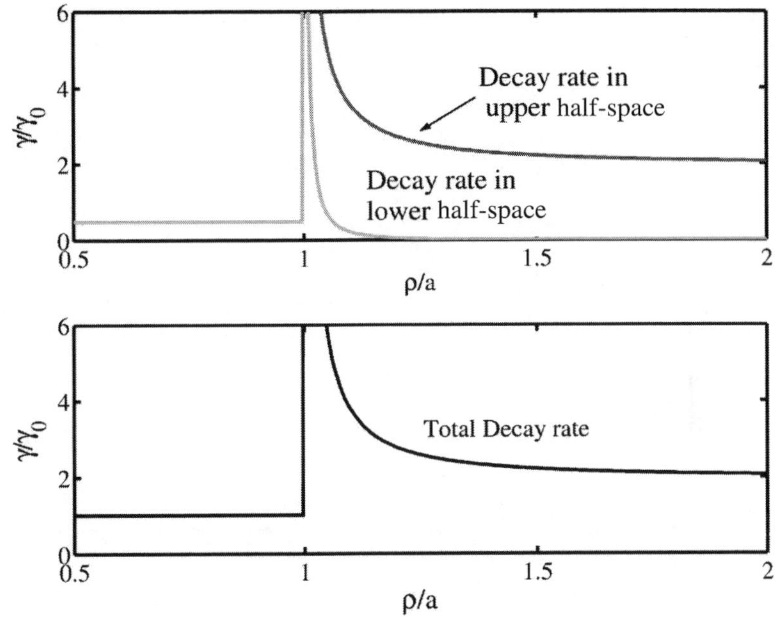

Figure 12.16 The decay rate of an atom with z orientation of the dipole moment, depending on its position in the aperture plane $z = 0$.

conducting plane. In Fig. 12.17, the total decay rate is shown, as well as the rates of the decay into the upper and lower half-spaces in the case of an atom situated in any position on the plane $z = 0$ and having the dipole moment orientation along the radius. From Eq. (12.31) and from the picture, it is seen that near the rim of the hole a substantial enhancement of the decay rate arises, and the emission propagates into both upper and lower half-spaces. When the finiteness of the screen width and the finiteness of its conductivity are taken into account, the spontaneous decay rate of an atom becomes limited near the nanohole rim.

Thus, in the present section analytical expressions allowing one to estimate relatively simply spontaneous decay rates of an atom or a molecule near a nanoaperture are presented. Unfortunately, more adequate expressions for real apertures have not been found yet.

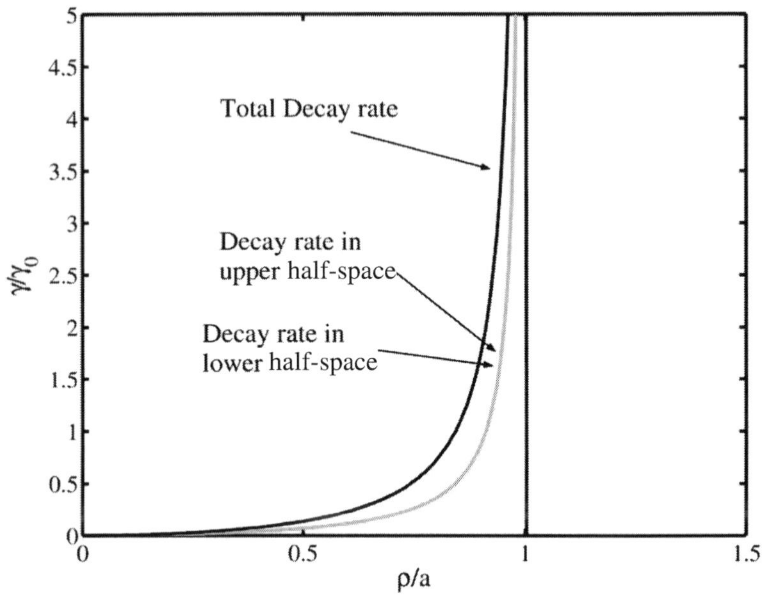

Figure 12.17 The decay rate of an atom with radial orientation of the dipole moment, depending on its position in the aperture plane.

The shift of the emission frequency of an atom situated near a nanohole can also be investigated with the help of the quasi-static approximation (Klimov, 2003).

12.5.2 *Experimental Investigations of Molecules' Emission Near Nanoapertures*

For the first time, the fluorescence of a single molecule in usual conditions with the help of an aperture near-field scanning microscope (SNOM) was observed by Betzig and Chichester (1993). Single molecules were fixed on a plane substrate and excited locally by emission from the microscope aperture. By changing the polarization of the emission incident on the molecules, detection of the dipole moment orientation of the molecule has appeared to be possible.

Then, analogous experiments were carried out with the help of substantially improved tips of aperture microscopes (Veerman et al.,

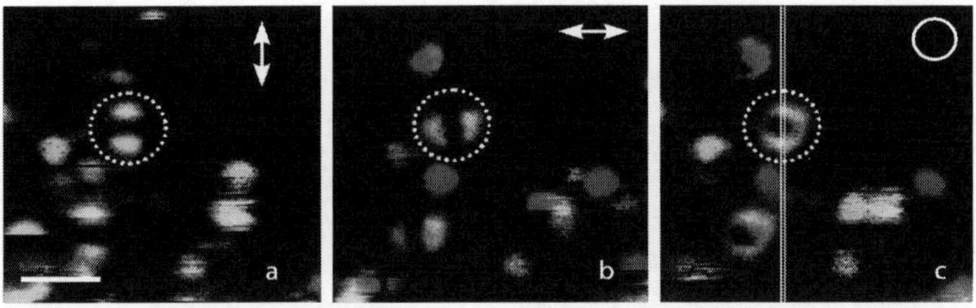

Figure 12.18 The consequence of three images of the same area of the size of 1.2 × 1.2 μm with molecules $DiIC_{18}$ in a PMMA layer of the thickness of 10 nm in a scanning microscope with an aperture of diameter 70 nm (SNOM). The emission polarization is linear vertical (a), linear horizontal (b), and circular (c). The image of the molecules changes as the polarization does. For example, the molecule image in a circle implies that its orientation is perpendicular to the film. The scale bar is 300 nm (Veerman et al., 1999). *Abbreviation*: PMMA, polymethyl methacrylate.

1999; Gersen et al., 2000, 2001). Owing to usage of FIBs for needles' production, they had a well-defined circular aperture with sharp edges. Another advantage of the used needles was the absence of metal grains near the aperture, resulting in an almost plane surface and allowing one to control the distance between the aperture and the surface with high precision.

In Fig. 12.18, images of several molecules are shown, and the color is used for indication of molecule emission polarization. The red color corresponds to the horizontal polarization in the plane of the image; the green color corresponds to the vertical one. An important peculiarity of these images of molecules is that the emission polarization of the fixed molecule depends on its orientation relative to the microscope tip. In a general case, this dependency is complex since the emission of the molecule near the hole is described by the total dipole moment of the "molecule + aperture" system, which does not coincide necessarily with the molecule orientation according to the theory stated above. In particular, in the case of a molecule near a nanoaperture the emission of the "molecule + microscope aperture" system is mainly determined by the dipole oriented perpendicularly to the surface. The modification of the molecule radiation pattern by the aperture

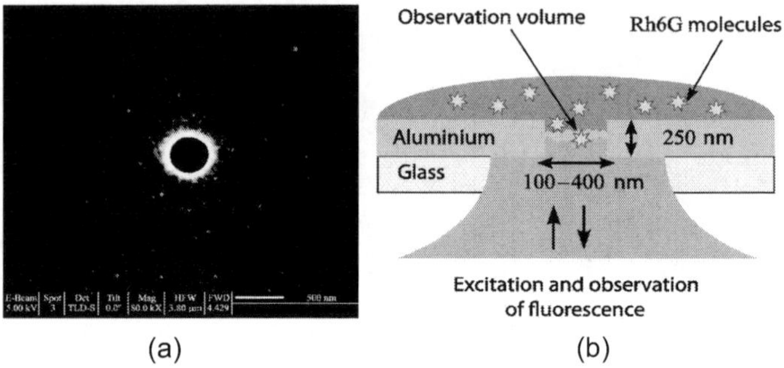

Figure 12.19 (a) The SEM image of a nanoaperture in an Al film coated on the microscope cover glass. (b) The scheme of the experiment on measurement of single molecules' fluorescence in a nanoaperture (Rigneault et al., 2005).

of the scanning microscope was confirmed experimentally (Gersen et al., 2000, 2001).

Quite recently, a range of interesting experiments focused on investigation of fluorescence of Rh6G molecules in a circular and a rectangular nanoaperture in an Al film of the width of 300 nm was carried out (Wenger et al., 2005; Rigneault et al., 2005). This aperture was irradiated by a light beam with a waist of 250 nm at the wavelength of 488 nm. Above the nanoapertures, an aqueous solution of the dye molecule Rh6G was placed. The geometry of this experiment is shown in Fig. 12.19. Using the method of fluorescence correlation spectroscopy (FCS), (Elson and Magde, 1974; Wenger et al., 2005; Rigneault et al., 2005) succeeded in finding a substantial increase of the fluorescence intensity and the total spontaneous decay rate per molecule.

In Fig. 12.20, results of the experiment with a circular aperture are shown.

From this picture, it is seen that inside the aperture of 150 nm an enhancement of the fluorescence intensity exceeds the fluorescence in an aqueous solution by 6.5 times in the linear regime (the weak intensity of the exciting light) and by more than 12 times in a nonlinear regime (the strong intensity of the exciting light), where the fluorescence intensity has the saturation in the aqueous solution. From the insertion in this picture, it is seen that a substantial

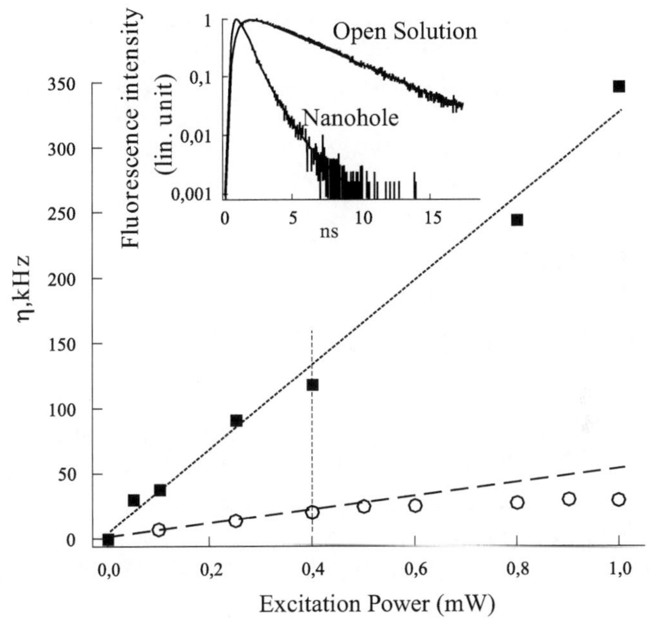

Figure 12.20 The measurement of the fluorescence rate η of molecules, depending on the excitation intensity. The case when the molecules are situated in an aperture of the diameter of 150 nm is shown by the symbols (■). The case when molecules are situated in a homogeneous fluid is shown by the symbols (o). Insertion: The curves of the fluorescence decay in a homogeneous fluid and in a nanoaperture of the diameter of 150 nm (Rigneault et al., 2005).

increase of the total rate of the spontaneous decay takes place in this situation. Certainly, due to the stochastic nature of the FCS method, these results have an averaged character referring both to the molecule position inside the aperture and to its dipole moment orientation. The main increase of the fluorescence intensity can be explained qualitatively by enhancement of the exciting field near the nanoaperture rim. This factor (taking into account the averaging over the aperture size) reaches its maximum value for apertures of a diameter of the order of 150 nm. For smaller sizes of the aperture, the averaged intensity decreases, possibly explaining the results of (Levene et al., 2003), where the changes in fluorescence intensity were not found. However, for an exhausting explanation of this effect, the factor of spontaneous decay enhancement is to be invoked, which results in an increase of excitation saturation intensity, and, possibly, the factor of change of the molecule radiation pattern, which results in an increase of the efficiency of photons collection.

12.6 Conclusion

Thus, in this chapter optical properties of a single nanoaperture and of an array of nanoapertures in a thin metal film are considered. It was shown that plasmons influence substantially both the coefficient of light transmission through nanoapertures and the outgoing radiation pattern. Besides, it was shown that nanoapertures influence substantially the emission of atoms and molecules situated near them, and this fact can be used for the development of optical nanodevices and nanosensors in the future.

References

Aigouy, L., de Wilde, Y., Mortier, M. (2003) Local optical imaging of nanoholes using a single fluorescent rare-earth-doped glass particle as a probe, *Appl. Phys. Lett.*, **83**, 147.

Abramowitz, M., Stegun, I.A. (1965) *Handbook of Mathematical Functions* (Dover, New York).

Ambrose, W.P., Goodwin, P.M., Martin, J.C., Keller, R.A. (1994) Alterations of single molecule fluorescence lifetimes in near-field optical microscopy, *Science*, **265**, 364.

Andrejewski, W. (1953) Die Beugung electromagnetischer Wellen an der leitenden Kreisscheibe und an der Kreisformingen Oeffnung in leitenden ebenen Schirm, *Z. Angew. Phys.*, **5**, 178–186.

Baida, F.I., van Labeke, D. (2002) Light transmission by subwavelength annular aperture arrays in metallic films, *Opt. Commun.*, **209**, 17–22.

Balykin, V.I., Klimov, V.V., Letokhov, V.S. (2003) Atom nanooptics based on photon dots and photon holes, *JETP Lett.*, **78**, 8.

Balykin, V.I., Klimov, V.V., Letokhov, V.S. (2005) Atom nano-optics, *Opt. Photonic News*, **March**, 44–48.

Barnes, W.L., Murray, W.A., Dintinger, J., Devaux, E., Ebbesen, T.W. (2004) Surface plasmon polaritons and their role in the enhanced transmission of light through periodic arrays of subwavelength holes in a metal film, *Phys. Rev. Lett.*, **92**(10), 107401.

Becker, R.S., Anderson, V.E., Birkhoff, R.D., Ferrell, T.L., Ritchie, R.H. (1981) Surface plasmon dispersion on a single-sheeted hyperboloid, *Can. J. Phys.*, **59**, 521–529.

Belkina, M.G. (1957) Diffraction of electromagnetic waves on a disc. Collected works: Diffraction of electromagnetic waves on some bodies of rotation (Soviet radio).

Bethe, H.A. (1944) Theory of diffraction by small holes, *Phys. Rev.*, **66**, 163.

Betzig, E., Chichester, R. (1993) Single molecules observed by near-field scanning optical microscopy, *Science*, **262**, 1422–1425.

Bian, R.X., Dunn, R.C., Xie, X.S., Leung, P.T. (1995) Single molecule emission characteristics in near-field microscopy, *Phys. Rev. Lett.*, **75**, 4772–4775.

Bonod, N., Popov, E., Nevière, M. (2005) Differential theory of diffraction by finite cylindrical objects, *J. Opt. Soc. Am. A*, **22**, 481–490.

Bouwkamp, C.J. (1950) On the diffraction of electromagnetic waves by small circular discs and holes, *Philips Res. Rep.*, **5**, 401.

Butter, J.Y.P., Hecht, B. (2006) Aperture scanning near-field optical microscopy and spectroscopy of single terrylene molecules at 1.8 K, *Nanotechnology*, **17**, 1547–1550.

Chang, Shih-Hui, Gray, S.K., Schatz, G.C. (2005) Surface plasmon generation and light transmission by isolated nanoholes and arrays of nanoholes in thin metal films, *Opt. Express*, **13**(8), 3150–3165.

de Abajo, G.F.J., Sáenz, J.J., Campillo, I., Dolado, J.S. (2006) Site and lattice resonances in metallic hole arrays, *Opt. Express*, **14**(1), 7–18.

Decca, R.S., Drew, H.D., Empson, K.L. (1997) Investigation of the electric-field distribution at the subwavelength aperture of a near-field scanning optical microscope, *Appl. Phys. Lett.*, **70**, 15, 1932–1934.

Degiron, A., Ebbesen, T.W. (2005) The role of localized surface plasmon modes in the enhanced transmission of periodic subwavelength apertures. *J. Opt. A: Pure Appl. Opt.*, **7**, S90–S96.

Degiron, A., Lezec, H.J., Yamamoto, N., Ebbesen, T.W. (2004) Optical transmission properties of a single subwavelength aperture in a real metal, *Opt. Commun.*, **239**, 61–66.

Degiron, E.T. (2004) Analysis of the transmission process through single apertures surrounded by periodic corrugations, *Opt. Express*, **12**, 3694–3700.

Degiron, H.J., Lezec, N., Yamamoto, Ebbesen, T.W. (2004) Optical transmission properties of a single subwavelength aperture in a real metal, *Optics Commun.*, **239**, 61–66.

Drezet, A., Nasse, M.J., Huant, S., Woehl, J.C. (2004) The optical near-field of an aperture tip, *Europhys. Lett.*, **66**, 41.

Drezet, A., Woehl, J.C., Huant, S. (2001) Far-field emission of a tapered optical fiber tip: a theoretical analysis, *J. Microsc.*, **202**, 359–361.

Drezet, A., Woehl, J.C., Huant, S. (2002) Diffraction by a small aperture in conical geometry: application to metal-used in near-field scanning optical microscopy, *Phys. Rev. E*, **65**, 046611.

Dunn, R.C. (1999) Near-field scanning optical microscopy, *Chem. Rev.*, **99**, 2891–2927.

Ebbesen, T.W., Lezec, H.J., Ghaemi, H.F., Thio, T., Wolff, P.A. (1998) Extraordinary optical transmission through sub-wavelength hole arrays, *Nature*, **391**, 667.

Elson, E., Magde, D. (1974) Fluorescence correlation spectroscopy. II. An experimental realization, *Biopolymers*, **13**, 29–61.

de Abajo, G.F.J. (2002) Light transmission through a single cylindrical hole in a metallic film, *Opt. Express*, **10**, 1475.

García-Vidal, F.J., Lezec, H.J., Ebbesen, T.W., Martín-Moreno, L. (2003) Multiple paths to enhance optical transmission through a single subwavelength slit, *Phys. Rev. Lett.*, **90**(21), 213901.

Genet, C., Ebbesen, T.W. (2007) Light in tiny holes, *Nature*, **445**, 39–46.

Gersen, H., Garciâa-Parajoâ, M.F., Novotny, L., Veerman, J.A., Kuipers, L., van Hulst, N.F. (2001) Near-field effects in single molecule emission, *J. Microsc.*, **202**(Pt 2), 374–378.

Gersen, H., García-Parajó, M.F., Novotny, L., Veerman, J.A., Kuipers, L., van Hulst, N.F. (2000) Influencing the angular emission of a single molecule, *Phys. Rev. Lett.*, **85**, 5312.

Ghaemi, H.F., Thio, T., Grupp, D.E., Ebbesen, T.W., Lezec, H.J. (1998) Surface plasmon enhance optical transmission through subwavelength holes, *Phys. Rev. B*, **58**(11), 6779–6782.

Gordon, R. and Brolo, A. (2005) Increased cut-off wavelength for a subwavelength hole in a real metal, *Opt. Express*, **13**, 1933–1938.

Grupp, D., Lezec, H.J., Thio, T., Ebbesen, T.W. (1999) Tunable enhanced light transmission through a single subwavelength aperture, *Adv. Mater.*, **11**, 860–862.

Grupp, D.E., Lezec, H.J., Thio, T., Ebbesen, T.W. (1999) Beyond the Bethe limit: tunable enhanced light transmission through a single sub-wavelength aperture, *Adv. Mater.*, **11**(10), 860–862.

Höppener, C., Molenda, D., Fuchs, H., Naber, A. (2002) Simultaneous topographical and optical characterization of near-field optical aperture probes by way of imaging fluorescent nanospheres, *Appl. Phys. Lett.*, **80** 1331–1333.

Hosaka, N., Saiki, T. (2001) Near-field fluorescence imaging of single molecules with a resolution in the range of 10 nm, *J. Microsc.*, **202**(Pt 2), 362–364.

Klimov, V.V., Letokhov, V.S. (1994) A simple theory of the near field in diffraction by a round aperture, *Opt. Commun.*, **106**, 151.

Klimov, V.V., Letokhov, V.S. (1995) New atom trap configuration in the near field of laser radiation, *Opt. Commun.*, **121**, 130.

Klimov, V.V. (2003) Spontaneous emission of an atom placed near the aperture of scanning microscope, *JETP Lett.*, **78**, 471–475.

Klyuchnik, V., Kurganov, S. Yu., Lozovik, Yu. E. (2003) Plasmons at a hole in a screen, *Phys. Solid State*, **45,** 1793–1797.

Krishnan, Thio, T., Kim, T.J., Lezec, H.J., Ebbesen, T.W., Wolff, P.A., Pendryd, J., Martin-Moreno, L., Garcia-Vidal, F.J. (2001) Evanescently coupled resonance in surface plasmon enhanced transmission, *Opt. Commun.*, **200**, 1–7.

Levene, M.J., Korlach, J., Turner, S.W., Foquet, M., Craighead, H.G., Webb, W.W. (2003) Zero-mode waveguides for single-molecule analysis at high concentrations, *Science*, **299**, 682–686.

Lezec, H.J., Degiron, A., Devaux, E., Linke, R.A., Martin-Moreno, L., Garcia-Vidal, F.J., Ebbesen, T.W. (2002) Beaming light from a subwavelength aperture, *Science*, **297**, 820–822.

Meixner, J., Andrejewski, W. (1950) Strenge Theorie der Beugung ebener elektromagnetisher Wellen an der vollkommen leitenden Kreisscheibe und an der kreisformigen Offnung im vollkommen leitenden ebenen Schirm, *Ann. Phys.*, **B.7**, 157.

Naber, A., Molenda, D., Fischer, U.C., Maas H-J Höppener, C., Lu, N., Fuchs, H. (2002) Enhanced light confinement in a near-field optical probe with a triangular aperture, *Phys. Rev. Lett.*, **89,** 210801.

Nomura, Y., Katsura, S. (1985) Diffraction of electromagnetic waves by circular plate and circular hole, *J. Phys. Soc. Japan*, **10**, 285.

Obermüller, Karrai, K. (1995) Far field characterization of diffracting circular apertures, *Appl. Phys. Lett.*, **67**, 3408.

Pohl, D.W., Courjon, D. (1992) *Near Field Optics* (Kluwer Academic Publishers, Dordrecht).

Popov, E., Bonod, N., Nevière, M., Rigneault, H., Lenne, P.-F., Chaumet, P. (2005) Surface plasmon excitation on a single subwavelength hole in a metallic sheet, *Appl. Opt.*, **44**, 2332–2337.

Porto, J.A., Garcia-Vidal, F.J., Pendry, J.B. (1999) Transmission resonances on metallic gratings with very narrow slits, *Phys. Rev. Lett.*, **83**(14), 2845–2848.

Rigneault, H., Capoulade, J., Dinitnger, J., Wenger, J., Bonod, N., Popov, E., Ebbesen, T.W., Lenne, P.F. (2005) Enhancement of single-molecule fluorescence detection in subwavelength apertures, *Phys. Rev. Lett.*, **95**, 117401.

Roberts (1987) Electromagnetic theory of diffraction by a circular aperture in a thick, perfectly conducting screen, *J. Opt. Soc. Am. A*, **4**, 1970.

Roberts (1989) Near-zone fields behind circular apertures in thick, perfectly conducting screens, *J. Appl. Phys.*, **65,** 2896–2899.

Sekatskii S.K., Dietler, G., Mironov, B.N., Kompanets, V.O., Matveets Yu. A., Chekalin, S.V., Letokhov, V.S. (2003) Nonperturbing observation of optical near field, *Appl. Phys. Lett.*, **83**, 15, 4900.

Shi, X., Hesselink, L., Thornton, R.L. (2003) Ultrahigh light transmission through a C-shaped nanoaperture, *Opt. Lett.*, **28**(15), 1320–1322.

Shin, D.J., Chavez-Pirson, A., Kim, S.H., Jung, S.T., Lee, Y.H. (2001) Diffraction by a subwavelength-sized aperture in a metal plane, *J. Opt. Soc. Am. A*, **18**, 1477.

Shin, D.J., Chavez-Pirson, A., Lee, Y.H. (2002) Multipole analysis of the radiation from near-field optical probes, *Opt. Lett.*, **25**, 171–173.

Shin, H., Catrysse, P.B., Fan, S. (2005) Effect of the plasmonic dispersion relation on the transmission properties of subwavelength cylindrical holes, *Phys. Rev. B*, **72**, 085436.

Shou, X., Agrawal, A., Nahata, A. (2005) Role of metal film thickness on the enhanced transmission properties of a periodic array of subwavelength apertures, *Opt. Express*, **13**(24), 9834–9840.

Shubeita, G.T., Sekatskii, S.K., Dietler, G., Letokhov, V.S. (2002) Local fluorescent probes for the fluorescence resonance energy transfer scanning near-field optical microscopy, *Appl. Phys. Lett.*, **80**, 2625–2627.

Shubeita, G.T., Sekatskii, S.K., Dietler, G., Potapova, I., Mews, A., Basché T. (2003) Scanning near-field optical microscopy using semiconductor nanocrystals as a local fluorescence and fluorescence resonance energy transfer source, *J. Microsc.*, **210**(Pt 3), 274–278.

Smythe, W.R. (1969) *Static and Dynamic Electricity*, 3rd ed. (McGraw-Hill, New York).

Stratton, J.A., Morse, P.M., Chu, L.J., Hutner, R.A. (1941) *Elliptic Cylinder and Spheroidal Wave Functions* (John Wiley, New York).

Thio, T., Pellerin, K.M., Linke, R.A., Lezec, H.J., Ebbesen, T.W. (2001) Enhanced light transmission through a single subwavelength aperture, *Opt. Lett.*, **26**(24), 1972–1974.

Trautman, J.K., Macklin, J.J., Brus, L.E., Betzig, E. (1994) Near-field spectroscopy of single molecules at room temperature, *Nature (London)*, **369**, 40.

Vainstein, L.A. (1988) Electromagnitnie volni (electromagnetic waves) (Radio i Sviaz, Moscow).

van der Molen, K.L., Segerink, F.B., van Hulst, N.F., Kuipers, L. (2004) Influence of hole size on the extraordinary transmission through subwavelength hole arrays, *Appl. Phys. Lett.*, **85**(19), 4316–4318.

Veerman, J.A., Garcia-Parajo, M.F., Kuipers, L., van Hulst, N.F. (1999) Single molecule mapping of the optical field distribution of probes for near-field microscopy, *J. Microsc.*, **194**, 477–482.

Wang, Qian-jin, Li, Jia-qi, Huang, Cheng-ping, Zhang, C., Zhu, Y.-Y. (2005) Enhanced optical transmission through metal films with rotation-symmetrical hole arrays, *Appl. Phys. Lett.*, **87**, 091105.

Wannemacher, R. (2001) Plasmon supported transmission of light through nanometric holes in mettallic thin film, *Opt. Commun.*, **195**, 107–118.

Webb, K.J., Li, J. (2006) Analysis of transmission through small apertures in conducting films, *Phys. Rev. B*, **73**, 033401.

Wenger, P.-F., Lenne, E., Popov, H., Rigneault, J., Dintinger, Ebbesen, T. (2005) Single molecule fluorescence in rectangular nano-apertures, *Opt. Express*, **13**, 7035–7044.

Xie, X.S., Dunn, R.C. (1994) Probing single molecule dynamics, *Science*, **265**, 361.

Chapter 13

Applications of Nanoplasmonics

13.1 Tumor Therapy and Visualization with the Help of Nanoparticles

Now, there is an enormous clinical demand for new methods of detection, visualization, and treatment of cancer, which would have good sensitivity at a reasonable cost. Standard chemotherapy kills both cancerous and healthy cells. At that, often patients lose their hair and suffer from a variety of side effects (blood cells and the immune system suffer). Therefore, the possibility of selective eradication of cancer cells will provoke a considerable breakthrough in the treatment of this fatal illness.

In recent years, it has been demonstrated in several laboratories that optical technologies could be an important step toward the solution of this problem. Optical technologies promise high-resolution and nondestructive functional visualization of tumors at a reasonable price. However, in many cases these technologies are restricted both by small intensities of optical signals from endogenous (intracellular) chromophores and by a weak spectral contrast between normal and diseased tissue. Recently, considerable interest has appeared for a combination of optical technologies and exogenous (external) contrast agents (nanoparticles) aimed to

Nanoplasmonics
Vasily Klimov

ISBN 978-981-4267-16-8 (Hardcover), 978-981-4267-42-7 (eBook)
www.panstanford.com

detect specific molecular signs of cancer, increase the detection level, and increase the clinical efficiency of visualization and even for cancer therapy. The selectiveness is achieved by usage of antibodies in contrast agents that recognize specific cancer cell antigens. As a result of binding of antibodies and antigens, contrast agents, which are metal nanoparticles, attach to the surfaces of cancer cells or penetrate them. Then, the diseased tissue is exposed to near-infrared radiation. Scattered radiation can be detected, and a tumor can be visualized. On the other hand, rather intensive radiation heats nanoparticles to a temperature of 70°C or more, which kills a cancerous cell. At that, the radiation almost does not affect healthy cells, in which the nanoparticles are absent.

Layered nanoparticles (nanoshells or nanomatryoshkas) appear to be especially promising contrast agents. Usually, these nanoparticles consist of a dielectric core (dioxide of silicon or ferrum) and a gold shell. Physical properties of gold nanoshells are similar to those of solid-gold nanoparticles; however, their absorption and scattering cross sections can change in a much wider frequency range, depending on the ratio between the core and shell radii. In Fig. 13.1, dependencies of the extinction cross section of a particle of a radius of 60 nm on the gold shell thickness is shown. From this picture, it is seen that even minor changes of the shell thickness result in a substantial shift of the plasmon resonance. The theoretical estimation of this shift can be made on the basis of formulas stated in Chapters 6, 7, and 8 for nanoshells of different geometries.

At the present time, there is already a technology of nanoparticle production that allows one to vary core and shell sizes in a wide range and, consequently, to control scattering and absorption cross sections (Oldenburg et al., 1998; Wang et al., 2006).

Living tissues have an increased penetration depth in the near-infrared range (700–1100 nm), and therefore it is necessary to optimize the properties of nanoparticles in this range exactly. Good scattering and small absorption of light by tumor cells, in which nanoparticles are situated, are necessary for tumor visualization.

For example, gold nanoshells with a core of 120 nm radius coated with a nanoshell of 35 nm thickness have a high scattering cross section in the near-infrared range (Fig. 13.2). This

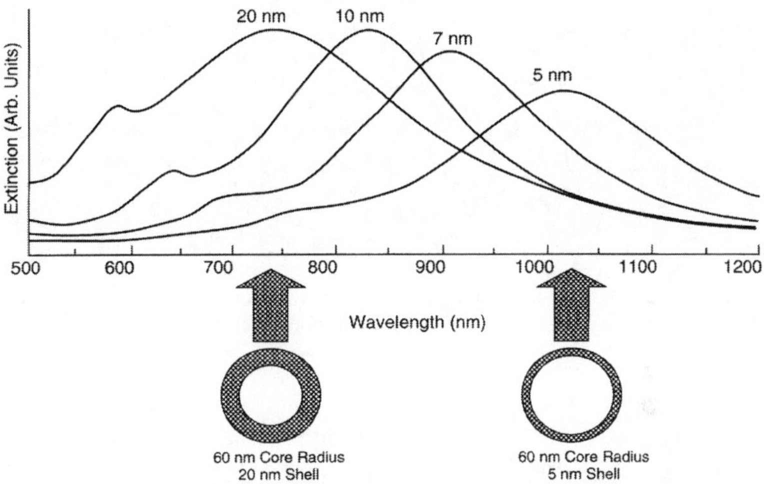

Figure 13.1 Dependencies of the extinction cross section of gold shell–silica core nanoshells with a nucleus of 60 nm radius on the shell thickness (Loo et al., 2004).

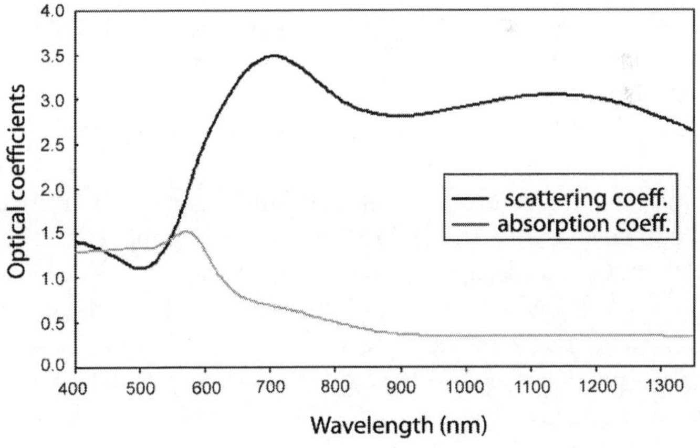

Figure 13.2 Normalized scattering and absorption cross sections of a layered nanoparticle with a silica core of 120 nm radius of, coated with a gold shell of 35 nm thickness. The scattering maximum (705–710 nm) is 2.5 times greater than the absorption maximum (570 nm) and is extended to the region of the near-infrared band (Loo et al., 2004).

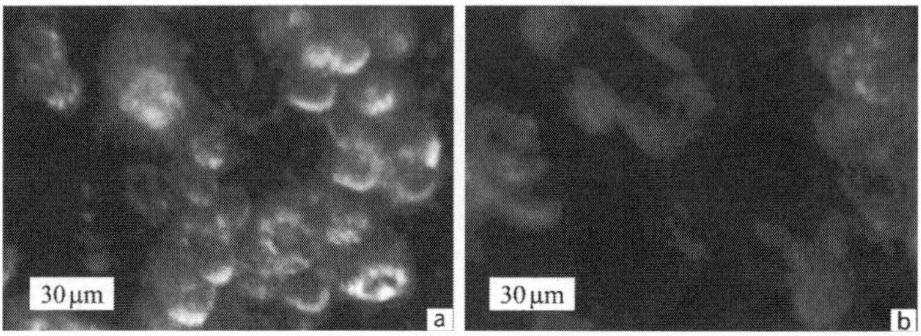

Figure 13.3 A dark-field image of cancer cells treated by nanoparticles functionalized by specific antibodies (a) and nonspecific antibodies IgG (Loo et al., 2004). *Abbreviation*: IgG, immunoglobulin G.

property was used for the visualization of breast tumor (SKBR3) (Loo et al., 2004, 2005). In Fig. 13.3, results of this visualization with the help of gold nanoshells of the sizes described above and coated with specific cancer antibodies HER2 are shown. From this picture, it is seen that the cancer cells become visible after they have been treated with nanoparticles of this kind (Fig. 13.3a). And, vice versa, when nanoparticles are coated with nonspecific, neutral antibodies, they do not attach to cancerous cells, and as a result the cancerous cells become barely visible (Fig. 13.3b). Thus, gold nanoparticles help to visualize tumors.

Loo et al. (2004), Hirsch et al. (2003) used smaller gold nanoshells with a core of a radius of 60 nm and a shell width of 10 nm for cancer therapy. These nanoshells have high absorption in the near-infrared range. In Fig. 13.4, preliminary results of breast carcinoma therapy by laser irradiation at the wavelength of 800 nm are shown. From this picture, the high efficiency of the nanoparticles' usage for cancer cell eradication is seen.

At the present time, other methods of tumor visualization with the help of gold nanoparticles are also being developed, in particular the optoacoustic method. Nanoparticle penetration and heating are performed by the same way as described above. However, in this case a scattered optical signal is not measured but an acoustic one, which arises in the moment of the start of water boiling near the overheated nanoparticle. A relatively small sound absorption in a living tissue is the advantage of this method (Copland et al., 2004).

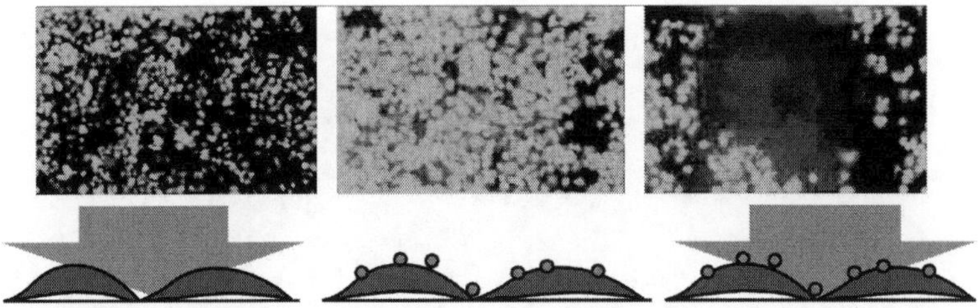

Figure 13.4 The image of cancer cells stained with the help of calcein AM (the green color shows living cells). (Left): Cells irradiated by laser only (no nanoparticles). (Middle): Cells treated by nanoparticles but not irradiated by laser. (Right): Cells treated by nanoparticles and irradiated by laser. The black spot corresponds to dead cells, which were killed as a result of the nanoparticles' heating by irradiation (Loo et al., 2004).

13.2 Biosensors on Surface Plasmons

Possibly, biosensors of different manufacturers, based on surface plasmon resonances, are most commonly used and most available on the market. There is a range of excellent reviews of this subject (Liedberg et al., 1995; Schuck, 1997; Homola et al., 1999; Myszka and Rich, 2000). The biosensors based on surface plasmons are used for biochemical analyses of glucose and blood urea nitrogen; immunoanalysis of proteins, hormones, steroids, and viruses; DNA analyses; and, finally, investigation of drugs' actions kinetics in real time. All biosensors based on surface plasmons use the fact of their strong localization and high sensitivity to a change of medium dielectric properties near the interface.

In nanosensors of this kind, the Kretschmann geometry of the surface plasmons' excitation (Chapter 4, Fig. 4.12) is used, and measurements of the angle determining the minimal reflection coefficient, the value of the reflection coefficient, and the resonance curve width are performed. This data is used for computer-based determination of real and imaginary parts of the refractive index and of the width of the analyzing layer on the basis of the Fresnel formulas for refraction coefficients.

Equations stated in Chapter 4 about surface plasmons imply that a change of the angle of minimal total internal reflection $\delta\theta$ as a variation function of dielectric permittivities of metal $\delta\varepsilon_{\mathrm{m}}$ and

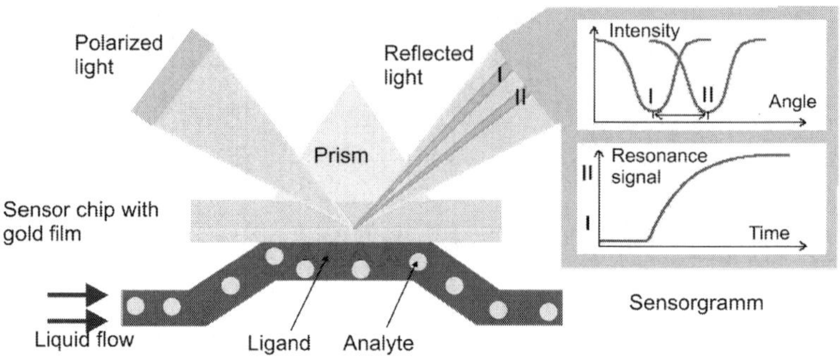

Figure 13.5 The scheme of a typical biosensor based on surface plasmons.

dielectric $\delta\varepsilon_{\mathrm{d}}$ is determined by the equation (Nunzi and Ricard, 1984)

$$\cot\theta\delta\theta = (2\varepsilon_{\mathrm{m}}\varepsilon_{\mathrm{d}}(\varepsilon_{\mathrm{m}} + \varepsilon_{\mathrm{d}}))^{-1}(\varepsilon_{\mathrm{m}}^2\delta\varepsilon_{\mathrm{d}} + \varepsilon_{\mathrm{d}}^2\delta\varepsilon_{\mathrm{m}}). \qquad (13.1)$$

Since $|\varepsilon_{\mathrm{m}}| \gg |\varepsilon_{\mathrm{d}}|$, the change of θ is more sensitive to the variation of electric permittivity of the sensitive layer ε_d than to the variation of electric permittivity of metal ε_m. Thus, this method is ideally suitable for measurement of $\delta\varepsilon_d$ and, consequently, for determination of the bound analyte quantity.

In commercial biosensors on surface plasmons, a charge-coupled device (CCD) camera or a set of photodiodes is usually used for measurement of reflected light properties change reflected light properties (Fig. 13.5). These facilities can be used for real-time measurement of kinetics of analyte binding with a functionalized substrate. To obtain this information, the investigated analyte solution moves near the sensor with a surface coated with corresponding ligands. During the binding phase, the analyte is being accumulated in the ligands, resulting in a gradual increase of the angle of minimal reflection until the saturation is achieved, which is determined by the quantity of the used ligand. During the dissociation phase, the flow of the analyte solution is replaced by a buffer solution washing out the analyte, and as a result the signal decreases. The curve obtained as a result of the association and dissociation cycle of an analyte is often called "sensorgram." The part of the sensorgram determined by the processes of analyte binding is shown in Fig. 13.5, on the right.

In conclusion of this section, let us mention again main advantages of biosensors making use of surface plasmons:

- No fluorescent or other markers are required, and this allows one to investigate a large class of biomolecular systems.
- There is a possibility of real-time measurement, which allows one to determine the kinetics of analyte binding with ligands in real time and the rate of drug action, in particular.
- For typical analyses, small amounts of substance are sufficient.

13.3 Biosensors Based on Localized Plasmons in Nanopaticles

Such devices are also very sensitive to dielectric properties of the environment. Two types of nanosensors are developed on the basis of this principle. According to the first one, an analyzed substance results in agglomeration of plasmon particles and changes in their absorption spectrum (Fig. 13.6 and Fig. 13.7). According to the other, an analyzed substance accumulates on the nanoparticles' surface, resulting in a change of the local electric permittivity and in a change of the solution absorption spectrum (Fig. 13.9).

13.3.1 *The Method of Nanoparticle Agglomeration*

In 1996, the Mirkin and Alivisatos groups published two papers simultaneously, where it was shown that gold nanoparticles can be organized into three-dimensional structures with the help of oligonucleotides attached to them (Alivisatos et al., 1996; Mirkin et al., 1996). Correspondingly, on the basis of a change of spatial organization of nanoparticles the oligonucleotides can be detected, that is, sensors can be developed.

In (Elghanian et al., 1997), one or other probe nucleotide sequence was attached to nanoparticles, and their combination corresponded to the nucleotide to be detected. In Fig. 13.6, nanoparticles with attached probe nucleotides, complementary

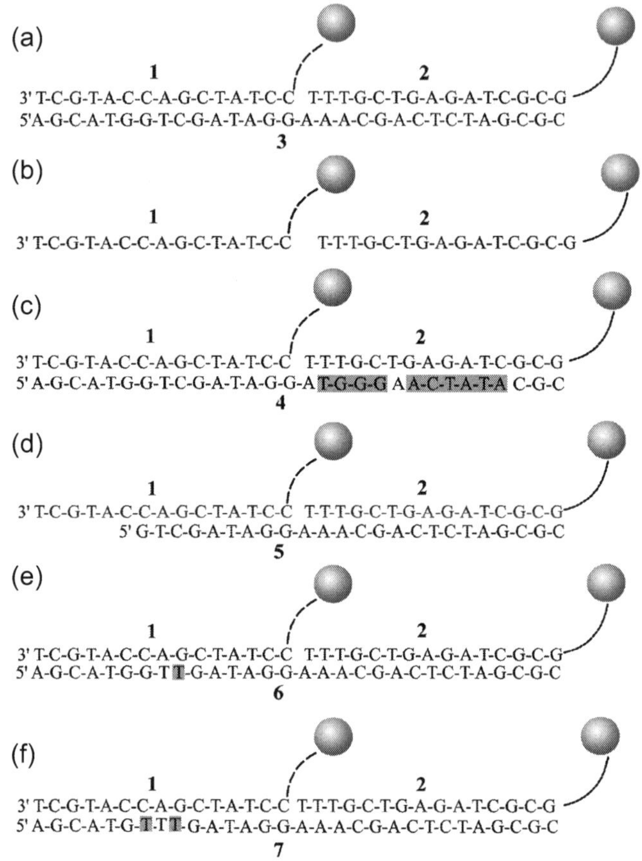

Figure 13.6 13 nm gold nanoparticles with special (probe) nucleotide sequences attached. In the upper line, the probe nucleotides are shown, each of which contains 15 bases denoted by the numbers 1 and 2, a combination of which corresponds to target molecules. In the lower line, different molecules, including the target ones, denoted by the number 3–7 are shown. (a) The complementary geometry—the ideal detection. (b) The probes without targets. (c) The semicomplementary geometry—the lower oligonucleotide has many mismatches relative to the target. (d) The absence of six nucleotides in the chain. (e) The mismatch of one nucleotide. (f) The mismatch of two nucleotides. For simplicity, only two nanoparticles are shown. In practice, the polymer aggregates of nanoparticles have a more complicated form, as it is shown in Fig. 13.7. The dashed line shows the flexible and nonactive connection between a nanoparticle and a "functional" part of the sensor (Elghanian et al., 1997).

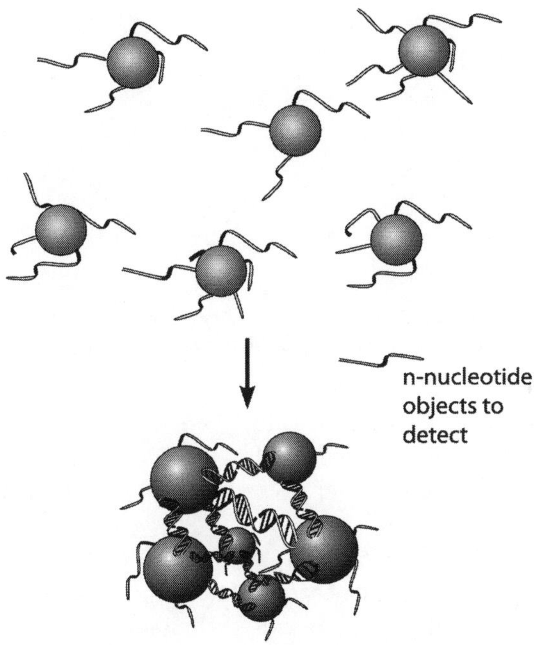

Figure 13.7 The schematic representation of nanoparticles' agglomeration process during target molecules addition to the solution. Connections between the nanoparticles and the oligonucleotides are not to scale, and a much greater number of oligomers is related to one nanoparticle than it is shown in the figure.

target molecules (Fig. 13.6a), and similar molecules are shown, which, however, do not agree completely with the structure of the probe nucleotides attached to the nanoparticles. In practice, much more "signal tails" are attached to one nanoparticle. After nanoparticles prepared in this manner have been added to the solution containing target DNA molecules, agglomeration of the nanoparticles occurs (Fig. 13.7) and the solution color changes from red to dark purple. This system can detect 10 femtomoles of oligonucleotides.

Hirsch et al. (2003) proposed to use nanoparticles having the plasmon resonance in the near-infrared region for prompt and effective immune analyses of blood. In commonly accepted methods of immune analysis, the visible frequency range was used. In

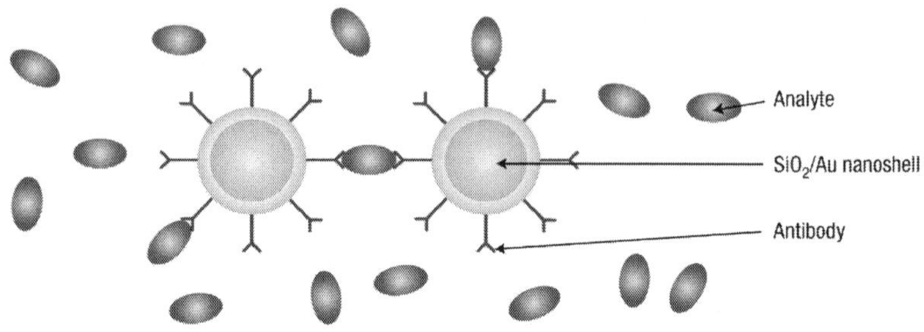

Figure 13.8 Forming of dimers from nanoshells by means of interaction between the antibodies located on them and the analyte (immunoglobulin).

this frequency range, a blood component exists that absorbs light strongly, and the blood is to be preliminarily cleared from absorbing components, and some other procedure is to be performed before carrying out the immune analysis.

In the procedure proposed by Hirsch et al. (2003), layered nanoparticles with a silica core coated with a gold layer, to which antibodies are attached that can be bound with the help of a detectable analyte (immunoglobulin), are used. The analyte addition to the solution of gold nanoparticles results in formation of dimers (Fig. 13.8) having an absorption spectrum different from that of single nanoparticles (Fig. 13.9). Then, the presence of the analyte (immunoglobulin) is detected promptly by the absorption spectrum measurement in the near-infrared range and thus not requiring a long-lasting procedure of blood purification from absorbing (in the visible light) components.

13.3.2 *The Method of Change of Local Dielectric Permittivity*

To describe principles of this method, let us assume that gold nanoparticles are coated (functionalized) with a layer of antibodies of the thickness of 3–7 nm. Detected proteins get bound with antibodies and form a shell of a thickness from 15 nm to 30 nm (Fig. 13.10). This additional protein layer shifts the plasmon resonance frequency (see Chapter 6, Section 6.6), provoking corresponding changes in the absorption spectrum.

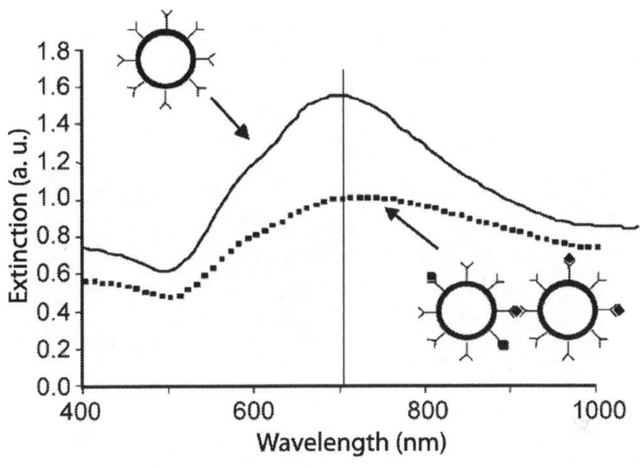

Figure 13.9 A spectrum of nanoshells with a core of 96 nm and a shell of 22 nm thickness and a spectrum of dimers of the same nanoshells bound by an analyte and antibodies (Hirsch et al., 2003).

In Fig. 13.11, plasmon resonance frequency dependence on the thickness of a protein layer coating a gold nanoparticle is shown. From this image, it is seen that a change of the plasmon resonance wavelength has the saturation in the region where the shell thickness exceeds the gold core diameter approximately by two times. It is very important for providing of selectiveness of this method, and, namely, other proteins, situated in the solution and

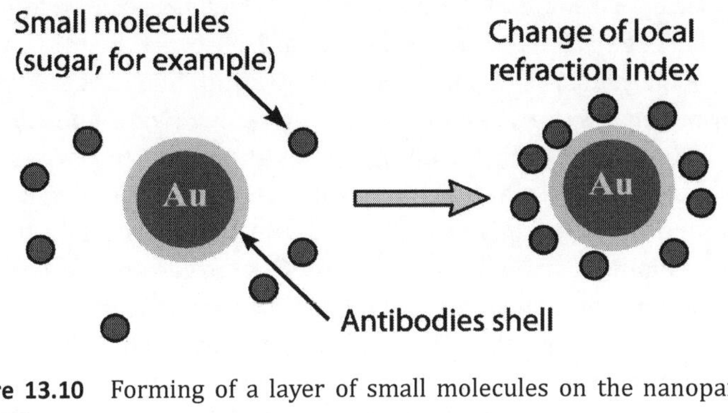

Figure 13.10 Forming of a layer of small molecules on the nanoparticle surface.

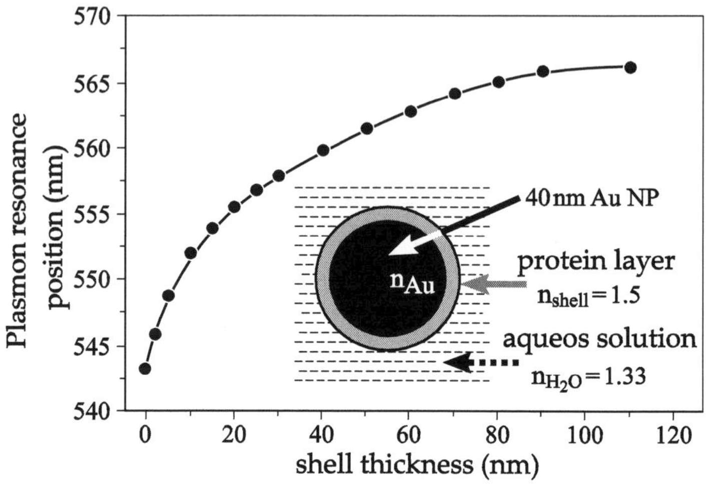

Figure 13.11 Dependence of plasmon resonance peak wavelength of a gold nanoparticle of 40 nm diameter on the thickness of a protein with the refractive index $n_{\text{shell}} = 1.5$. Calculations were performed in the context of Mie's theory for spherically layered systems (Raschke et al., 2003; see also Chapter 6).

not interesting for us, are not expected to change the position of plasmon resonance. From Fig. 13.11, it is seen that the proteins situated at the distance of 100 nm from the nanoparticles will not cause spectrum distortion and appearance of a false signal. More detailed investigation of the relation between the plasmon resonance position and the protein layer thickness can be found in (Malinsky et al., 2001; Haes et al., 2004a, 2004b).

This method of nanoparticles' usage was applied for detection of small molecules, for example, sugar molecules (Nath and Chilkoti, 2002, 2004). Stabilizing coatings of gold nanoparticles (ligands) are selected in the manner that ensures that target molecules can be bound with them. As a result of the nanoparticles' addition to the solution, gradual binding of the target molecules with the nanoparticles takes place, and the electric permittivity of the surface adjacent layer and the plasmon spectrum change. It is rather important that a spectrometric sensor have the detection limit from 16 nanomoles for streptavidin. This sensor provides well-reproducible results, it can be constructed easily (only a spectrophotometer in the ultraviolet and visible range is needed), and it can be used for real-time observation of molecules to be detected for many samples simultaneously.

13.4 Spectroscopy of Single Plasmon Nanoparticles

Biosensors considered in Section 13.3 are effective for solution of many problems. Their main disadvantage is nonhomogeneous plasmon resonance broadening since now it is difficult to ensure equal sizes and shapes of millions of nanoparticles. Biosensors using spectroscopy of a single nanoparticle are free from this disadvantage.

Methods of this kind were proposed by Klar et al. (1998) and confirmed experimentally by Mock et al. (2003). In an early investigation by Klar et al. (1998), a near-field microscope was used for observation of nanoparticles. However, dark-field spectroscopy and related methods are more convenient for investigation of light scattered by gold nanoparticles (Yguerabide and Yguerabide, 1998a, 1998b; Schultz et al., 2000; Sonnichsen et al., 2000, 2002; Taton et al., 2001).

The first biosensors based on usage of single gold nanoparticles were realized by Raschke et al. (2003). In these experiments, a gold nanoparticle functionalized (coated) by biotin-modified albumin was used for directed detection of streptavidin. The main result of this experiment is shown in Fig. 13.12.

At first, during 15 minutes a functionalized particle without addition of an analyte (streptavidin) was observed. At that the peak of the scattered light was not changed to an accuracy of ± 0.5 meV. After streptavidin was added, the peak began to move in the red region. The displacement of the peak is saturated for a shift of 5 meV. This shift value corresponds to approximately 200 molecules of streptavidin on the nanoparticle surface. Addition of other solutions of proteins containing no streptavidin does not make any change, and this confirms selectiveness of this method.

Nonspecific detection of 60,000 molecules of hexadecanethiol with usage of one gold nanoparticle was demonstrated by

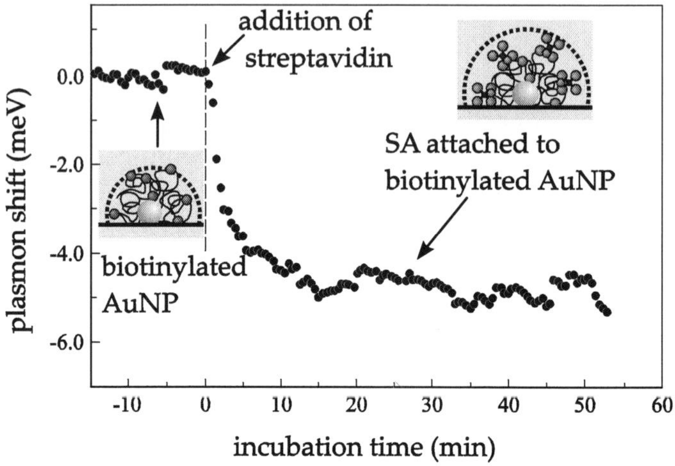

Figure 13.12 The illustration of operation of a nanosensor based on a single gold nanoparticle. The gold nanoparticle was functionalized by biotin-modified albumin. At first, during 15 minutes the functionalized particle without addition of an analyte (streptavidin) was observed. The peak of the scattered light was not changed. After streptavidin was added, the peak began to move to the red region. The displacement achieved its maximum for a shift of 5 meV. This shift value corresponds to approximately 200 molecules of streptavidin on the nanoparticle surface (Raschke et al., 2003).

McFar-land and Van Duyne (2003). Quite recently, Liu et al. (2005) were able to measure plasmon spectra of more than 1,000 separated gold colloidal nanoparticles on a glass slide. Since nanoparticles are rather small, the biochip's area for detection of one sort of proteins is restricted only by a microscope's resolution. Therefore, the information density of sensors based on single nanoparticles is of the order of $1/\mu m^2$. This offers a big hope for further miniaturization of the biochip technology.

Separate metal nanoparticles as markers. Methods using nanoparticles for precise and prompt analyses can be applied also in cases when detection and analysis of small amounts of molecules are to be performed. In this case, nanoparticles appear as a substitute of fluorescence markers. Main advantages of gold particles is their (in comparison with fluorophores) large scattering cross sections and high photostability.

At the first time, detection of small amounts of molecules by counting of single metal nanoparticles was performed by Schultz et al. (2000). They realized an immunocytological analyzer and even a "sandwich" of such analyzers. In these analyzers, a number of nanoparticles visible with the help of dark-field spectroscopy was treated as a signal. Results of more detailed investigations of possibilities of single nanoparticles to act as markers in histological, immune, and DNA research were stated by (Yguerabide and Yguerabide, 2001). Another strategy is to use oligonucleotides bound with biotin molecules. Large silver nanoparticles functionalized by antibiotin antibodies are used after a biomolecular "sandwich" with analyzed oligonucleotides has been formed on the substrate (Oldenburg et al., 2002) and again can be detected by dark-field microscopy.

Detection of metal nanoparticles of a diameter of less than 20 nm is a difficult problem since they have small scattering cross sections. Nevertheless, sometimes it is necessary to use such particles for attachment of antibodies or specific oligonucleotides since larger nanoparticles do not allow antibodies to approach and recognize corresponding antigens or DNA. To count the amount of small gold nanoparticles, several strategies are developed. For example, emission of small gold nanoparticles can be enhanced with the help of silver (Schultz et al., 2000; Storhoff et al., 2004b). In this case, silver is precipitated on the surface of gold nanoparticles, and the larger nanoparticles can be easily detected by means of dark-field spectroscopy.

Recently, other interesting ways were proposed to increase sensitivity of methods of analysis using gold nanoparticles. Gold nanoparticles of a diameter of less than 10 nm were detected when interferential methods of confocal microscopy were used (Lindfors et al., 2004).

13.5 Element Base for Plasmonic Integrated Circuit

At the present time, there is a necessity for development of devices allowing fast data transmission and processing. These necessities have already led to development of silicon electronics, and for the moment in this area we observe further progress, which results in reduction of sizes and an increase of operation frequency and efficiency of computers. In turn, the reduction of the sizes of processing and transmission devices results in a further fantastic growth of the number of possible applications of these devices. Unfortunately, interconnections' heating in microcircuits makes difficulties for further progress (Bohr, 1995; Chiang et al., 2002; Banerjee et al., 1996). Optical interconnections have almost unlimited informational capacities, and on their basis solutions can be found, allowing one to avoid existing problems of microcircuits (Miller, 2000; Chen et al., 2005).

The optical alternative can be especially attractive with respect to future microcircuits with more distributed architecture, in which a multitude of fast silicon cores are to be connected by high-speed optical lines. Unfortunately, development of such kind of devices is complicated by a great difference in sizes of electronic components, which are to be 100 nm or less, and dielectric optical components, which are limited in size owing to diffraction laws (1 μm or greater). This evident mismatch of sizes of electronic and photonic devices results in significant problems in their coupling. Further progress requires development of brand-new technologies that provide data transmission between nanoscale electronic cores at optical frequencies and thus fill the gap between nanoscale electronics and microscale photonics. In Fig. 13.13, the place of nanoplasmonics among existing technologies of data processing and transmission is shown schematically.

Devices based on surface and localized plasmons appear to be evident candidates for a successful solution of this problem. At the present time, active investigations of nanoplasmonics' application to solution of the problem of data transmission between nanoscale electronic cores at optical frequencies are in progress. In this section, both passive (waveguides, splitters, mirrors) and active (shutters and switchers) elements are considered.

13.5.1 *Passive Elements*

Metal striplines, which resemble usual metal conductors to a great extent, appear to be the most prospective and investigated plasmon

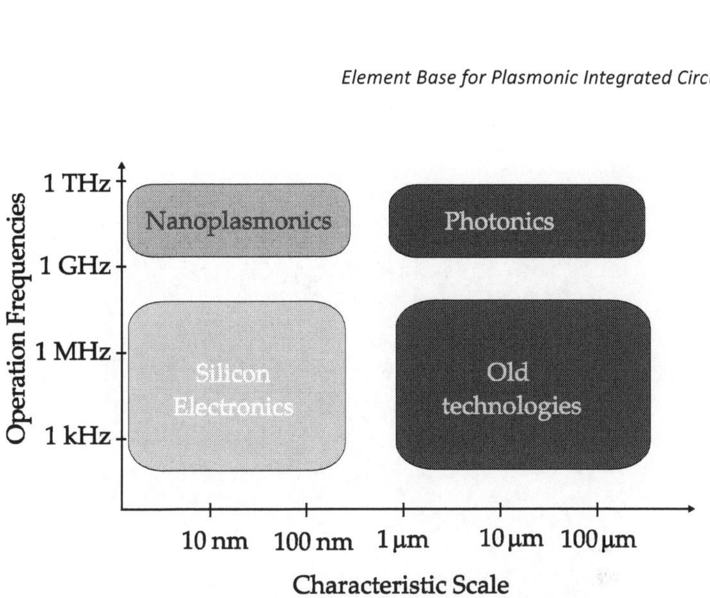

Figure 13.13 The operation rate and critical sizes of different technologies of data processing and transmission.

waveguides. In particular, Zia et al. (2004) formed gold nanostrips of 55 nm thickness and of the width from 50 nm to 5 μm on a silica substrate with the help of electron beam lithography. Gold stripes are ideal candidates for investigation of plasmon propagation since they can be produced easily, are not exposed to oxidation, and have properties similar to the properties of cuprum and aluminum (Zia et al., 2004).

Fig. 13.14a shows the optical microphotograph of a typical plasmon device that consists of a "large" gold pad, where plasmons are excited with the help of methods described in Chapter 4 and then directed to plasmon waveguides of different geometry. In the insertion, an enlarged scanning electron microscopy (SEM) image of one of the gold stripes is shown. The red arrow shows how plasmons enter a waveguide. Fig. 13.14b,c,d shows intensities' distribution of plasmon fields excited at the wavelength of 780 nm and propagating along waveguides of 3 μm, 1.5 μm, and 0.5 μm width. A 3-micron stripe waveguide supports plasmon signal propagation at the distance of several tens of microns. These pictures and theoretical calculations imply that the propagation length decreases as the nanostrip width does. More fundamental understanding of this behavior and optimization of

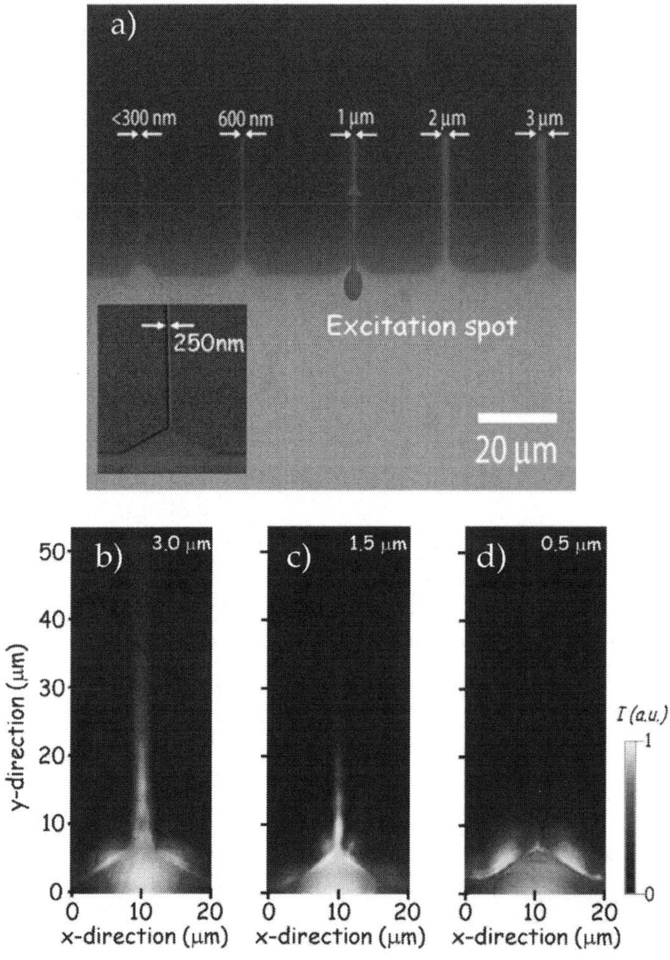

Figure 13.14 Visualization of process of plasmon propagation along gold nanostrips of different widths. (a) The image of gold nanostrips on a substrate of silicon dioxide formed with the help of FIB lithography. The red arrow shows the plasmon launch to a nanostrip of 1 μm width. (b, c, d) SNOM photographs of the process of plasmon propagation along strips of 3 μm, 1.5 μm, and 0.5 μm thickness (Zia et al., 2006). *Abbreviations*: SNOM, scanning near-field optical microscopy; FIB, focused ion beam.

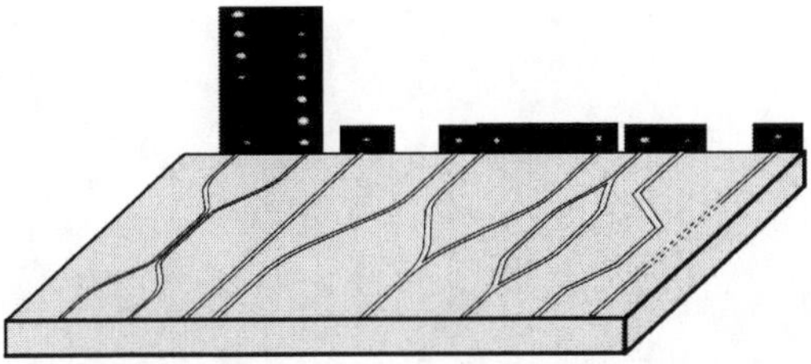

Figure 13.15 The schematic representation of several optical devices based on surface plasmons. On the top, optical images are shown that arise at the output of really functioning devices. The Bragg grating formed in the middle of the waveguide is shown on the right. Different signals ratios at the output of the four-port coupler (on the left) are related to the distance between straight sections of the waveguides. This distance was changed by 1 μm steps from 8 μm to 2 μm (below), and plasmons were launched from the left input arm (Charbonneau et al., 2005).

these waveguides can be obtained on the basis of a solution of the total system of Maxwell's equations in this geometry.

Straight and bent waveguides are the basis for development of more complex nanodevices, such as S-bend, four-port divider, Y-divider, and Mach–Zehnder interferometers. Functioning of these devices was demonstrated experimentally with the help of gold strips inside SiO_2 at the wavelength of $\lambda = 1{,}550$ nm (Charbonneau, 2001) and recently with the help of gold nanostrips on the surface of SiO_2, coated with a layer of polymer for matching of the refractive indices, also at the wavelength of $\lambda = 1{,}550$ nm (Charbonneau et al., 2005). In Fig. 13.15, measured intensities at the output of these devices are shown. On the right of this picture, the Bragg grating formed in the middle of the waveguide is shown. Different correlations at the outputs of four-port couplers (on the left) are related only to the distance between straight sections of the waveguides. This distance was changed with 1 μm intervals from 8 μm to 2 μm (below), and plasmons were launched from the left input arm.

An element that turns a wave through 90 degrees is the most important part of any nontrivial waveguide. In plasmonic stripline waveguides, this turn can be performed, for example, with the help of the Bragg mirror (Weeber et al., 2004).

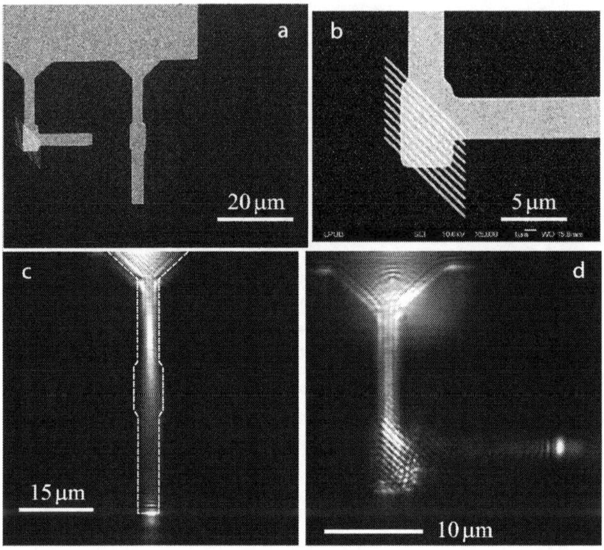

Figure 13.16 The illustration of possibility to turn a waveguide through 90 degrees. (a, b) SEM images of the turning element with an inclined Bragg lattice. The Bragg mirror consists of gold combs of 60 nm height and with 550 nm distance between them. Losses related to the 90 degrees turn were compared with the case of a straight waveguide of analogous length. (c, d) SNOM images of plasmon intensity in a straight waveguide and a waveguide with a 90 degrees bend (Weeber et al., 2004).

In Fig. 13.16a,b, the SEM image of a turning element with an inclined Bragg grating is shown. These nanostructures were formed with the help of double electron beam lithography: initially a mirror was formed, and then gold stripes were formed. The Bragg mirror consists of gold combs of 60 nm height, 150 nm width, and 550 nm period. To estimate the turning element effectiveness, losses related to a turn trough 90 degrees were compared with the case of a straight waveguide of analogous length. In Fig. 13.16c,d, SNOM images of plasmon intensity in straight waveguides and waveguides with 90 degrees bend are shown. Comparing intensities of straight and bent waveguides, (Weeber et al., 2004) have made the conclusion that the bend induces the losses of about 2 dB. The turn through 90 degrees is the simplest application of the Bragg mirror.

With the help of it, other elements, in particular dividers, can be formed also.

So far, we have given examples of plasmonic stripline nanowaveguides that enable long-range propagation of plasmons. Certainly, at that, modes' localization in the transverse direction was not too strong. However, there are other possibilities of plasmon channeling. In particular, for this purpose a "metal-dielectric-metal" system can be used. We have considered the theory of these modes in Chapter 4. In this case, the transverse superlocalization of plasmons is possible for a rather long range of propagation. For example, when the dielectric gap thickness between the metals is 50 nm, the area of localization is of the order of 100 nm, and the propagation length exceeds 1 μm, which cannot be achieved with the help of usual dielectric waveguides (Zia et al., 2004). More complex devices can also be considered on this basis (Tanaka and Tanaka, 2003; Veronis and Fan, 2005; Bozhevolnyi et al., 2006).

Besides, to control plasmon propagation, one can use nano-waveguides, which are V-shaped nanogrooves milled in a metal film. The analytical (Novikov and Maradudin, 2002) and numerical (Pile and Gramotnev, 2004) investigations have shown that localized modes (channel plasmons) exist on the grooves' bottom. Since the dispersion laws of channel and usual surface plasmons differ, they do not interact, and channel plasmons can propagate without a decrease of localization. The law of the dispersion of these modes is qualitatively similar to that in planar nanostructures (Bozhevolnyi et al., 2005b). It was shown experimentally that in grooves of 0.6 μm width and 1 μm depth, which were milled by FIBs on a gold surface, plasmon propagation is possible at the wavelength of $\lambda =$ 1,500 nm for a distance of more than 100 μm with the localization of the order of 1.1 μm (Bozhevolnyi et al., 2005b). Examples of plasmon propagation in different passive elements based on channel plasmons are shown in Fig. 13.17 (splitters and interferometers) and in Fig. 13.18 (ring resonator) (Bozhevolnyi et al., 2006). It should be noted that in this investigation the grooves' sizes and the localization of modes are not substantially less than the wavelength, and this explains the great length of the plasmon propagation.

Of course, the examples of plasmonic nanowaveguides and related plasmonic components considered above do not exhaust

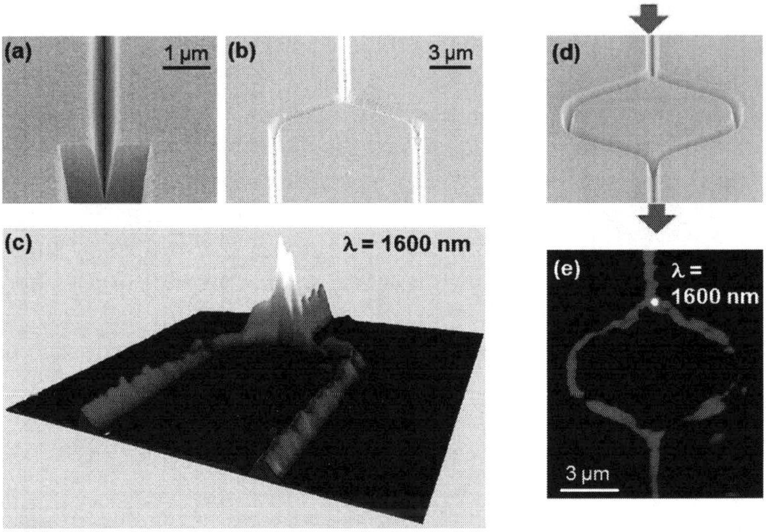

Figure 13.17 The plasmon Y-splitter and Mach–Zehnder interferometers. (a) The SEM image of the the groove profile along which the plasmons propagate, (b) the SEM image of Y-splitter, (c) the SNOM image of the plasmon field in the Y-splitter at a wavelength of 1,600 nm. (d) The SEM image image of the Mach–Zehnder interferometer (e) the SNOM image of the plasmon field in the Y-splitter at a wavelength of 1,600 nm (by the courtesy of Sergey Bozhevolnyi).

all possibilities to provide long-range propagation of highly localized plasmons. Now, new possibilities to improve plasmonic nanowaveguides are under investigation. In this direction, one should first of all mention dielectric-loaded surface plasmon polariton waveguides on a metal film (Steinberger et al., 2006, 2007). The geometry of such a waveguide is shown in Fig. 13.19. A waveguide of such a geometry can be very effective for long-range propagation of highly localized plasmons. Besides, such a geometry seems to be very technological. From a physical point of view, such a geometry of a waveguide results in lower losses in comparison with the above-considered waveguides because now the main energy of the electromagnetic field is localized in a dielectric.

Even lower losses can be achieved if one makes use of hybridization of plasmonic modes and modes in a semiconductor waveguide

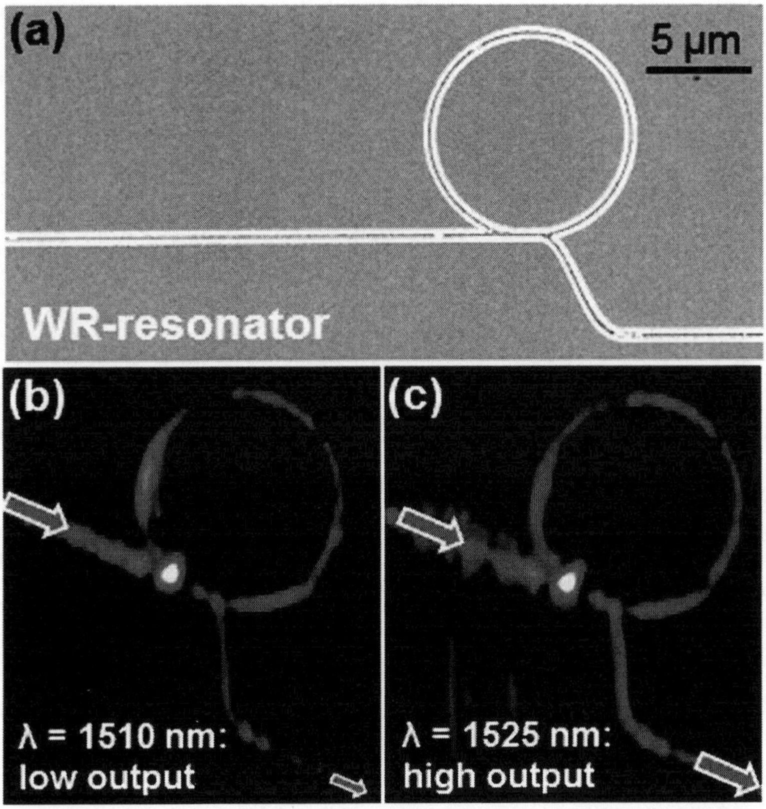

Figure 13.18 The plasmon ring resonator based on channel plasmon technology. (a) The SEM image of the ring resonator, (b) and (c) the SNOM image of the plasmon field in the ring resonator at a wavelength of 1510 and 1525 nm. (By the courtesy of Sergey Bozhevolnyi).

separated by a gap from the metal surface (see Fig. 13.20). Coupling between the plasmonic and the waveguide modes across the gap enables "capacitor-like" energy storage that allows effective subwavelength transmission in nonmetallic regions. In this way, surface plasmon polaritons can travel over large distances (40–150 μm) with strong mode confinement (ranging from $\lambda^2/400$ to $\lambda^2/40$). This approach is fully compatible with semiconductor fabrication techniques and could lead to truly nanoscale semiconductor-based plasmonics and photonics (Oulton et al., 2008).

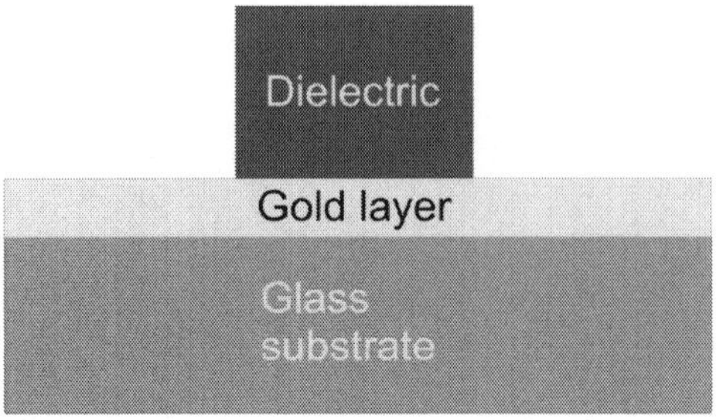

Figure 13.19 The dielectric-loaded surface plasmon polariton waveguide on a metal film.

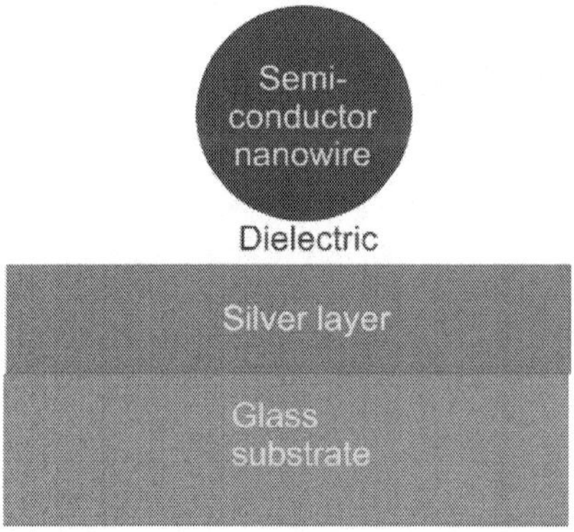

Figure 13.20 The semiconductor waveguide separated by a dielectric gap from the metal surface. Great propagation length is provided by hybridization of surface plasmon modes and modes inside the semiconductor nanowire.

13.5.2 *Active (Dynamic) Elements*

In the previous section, it was demonstrated that there are some achievements in development of data transmission line based on plasmons in metal or metal-dielectric nanostructures, that is, in development of passive elements. However, active elements, which can change their state, are also required for effective control over optical data flows. Such active elements can be used to match silicon cores and optical transmission lines. Moreover, purely plasmonic data processors and computers can be developed on the basis of such active elements.

Active elements of this kind are to be nonlinear necessarily. Electro-optical, magneto-optical, and thermo-optical properties of different parts of nanodevices can be used as physical processes providing nonlinearity. Here, we will restrict ourselves to a brief consideration of some active elements on the basis of thermo-optical effects, that is, on the basis of the refractive index modification by temperature.

The most important property of plasmon waveguides on the basis of striplines is that they can transmit both plasmons, that is, an information optical signal, and control electric currents of low frequency. Recently, such active plasmon components as different optical modulators and switches in the range of the telecom wavelengths of $\lambda \approx 1{,}500$ nm were proposed and realized on the basis of thermo-optical effects (Nikolajsen et al., 2004). In nanodevices of this kind, the control over plasmon transmission through a waveguide is provided by changing of the refractive index of the substance surrounding the waveguide (plastic) due to the plasmon waveguide heating by an electric current. It was also demonstrated that these striplines can be used as sensors of the energy transmitted by plasmons by measurement of changes in waveguide electrical resistance (Bozhevolnyi et al., 2005a).

In Fig. 13.21a,b, principal schemes of a Mach–Zehnder interferometer and a directional-coupler switch with one arm heated by an electric current are shown. In Fig. 13.21c, the image of a Mach–Zehnder interferometer with the maintained proportion is shown. In this figure, one can see a relatively great arms' curvature (the curvature radius is 20 mm) needed for losses reduction during plasmon propagation. Slits in the waveguides (10 μm) are necessary for electric isolation of the heated section from the rest of the nanodevice's sections. The slits presence does not add any essential losses to plasmon propagation. In Fig. 13.21d, a zoomed image of "big" electrodes (20 × 20 μm) connected to the controllable arm of the interferometer by means of a gold wire of 20 μm length and 20 nm thickness is shown.

Another possibility to develop active plasmon nanodevices is making use of nonlinear elements on the same substrate where the

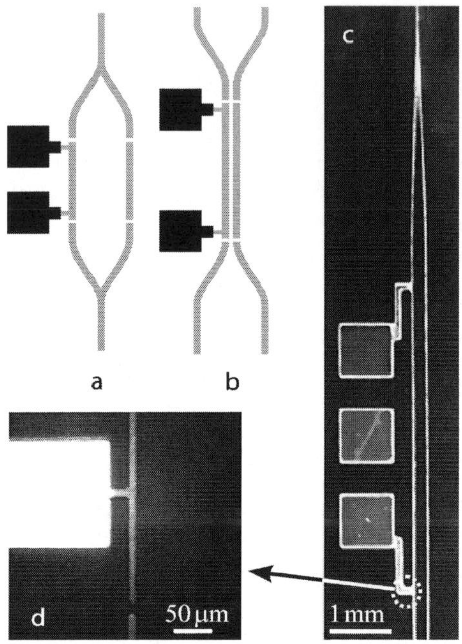

Figure 13.21 Principal schemes of a plasmonic Mach–Zehnder interferometer and a directional coupler switch, one of the arms of which is heated by an electric current (a, b). The optical microscope image of an optical Mach–Zehnder interferometer (c). The magnified image of "big" electrodes (20 × 20 μm) that are connected to the operating interferometer arm by means of a gold wire of 20 μm length and 20 nm thickness (d) (Nikolajsen et al., 2004).

plasmon nanowaveguide is situated. A metal stripe of gallium, where a structural phase transition can occur, can be used as an element of this kind. Gallium is remarkable for it melts at 29.8°C and its electric permittivity varies sevenfold at that! At the present time, it has been already shown that making use of such phase transitions can form the basis for development of purely optical switches of milliwatt power in thin films and in monolayers of nanoparticles (Bennet et al., 1998; Albanis et al., 1999; 2001; Petropoulos et al., 1999; 2001; MacDonald et al., 2001a, 2001b, 2003, 2004; Zheludev, 2002; Fedotov et al., 2003; Pochon et al., 2004).

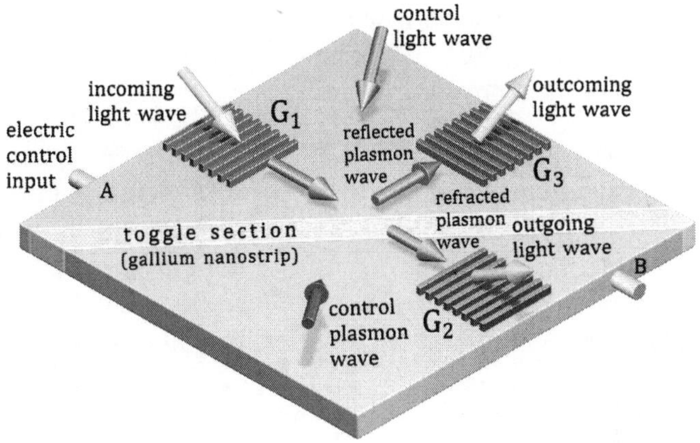

Figure 13.22 The generalized scheme of the active switch of the surface plasmon propagation direction.

In Fig. 13.22, the generalized scheme of an active switcher of the surface plasmon propagation direction is shown (Krasavin et al., 2007). The main element of this scheme is a metal-dielectric plasmon waveguide containing a section of gallium and Bragg gratings for plasmons input and output, that is, for optical signals. In a real application several switches of such kind can be connected together, resulting in development of more complex devices for plasmon signals transformation. Nevertheless, the general scheme of the switch shown in this figure contains all fundamental elements and is sufficient for the demonstration of its potential. The incoming light signal generates surface plasmons on the grating G1. This signal propagates along the planar waveguide and transmits through the switching gallium section or reflects from it and then with the help of the gratings G2 and G3 transforms again into a free propagating optical signal. The power of the optical signal outgoing from one or other channel depends on properties of the switching section, which depend on the structural phase of gallium. In turn, the gallium structural phase can change reversibly with the help of plasmons themselves, external optical irradiation, and heating by the current running through the gallium section. It is evident that any other material having different structural phases can be used for development of analogous switches, but for the moment gallium is the best known for its polymorphous properties (Defrain, 1977; Bosio, 1978) and the most suitable for development of the active elements of nanoplasmonics.

13.6 Applications Based on the Nanoparticles' Influence on the Radiation and Fluorescence of Atoms and Molecules

Control over the optical properties of single molecules, nanocrystal quantum dots, and nitrogen vacancy (NV) centers in diamond nanocrystals is also one of the most important directions of modern science and technology development.

At the present time, fluorescence is the key technology in medicine and biology. For example, decoding of DNA structure with the help of fluorescent methods was carried out as long ago as in 1987 (Smith et al., 1986; Prober et al., 1987), resulting in almost total determination of the structure of the human genome (Human Genome, 2001; Pennisi, 2001). Fluorescent methods have replaced radioactive ones in majority of medical tests (Kricka et al., 1997; Ichinose et al., 1987). Recently, the fluorescent methods were used for development of gene chips containing more than 30,000 individual sequences of DNA and suitable for high-performance gene investigations (Lipschutz et al., 1999; Brown and Botstein, 1999). All these important applications were based on spectral properties of free fluorophores, and the main task was to develop and to select the proper fluorophore. Owing to development of nanotechnologies and nanoplasmonics, at the present time a possibility has appeared to extend application area of fluorescent methods and to increase their effectiveness crucially by using plasmonic nanoparticles and nano-objects.

First of all, observation of single molecules should be noted. The data on a single molecule can provide information about the kinetics of some process that cannot be obtained by means of an investigation of molecule ensembles (Xie Lu, 1999; Lu Xun Xie, 1998).

Scanning near-field optical microscopes providing high spatial resolution are one of the most effective instruments for observation of a fluorescent molecule (see, e.g., Dunn, 1999). In investigations of this kind, the needle of a scanning microscope (which can be both metal and dielectric) is situated in the immediate vicinity of the observed molecule and therefore influences both excitation and emission rates of the molecule and its spectral properties substantially due to the emission frequency shift. As a result, the scan image of the molecule is related to its position relative to the needle and to its emission spectrum only indirectly. To find the latter, the influence of the microscope tip on the excitation rate and the emission characteristics of the molecule are to be taken into account.

With the help of microscopes of this kind, a range of investigations of single molecules was carried out (Betzig and Chichester, 1993; Trautman et al., 1994; Xie and Dunn, 1994; Ambrose et al., 1994; Bian et al., 1995; Veerman et al., 1999; Gersen et al., 2000, 2001). For a qualitative description of the influence of such a microscope on the observed molecule, the theory developed in Chapter 12 can be used. Typical images of the molecules are shown in Fig. 12.18, Chapter 12.

Obtained scan images should be interpreted with precaution since they have sizes of the order of the microscope aperture size, and additional data is required to confirm that a single molecule is observed, in fact. There are many investigations, and even in the first of them the scan images agree qualitatively with the theory described above; however, there are some apprehensions that, in fact, clusters of molecules, which have approximately similar scan images, are observed. In this case, results of the theory stated above are to be averaged with respect to the cluster volume.

There are also other methods of observation of single molecules on the basis of their fluorescence. In particularly, (Levene et al., 2003; Rigneault et al., 2005) proposed to use an array of nanoholes in a metal film for the observation of single molecules' dynamics. In devices of this kind, the observation volume can be very small (of the order of 1 zeptoliter $= 10^{-21}$ L), and parallel observation of the molecules' dynamics in a million nanoapertures is possible. Nanoapertures influence substantially the optical properties of fluorophores (Rigneault et al., 2005), and this influence can be estimated with the help of approaches given in Chapter 12.

Figure 13.23 The nanoantenna on the tip of a scanning microscope (Farahani et al., 2005).

An apertureless scanning microscope is another instrument for the observation of single molecules and nanocrystal quantum dots. To describe the influence of the tip of this microscope on the emission of the examined molecule, a theoretical model where the tip of the microscope is described with the help of a conical surface is quite suitable (Klimov, 1998; Klimov and Perventsev, 1999). Moreover, on the end of the tip a (metal) nanobody of a certain structure ("nanoantenna") can be placed, which allows one to control additionally the emission of the examined molecule or the quantum dot. Farahani et al. (2005) used nanoantennas of the shape of a bowtie for that purpose (Fig. 13.23).

Quite recently, excellent research (Anger et al., 2006) appeared, in which the fluorescence intensity change of a dye, "Nile blue," molecule was measured, depending on the distance to a gold nanosphere of 80 nm diameter attached to the tip of an apertureless scanning microscope. In Fig. 13.24, results of measurements and calculations based on the theory presented in Chapter 6 are stated. It is seen from the picture that as the nanosphere approaches the molecule, at first the fluorescence intensity increases by 30 times, and then the fluorescence quenches due to prevalence of nonradiative processes. The measurements performed by Anger et al. (2006) agree completely both with the theory and with other measurements performed for fixed distances between the molecule and the nanosphere.

In general, these experiments show a possibility of flexible control of both radiative and nonradiative characteristics of the

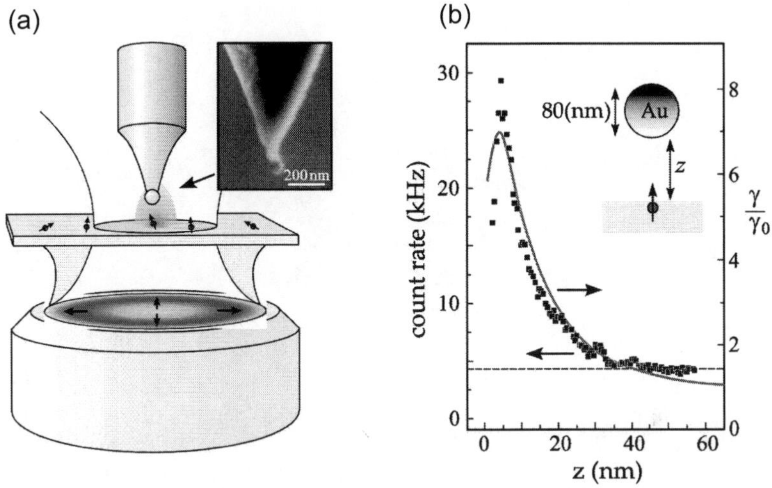

Figure 13.24 The influence of a gold nanoparticle located on a scanning microscope tip on the fluorescence of a single molecule of the Nile blue dye. (a) The geometry of the experiment. A microscope objective (at the bottom) was used to focus a radially polarized laser beam on the molecule and to collect fluorescence photons from it. (b) Results of fluorescence intensity measurement and their comparison with the theory (Anger et al., 2006).

molecules' emission. Let us emphasize that this control is provided not only by means of local enhancement or weakening of the exciting optical field but also by means of change of molecules' spontaneous decay rates due to the presence of the tip (see Appendix A1).

Detection of small amounts of molecules, microorganisms, and viruses is also an important application of plasmon nanoparticles. First of all, this problem is significant in the context of prevention of natural and anthropogenic dangers (terrorism and bioterrorism). In this direction, the problem is to detect promptly and effectively small amounts of molecules, single molecules, and pathogen viruses in a given volume. Here, optical methods seem to be quite effective and operative, but intensity of single molecules' fluorescence can be rather small, requiring a long time for accumulation of information. Insertion of nanoparticles that can be attached to a detected molecule or a virus with the help of one or other chemical method (the "antibody-antigen" method) in the examined volume allows one

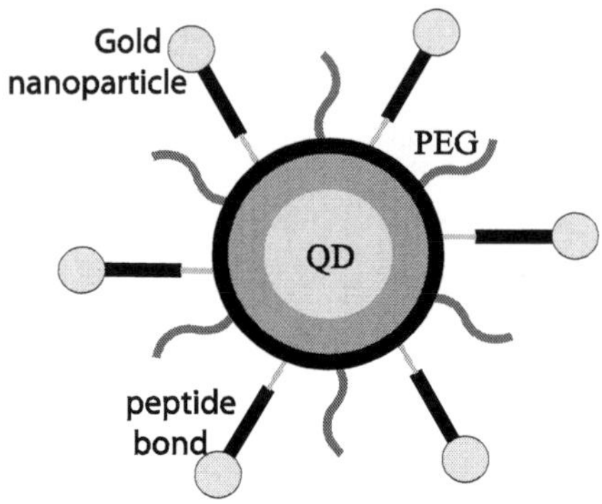

Figure 13.25 The structure of a sensor based on quantum dots. PEG provides improvement of water solubility and an increase of the quantum dots' stability (Chang et al., 2005). *Abbreviation*: PEG, polyethyleneglycol.

to enhance the fluorescence intensity considerably and to detect small amounts of molecules or viruses. In turn, it allows one to solve the problem of prevention of terroristic and others threats. In this case, the theory stated in Chapters 6, 7, and 8, describing properties of spheroidal and ellipsoidal nanoparticles and their influence on the emission of atoms and molecules, is effective.

Gold nanoparticles can be also used in devices based on fluorescence quenching. In particular, Chang et al. (2005) elaborated a nanosensor based on the combination of a nanocrystal quantum dot and gold nanoparticles. In the initial state, gold nanoparticles are attached to the quantum dot by means of chemical bonds (peptide linkers), which are selected on the basis of what protease is to be detected (Fig. 13.25).

The distance between the quantum dot and the gold nanoparticles is so small that nonradiative decay processes prevail, and in the initial state the quantum dot does not emit at all or emits weakly. The detected protease splits the peptide linkers, and the gold nanoparticles move away from the quantum dot, which now can emit efficiently and thus indicate the presence of the detected protease. The operation principle of such a sensor is shown in Fig. 13.26.

Devices for DNA structure decoding are one of the most significant applications of fluorescence enhancement in the vicinity

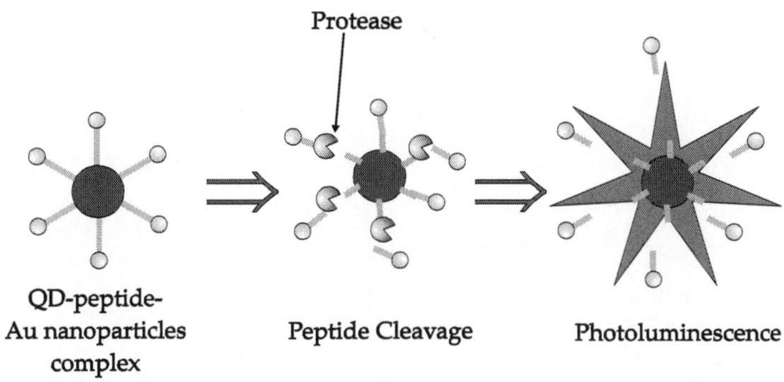

Figure 13.26 The illustration of activation of a sensor based on quantum dots. Target protease splits peptide bond and unbinds gold nanoparticles, which quenched quantum dot photoluminescence (Chang et al., 2005).

of nanoparticles. The problem of DNA structure sequencing is of crucial importance for development of biotechnologies, and a range of approaches to treat this problem now exists. In DNA, each nucleotide residue has a base absorbing ultraviolet light at $\lambda \approx 260$ nm. A corresponding fluorescence could be expected. However, in fact, the DNA bases' fluorescence or the fluorescence of isolated bases is extremely weak and cannot be used effectively for determination of DNA structure (Georghiou et al., 1985; Georghiou et al., 1996). To decode the DNA structure, a set of fluorophores has been developed that fluoresce intensively in the range from ultraviolet to infrared when attached to a corresponding nucleotide. Now, one nucleotide can be cleaved from the DNA, and then it can be identified on the basis of attached fluorescent marker spectrum, and thus the DNA structure can be determined step by step.

Lacowicz et al. (2003) proposed to use silver (or other) nanoparticles to sequence the DNA structure without usage of fluorescent markers. The idea of this approach is shown in Fig. 13.27. Within this approach, a nucleotide sequence or even a whole single strand of DNA passes through a flow chamber with plasmon particles that are resonant with the emission of some or other nucleotides deposited on its walls. The nucleotides have weak fluorescence in free state; however, in the presence of nanoparticles the emission at one or other frequency increases substantially, and one can say

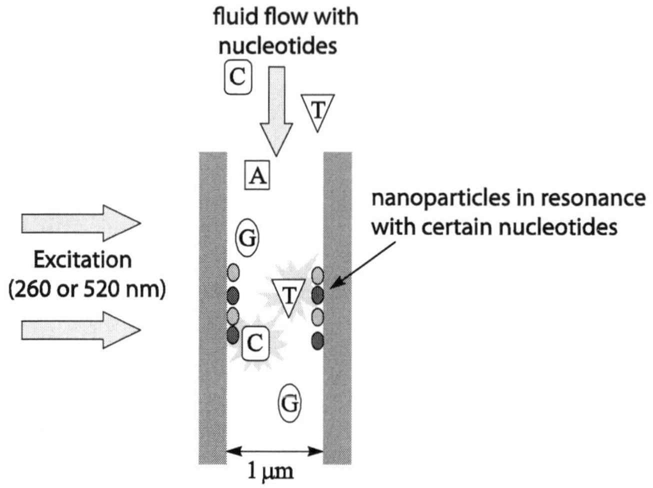

Figure 13.27 Detection and identification of separate DNA nucleotides with the help of resonance plasmon nanoparticles (Lakowicz et al., 2003).

that in this point of space one or other nucleotide is situated. Thus, in the proposed scheme sequential decoding of the DNA structure takes place. In this case, influence of nanoparticles on the spontaneous emission is decisive, and the theory of molecules' emission in the vicinity of nanoparticles, presented in Chapters 6, 7, and 8 (spherical, spheroidal, and ellipsoidal nanoparticles), fits this problem perfectly.

The interaction between fluorophores and nanoparticles has great significance also in usual fluorescent investigations, in which a great number of fluorophores is involved simultaneously. For example, in Fig. 13.28 a photograph of a protein (albumin) labeled by a standard fluorescein dye is shown. The protein labeled in this way was placed on a silica substrate or on a silica substrate coated with silver nanoparticles. The fluorescence of the protein drop is almost invisible on the silica substrate (Fig. 13.28a) and is brightly visible on the substrate coated with silver nanoparticles (Fig. 13.28c). In Fig. 13.28b, the drop of protein is on the boundary of the area coated with silver nanoparticles. The difference between the fluorescence intensity is due to an increase of the radiative

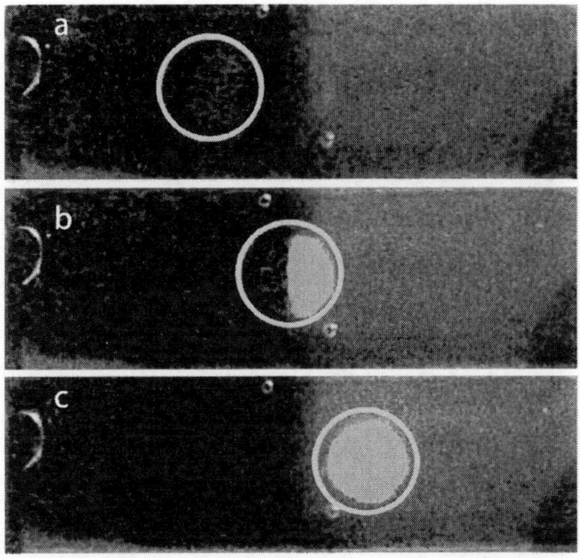

Figure 13.28 A photograph of HSA labeled with the fluorescein on a silica substrate and on a silica substrate coated with silver nanoparticles. (a) The protein drop on the silica; (b) the protein drop located on the boundary of the region coated by silver nanoparticles; and (c) the protein drop on the substrate coated with silver nanoparticles (Lakowicz et al., 2003). *Abbreviation*: HSA, human serum albumin.

part of the spontaneous decay rate, resulting in an increase of the quantum yield of fluorescein fluorescence, in turn.

An even greater fluorescence increase can be achieved by the usage of nanoparticles of a more complex shape, such as nanotriangles and nanorods (Aslan et al., 2005).

A plasmon nanoparticle can be used for fluorescence enhancement of intracellular fluorophores, fluorophores introduced into a cell, and fluorophores situated on the nanoparticle itself (Lakowicz et al., 2003). The nanoparticle can be moved inside the cell with the help of a micromanipulator or an atomic force microscope. This method requires a special selection of wavelength to reduce heating of the examined cell when manipulating the nanoparticles' positions (Fig. 13.29).

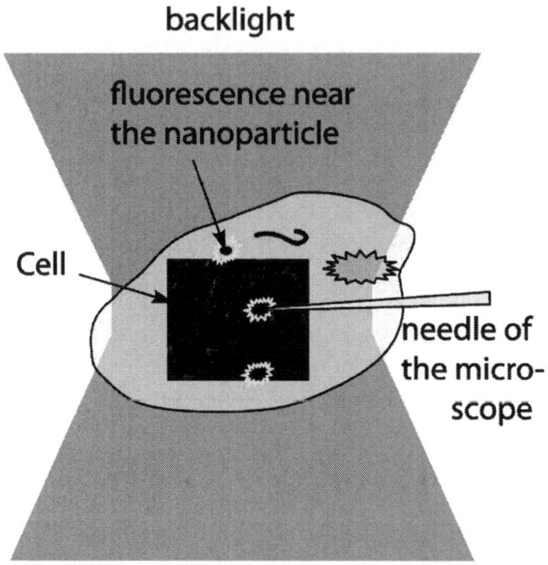

Figure 13.29 Usage of plasmonic nanoparticles for a fluorescent analysis of cells. The plasmonic nanoparticle is moved with the help of a scanning microscope or other manipulator, and fluorophores located near it become more perceptible (Lakowicz et al., 2003).

The nanoparticles' influence on the molecules' emission can be used for increase of existing fluorophores' efficiency and for development of new fluorophores. Complex nanoparticles with several plasmon resonances suit this purpose especially (Guzatov and Klimov, 2005). For example, in ellipsoidal nanoparticles, which are described in Chapter 8, there are three plasmon resonances. Selecting a shape of an ellipsoidal (or other) nanoparticle, one can make two of these plasmon resonances coincident with bands of absorption and emission of the initial fluorophore. This results in increase of rates of both molecule excitation and light emission, leading to a substantial fluorescence intensity enhancement and thus to development of highly effective fluorophores (Fig. 13.30).

Sources of single photons (Greulich and Thiel, 2001) are the other important applications of "nanoparticle+molecule" systems. These sources of light have great significance in experiments

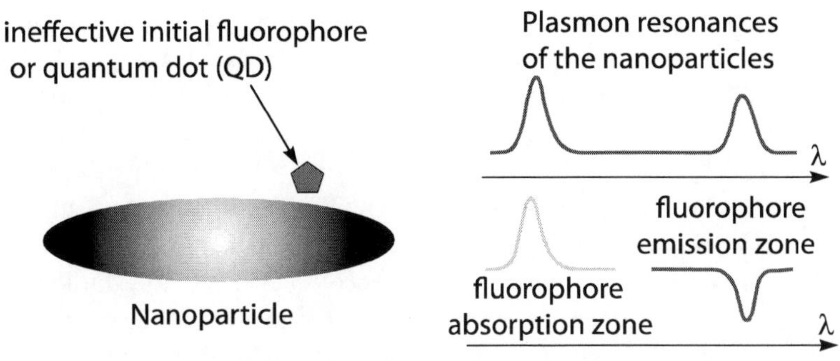

Figure 13.30 The operation principle of an artificial fluorophore with improved characteristics both of absorption and of emission of light. Effective light absorption by the fluorophore takes place at one plasmon resonance frequency and effective emission on the other. As a result, the fluorophore's "brightness" increases substantially (Guzatov and Klimov, 2005).

in quantum mechanics and electrodynamics and in devices for quantum calculations and quantum cryptography.

Quite recently, a true single-photon light source was created (Lounis and Moerner, 2000). In this device, separate terrylene molecules were placed in the *p*-terphenyl matrix in the ratio of $1:10^{11}$ so that in the *p*-terphenyl film of 10 μm thickness of terrylene molecules were situated at a distance of more than 1 μm from one another and excited by radiation with wavelength of 532 nm and impulse time of 35 ps every 160 ms. Each impulse had the energy of 22 pJ (5×10^7 photons!) and was focused onto a single molecule with the help of a microscope. This number of photons was sufficient to excite a molecule of terrylene and to obtain a photon of fluorescence with probability close to 1. The lifetime of a terrylene molecule in an excited state is about 3.8 ns in the matrix of *p*-terphenyl. When the molecule is already excited, it cannot absorb a photon for the second time, so when the irradiation is continuous, photons are emitted nearly every 3.8 ns. The exciting impulse time was so small that a probability of double excitation was less than 8×10^{-4}, that is, practically every 160 ms only one photon was emitted.

Necessity to use 5×10^7 pumping photons for generation of one fluorescence photon is an important disadvantage of the proposed device. Evidently, it is an essential barrier to creation of quantum computers and encryption devices since their operation will require huge energy consumption for 1 qubit of information. To increase the rate and probability of the molecule excitation, such a nanoparticle can be applied, in the vicinity of which a substantial local field enhancement and an acceleration of the molecule excitation take place (receiving "nanoantenna"). However, in this case the interaction between the nanoparticle and the molecule should be taken into account to provide effective interaction and minimization of nonradiative losses (see Chapter 10).

On the other hand, a single molecule or a quantum dot can be used as a nanoscale light source (Michaelis et al., 2000). A source of this kind can be used both for observation of other molecules with a higher resolution and in optoelectronic nanodevices. In this case, the combination of the molecule with the nanoparticle also increases efficiency of its operation. In this case, the nanoparticle plays a role of a radiating "nanoantenna."

And finally, variants of nanolasers' development on the basis of plasmon nanoparticles are suggested. This kind of laser is based on surface plasmon amplification by stimulated emission of radiation (SPASER) (Bergman and Stockman, 2003; Vinogradov et al., 2012; Lisyansky et al., 2012).

In a usual laser, amplification is achieved inside an optical cavity formed by mirrors, and its size cannot be smaller than the characteristic wavelength that is about 1 μm. In a SPASER-based nanolaser, resonant plasmonic nanoparticles, whose sizes can be substantially smaller than the light wavelength, are used for amplification of localized plasmons. It is possible because, in fact, the energy is stored in free-electron oscillations with optical frequency. These oscillations then emit light in outside space. It is important to note that the radiation pattern of a single-nanoparticle nanolaser is of dipole type in contrast to high directivity of usual lasers.

Recently, a SPASER-based nanolaser was realized experimentally (Noginov et al., 2009). In this excellent experiment, a gold nanoparticle (diameter 14 nm) was used as a plasmonic cavity, while

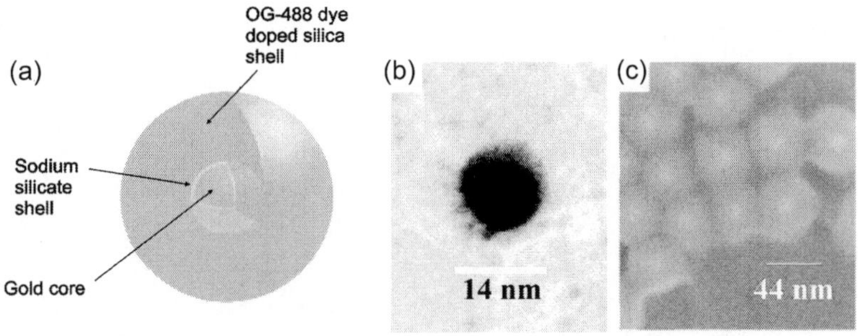

Figure 13.31 SPASER design. (a) Diagram of the hybrid nanoparticle architecture (not to scale), indicating dye molecules throughout the silica shell. (b) TEM image of Au core. (c) SEM image of Au/silica/dye core–shell nanoparticles (Noginov et al., 2009).

dye molecules on its surface (outer diameter 44 nm) were an active medium (Fig. 13.31).

Noginov et al. (2009) were able to detect threshold narrowing of the emission line with increasing of pumping (see Fig. 13.32). This fact supports existence of the laser effect, but special measurements of emitted light coherence are needed to prove it finally.

"Atom+plasmonic particle" systems have great significance in application of atom nano-optics. Atom nano-optics is an area of atom optics where atom motion is controlled with nanometer accuracy. These applications include atom-focusing devices, atom traps, and nanopincers (nanotweezers). To obtain atoms localization in nanoscale regions of space, combinations of optical fields and a nanobody are required (Balykin et al., 2006). At that, the presence of nanobodies influences both the fields confining atoms and the atoms' inner structure substantially. As a result, the theory described in previous chapters is required both for the atoms' localization and for their visualization (by means of fluorescence). As an example, let us consider an atomic trap near a nanowire with a current that is placed in a homogeneous magnetic field perpendicular to the wire (Denschlag et al., 1998, 1999a, 1999b). In Fig. 13.33, the fluorescence of atoms situated in such a nanotrap for different values of the current flowing through the nanowire is shown (the more the current is, the further from the axis the trap is situated).

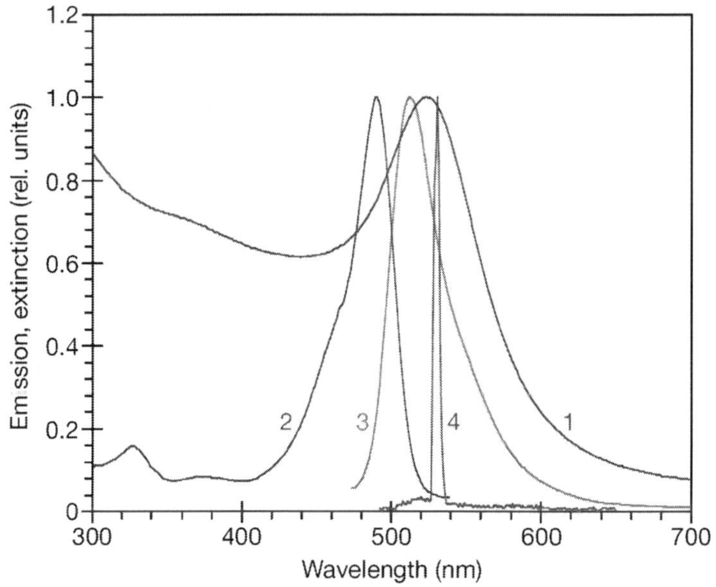

Figure 13.32 Normalized extinction (1), excitation (2), spontaneous emission (3), and stimulated emission (4) spectra of Au/silica/dye nanoparticles. The peak extinction cross section of the nanoparticles is 1.1310^{-12} cm^2. The emission and excitation spectra were measured in a spectrofluorometer at low fluence (Noginov et al., 2009).

Until now, we have considered fluorescence of ordinary molecules near plasmonic nanoparticles. One can expect even more interesting effects, considering spontaneous emission of optically active (chiral) molecules placed near resonant chiral nanoparticles. It was shown that chiral nanoparticles with specially tuned permittivity and permeability can enhance the decay rate of right or left molecules selectively (Klimov et al., 2011; see also Chapter 11).

This effect can be used in many applications. Here, we mention only two, the most obvious ones. The first application is observing separately right or left molecules with a scanning microscope (see Fig. 13.34).

To do so, specially tuned (see Chapter 11, Eq. (11.67)) nanoparticles are attached to the tip of a scanning microscope to increase the spontaneous emission of a molecule with a selected chirality (right

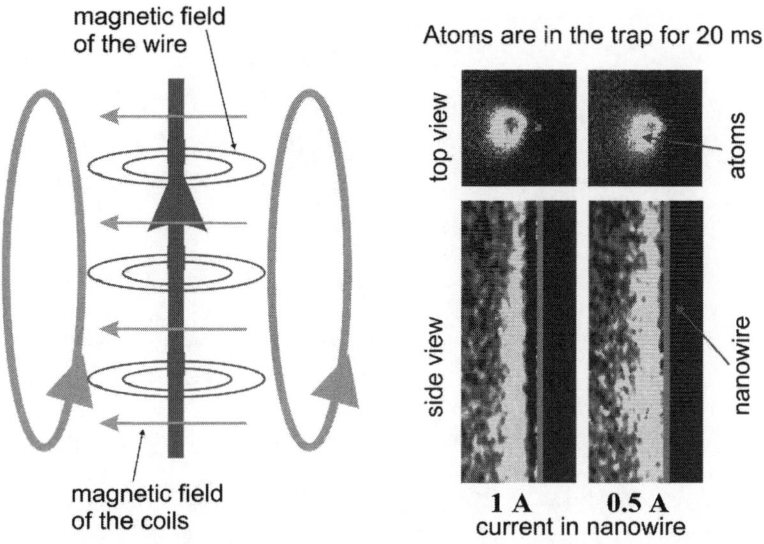

Figure 13.33 The fluorescence of atoms located in a cylindrical nanotrap near a metal wire within a 20 ms period. Magnetic field is perpendicular to the wire. (a) The magnetic field configuration and (b) the fluorescence distribution in the space for different values of the current in the nanowire (Denschlag et al., 1999).

or left). A short metallic spiral can be used as such a nanoparticle because it can possess chirality and a negative magnetic response. Indeed, the intensity of the fluorescence of a molecule is defined by the relation

$$I_{\text{fluo}} = \hbar\omega \frac{\Gamma_{\text{pump}}\Gamma_{\text{rad}}}{\Gamma_{\text{pump}} + \Gamma_{\text{nonrad}} + \Gamma_{\text{rad}}}, \qquad (13.2)$$

where Γ_{pump}, Γ_{nonrad}, and Γ_{rad} are the rate of excitation, nonradiative, and radiative decays of a molecule, respectively. In the presence of nanoparticles with specially tuned parameters (see Chapter 11, Eq. (11.67)), the radiative decay rate Γ_{rad} of one sort of molecules (e.g., right ones) increases, while the radiative decay rate of molecules with the opposite chirality (left) is inhibited. Due to this fact, the contrast between the brightness of the right and left molecules can reach a value of 10–100 or more and, consequently, one will see images of molecules with the chosen chirality only.

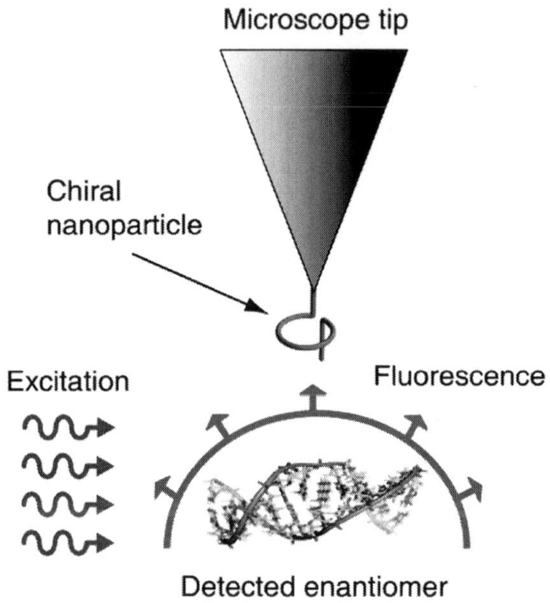

Figure 13.34 Illustration of the operation principle of a nanoscope, which is capable of distinguishing right and left enantiomers.

A significantly more important application of the theory presented in Chapter 11, Eq. (11.67), is separation of drug enantiomers in racemic mixtures (see Porter, 1991; Jaques et al., 1981; Li and Shapiro, 2010) for other methods of resolution). The operation principle of such a pure optical device is shown in Fig. 13.35.

The key element in this scheme is the reaction chamber, which contains arrays of nanoparticles prepared in accordance with Eq. (11.67). As an example of such an array, one can consider gold helix photonic metamaterials (Gansel et al., 2009) or so-called stereometamaterials (Liu et al., 2009). The racemic mixture of enantiomers is placed in this chamber and then excited by one or another way (e.g., photoexcitation). Due to the presence of chiral nanoparticles, one type of optically active enantiomers radiates efficiently and goes to the ground state quickly, while the remaining excited enantiomers can be ionized by a resonant field and then removed from the chamber. Other methods of removal

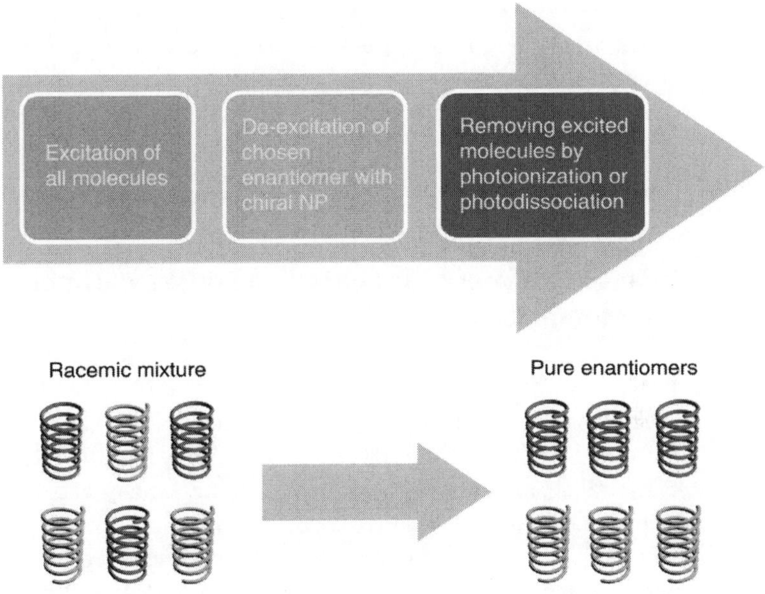

Figure 13.35 The operation principle of a pure optical system of separation of left and right enantiomers in racemic mixtures.

of excited molecules or their decay products are also possible. As a result, the desired pure enantiomer will be accumulated in the chamber. This approach for pure optical spatial separation of enantiomers has some advantages in comparison with other proposals (Li and Shapiro, 2010) because we have no need to use very low temperatures.

And finally, the nanoparticles' influence on the emission processes in semiconductor devices has great significance (Yamamoto et al., 2000). Now, there are high hopes that nanoparticles can increase the efficiency or the cost efficiency both of solar batteries and of light-emitting diodes substantially (Atwater, 2008; Bagnall and Boreland, 2008). However, here the level of the theory development remains behind the case of the interaction between nanoparticles and separate atoms and molecules, since emission process in semiconductors has substantially a many-particle character, and relatively simple models of dipole and quadrupole emissions are not quite adequate here.

Actually, the theory of the emission of atoms, molecules, and simple quantum systems in the presence of nanoparticles in general has great significance for description and optimization of existing optical nanodevices. However, this theory will have even greater significance for development of new nanodevices that will prevail in the 21st century.

13.7 Super- and Hyperlenses Based on Surface Plasmons and Metamaterials

In Chapter 11, we have already discussed a possibility of creation of a perfect lens (see Figs. 11.9 and 11.12) based on a layer of a substance with a negative refractive index. Due to difficulties of experimental realization of media with simultaneously negative electric permittivities and magnetic permeabilities, existence of such a lens has not been demonstrated yet in the optical region of frequencies. Moreover, as it was shown by (Klimov, 2008, 2009a, 2009b), the perfect lens is not realizable in the ordinary sense. However, almost-perfect lenses, which overcome the diffraction limit, have been already realized in different variants (Fang et al., 2005; Liu et al., 2007; Smolyaninov et al., 2007).

This possibility is based on surface plasmon properties in thin films. Sir John Pendry (Pendry, 2000) showed that under certain conditions near fields of nano-objects can be enhanced inside a metal film. For that case, the near-field intensity distribution in the plane perpendicular to the film is shown in Fig. 13.36. As a result of this enhancement of the near fields, the information on the object subwavelength features contained in them is not lost as it does in usual lenses, and an analogue of super-resolution can be achieved near the surface of the metal film.

Pendry (2000) proposed also a scheme of realization of such a lens (see Fig. 13.37). He suggested to use of a silver film of 40 nm thickness and with electric permittivity $\varepsilon = -1$ ($\lambda = 350$ nm) as a superlens so that the planes of the object and the image were situated at a distance of 40 nm from the midpoint of the film (Fig. 13.37a). In Fig. 13.37b, the near-field electric potential distribution in the plane of the object that consisted of two stripes situated at the distance of $d = 80$ nm, $d \ll \lambda = 350$ nm is shown. In

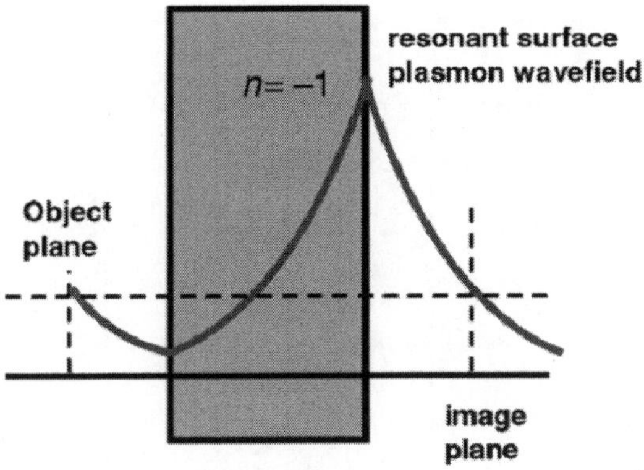

Figure 13.36 The illustration of the near fields' increase inside a metal film in the case of surface plasmon resonance on the right side of the film.

Fig. 13.37c, the potential distribution in the plane of the image in the case of silver film presence or absence is shown. It is well seen from the last picture that in the case of silver film presence, stripes remain resolved in contrast to the case of its absence. Let us note, however, that there is no true three-dimensional focus in lenses of this type since the same two-dimensional image is formed at any nanoscale distance from the surface of the film and the highest intensity of image will be observed exactly on the surface of the film and not at the image plane.

For the first time, this idea of Pendry was confirmed experimentally by Fang et al. (2005). In Fig. 13.38, the scheme of this experiment is shown. In the object plane on a silica substrate, stripes of chromium of 60 nm width and situated parallel to each other with a period of 120 nm are fabricated with the help of FIBs (see Chapter 2). In addition to the stripes, a more complex object was formed—the word "**NANO**." The chromium structure was coated by a 40 nm layer of a polymethyl methacrylate (PMMA) polymer, on which a silver film of 35 nm thickness and a photoresist of 120 nm thickness were deposited. The superlens prepared in that manner was irradiated from below by ultraviolet light with the wavelength $\lambda = 365$ nm, for which $\varepsilon_{\text{Ag}} = -1$ and the surface plasmon can be excited. The same system was used as the control device but with a silver layer replaced by a layer of PMMA polymer.

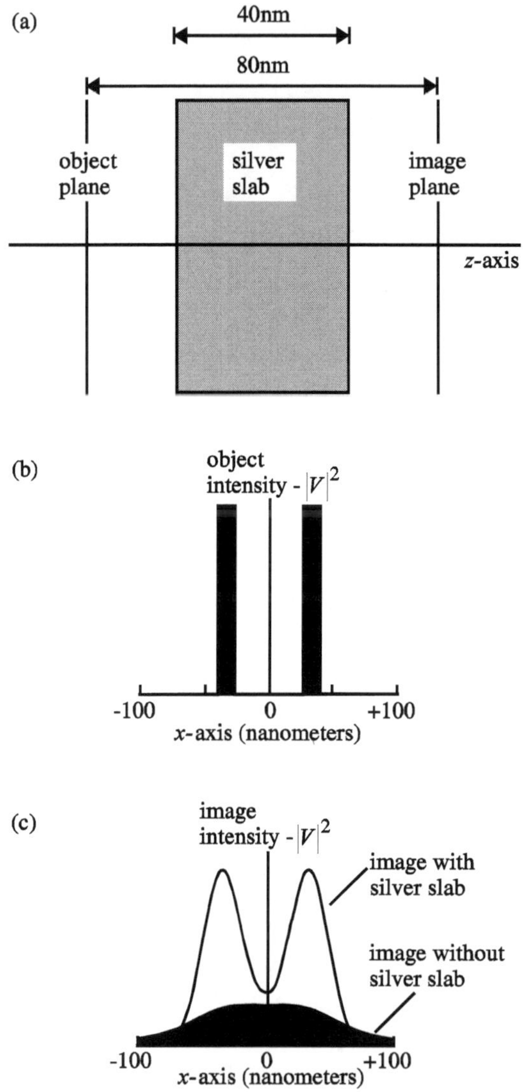

Figure 13.37 (a) The scheme of a superlens. Quasi-static potential in the object plane is transformed with the help of a silver nanofilm. (b) Distribution of electrostatic potential square in the form of two stripes in the object plane. (c) Distribution of electrostatic potential square in the image plane in the presence of the silver film and in the absence of it (Pendry, 2000).

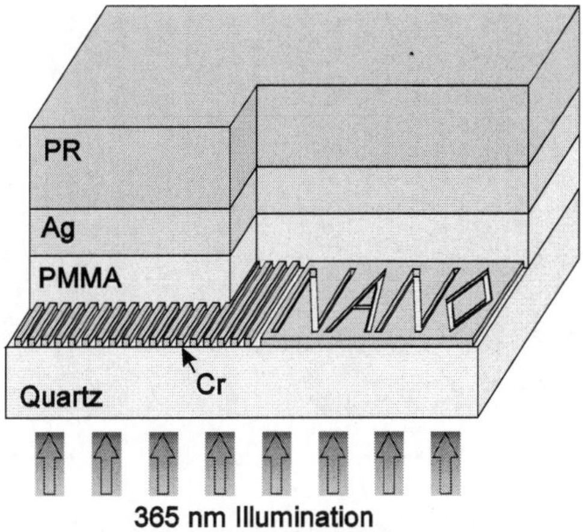

Figure 13.38 The scheme of an experiment demonctrting superlenses' abilities. The object is nanostructures milled from a chromium film of 50 nm thickness with the help of an FIB. On the left, the object is formed in the shape of a nanoarray with the stripe width of 60 nm and with the distance between them of 60 nm. On the right, the object is formed in the shape of the word "NANO." The silver film of 35 nm thickness is separated from the object by a PMMA polymer film of 40 nm thickness. The image is recorded on a photoresist lying on the silver film (Fang et al., 2005).

In Fig. 13.39, AFM images of the object are shown in the case of the presence (a, c) and in the case of the absence (b, d) of a silver film, that is, a superlens. From this picture it is clearly seen that the superlens forms an image with subwavelength resolution (Fig. 13.39a,c), indeed.

In Fig. 13.40a, the AFM image of a more complex nanostructure—the word "NANO"—is shown. As in the case of nanostrips, in the image of the word "NANO" (Fig. 13.40b), formed with the help of a superlens, some features are seen that are five times finer than those in the image obtained without the lens (Fig. 13.40c). The effect of the super-resolution is seen especially well in Fig. 13.40d, where the averaged surface profiles along the transverse cross section of the lines forming the letters are shown.

The superlens considered above deals only with the near fields that decay promptly when propagating in free space, and thus the super-resolution effects decrease promptly when moving away from

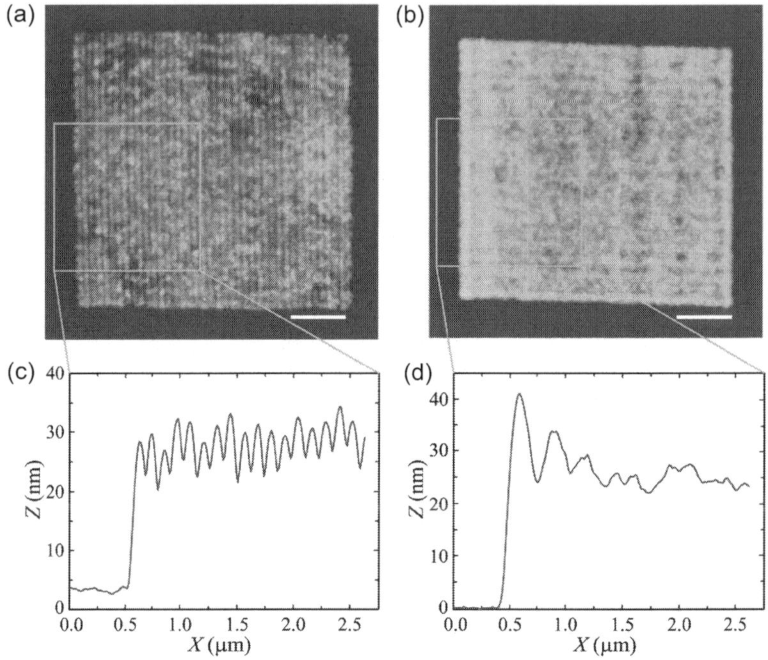

Figure 13.39 (a) The AFM image of an object in the shape of a nanoarray on a photoresist; the image is formed in the presence of a silver film (the scale bar is 1 mcm, the false color is from dark red to canary yellow, and it corresponds to the surface height range from 0 to 150 nm). (b) The AFM image of the object in the shape of a nanoarray on the photoresist in the control experiment, in which the silver film was replaced by a PMMA polymer layer of the same thickness. (c) The averaged profile of the photoresist surface cross section in the case of superlens action. (d) The averaged profile of the photoresist surface cross section in the control case (Fang et al., 2005).

the lens. Recently, Liu et al. (2007), Smolyaninov et al. (2007) have demonstrated experimentally that this problem can be avoided in so-called hyperlenses.

The basis of the hyperlenses' conception (Jacob et al., 2006; Salandrino and Engheta, 2006) is formed by a nanostructured material with permittivity tensor components of different signs in orthogonal directions. This anisotropy is necessary to remove the lower limit of the wavelength of the waves propa-

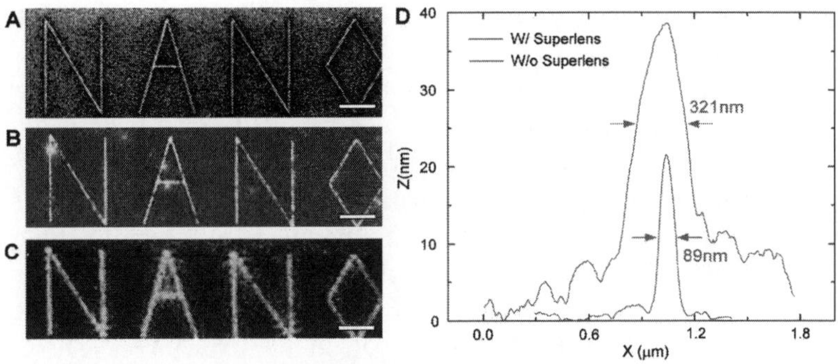

Figure 13.40 Forming of the image of a complex nano-object with the help of a superlens. (a) The FIB images of a complex nano-object (the word "NANO"). The width of the lines forming the letters is 40 nm, and the scale bar is 2 μm in the figures a, b, and c. (b) The AFM image of an arbitrary nano-object (the word "NANO") formed by a silver film on a photoresist. (c) The AFM image of the nano-object (the word "NANO") on the photoresist where the silver film is replaced by a PMMA polymer layer of the same thickness. (d) The averaged profile of the letter "A" cross section on the photoresist surface in the case of superlens action (the blue line; the width is 89 nm) and in the control case (the red line; the width is 321 nm (Fang et al., 2005).

gating in usual isotropic dielectrics, where the dispersion law for a wavevector having, for example, two components has the form

$$\frac{\mathbf{k}_{\perp}^2 + \mathbf{k}_{\parallel}^2}{\varepsilon} = \left(\frac{\omega}{c}\right)^2. \tag{13.3}$$

In the case of large transverse components of the wavevector $k_{\perp} \gg \omega/c$ bearing the subwavelength features of the object, the longitudinal component $k_{\parallel}$ becomes imaginary, and the field decays exponentially in the longitudinal direction, which is considered as a propagation direction. It is this circumstance that restricts resolution of usual lenses.

In the case of an anisotropic medium with permittivity tensor components of different signs $\varepsilon_{\parallel} > 0, \varepsilon_{\perp} = -|\varepsilon_{\perp}| < 0$, the dispersion equation has the form

$$\frac{\mathbf{k}_{\parallel}^2}{\varepsilon_{\parallel}} - \frac{\mathbf{k}_{\perp}^2}{|\varepsilon_{\perp}|} = \left(\frac{\omega}{c}\right)^2, \tag{13.4}$$

and for arbitrary large values of the transverse wavevector $\mathbf{k}_{\perp}$ the longitudinal wavevector $\mathbf{k}_{\parallel}$ remains real and the wave propagates in this direction without attenuation. The hyperbolic form of Eq. (13.4) conditions the name "hyperlens."

In the absence of the upper limit for the transverse wavevector (or lower limit for wavelength), there is also no diffraction limit, and therefore an arbitrary high resolution is possible. However, since only short-wavelength radiation goes out from the optical "hyper-space" to air, it cannot propagate freely and decays exponentially. To transform these fields into propagating far fields effectively, first their wavelength is to be increased to make them propagate freely without decay in the air. To realize such an adiabatic increase of the wavelength, cylindrical geometry can be used (Jacob et al., 2006). In Fig. 13.41, results of experiments by Liu et al. (2007) on the hyperlens realization are shown. In this experiment, a semicylinder of layered metamaterials with the electric permittivity strongly anisotropic in the radial and tangential directions was used. The nano-object whose image should be obtained in the far field was situated in the hollow center and irradiated from the top of the cylinder.

On the other hand, Smolyaninov et al. (2007) combined the idea of the hyperlens with the concept of the superlens of Pendry (2000). As it was mentioned above, the superlens consists of material with electric permittivity equal to -1. Instead of immediate transformation of the near fields into the far ones, the superlens enhances the near fields by means of resonance interaction with surface plasmons (Fig. 13.36). In the device developed by Smolyaninov et al. (2007), consisting of concentric rings of PMMA plastic on a gold film, the rings were playing the role of regions enhancing the near field and forming strong anisotropy of the constituent metamaterial. Amplification ability of this lens was demonstrated by an example of forming an image of two or three PMMA nanodots situated near the inner ring.

Strictly speaking, the above-considered superlenses and hyper-lenses are not lenses in the usual sense because they have no three-dimensional focus spot at all and the notion "depth of focus" cannot be applied for them. In fact, superlenses transfer information from one surface of a thin metal film to another, while hyperlenses transfer information from one fixed plane to another. In both cases, the image has a "flat" nature. As it was shown in Chapter 11, it is

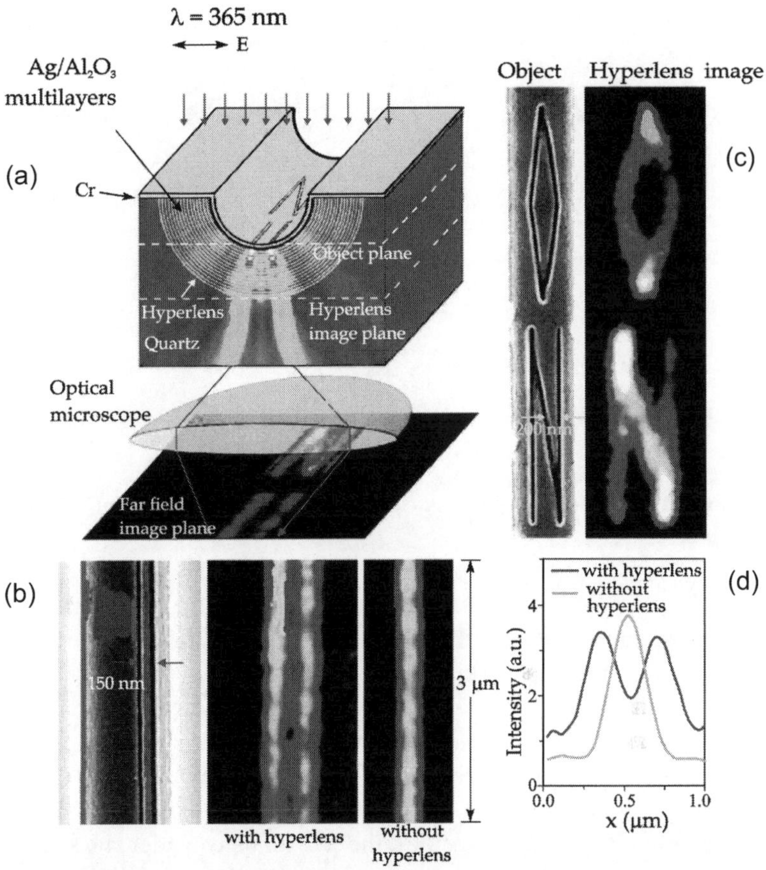

Figure 13.41 A magnifying optical hyperlens. (a) The hyperlens operation scheme and results of numerical calculation of forming of an image of an object of subway sizes. (b) The image of an object formed by the hyperlens in the form of two lines of 35 nm width and with 150 nm distance between them. (Left): The SEM image of the object fabricated on the internal surface of the hyperlens; (middle) the image magnified by the hyperlens where both lines are resolved clearly; (right): the control image without the hyperlens where the lines are not resolved. (c) The image of an arbitrary object (the word "ON") obtained with the help of the hyperlens. The line width is 40 nm. (d) The averaged profile of the image in the presence of the hyperlens (the red color) and in the control case (the green color). The hyperlens is made of 16 layers of Ag/Al_2O_3 (Liu et al., 2007).

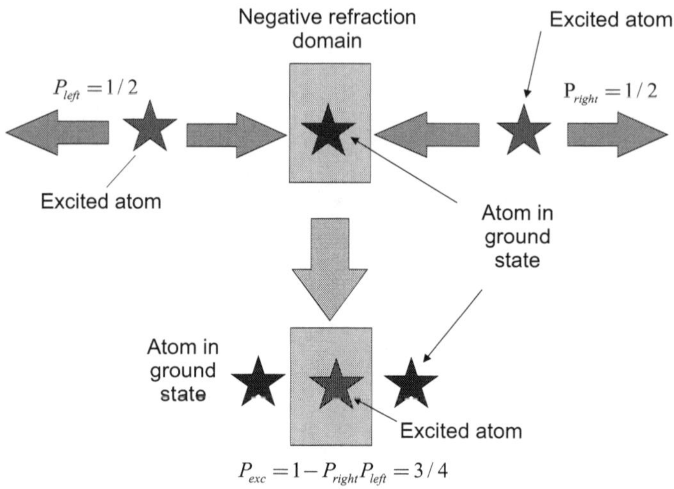

Figure 13.42 Effective atom excitation with the help of perfect lens geometry.

not possible to build a true three-dimensional perfect lens with a well-defined focus spot in the case of a negative refraction slab. However, it is possible to use a negative refraction slab and other geometrical shapes for other applications if one admits the approach suggested by Klimov (2009a, 2009b) and by Klimov et al. (2011).

For example, one can use a negative refraction slab for effective excitation of atoms or molecules. Indeed, let us consider the system presented in Fig. 13.42. The ray picture for this system is shown in Fig. 11.12. At the initial instant of time (top part of figure), two atoms (the red stars) are in the exited state, while the atom inside the negative refraction slab is in the ground state (the blue star). The probability that a photon will be emitted to the left by the left exited atom is the same as the probability that a photon will be emitted to the right by the right excited atom and equals $1/2$. It means that the probability to emit at least one photon inside the slab and to excite the atom in the ground state is equal to $3/4$ (the lower part of the figure). It is very good efficiency because usually to excite an atom effectively, one needs about 10^6 or more photons rather than 2, as we have in the proposed scheme.

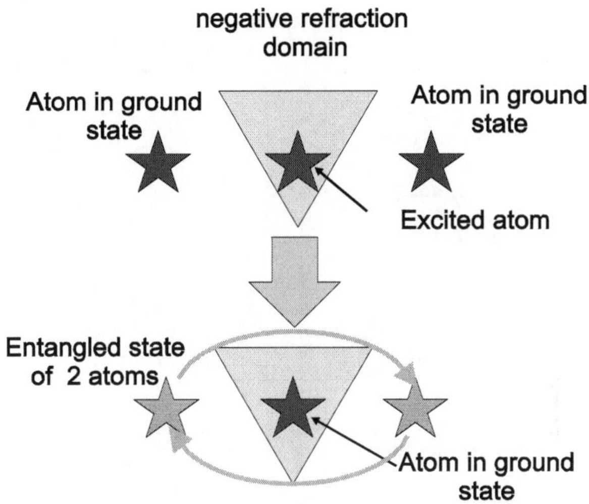

Figure 13.43 Preparation of atomic entangled states.

Reversing the geometry in Fig. 13.42, one can prepare an entangled state of two excited atoms (Fig. 13.43). Indeed, if now an excited atom is placed inside the negative refraction domain, it can emit photons both to the left and to the right, and in any case it excites an atom in the ground state with a probability equal to 1 due to perfect focusing of light in this geometry. However, according to quantum mechanics principles, nobody knows before measurements which direction an emitted photon will choose. Therefore, the entangled state of excited and unexcited atoms will be prepared. Such states can be useful in quantum computations.

Moreover, if instead of a single negative refraction domain one uses a chessboard with negative refraction boxes (blue) and properly prepare initial states of excited and unexcited atoms, one can perform distributed quantum computations (Fig. 13.44).

Thus, owing to invention of super- and hyperlenses, investigations in the area of nanotechnologies and construction of nanodevices will become easier, more convenient, and, above all, cheaper. First of all, superlenses will be useful in optical microscopy, medicine, and biology. With the help of modern optical microscopes, scientists are able to observe large cell components, such as

Figure 13.44 More complicated geometry for distributed quantum computing with a Nim chessboard.

mitochondria and nuclei, while the motion of proteins along the cytoskeleton can be observed by means of hyperlens microscopy. With the help of scanning electronic microscopy (SEM) and AFM, objects of sizes of several atoms can be distinguished. But acquiring such images takes a long time since these types of microscopy are restricted by movements of the probes scanning the surface. In addition, a living cell sample cannot be examined with the help of them since specimens for scanning tunneling microscopy (STM), SEM, and AFM are to be prepared specifically. In the future, functioning biological nanosystems can be investigated in real time with the help of hyperlenses, and one can understand the nature of different diseases better. Development of these technologies can form the basis for increasing of digital video disc (DVD) capacity since the information can be read from nanoscale regions with the help of hyperlenses. As a result, the whole library of the US Congress can be stored on such ultrahigh-density DVDs, and this is really a fantastic capacity for a 12 cm disc (Srituravanich et al., 2008).

13.8 Invisibility Cloaks Based on Metamaterials

As it was mentioned in Chapter 6, a nanoparticle coated with a layer of material with corresponding properties (see Eqs. (6.82) and

(6.83)) has zero polarizability and does not scatter or absorb light and thus stays invisible.

If it was possible to make big objects invisible with the help of any coating, it would have enormous practical significance, first of all, in military applications.

At first sight, it seems that by a slight change of parameters of the layer that makes a small particle invisible, an arbitrary big object can also be made invisible. However, it is not true. Big objects have not only dipole but also quadrupole, etc., polarizabilities, and to suppress the emission of all multipoles, a system with an infinite number of control parameters is required in a general case. On the other hand, there is a range of purely mathematical theorems proving that there are no simple methods of creating invisibility cloaks. Nachman (1988), Wolf and Habashy, (1993) have proved that total invisibility can be achieved only in the case of a finite number of directions where the object occurs to be infinitely thin. Besides, as we have seen before, objects that are small in comparison with the emission wavelength can be made invisible (Kerker, 1975; Alu and Engheta, 2005).

Nevertheless, there are ways and approaches allowing one to develop cloaks inside which any object can be hidden. The point is that mathematical theorems mentioned above are inapplicable in the case of media with a negative refractive index and in the case of strongly anisotropic media. Besides, these theorems tell that creation of the *perfect* invisibility cloak is impossible, but they do not deny the possibility of creation of *almost*-invisible cloaks.

However, the modern development of nanotechnologies and nanoplasmonics allows one to create artificial media, that is, metamaterials, both with a negative refractive index and with any degree of anisotropy. Examples of media with a negative refractive index and with strong anisotropy were discussed in Chapter 11 and in the previous section. Thus, the technological basis allowing one to investigate the possibilities of the creation of invisibility cloaks exists.

To create invisibility cloaks Pendry et al. (2006) proposed a method based on coordinate transformation:

$$(x, y, z) \to (u(x, y, z), v(x, y, z), w(x, y, z)), \qquad (13.5)$$

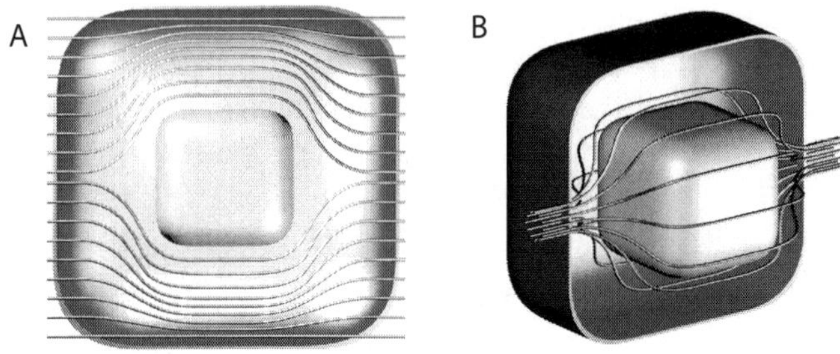

Figure 13.45 The illustration of rays bypassing the region the content of which remains invisible.

as a result of which the regular ray picture in the (x, y, z) space with homogeneous and isotropic characteristics is transformed to the ray picture in the (u, v, w) space with nonhomogeneous and nonisotropic characteristics, having a region that no ray penetrates. In other words, in the (u, v, w) space the rays "bypass" the region where the concealed object can be placed (Fig. 13.45).

Since Maxwell's equations are invariant in form relative to any coordinate transformations, the coordinate transformation will be attended by a change of both fields and the electromagnetic characteristics of the space.

For example, if the initial spherical region $r < R_2$ in the (x, y, z) space was homogeneous and isotropic, and rays propagated there without changing direction, then after the transformation

$$r' = R_1 + r(R_2 - R_1)/R_2; \theta' = \theta; \phi' = \phi, \tag{13.6}$$

no rays can penetrate the sphere of radius R_1 in the (r', θ', ϕ') space, and thus this region becomes invisible.

The coordinate transformation, Eq. (13.6), is attended by a change of electric permittivity and magnetic permeability tensors in the region $R_2 > r > R_1$:

$$\begin{aligned} \varepsilon'_{r'} &= \mu'_{r'} = \frac{R_2}{R_2 - R_1}\frac{(r' - R_1)^2}{r'^2}, \\ \varepsilon'_{\theta'} &= \mu'_{\theta'} = \varepsilon'_{\phi'} = \mu'_{\phi'} = \varepsilon_t = \mu_t = \frac{R_2}{R_2 - R_1}. \end{aligned} \tag{13.7}$$

In the untransformed region $r > R_2$, the ray paths remain straight, and the material tensors do not change:

$$\varepsilon'_{r'} = \mu'_{r'} = \varepsilon'_{\theta'} = \mu'_{\theta'} = \varepsilon'_{\phi'} = \mu'_{\phi'} = 1. \tag{13.8}$$

In the region $r < R_1$, one can choose any values of permittivity and permeability because no rays enter this domain.

Let us note once again that no rays can penetrate or come out from the region with the material tensors Eqs. (13.7) and (13.8). On the external surface of the invisibility coating ($r = R_2$), the tensor components of the dielectric and magnetic permittivities are related by the ratios

$$\begin{aligned} \varepsilon'_{\theta'} &= \varepsilon'_{\phi'} = 1/\varepsilon'_{r'}, \\ \mu'_{\theta'} &= \mu'_{\phi'} = 1/\mu'_{r'}, \end{aligned} \tag{13.9}$$

which are conditions for a perfectly matched layer (PML) and ensure the absence of reflection on this boundary. Thus, a natural connection between the considered invisibility cloak and the thoroughly studied reflectionless interfaces (Berenger, 1994, 1996) arises.

Let us consider the case $R_2 \gg \lambda$, where λ is a typical light wavelength and the ray paths can be used for representation of the Poynting vectors, as an illustration. In Fig. 13.45, the paths of the rays incident from infinity on the region with the characteristics Eqs. (13.7) and (13.8) are shown. In this figure, the rays are obtained as a result of the integration of Hamilton's equation obtained from Maxwell's equations in the ray approximation. Indeed, it is seen from the picture that all rays deflected from the region $r < R_1$, in which the concealed object can be placed, therefore.

A detailed consideration of electromagnetic waves scattering on a spherically layered system with the characteristics

$$\begin{aligned} &\varepsilon_r = \varepsilon_t \frac{(r - R_1)^2}{r^2}, \ \mu_r = \mu_t \frac{(r - R_1)^2}{r^2}, \\ &\varepsilon_\theta = \varepsilon_\phi = \varepsilon_t = \varepsilon_2 \frac{R_2}{R_2 - R_1}, \\ &\mu_\theta = \mu_\phi = \mu_t = \mu_2 \frac{R_2}{R_2 - R_1}, \end{aligned} \tag{13.10}$$

inside the region $R_1 < r < R_2$ and with the isotropic characteristics ε_2, μ_2 outside the layer was carried out by (Chen et al., 2007) within the complete system of Maxwell's equations.

In general, this solution follows the known Mie solutions with isotropic spherical layers that use the spherical harmonics expansion and the solutions match in different regions on their boundaries (see Chapter 6). However, anisotropy and spatial inhomogeneity of Eq. (13.10) require a special approach. To solve Maxwell's equations in the spherical layer $R_1 < r < R_2$ with the characteristics Eq. (13.10), Chen et al. (2007) introduced scalar potentials for transverse electric (TE) and magnetic (TM) waves Φ_{TM} and Φ_{TE}:

$$\begin{aligned} &\mathbf{B}_{\mathrm{TM}} = \mathrm{rot}\left(\hat{\mathbf{r}}\Phi_{\mathrm{TM}}\right); \mathbf{D}_{\mathrm{TM}} = \tfrac{i}{k}\mathrm{rot}\left(\overleftrightarrow{\mu}^{-1}\mathbf{B}_{\mathrm{TM}}\right), \\ &\mathbf{D}_{\mathrm{TE}} = -\mathrm{rot}\left(\hat{\mathbf{r}}\Phi_{\mathrm{TE}}\right); \mathbf{B}_{\mathrm{TE}} = -\tfrac{i}{k}\mathrm{rot}\left(\overleftrightarrow{\varepsilon}^{-1}\mathbf{D}_{\mathrm{TE}}\right), \end{aligned} \tag{13.11}$$

where $\hat{\mathbf{r}}$ is the unit vector in the radial direction and $\overleftrightarrow{\mu}^{-1}$ and $\overleftrightarrow{\varepsilon}^{-1}$ are tensors inverse to the tensors of magnetic permeability and electric permittivity.

By substituting Eq. (13.11) in Maxwell's equation with the material tensors Eq. (13.7), one can obtain wave equations for the Φ_{TM} and Φ_{TE} potentials in the spherical system of coordinates:

$$\begin{aligned} &\left\{\frac{1}{(\mathrm{SR})}\frac{\partial^2}{\partial r^2} + \frac{1}{r^2 \sin\theta}\frac{\partial}{\partial\theta}\left(\sin\theta\frac{\partial}{\partial\theta}\right)\right. \\ &\left. + \frac{1}{r^2\sin^2\theta}\frac{\partial^2}{\partial\varphi^2} + \frac{1}{(\mathrm{SR})}k_t^2\right\}\Phi = 0, \end{aligned} \tag{13.12}$$

where $k_t = \frac{\omega}{c}\sqrt{\mu_\theta\varepsilon_\theta} = \frac{\omega}{c}\sqrt{\mu_\varphi\varepsilon_\varphi}$, SR denotes the anisotropy order, and for the TM and TE waves we have $\mathrm{SR}_{\mathrm{TM}} = \varepsilon_t/\varepsilon_r$ and $\mathrm{SR}_{\mathrm{TE}} = \mu_t/\mu_r$, correspondingly. When the tensors of permittivity and permeability are given by the expressions in Eq. (13.7), the anisotropy degree is the same for both polarizations:

$$\mathrm{SR}_{\mathrm{TM}} = \mathrm{SR}_{\mathrm{TE}} = \mathrm{SR} = \frac{r^2}{(r-R_1)^2}. \tag{13.13}$$

It is natural to search for a solution of Eq. (13.12) in the form of

$$\Phi = Y_n^m(\theta,\varphi) f_n(r), \tag{13.14}$$

where $Y_n^m(\theta,\varphi)$ are usual spherical harmonics.

The tensors of electric permittivity and magnetic permeability, Eq. (13.7), have a wonderful property: the solutions of the equation for the radial function f_n

$$\left\{\frac{\partial^2}{\partial r^2} + \left[k_t^2 - \frac{n(n+1)}{(r-R_1)^2}\right]\right\} f_n = 0, \tag{13.15}$$

are arbitrary Riccati–Bessel functions with a shifted argument:

$$f_n(r) = z_n(k_t(r - R_1)). \tag{13.16}$$

Thus, potentials that are finite inside the region $R_1 < r < R_2$ can be written in the form:

$$\begin{aligned}\Phi^c_{\rm TM} &= \frac{\cos\varphi}{\omega}\sum_{n=1}^{\infty} d_n^M \psi_n(k_t(r - R_1))P_n^1(\cos\theta),\\ \Phi^c_{\rm TE} &= \frac{\sin\varphi}{\omega}\sum_{n=1}^{\infty} d_n^E \psi_n(k_t(r - R_1))P_n^1(\cos\theta),\end{aligned} \tag{13.17}$$

where $\psi_n(z) = zj_n(z)$, and $j_n(z)$ is Bessel's spherical function.

Let us emphasize that the potentials in Eq. (13.17) vanish on the boundary $r = R_1$, and there is only one independent coefficient for each harmonic. In the case of a usual shell, there would be two coefficients of this kind. This is the circumstance that allows one to create an invisibility cloak.

In a general case, in the region $r > R_2$ fields consist of the incident $\Phi^{\rm in}_{\rm TM}$ and $\Phi^{\rm in}_{\rm TE}$ and scattered $\Phi^{\rm s}_{\rm TM}$ and $\Phi^{\rm s}_{\rm TE}$ parts, which are expressed by means of nonshifted Ricatti–Bessel functions. In the case of the x polarized plane unit wave incident along the z axis, $\mathbf{E}^{\rm in} = (1, 0, 0)e^{ik_2z}$ they can be written in the following form:

$$\begin{aligned}\Phi^{\rm in}_{\rm TM} &= \frac{\cos\varphi}{\omega}\sum_{n=1}^{\infty} a_n\psi_n(k_2r)P_n^1(\cos\theta),\\ \Phi^{\rm in}_{\rm TE} &= \frac{\sin\varphi}{\omega}\sum_{n=1}^{\infty} a_n\psi_n(k_2r)P_n^1(\cos\theta),\end{aligned} \tag{13.18}$$

$$\begin{aligned}\Phi^{\rm s}_{\rm TM} &= \frac{\cos\varphi}{\omega}\sum_{n=1}^{\infty} a_nT_n^M\zeta_n(k_2r)P_n^1(\cos\theta),\\ \Phi^{\rm s}_{\rm TE} &= \frac{\sin\varphi}{\omega}\sum_{n=1}^{\infty} a_nT_n^E\zeta_n(k_2r)P_n^1(\cos\theta),\end{aligned} \tag{13.19}$$

where $\zeta_n(z) = zh_n^{(1)}(z)$, $h_n^{(1)}(z)$ is the spherical Hankel function, $a_n = \frac{(-1)^n(2n+1)}{n(n+1)}$, and $k_2 = \frac{\omega}{c}\sqrt{\varepsilon_2\mu_2}$ is the wavevector in the outer region.

The fields inside the region $r < R_1$ are also expressed through nonshifted Ricatti–Bessel functions:

$$\begin{aligned}\Phi^1_{\rm TM} &= \frac{\cos\varphi}{\omega}\sum_{n=1}^{\infty} c_n^M\psi_n(k_0r)P_n^1(\cos\theta),\\ \Phi^1_{\rm TE} &= \frac{\sin\varphi}{\omega}\sum_{n=1}^{\infty} c_n^E\psi_n(k_0r)P_n^1(\cos\theta),\end{aligned} \tag{13.20}$$

where $k_0 = \frac{\omega}{c}\sqrt{\varepsilon_0\mu_0}$ is the wavevector in this region.

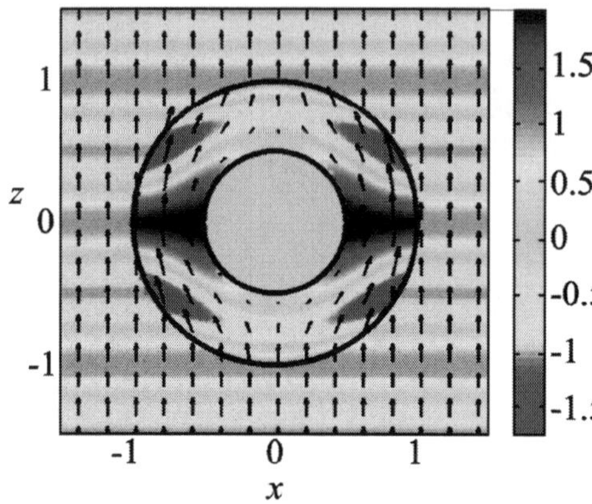

Figure 13.46 Distribution of the $\mathbf{E}_x$ component of the electric field of a plane wave incident along the z axis and corresponding Poynting vector in the xz plane. Plane-polarized waves are incident on the invisibility cloak with the internal radius $R_1 = 0.5\lambda_0$ and the external radius $R_2 = \lambda_0$ (Chen et al., 2007).

The condition of the tangential components of **E** and **H** continuity on the surface implies that the coefficients $T_n^{M,E}$, characterizing the scattered field, and the coefficients $c_n^{M,E}$, characterizing the field inside the region $r < R_1$, are equal to zero. The connection between the incident field and the field inside the region $R_1 < r < R_2$ is described by the relations:

$$d_n^E = \frac{\varepsilon_t}{\varepsilon_2} a_n, \; d_n^M = \frac{\mu_t}{\mu_2} a_n. \tag{13.21}$$

In Fig. 13.46, the distribution of the electric field component E_x of the plane wave incident along the z axis in the xz plane is shown in pseudocolor, and the distribution of the corresponding Poynting vector is shown by the arrows (the plane x polarized wave is incident on the ideal invisibility cloak with the characteristics Eq. (13.7) and with the internal radius $R_1 = 0.5\lambda_0$ and the external radius $R_2 = \lambda_0$). Indeed, from this picture it is seen that the scattered field is absent. Moreover, from this figure it is seen that a so-called problem of the axial ray does not exist since the Poynting vector is equal to zero on the poles of the inner surface, where the axial ray enters the invisibility region.

Recently, the ideas of Pendry et al. (2006) were realized experimentally by Schurig et al. (2006). However, they have tried to realize not a spherical but a cylindrical invisibility cloak, in which the following coordinate transformation is to be used to create a totally "invisible" region $r < a$:

$$r' = \frac{b-a}{a} r + a;\ \theta' = \theta;\ z' = z, \tag{13.22}$$

resulting in the following values of the diagonal components of the tensors of permittivity and permeability:

$$\varepsilon_r = \mu_r = \frac{r-a}{r};\ \varepsilon_\theta = \mu_\theta = \frac{r}{r-a};\ \varepsilon_z = \mu_z = \left(\frac{b}{b-a}\right)^2 \frac{r-a}{r}, \tag{13.23}$$

(nondiagonal components are equal to zero).

From Eq. (13.23), it is seen that for realization of the perfect cylindrical invisibility cloak the metamaterial with ε and μ that not only nonhomogeneous and anisotropic but also take infinitely great values for $r = a$, is to be created (let us note that in this respect the case of the spherical invisibility cloak is simpler since in this case there are no infinities).

Due to these difficulties, Schurig et al. (2006) have tried to realize a system in which

$$\varepsilon_z = \left(\frac{b}{b-a}\right)^2;\ \mu_r = \left(\frac{r-a}{r}\right)^2;\ \mu_\theta = 1, \tag{13.24}$$

the rest of diagonal components are to be equal to 1, and non-diagonal components are to be equal to 0 also. By this substitution, equations for the electric field that is polarized along the cylinder axis remain the same, but for $r = b$ the boundary condition ceases to be agreed, and therefore the reflection appears, though the greater the external radius b in comparison with the wavelength, the less this reflection.

To realize the dependencies of Eq. (13.24), Schurig et al. (2006) developed and created a metamaterial, whose photograph is shown in Fig. 13.47. In the same picture, the values of ε_z, μ_r, and μ_θ that were realized in the experiment are shown. The elaborated invisibility cloak consisted of 10 cylindrical layers on which SRRs were situated, and the geometry of the latter was changed depending on the layer on which they were situated.

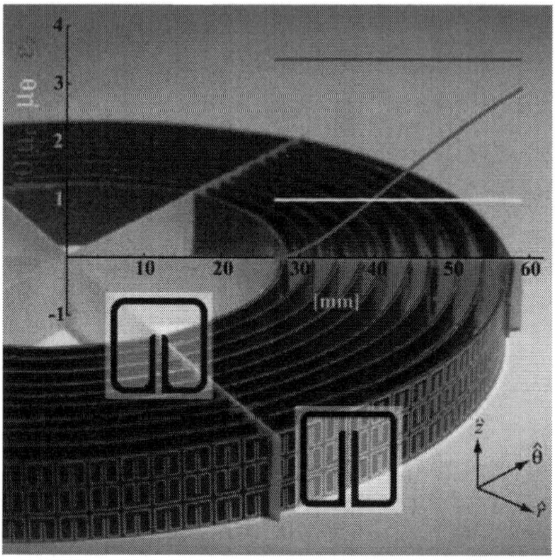

Figure 13.47 Experimental realization of a cylindrical invisibility cloak in UHF band in the form of a system of SRRs placed on10 cylinders. Colour lines show constitutive parameters realized in the experiment: the red line stands for 10 μ_r, the green line stands for $\mu_\theta = 1$, and, finally the blue line stands for $\varepsilon_z = 3.423$. The SRRs located on the 1st (internal) and the 10th (outer) cylinders are shown in the transparent insertions (Schurig et al., 2006). *Abbreviations*: UHF, ultrahigh frequency; SRR, split-ring resonator.

In Fig. 13.48, results of computer modeling (Fig. 13.48a and Fig. 13.48b) and experimental measurements of the z component of the electric field in the absence of an invisibility coating (Fig. 13.48c) and in the presence of it (Fig. 13.48d) are shown. In an ideal case (see Eq. (13.23)), computer modeling shows almost total absence of the scattered field, that is, an almost ideal invisibility coating. By computer calculations, taking into account the approximate expressions in Eq. (13.24) and losses that correspond to independent calculations of SRRs, increases scattering, but anyway the effect of the invisibility coating gives much less scattering than experimentally measured values of a metal cylinder without a cloak (Fig. 13.48c). Experimental measurements of the metal cylinder with a cloak (Fig. 13.48d) also show an essential reduction of the

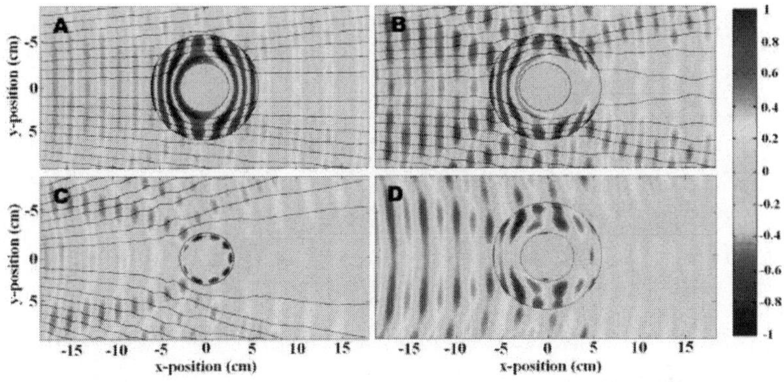

Figure 13.48 An instant image of steady-state and time-dependent electric field distribution. In figures a–c, black lines show the energy flow direction, that is, the Poynting vector. The invisibility cloak is located in the ring region between the black circles and surrounds the conducting copper cylinder. (a) The numerical simulation making use of precise constitutive equations, Eq. (13.23). (b) The numerical simulation making use of simplified constitutive conditions, Eq. (13.24) and taking losses in SRRs into account. (c) Results of experimental measurement of the fields near the conducting copper cylinder. (d) Results of experimental measurement of the fields near the conducting copper cylinder surrounded by the invisibility cloak. The right column shows the correspondence between the color and the field amplitude (Schurig et al., 2006).

scattering and agree with results of realistic modeling of the same system (Fig. 13.48d).

By now, we have considered situations in which an invisibility cloak was realized on the basis of anisotropic and nonhomogenous metamaterials with positive values of electric permittivity and magnetic permeability tensors. However, at least in principle, it is possible to create invisibility cloaks based on isotropic materials with a negative refractive index. As it was mentioned in Chapter 11, there is a significant advance in creation of such materials, even for optical frequencies. An example of the theory of a cylindrical invisibility cloak using isotropic materials with a negative refractive index was proposed by Ochiai et al. (2008). In Fig. 13.49, the transverse cross section of such a cylinder and corresponding ray paths are shown (Nozawa, 2008).

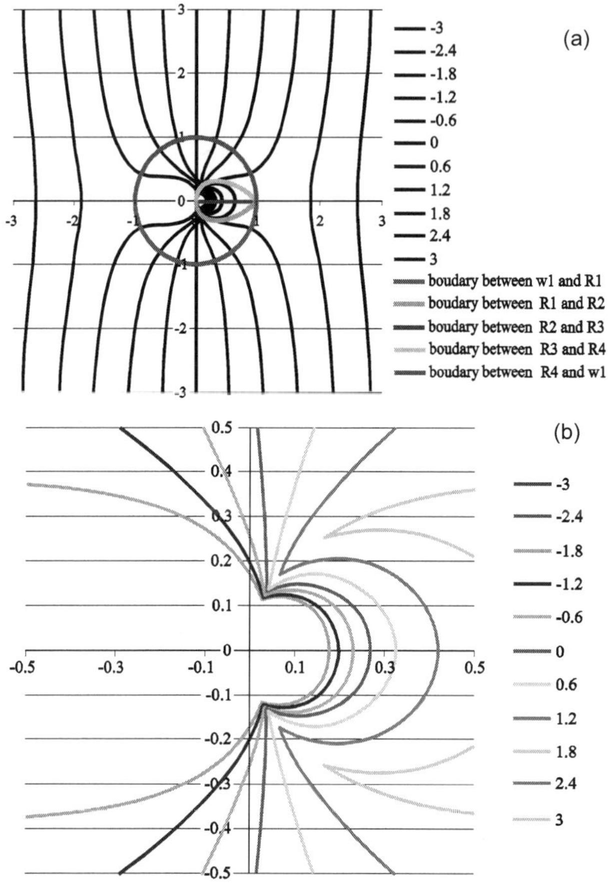

Figure 13.49 (a) Transverse cross section of a cylindrical invisibility cloak with an isotropic negative refraction material. Black lines correspond to rays of light, and lines of other colors show boundaries between five regions forming the invisibility cloak. The region outside the circle of the unit radius, w1, has the refractive index $n > 0$. The region constrained by the red and blue lines and the x axis, R1, has a positive refractive index $n > 0$; the region constrained by the brown and blue lines, R2, has a negative refractive index $n < 0$. The region constrained by the brown and green lines, R3, has a positive refractive index $n > 0$. The region constrained by the purple and green lines and the x axis, R4, has a negative refractive index $n < 0$. (b) A magnified view of the central region of (a). The color lines now show the ray paths only. The region in the center cannot be accessed by the rays, and any cylindrical object can be concealed there (Nozawa, 2008).

The whole system consists of five regions with negative and positive refractive indices, absolute values of which increase when approaching the cylinder axis. For any ray incident on this system, in the center of it a region is present that cannot be reached by the ray, that is, this system is really an invisibility cloak for this region.

Thus, anisotropic materials can really form the basis for creation of invisibility cloaks at least in the UHF region (experiments of Schurig et al. (2006). Creation of invisibility cloaks in regions of shorter frequency is a more complex task, which nevertheless appears to be resolvable. At that, the problem of the frequency bandwidth for which this cloak is effective arises always.

References

Albanis, V., Dhanjal, S., Emel'yanov, V.I., Fedotov, V.A., MacDonald, K.F., Petropoulos, P., Richardson, D.J., Zheludev, N.I. (2001) Nanosecond dynamics of a gallium mirror's light-induced reflectivity change, *Phys. Rev. B*, **63**, 165207.

Albanis, V., Dhanjal, S., Zheludev, N.I., Petropoulos, P., Richardson, D.J. (1999) Cross-wavelength all-optical switching using nonlinearity of liquefying gallium, *Opt. Express*, **5**, 157.

Alivisatos, A.P., Peng, X., Wilson, T.E., Johnsson, K.P., Loweth, C.J., Bruchez, M.P., Jr., Schultz, P.G. (1996) Organization of nanocrystal molecules, using DNA, *Nature*, **382,** 609–611.

Alu, A., Engheta, N. (2005) Achieving transparency with plasmonic and metamaterial coatings, *Phys. Rev. E*, **72**, 016623.

Ambrose, W.P., Goodwin, P.M., Martin, J.C., Keller, R.A. (1994) Alterations of single molecule fluorescence lifetimes in near-field optical microscopy, *Science*, **265**, 364.

Anger, P., Bharadwaj, P., Novotny, L. (2006) Enhancement and quenching of single-molecule fluorescence, *Phys. Rev. Lett.*, **96**, 113002.

Aslan, K., Lakowicz, J.R., Geddes, C.D. (2005) Metal-enhanced fluorescence using anisotropic silver nanostructures: critical progress to date, *Anal. BioAnal. Chem.*, **382**, 926–933.

Atwater, H.A. (2008) Nanostructures for high-efficiency photovoltaics, *Proc. SPIE*, **7047**, 704702.

Bagnall, D.M., Boreland, M. (2008) Photovoltaic technologies, *Energy Policy*, **36**, 4390–4396 (and references therein).

Balykin, V.I., Klimov, V.V., Letokhov, V.S. (2006) Atom nanooptics, in *Handbook of Theoretical and Computational Nanotechnology*, ed. Michael Rieth, Wolfram Schommers, **7**, 1–78.

Banerjee, K., et al. (1996) The effect of interconnect scaling and low-k dielectric on the thermal characteristics of the IC metal, *Tech. Dig. IEDM*, 65.

Bennet, P.J., Dhanjal, S., Petropoulos, P., Richardson, D.J., Zheludev, N.I., Emel'yanov, V.I. (1998) A photonic switch based on a gigantic, reversible optical nonlinearity of liquefying gallium, *Appl. Phys. Lett.*, **73**, 1787.

Berenger, J.P. (1994) A perfectly matched layer for the absorption of electromagnetic waves, *J. Comput. Phys.*, **114**, 185–200.

Berenger, J.P. (1996) Three-dimensional perfectly matched layer for the absorption of electromagnetic waves, *J. Comput. Phys.*, **127**, 363–379.

Bergman, D.J., Stockman, M.I. (2003) Surface plasmon amplification by stimulated emission of radiation: quantum generation of coherent surface plasmons in nanosystems, *Phys. Rev. Lett.*, **90**, 027402.

Betzig, E., Chichester, R. (1993) Single molecules observed by near-field scanning optical microscopy, *Science*, **262**, 1422–1425.

Bian, R.X., Dunn, R.C., Sunney, X.X., Leung, P.T. (1995) Single molecule emission characteristics in near-field microscopy, *Phys. Rev. Lett.*, **75**, 4772–4775.

Bohr, M.T. (1995) Interconnect scaling-the real limiter to high performance ULSI, *Tech. Dig. IEDM*, 241.

Bosio, L. (1978) Crystal structures of Ga(II) and Ga(III), *J. Chem. Phys.*, **68**, 1221–1223.

Bozhevolnyi, S.I., Nikolajsen, T., Leosson, K. (2005a) Integrated power monitor for long range surface plasmon polaritons, *Opt. Commun.*, **255**, 51–56.

Bozhevolnyi, S.I., Volkov, V.S., Devaux, E., Ebbesen, T.W. (2005b) Channel plasmon polariton guiding by subwavelength metal grooves, *Phys. Rev. Lett.*, **95**, 046802.

Bozhevolnyi, S.I., Volkov, V.S., Devaux, E., Laluet, J.-Y., Ebbesen, T.W. (2006) Channel plasmon subwavelength waveguide components including interferometers and ring resonators, *Nature*, **440**, 508–511.

Brown, P.O., Botstein, D. (1999) Exploring the new world of the genome with DNA microarrays, *Nat. Genetics Suppl.*, **21**, 33–37.

Chang, E., et al. (2005) Protease-activated quantum dot probes, *Biochem. Biophys. Res. Commun.*, **334**, 1317–1321.

Charbonneau, R. (2001) *Demonstration of Passive Integrated Optics Technology Based on Plasmonics*, MASc thesis (University of Ottawa, Ottawa).

Charbonneau, R., Lahoud, N., Mattiuse, G., Berini, P. (2005) Demonstration of integrated optics based on long ranging surface plasmon polaritons, *Opt. Express*, **13**, 977–984.

Chen, G., Chen, H., Haurylau, M., Nelson, N., M. Fauchet, P., Friedman, E.G., Albonesi, D. (2005) *Predictions of CMOS Compatible On-Chip Optical Interconnect, SLIP '05: Proceedings of the 2005 International Workshop on System Level Interconnect Prediction*, 13–20.

Chen, H., Wu, B.I., Zhang, B., Kong, J.A. (2007) Electromagnetic wave interactions with a metamaterial cloak, *Phys. Rev. Lett.*, **99**(6), 063903.

Chiang, T.-Y., et al. (2002) Impact of joule heating on scaling of deep submicron Cu/low- k interconnects, *IEEE Symp. VLSI Circuits, Dig. Tech. Papers*, 38.

Copland, J.A., Eghtedari, M., Popov, V.L., Kotov, N., Mamedova, N., Motamedi, M., Oraevsky, A.A. (2004) Bioconjugated gold nanoparticles as a molecular based contrast agent: implications for imaging of deep tumors using optoacoustic tomography, *Mol. Imaging Biol.*, **6**, 341–349.

Defrain, A. (1977) Metastable states of gallium. Undercooling and polymorphism, *J. Chem. Phys.*, **74**, 851–862.

Denschlag, J., Cassettari, D., Schmiedmayer, J. (1999b) Guiding neutral atoms with a wire, *Phys. Rev. Lett.*, **82**, 2014.

Denschlag, J., Cassettari, D., Chenet, A., Schneider, S., Schmiedmayer, J. (1999a) A neutral atom and a wire: towards mesoscopic atom optics, *Appl. Phys. B*, **69**, 291–301.

Denschlag, J., Umshaus, G., Schmiedmayer, J. (1998) Probing a singular potential with cold atoms: a neutral atom and a charged wire, *Phys. Rev. Lett.*, **81**, 737.

Dunn, R.C. (1999) Near-field scanning optical microscopy, *Chem. Rev.*, **99**, 2891–2927 (and reference therein).

Elghanian, R., Storhoff, J.J., Mucic, R.C., Letsinger, R.L., Mirkin, C.A. (1997) Selective colorimetric detection of polynucleotides based on the distance dependent optical properties of gold nanoparticles, *Science*, **277**, 1078–1081.

Fang, N., Lee, H., Sun, C., Zhang, X. (2005) Sub–diffraction-limited optical imaging with a silver superlens, *Science*, **308**, 534.

Farahani, J.N., Pohl, D.W., Eisler, H.-J., Hecht, B. (2005) Single quantum dot coupled to a scanning optical antenna: a tunable superemitter, *Phys. Rev. Lett.*, **95**, 017402.

Fedotov, V.A., MacDonald, K.F., Stevens, G.C., Zheludev, N.I. (2003) Nanoscale light-induced melting as a mechanism for broadband optical switching, *Nonlinear Opt., Quant. Opt.*, **30,** 53.

Gansel, J.K., Thiel, M., Rill, M.S., Decker, M., Bade, K., Saile, V., Freymann, G. von, Linden, S., Wegener, M. (2009) Gold helix photonic metamaterial as broadband circular polarizer, *Science*, **325**, 1513–1515.

Georghiou, S., Braddick, T.D., Philippetis, A., Beechem, J.M. (1996) Large-amplitude picosecond anisotropy decay of the intrinsic fluorescence of double-stranded DNA, *Biophys. J.*, **70**, 1909–1922.

Georghiou, S., Nordlund, T.M., Saim, A.M. (1985) Picosecond fluorescence decay time measurements of nucleic acids at room temperature in aqueous solution, *Photochem. Photobiol.*, **41**, 209–212.

Gersen, H., García-Parajó, M.F., Novotny, L., Veerman, J.A., Kuipers, L., Van Hulst, N.F. (2001) Near-field effects in single molecule emission, *J. Microsc.*, **202**, 374–378.

Gersen, H., García-Parajó, M.F., Novotny, L., Veerman, J.A., Kuipers, L., van Hulst, N.F. (2000) Influencing the angular emission of a single molecule, *Phys. Rev. Lett.*, **85**, 5312.

Greulich, K.O., Thiel, E. (2001) Single photon light sources, *Single Mol.*, **2**, 1.

Guzatov, D.V., Klimov, V.V. (2005) Radiative decay engineering by triaxial nanoellipsoids, *Chem. Phys. Lett.*, **412**, 341–346.

Haes, A.J., Zou, S., Schatz, G.C., Van Duyne, R.P. (2004a) A nanoscale optical biosensor: the long range distance dependence of the localized surface plasmon resonance of noble metal nanoparticles, *J. Phys. Chem. B*, **108**(1), 109–116.

Haes, A.J., Zou, S., Schatz, G.C., Van Duyne, R.P. (2004b) A nanoscale optical biosensor: the short range distance dependence of the localized surface plasmon resonance of noble metal nanoparticles, *J. Phys. Chem. B*, **108**(22), 6961–6968.

Hirsch, L.R., Stafford, R.J., Bankson, J.A., Sershen, S.R., Rivera, B., Price, R.E., Hazle, J.D., Halas, N.J., West, J.L. (2003) Nano-shell mediated near-infrared thermal therapy of tumors under magnetic resonance guidance, *PNAS*, **100**, 13549–13554.

Hirsch, L., Jackson, J., Lee, A., Halas, N.J., West, J. (2003) A rapid whole blood immunoassay using gold nanoshells, *Anal. Chem.*, **75**, 2377–2381.

Homola, J., Yee, S., Gauglitz, G. (1999) Surface plasmon resonance sensors: review, *Sens. Actuators*, **B54**, 3–15.

Human Genome (2001) The human genome, *Nature*, **409**, 818–958.

Ichinose, N., Schwedt, G., Schnepel, F.M., Adachi, K. (1987) *Fluorometric Analysis in Biomedical Chemistry* (Wiley, New York), 225 p.

Jacob, Z., Alekseyev, L.V., Narimanov, E. (2006) Optical hyperlens: far-field imaging beyond the diffraction limit, *Opt. Express*, **14**, 8247–8256.

Jaques, J., Wilen, S.H., Collet, A. (1981) *Enantiomers Racemates and Resolution* (Wiley-Interscience, New York).

Kerker, M. (1975) Invisible bodies, *J. Opt. Soc. Am.*, **65**, 376.

Klar, T., Perner, M., Grose, S., Spirkl, W., von Plessen, G., Feldman, J. (1998) Surface-plasmon resonances in single metallic nanoparticles, *Phys. Rev. Lett.*, **80**, 4249–4252.

Klimov, V.V., Perventsev, Ya. A. (1999) Variation of the emission characteristics of an atom located near an ideally conducting conical surface, *Quantum Electron.*, **29**, 847–851.

Klimov, V.V. (1998) Spontaneous emission of an atom oscillator located near an ideally conducting conical surface, *JETP Lett.*, **68**, 610–613.

Klimov, V.V. (2008) Novel approach to a "perfect" lens, arXiv:0811.4381.

Klimov, V.V. (2009a) Novel approach to a perfect lens, *JETP Lett.*, **89**, 270–273.

Klimov, V.V. (2009b) Energy flows in a "perfect" lens, http://demonstrations.wolfram.com/EnergyFlowInANegativeIndexMaterial

Klimov, V.V., Baudon, J., Ducloy, M. (2011) Comparative focusing of Maxwell and Dirac fields by negative refraction half-space, *Europhys. Lett.*, **94**, 20006.

Klimov, V.V., Guzatov, D.V., Ducloy, M. (2011) Engineering of radiation of optically active molecules with chiral nano-meta-particles, *Europhys. Lett.*, **97**, 47004.

Klimov, V.V., Perventsev, Y.A. (1999) Variation of the emission characteristics of an atom located nearly an ideally conducting conical surface, *Quantum Electron.*, **29**, 9–13.

Krasavin, A.V., MacDonald, K.F., Zheludev, N.I. (2007) Active plasmonics, in *Nanophotonics with Surface Plasmons*, eds. V.M. Shalaev, S. Kawata (Elsevier).

Kricka, L.J., Skogerboe, K.J., Hage, D.A., Schoeff, L., Wang, J., Sokol, L.J., Chan, D.W., Ward, K.M., Davis, K.A. (1997) Clinical chemistry, *Anal. Chem.*, **69**, 165R–229R.

Lakowicz, J.R., Malicka, J., Gryczynski, I., Gryczynski, Z., Geddes, C.D. (2003) Radiative decay engineering: the role of photonic mode density in biotechnology, *J. Phys. D: Appl. Phys.*, **36**, R240.

Levene, M.J., Korlach, J., Turner, S.W., Foquet, M., Craighead, H.G., Webb, W.W. (2003) Zero-mode waveguides for single-molecule analysis at high concentrations, *Science*, **299**, 682–686.

Li, X., Shapiro, M. (2010) Spatial separation of enantiomers by coherent optical means, *J. Chem. Phys.*, **132**, 041101.

Liedberg, B., Nylander, C., Landstrom, I. (1995) Biosensing with surface plasmon resonance, how it all started, *Biosens. Bioelectron.*, **10**, i–ix.

Lindfors, K., Kalkbrenner, T., Stoller, P., Sandoghdar, V. (2004) Detection and spectroscopy of gold nanoparticles using supercontinuum white light confocal microscopy, *Phys. Rev. Lett.*, **93**, 037401.

Lipschutz, R.J., Fodor, S.P.A., Gineras, T.R., Lockhart, D.J. (1999) High density synthetic oligonucleotide arrays, *Nat. Genetics*, **21**(Suppl 1), 20–24.

Lisyansky, A.A., Andrianov, E.S., Dorofeenko, A.V., Pukhov, A.A., Vinogradov, A.P. (2012) Forced Spaser Oscillations, *Proc. of SPIE*, **8457**, 84570X (16 p).

Liu, G.L., Doll, J.C., Lee, L.P. (2005) High-speed multispectral imaging of nanoplasmonic array, *Opt. Express*, **13**, 8520–8525.

Liu, N., Liu, H., Zhu, S., Giessen, H. (2009) Stereometamaterials, *Nat. Photonics*, **3**(March), 157.

Liu, Z., Lee, H., Xiong, Y., Sun, C., Zhang, X. (2007) Far-field optical hyperlens magnifying sub-diffraction-limited objects, *Science*, **315**, 1686.

Loo, C.H., Hirsch, L.R., West, J.L., Lee, M., Chang, E., Halas, N.J., Drezek, R.A. (2005) Gold nanoshell bioconjugates for molecular imaging in living cells, *Opt. Lett.*, **30**, 1012–1014.

Loo, C., Lin, A., Hirsch, L., Min-Ho Lee, Barton, J., Halas, N., West, J., Drezek, R. (2004) Nanoshell-enabled photonics-based imaging and therapy of cancer, *Technol. Cancer Res. Treat.*, **3**, 33–40.

Lounis, B., Moerner, W.E. (2000) Single photons on demand from a single molecule at room temperature, *Nature*, **407**, 491–493.

Lu, H.P., Xun, L., Xie, X.S. (1998) Single-molecule enzymatic dynamics, *Science*, **282**, 1877.

MacDonald, K.F., Fedotov, V.A., Eason, R.W., Zheludev, N.I., Rode, A.V., Luther-Davies, B., Emelyanov, V.I. (2001a) Nanosecond dynamics of a gallium mirror's light-induced reflectivity change, *J. Opt. Soc. Am. B*, **18**, 331.

MacDonald, K.F., Fedotov, V.A., Pochon, S., Stevens. G.C., Kusmartsev, F.V., Emel'yanov, V.I., Zheludev, N.I. (2004) Controlling the coexistence of structural phases and the optical properties of gallium nanoparticles with optical excitation, *Europhys. Lett.*, **67**, 614.

MacDonald, K.F., Fedotov, V.A., Zheludev, N.I., Zhdanov, B.V., Knize, R.J. (2001b) Structural phase transition as a mechanism for broadband, low-threshold reflectivity switching in gallium, *Appl. Phys. Lett.*, **79**, 2375.

MacDonald. K.F., Fedotov, V.A., Zheludev. N.I. (2003) Optical nonlinearity resulting from a light-induced structural transition in gallium nanoparticles, *Appl. Phys. Lett.*, **82**, 1087.

Malinsky, M.D., Kelly, K.L., Schatz, G.C., Van Duyne, R.P. (2001) Chain length dependence and sensing capabilities of the localized surface plasmon resonance of silver nanoparticles chemically modified with alkanethiol self-assembled monolayers, *J. Am. Chem. Soc.*, **123**, 1471–1482.

McFarlad, A.D., Van Duyne, R.P. (2003) Single silver nanoparticles as real-time optical sensors with zeptomole sensitivity, *Nano Lett.*, **3**, 1057–1062.

Michaelis, J.M., Hettich, C., Mlynek, J., Sandoghdar, V. (2000) Optical microscopy using a single-molecule light source, *Nature*, **405**, 325–328.

Miller, D.A. (2000) Rationale and challenges for optical interconnects to electronic chips, *Proc. IEEE*, **88**, 728.

Mirkin, C.A., Letsinger, R.L., Mucic, R.C., Storhoff, J.J. (1996) A DNA-based method for rationally assembling nanoparticles into macroscopic materials, *Nature*, **382,** 607–609.

Mock, J.J., Smith, D.R., Schultz, S. (2003) Local refractive index dependence of plasmon resonance spectra from individual nanoparticles, *Nano Lett.*, **3**, 485–491.

Myszka, D.G., Rich, R.L. (2000) Implementing surface plasmon resonance biosensors in drug discovery, *Pharm. Sci. Technol. Today*, **3**, 310–317.

Nachman, A.I. (1988) Reconstructions from boundary measurements, *Ann. Math.*, **128**, 531.

Nath, N., Chilkoti, A. (2002) A colorimetric gold nanoparticle sensor to interrogate biomolecular interactions in real time on a surface, *Anal. Chem.*, **74**, 504–509.

Nath, N., Chilkoti, A. (2004) Label-free biosensing by surface plasmon resonance of nanoparticles on glass: optimization of nanoparticle size, *Anal. Chem.*, **76**, 5370–5378.

Nikolajsen, T., Leosson, K., Bozhevolnyi, S.I. (2004) Surface plasmon polariton based modulators and switches operating at telecom wavelengths, *Appl. Phys. Lett.*, **85**, 5833–5836.

Noginov, M.A., Zhu, G., Belgrave, A.M., Bakker, R., Shalaev, V.M., Narimanov, E.E., Stout, S., Herz, E., Suteewong, T., Wiesner, U. (2009) Demonstration of a spaser-based nanolaser, *Nature*, **460**, 1110–1112.

Novikov, I.V., Maradudin, A.A. (2002) Channel polaritons, *Phys. Rev. B*, **66**, 035403.

Nozawa, T. (2008) Invisible cloak realized in theory, *Tech_On*, **4**, 18.

Nunzi, J.M., Ricard, D. (1984) Optical phase conjugation and related experiments with surface plasmon waves, *Appl. Phys. B*, **35**, 209–216.

Ochiai, T., Leonhardt, U., Nacher, J.C. (2008) A novel design of dielectric perfect invisibility devices, *J. Math. Phys.*, **49**, 032903.

Ochiai, T., Leonhardt, U., Nacher, J.C. (2008) A novel design of dielectric perfect invisibility devices, arXiv:0711.1122v2.

Oldenburg, S.J., Averitt, R.D., Westcott, S.L., Halas, N.J. (1998) Nanoengineering of optical resonances, *Chem. Phys. Lett.*, **288**, 243–247.

Oldenburg, S.J., Genick, C.C., Clark, K.A., Schultz, D.A. (2002) Base pair mismatch recognition using plasmon resonant particle labels, *Anal. Biochem.*, **309**, 109–116.

Oulton, R.F., Sorger, V.J., Genov, D.A., Pile, D.F.P., Zhang, X. (2008) A hybrid plasmonic waveguide for subwavelength confinement and long-range propagation, *Nat. Photonics*, **2**, 496.

Pendry, J.B. (2000) Negative refraction makes a perfect lens, *Phys. Rev. Lett.*, **85**, 3966.

Pendry, J.B., Schurig, D., Smith, D.R. (2006) Controlling electromagnetic fields, *Science*, **312**, 1780.

Pennisi, E. (2001) The human genome, *Science*, **291**, 1177–1180.

Petropoulos, P., Kim, H.S., Richardson, D.J., Fedotov, V.A., Zheludev, N.I. (2001) Light-induced metallization at the gallium-silica interface, *Phys. Rev. B*, **64**, 193312.

Petropoulos, P., Offerhaus, H.L., Richardson, D.J., Dhanjal, S., Zheludev, N.I. (1999) Passive Q-switching of fiber lasers using a broadband liquefying gallium mirror, *Appl. Phys. Lett.*, **74**, 3619.

Pile, D.F.P., Gramotnev, D.K. (2004) Channel plasmon polariton in a triangular groove on a metallic surface, *Opt. Lett.*, **29**, 1069.

Pochon, S., MacDonald, K.F., Knize. R.J., Zheludev, N.I. (2004) Phase coexistence in gallium nanoparticles controlled by electron excitation, *Phys. Rev. Lett.*, **92**, 145702.

Porter, W.H. (1991) Resolution of chiral drugs, *Pure Appl. Chem.*, **63**, 1119.

Prober, J.M., Trainor, G.L., Dam, R.J., Hobbs, F.W., Robertson, C.W., Zagursky, R.J., Cocuzza, A.J., Jensen, M.A., Baumeister, K. (1987) A system for rapid DNA sequencing with fluorescent chain-terminating dideoxynucleotides, *Science*, **238,** 336–343.

Raschke, G., Kowarik, S., Franzl, T., So1nnichsen, C., Klar, T.A., Feldmann, J., Nichtl, A., Ku1rzinger, K. (2003) Biomolecular recognition based on single gold nanoparticle light scattering, *Nano Lett.*, **3**, 935–938.

Rigneault, H., Capoulade, J., Dintinger, J., Wenger, J., Bonod, N., Popov, E., Ebbesen, T.W., Lenne, P.F. (2005) Enhancement of single-molecule fluorescence detection in subwavelength apertures, *Phys. Rev. Lett.*, **95**, 117401.

Salandrino, A., Engheta, N. (2006) Far-field subdiffraction optical microscopy using metamaterial crystals: theory and simulations, *Phys. Rev. B*, **74**, 075103.

Schuck, P. (1997) Use of surface plasmon resonance to probe the equilibrium and dynamic aspects of interaction between biological macromolecules, *Annu. Rev. Biophys. Biomol. Struct.*, **26**, 541–566.

Schultz, S., Smith, D.R., Mock, J.J., Schultz, D.A. (2000) Single-target moleculedetection with nonbleaching multicolor optical immunolabels, *Proc. Natl. Acad. Sci.*, **97**(3), 996–1001.

Schurig, D., Mock, J.J., Justice, B.J., Cummer, S.A., Pendry, J.B., Starr, A.F., Smith, D.R. (2006) Metamaterial electromagnetic cloak at microwave frequencies, *Science*, **314**, 977–980.

Smith, L.M., Sanders, J.Z., Kaiser, R.J., Hughes, P., Dodd, C., Connell, C.R., Heiner, C., Kent, S.B.H., Hood, L.E. (1986) Fluorescence detection in automated DNA sequence analysis, *Nature*, **321**, 674–679.

Smolyaninov, I.I., Hung, Y.-J., Davis, C.C. (2007) Magnifying superlens in the visible frequency range, *Science*, **315,** 1699–1701.

Sönnichsen, C., Franzl, T., Wilk, T., von Plessen, G., Feldmann, J., Wilson, O., Mulvaney, P. (2002) Suppression of interband damping of plasmons in gold nanorods, *Phys. Rev. Lett.*, **88**, 077402.

Sönnichsen, C., Geier, S., Hecker, N.E., von Plessen, G., Feldmann, J., Ditlbacher, H., Lamprecht, B., Krenn, J.R., Aussenegg, F.R., Chan, V.Z.-H., Spatz, J.P., Möller, M. (2000) Spectroscopy of single metallic nanoparticles using total internal reflection microscopy, *Appl. Phys. Lett.*, **77**, 2949.

Srituravanich, W., Pan, L., Wang, Y., Sun, C., Bogy, D.B., Zhang, X. (2008) Flying plasmonic lens in the near field for high-speed nanolithography, *Nat. Nanotechnol.*, **3**, 733–737.

Steinberger, B., Hohenau, A., Ditlbacher, H., Aussenegg, F.R., Leitner, A., Krenn, J.R. (2007) Dielectric stripes on gold as surface plasmon waveguides: bends and directional couplers, *Appl. Phys. Lett.*, **91**, 081111.

Steinberger, B., Hohenau, A., Ditlbacher, H., Stepanov, A.L., Drezet, A., Aussenegg, F.R., Krenn, J.R. (2006) Dielectric stripes on gold as surface plasmon waveguides, *Appl. Phys. Lett.*, **88**, 094104.

Storhoff, J., Marla, S., Bao, P., Hagenow, S., Mehta, H., Lucas, A., Garimella, V., Patno, T., Buckingham, W., Cork, W., Müller, U. (2004) Gold nanoparticle probe-based gene expression analysis with unamplified total human RNA, *Biosens. Bioelectron.*, **19**, 875–883.

Tanaka, K., Tanaka, M. (2003) Simulation of nanometric optical circuits based on surface plasmon polariton gap waveguide, *Appl. Phys. Lett.*, **82**, 1158.

Taton, T.A., Lu, G., Mirkin, C.A. (2001) Two-color labeling of oligonucleotide arrays via size-selective scattering of nanoparticle probes, *J. Am. Chem. Soc.*, **123**, 5164–5165.

Trautman, J.K., Macklin, J.J., Brus, L.E., Betzig, E. (1994) Near-field spectroscopy of single molecules at room temperature, *Nature* (London), **369**, 40.

Veerman, J.A., Garcia-Parajo, M.F., Kuipers, L., Van Hulst, N.F. (1999) Single molecule mapping of the optical field distribution of probes for near-field microscopy, *J. Microsc.*, **194**, 477–482.

Veronis, G., Fan, S. (2005) Bends and splitters in metal-dielectric-metal subwavelength plasmonic waveguides, *Appl. Phys. Lett.*, **87**, 131102.

Vinogradov, A.P., Andrianov, E.S., Pukhov, A.A., Dorofeenko, A.V., Lisyansky, A.A. (2012) Quantum plasmonics of metamaterials: loss compensation using spasers, *Phys. Usp.*, **55**, 1046.

Wang, H., Wu, Y., Lassiter, B., Nehl, C.L., Hafner, J.H., Nordlander, P., Halas, N.J. (2006) Symmetry breaking in individual plasmonic nanoparticles, *PNAS*, **103**, 10856–10860.

Weeber, J.C., Lacroute, Y., Dereux, A., Devaux, E., Ebbesen, T., Girard, C., Gonsalez, M.U., Baudrion, A.L. (2004) Near field characterization of Bragg mirrors engraved in surface plasmon waveguides, *Phys. Rev. B*, **70**, 235406.

Wolf, E., Habashy, T. (1993) Invisible bodies and uniqueness of the inverse scattering problem, *J. Mod. Opt.*, **40**, 785.

Xie, X.S., Lu, H.P. (1999) Single-molecule enzymology, *J. Biol. Chem.*, **274**, 15967.

Xie, X.S., Dunn, R.C. (1994) Probing single molecule dynamics, *Science*, **265**, 361.

Yamamoto, Y., Tassone, F., Cao, H. (2000) *Semiconductor Cavity Quantum Electrodynamics* (Springer-Verlag, Berlin, Heidelberg).

Yguerabide, J., Yguerabide, E.E. (2001) Resonance light scattering particles as ultrasensitive labels for detection of analytes in a wide range of applications, *J. Cell. Biochem.*, **37**(Suppl), 71–81.

Yguerabide, J., Yguerabide, E.E. (1998a) Light-scattering submicroscopic particles as highly fluorescent analogs and their use as tracer labels in clinical and biological applications: I. Theory, *Anal. Biochem.*, **262**, 137–156.

Yguerabide, J., Yguerabide, E.E. (1998b) Light-scattering submicroscopic particles as highly fluorescent analogs and their use as tracer labels in clinical and biological applications: II. Experimental characterization, *Anal. Biochem.*, **262**, 157–176.

Zheludev, N.I. (2002) Nonlinear optics on the nanoscale, *Contemp. Phys.*, **43**, 365.

Zia, R., Schuller, J.A., Chandran, A., Brongersma, M.L. (2006) Plasmonics: the next chip-scale technology, *Mater. Today*, **9**, 20–27.

Zia, R., Selker, M.D., Catrysse, P.B., Brongersma, M.L. (2004) Geometries and materials for subwavelength surface plasmon modes, *J. Opt. Soc. Am.* A, **21**, 2442.

Chapter 14

Conclusion

Nanoplasmonics is an amazing and fast-developing research area, and I hope that after reading this book completely or partially, an interested reader will find here not only an overview of the current situation in this field but also a profound basis for his or her own future investigations. It is quite obvious that a diversity of developing and prospective applications of nanoplasmonics and nano-optics determined by effects demonstrated above provides an interest in this branch for many years ahead. I have tried to make the book as self-sufficient as possible from the theoretical point of view, and I expect that many formulas can be used immediately by both theorists and experimentalists in nanoplasmonics and related areas. Theoretical methods and approaches presented in the book can be also applied in further original investigations. As to experimental aspects of nanoplasmonics, this work contains rather an overview of fundamental experiments and applications having reputable substantiation and interpretation. For a more detailed study of experimental aspects, one should consider specific literature listed in each chapter exhaustively. Moreover, due to fast development of nanoplasmonics and nano-optics, new ideas and directions arise persistently and cannot be represented in this book, naturally. Nevertheless, an acquaintance with methods and approaches of this

Nanoplasmonics
Vasily Klimov

ISBN 978-981-4267-16-8 (Hardcover), 978-981-4267-42-7 (eBook)
www.panstanford.com

work will be sufficient for a quick understanding of the essence of new ideas and experiments.

I hope that this book will be helpful also for educational purposes and will attract young specialists to the exciting field of nano-optics and nanophotonics, especially as development of nanotechnologies is becoming a national priority in many countries.

And, finally, I am deeply grateful to Letokhov Vladilen Stepanovich (1939–2009). Twenty years ago he drew my attention to a new area of science that did not exist at that time—to nano-optics that contains nanoplasmonics. It was the impulse that resulted in the appearance of the book.

Due to a variety of interesting effects, approaches, and applications that I regarded obligatory in this work, the book is rather voluminous, and some roughness and inaccuracies are admissible here. I will appreciate any corrections, comments, and criticism that can be sent to my email address: vklim@sci.lebedev.ru.

Appendix A1

Short Theory of Spontaneous Emission and Fluorescence of Atoms and Molecules in the Presence of Nanobodies

One of the most important problems of nano-optics and nanoplasmonics is the atoms' and molecules' emission control that allows one to create many types of complex nanodevices from effective light sources to nanolasers and quantum computers. In this appendix, main statements of the theory of atoms and molecules' radiation in the vicinity of nanobodies are presented.

A1.1 The Nanobodies' Influence on the Rate of an Atom's or a Molecule's Spontaneous Emission

For a weak interaction between an atom and a nanobody, that is, in the case when the spontaneous emission is exponentially decreasing in time, an expression for the linewidth γ_a of the excited state a of an atom located at the point $\mathbf{r} = \mathbf{r}'$ (hereafter primed coordinates will

Nanoplasmonics
Vasily Klimov

ISBN 978-981-4267-16-8 (Hardcover), 978-981-4267-42-7 (eBook)
www.panstanford.com

denote the location of an atom) is the sum over all atomic states n (Wylie and Sipe, 1984, 1985):

$$\gamma_a = \gamma_{a,0} + \frac{2}{\hbar}\sum_n d_\alpha^{an} d_\beta^{na} \mathrm{Im}G_{\alpha\beta}(\mathbf{r}', \mathbf{r}'; \omega_{na})\Theta(\omega_{an}), \quad \text{(A1.1)}$$

where $\gamma_{a,0}$ is the linewidth in a free space in the absence of a nanobody, d_α^{an} is a matrix element of the operator of the electric dipole moment between atomic states a and n, $\omega_{na} = (W_n - W_a)/\hbar$ is the frequency of the transition between states n and a with energies W_n and W_a, $G_{\alpha\beta}$ is a part of Green's function of Maxwell's equations with the dipole source $\mathbf{d}_0$, and this part is related to the field reflected from the nanobody E_α^r by the relation

$$E_\alpha^r(\mathbf{r}) = G_{\alpha\beta}(\mathbf{r}, \mathbf{r}'; \omega_{na}) d_{0,\beta}, \quad \text{(A1.2)}$$

where $\alpha, \beta = 1, 2, 3$ are indices of the Cartesian system (summation over the repeated indices is assumed) and Θ is the Heaviside unit step function. Hereafter, we shall consider only one channel of the excited state decay, that is, only one term in sum (A1.1). To take into account the possibility of decay into different states, partial linewidths are to be summed. In the case of the partial linewidth, the expression that can be obtained in the context of the classical (Lorentz) theory of an atom for the relative linewidth (i.e., for the spontaneous decay rate) coincides with the corresponding quantum mechanical expression (A1.1).

The expression for the level shift δW_a of an arbitrary state a of an atom in the presence of a nanoparticle has the form:

$$\delta W_a = \delta W_a^r + \delta W_a^{vdW}, \quad \text{(A1.3)}$$

$$\delta W_a^r = -\sum_n d_\alpha^{an} d_\beta^{na} \mathrm{Re}G_{\alpha\beta}(\mathbf{r}', \mathbf{r}'; \omega_{na})\Theta(\omega_{an}),$$

$$\delta W_a^{vdW} = -\frac{\hbar}{2\pi}\int_0^\infty d\xi\, G_{\alpha\beta}(\mathbf{r}', \mathbf{r}'; \omega = i\xi)\Pi_{\alpha\beta}^a(\omega = i\xi), \quad \text{(A1.4)}$$

where $\Pi_{\alpha\beta}^a$ is a tensor of the polarizability of the atomic state a.

$$\Pi_{\alpha\beta}^a(\omega) = \frac{2}{\hbar}\sum_n \frac{\omega_{na} d_\alpha^{an} d_\beta^{na}}{(\omega_{na})^2 - (\omega + i0)^2}, \quad \text{(A1.5)}$$

where summing is performed over all intermediat atom states n.

Let us note that the expressions (A1.1) and (A1.3) are obtained in the context of most general assumptions and therefore they have a very wide application area.

Thus, the problem of founding of spectroscopic characteristics of an atom in the vicinity of any nanobody is reduced to founding of the reflected field or Green's function, Eq. (A1.2), and analysis of Eqs. (A1.1) and (A1.3).

Often, in the case of nanobodies the perturbation theory with respect to a small parameter $2\pi b/\lambda \ll 1$ can be used, where b is the typical size of the nanobody and λ is the emission wavelength (Rayleigh theory). In this case, the expression for Green's function of the reflected field can be presented as the series over the wavevector powers:

$$\begin{aligned} G_{\alpha\beta}(\mathbf{r}, \mathbf{r}'; \omega) &= G^{(0)}_{\alpha\beta}(\mathbf{r}, \mathbf{r}') + kG^{(1)}_{\alpha\beta}(\mathbf{r}, \mathbf{r}') + k^2 G^{(2)}_{\alpha\beta}(\mathbf{r}, \mathbf{r}') \\ &\quad + ik^3 G^{(3)}_{\alpha\beta}(\mathbf{r}, \mathbf{r}') + \ldots, \end{aligned} \tag{A1.6}$$

where $G^{(j)}_{\alpha\beta}$ $(j = 0, 1, 2, \ldots)$ are coefficients that can be found from the solution of corresponding quasi-static problems (Stevenson 1953a, 1953b) and $k = \omega/c$. Let us note that when building Eq. (A1.6), the dependence $\varepsilon(\omega)$ on the frequency should not be taken into account. The first three terms of (A1.6) describe the near fields, while beginning from the fourth term the emission fields arise. By substituting the expansion Eq. (A1.6) in Eq. (A1.1), we obtain the following expression for the total rate of transitions from the state e to g near a nanobody:

$$\begin{aligned} \gamma_{e\to g} &= \underbrace{\frac{2}{\hbar} d^{eg}_{\alpha} d^{ge}_{\beta} \operatorname{Im}\left\{ G^{(0)}_{\alpha\beta}(\mathbf{r}', \mathbf{r}') + \ldots \right\}}_{\text{nonradiative}} \\ &\quad + \underbrace{\gamma_{e\to g,0} + \frac{2}{\hbar} d^{eg}_{\alpha} d^{ge}_{\beta} \operatorname{Re}\left\{ k_0^3 G^{(3)}_{\alpha\beta}(\mathbf{r}', \mathbf{r}') + \ldots \right\}}_{\text{radiative}}. \end{aligned} \tag{A1.7}$$

Here, $k_0 = \omega_{eg}/c$. In Eq. (A1.7), the first term is nonzero only for absorbing media and describes nonradiative (heat) losses of the atom energy, while the rest of the terms are nonzero also in the case of the absence of absorption. Mainly, these terms describe radiative losses, that is, an observable emission of an atom. Thus, to find leading terms of radiative and nonradiative losses, $G^{(0)}_{\alpha\beta}(\mathbf{r}', \mathbf{r}')$ and $G^{(3)}_{\alpha\beta}(\mathbf{r}', \mathbf{r}')$ are to be found, correspondingly.

To find the term that describes nonradiative losses, it is sufficient to solve the quasi-static problem of the electrodynamics with a dipole source placed in the location of the emitting atom. Direct finding of the radiative losses, which are described by terms of the third order over k, appears to be a difficult problem. However, in the case of an atom located in the vicinity of a nanobody the emission pattern is of a dipole type, and again the total dipole moment of the "atom + nanobody" system can be found from the analysis of $G^{(0)}_{\alpha\beta}(\mathbf{r}, \mathbf{r}')$ at far distances from the system. Thus, in the case of nanobodies, the radiative linewidth of the transition of the excited state e to the ground state g will be described by the expression

$$\left(\frac{\gamma}{\gamma_0}\right)^{\text{radiative}} = \frac{|\mathbf{d}_{\text{total}}|^2}{d_0^2}, \tag{A1.8}$$

where γ_0 is the transition linewidth in the absence of a nanobody, $\mathbf{d}_{\text{total}} = \mathbf{d}_0 + \delta\mathbf{d}$ is the total dipole moment of the "atom + nanobody" system, and $\delta\mathbf{d}$ is the dipole moment induced in the nanobody. In the context of quantum mechanical approach, correctness of Eq. (A1.8) can be also shown (Klimov and Ducloy, 2004).

Analogously, by substituting the expansion Eq. (A1.6) in the expression for the level shift, Eq. (A1.3), of any state, the expression can be obtained:

$$\begin{aligned}\delta E_a = &-\frac{1}{2}\left\langle a\left|d_\alpha d_\beta\right| a\right\rangle \text{Re}\left\{G^{(0)}_{\alpha\beta}(\mathbf{r}', \mathbf{r}')\right\} \\ &-\frac{1}{2}\sum_n d^{an}_\alpha d^{na}_\beta \text{Im}G^{(0)}_{\alpha\beta}(\mathbf{r}', \mathbf{r}')\frac{\omega_{an}}{|\omega_{an}|} + \ldots\end{aligned} \tag{A1.9}$$

In the case of a substance with small losses, the second term in Eq. (A1.9) can be ignored in comparison with the first one. In this case, the following expression is obtained for the emission frequency shift of the transition between the states e and g in the vicinity of a nanobody:

$$\Delta\omega = -\frac{1}{2\hbar}\left(\left\langle e\left|d_\alpha d_\beta\right| e\right\rangle - \left\langle g\left|d_\alpha d_\beta\right| g\right\rangle\right) \text{Re}\left\{G^{(0)}_{\alpha\beta}(\mathbf{r}', \mathbf{r}')\right\} + \ldots \tag{A1.10}$$

As a result, to find the change of the spontaneous decay rate and the frequency shifts in the presence of any nano-object with small losses ($\varepsilon'' \ll \varepsilon'$) and having the small sizes in comparison with the emission wavelength, it is sufficient to solve the quasi-static problem of the dipole located near this object and to find $G^{(0)}_{\alpha\beta}(\mathbf{r}, \mathbf{r}')$. The value of this function in the coinciding points $\mathbf{r} = \mathbf{r}'$ will give the frequency shift and the nonradiative part of the transition rate according to Eqs. (A1.10) and (A1.7), correspondingly. The total dipole moment found from the far field asymptotic of $G^{(0)}_{\alpha\beta}(\mathbf{r}, \mathbf{r}')$ allows one to calculate the radiative decay rate according to Eq. (A1.8).

A1.2 Nano-Objects' Influence on Fluorescence of Molecules

Above, the theory of the nanoenvironment influence of properties of a single optical transition, that is, on the linewidth and the frequency shift, was presented. However, practically one deals with more complex processes, in which several different transitions related both to radiation absorption and to emission take place.

Often,transitions between different states and various radiative and nonradiative processes in organic molecules are described by means of a so-called Jablonski diagram, which is shown in Fig. A1.1. In this diagram, radiative processes are denoted by straight lines, while nonradiative ones are denoted by wavy lines.

In the ground state of the majority of molecules (in particular, in organic ones), electron spins are directed in opposite directions (paired) and therefore the molecule total spin is equal to zero $S = 0$. Correspondingly, the ground state is singlet. Usually, the singlet energy states are denoted as S_0, S_1, S_2, ... in the order of increasing energy. The Jablonski diagram for such molecules is shown in Fig. A1.1.

For molecules with the singlet ground state S_0, the excited state can by either singlet (for example, S_1) or triplet when the electron spins are parallel. Excited triplet energy states are denoted as T_1, T_2, ... in the order of increasing energy.

The quantum mechanical consideration (Landau and Lifshitz, 1989) shows that for the same orbital moment the triplet state (for example, the T_1 state) lies lower that the singlet state. In Fig. A1.1, possible processes in molecule that was excited to the singlet state S_2 at the initial time are shown.

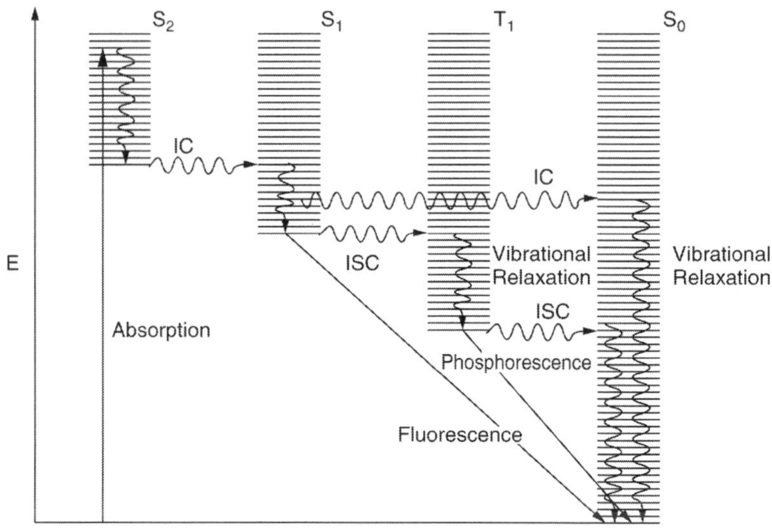

Figure A1.1 The Jablonski diagram describing the possible fates of molecule excitation.

Horizontal equidistant lines show vibrational sublevels. In a general case, the nonradiative transition from S_2 to S_1 is a predominant mechanism. Some molecules only admit the photon emission with the transition from the state S_2 to S_0. Transitions between states of the same spin multiplicity (e.g., from S_2 to S_1) are called an internal conversion (IC). Usually, the process of IC is followed by oscillation relaxation when the oscillation energy transforms to heat and the molecule turns out to be on the lowest vibrational level of the electron state S_1. From this state, the molecule can return to the ground state S_0 by emitting a photon. It is such transitions that are called *fluorescence*. In a general case, this transition is characterized by a short lifetime of the upper state, which has the order of one nanosecond usually.

On the other hand, the excitation energy can pass between states of different spins (from S_1 to T_1) by means of another nonradiative transition that is called intersystem crossing (ISC). This transition breaks the spin conservation law during a change of electronic state and therefore is called a spin-forbidden transition. This spin nonconservation is related to spin–orbital interaction. Then, a fast oscillation relaxation follows, the result of which is that the molecule turns out to be on the lower oscillation level of the triplet state T_1. The photon emission from this triplet state to the singlet ground state S_0 breaks again the spin conservation law and is called *phosphorescence*. This process is again related to the spin–orbital interaction and therefore is very weak and has a long lifetime of the upper state T_1. Moreover, nonradiative intersystem transitions from the state T_1 to S_0 are possible.

All the processes described above give important information on the molecule structure and form the basis of different types of spectroscopic investigations. One of the most important processes is the molecules' fluorescence in the presence of nano-objects (Lakowicz et al., 2003). In Fig. A1.2, a simplified scheme of levels for fluorescence description is shown.

With the help of this scheme, the fluorescence near nano-objects (the theory of free-space fluorescence is quite well developed; see,

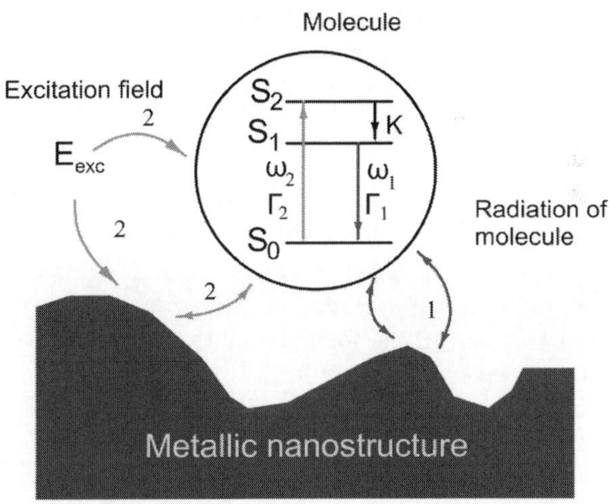

Figure A1.2 The simplified scheme of levels, which allows one to describe effectively fluorescence in an arbitrary environment. S_0, S_1, and S_2 are ground and excited singlet states of the molecule. K characterizes the vibrational relaxation of the state S_2. $\mathbf{E}_{ext}$ is an excitation field. Interaction between an incident field, a molecule, and the nanoenvironment is shown symbolically by the semicircular arrows at frequencies ω_1 and ω_2, correspondingly.

e.g., (Cohen-Tannoudji et al., 1989)) can be described effectively, and in the case of the stationary narrow-band resonance excitation the fluorescence intensity can be described by the expression

$$I_{fl} = \frac{\hbar\omega_1}{\frac{1}{\tilde{\Gamma}_1} + \frac{2}{K} + \frac{1}{\Gamma_{pump}}\left(1 + \frac{\Gamma_2}{K}\right)} Q_1, \qquad \text{(A1.11)}$$

where $\tilde{\Gamma}_1 = \Gamma_1 + \Gamma_{1,nonopt} = \Gamma_{1,rad} + \Gamma_{1,nonrad} + \Gamma_{1,nonopt}$ are the total rate of the optical and nonoptical transitions from the level S_1 to the level S_0, $1/K \approx 10^{-10}s$ is the rate of the vibrational relaxation of the state S_2, and Γ_2 is the rate of spontaneous transitions from the level S_2 to the level S_0. The pumping rate $\Gamma_{pump} = 4|\Omega|^2/\Gamma_2$ is determined by the Rabi frequency $\Omega = \frac{\mathbf{d}_{20}\mathbf{E}_{exc}}{\hbar}$, which, in turn, depends on the transition dipole moment $\mathbf{d}_{20}$ and the external electric field $\mathbf{E}_{ext}$. The factor

$$Q_1 = \frac{\Gamma_{1,rad}}{\Gamma_1 + \Gamma_{1,nonopt}}$$

determines the quantum efficiency of fluorescence, that is, a probability of the real photon emission by the transition from S_1 to S_0, and is expressed through the ratio between the radiative linewidth and its total width, which includes nonoptical relaxation processes also.

Generally speaking, in the case of a molecule in the nanoenvironment all the variables entering Eq. (A1.11) will depend on the molecule's position in the nanoenvironmen (see Fig. A1.2).

Apparently, nonoptical processes (K, $\Gamma_{1,nonopt}$), owing to their short-range nature, do not suffer a substantial influence of nano-objects, and as a result, the formula for the fluorescence intensity can be written as (Girard et al., 2005)

$$I_{fl} = \frac{\hbar\omega_1(\mathbf{r})}{\frac{1}{\Gamma_1(\mathbf{r})+\Gamma_{1,nonopt}} + \frac{2}{K} + \frac{1}{\Gamma_{pump}(\mathbf{r})}\left(1 + \frac{\Gamma_2(\mathbf{r})}{K}\right)} Q_1(\mathbf{r}), \qquad \text{(A1.12)}$$

$$Q_1(\mathbf{r}) = \frac{\Gamma_{1,rad}(\mathbf{r})}{\Gamma_1(\mathbf{r})+\Gamma_{1,nonopt}} = \frac{\Gamma_{1,rad}(\mathbf{r})}{\Gamma_{1,rad}(\mathbf{r})+\Gamma_{1,nonrad}(\mathbf{r})+\Gamma_{1,nonopt}}$$

where $\Gamma_1(\mathbf{r})$, $\Gamma_2(\mathbf{r})$*** $\omega_1(\mathbf{r})$ are determined by Eqs. (A1.1) and (A1.3), correspondingly. In Eq. (A1.12) and further, the rate $\Gamma_{1,nonrad}(\mathbf{r})$ is related to nonradiative losses in the nanoparticle material. By construction, this rate differs from the nonradiative rate $\Gamma_{1,nonopt}(\mathbf{r})$, which is nonzero even in the absence of a nanoparticle.

From Eq. (A1.12), it is seen that the spatial dependence of fluorescence intensity on a molecule's position is determined by corresponding linewidths and frequency shifts, which can be found with the help of the methods stated above. Furthermore, fluorescence intensity is defined by the Rabi local frequency value, which also experiences substantial changes in the presence of nanobodies.

In the case of a strong external field $\Gamma_{\text{pump}}(\mathbf{r}) \to \infty$ (the case of saturation), Eq. (A1.12) is simplified essentially as

$$I_{\text{fl}}(\mathbf{r}) = \frac{\hbar\omega_1(\mathbf{r})\Gamma_{1,\text{rad}}(\mathbf{r})}{1 + \frac{2}{K}\left(\Gamma_{1,\text{rad}}(\mathbf{r}) + \Gamma_{1,\text{nonrad}}(\mathbf{r}) + \Gamma_{1,\text{nonop}}\right)}. \tag{A1.13}$$

It is important that in the case of relatively small nonradiative losses $(\Gamma_{1,\text{nonrad}}(\mathbf{r}) + \Gamma_{1,\text{nonopt}})/K \leq 1$) the increase of the radiative transition rate near nanobodies $\Gamma_{1,\text{rad}}(\mathbf{r})$ (e.g., in the case of plasmon resonance in nanoparticles) results in a proportional fluorescence intensity increase. One can try to develop new types of nanosensors on the basis of this effect.

In the case of small excitation fields, we have the following expression instead of Eq. (A1.12), correspondingly:

$$I_{\text{fl}}(\mathbf{r}) = \frac{\hbar\omega_1(\mathbf{r})\Gamma_{\text{pump}}(\mathbf{r})}{1 + \frac{\Gamma_2(\mathbf{r})}{K}} Q_1(\mathbf{r}). \tag{A1.14}$$

In this case, the intensity is determined by a nonlinear combination of factors that are changing substantially in space and can be found with the help of the methods represented above. In a particular case when nonradiative losses are small in comparison with radiative ones (high quantum efficiency $Q_1(\mathbf{r}) = 1$), we have the following expression instead of Eq. (A1.14):

$$I_{\text{fl}}(\mathbf{r}) = \frac{\hbar\omega_1(\mathbf{r})4\Omega^2(\mathbf{r})}{\Gamma_2(\mathbf{r})\left(1 + \frac{\Gamma_2(\mathbf{r})}{K}\right)} = \frac{\hbar\omega_1(\mathbf{r})\Gamma_{\text{pump}}(\mathbf{r})}{\left(1 + \frac{\Gamma_2(\mathbf{r})}{K}\right)}. \tag{A1.15}$$

And, in general, fluorescence intensity's dependence on a molecule position is determined by spatial variations both of a pumping rate $\Gamma_{\text{pump}}(\mathbf{r})$ and of the linewidth $\Gamma_2(\mathbf{r})$.

References

Cohen-Tannoudji, C., Dupont-Roc, J., Grynberg, G. (1989) *Photons and Atoms: Introduction to Quantum Electrodynamics* (Wiley, New York).

Girard, C., Martin, O.J.F., Lévèque, G., Colas des Francs, G., Dereux, A. (2005) Generalized Bloch equations for optical interactions in confined geometries, *Chem. Phys. Lett.*, **404**, 44–48.

Klimov, V.V., Ducloy, M. (2004) Spontaneous emission rate of an excited atom placed near a nanofiber, *Phys. Rev. A*, **69**, 013812.

Lakowicz, J.R, Malicka, J., Gryczynski, I., Gryczynski, Z., Geddes, C.D. (2003) Radiative decay engineering: the role of photonic mode density in biotechnology, *J. Phys. D: Appl. Phys.*, **36**, R240–R249.

Landau, L.D., Lifshitz, E.M. (1989) *Quantum Mechanics*, 3rd ed. (Butterworth-Heinemann, Oxford).

Stevenson, A.F. (1953a) Solution of electromagnetic scattering problems as power series in the ratio (dimension of scatterer)/wavelength, *J. Appl. Phys.*, **24**, 1134–1142.

Stevenson, A.F. (1953b) Electromagnetic scattering by an ellipsoid in the third approximation, *J. Appl. Phys.*, **24**, 1143–1151.

Wylie, J.M., Sipe, J.E. (1984) Quantum electrodynamics near interface, *Phys. Rev. A*, **30**, 1185–1193.

Wylie, J.M., Sipe, J.E. (1985) Quantum electrodynamics near interface II, *Phys. Rev. A*, **32**, 2030–2043.

Appendix A2

Popular Numerical Methods in Nano-Optics and Nanoplasmonics

At the present time, a wide range of effective numerical approaches to solutions of Maxwell's equations in the presence of nanoparticles and nanostructures is developed. These approaches allow one to find plasmon resonances, a reflected field, or a Green's function and, consequently (see Appendix A1), spectroscopic characteristics of both nanoparticles and molecules located in the vicinity of them. The following methods may be considered the main ones:

- discrete dipole approximation (DDA)
- T-matrix method
- multiple multipole (MMP) methods
- finite-difference time domain (FDTD) methods
- direct space integral equation methods

All of these methods have both advantages and disadvantages. In particular, initially the DDA, T-matrix, and MMP methods were developed for solution of scattering problems. At that, it turns out to be that these methods do not describe a field near nanoparticles quite well. On the other hand, methods based on the solution of direct space integral equations turn out to be well suitable for

Nanoplasmonics
Vasily Klimov

ISBN 978-981-4267-16-8 (Hardcover), 978-981-4267-42-7 (eBook)
www.panstanford.com

description of near fields and, in particular, for the description of a Green's function in the vicinity of a source. Hence, apparently, rates of atom spontaneous decay and molecule fluorescence in a complex environment were investigated numerically with the help of this approach mainly.

In all of these methods, positions of resonance peaks and, therefore, plasmon oscillation spectra can be found after the problem for a large number of frequencies (or wavelengths) has been solved. When more detailed characteristics of plasmon spectra are to be found, apparently, ε-methods based on the surface integral equations method, which was described in Chapter 5, are most suitable.

A2.1 Discrete Dipole Approximation

The DDA method is a numerical method in which any object is represented as a set of a large number of N polarizable cubes located at nodes of the cubic grid of a period d (Purcell and Pennypacker, 1973). Sometimes, this method is called a coupled dipole approximation (CDA) method. Within the context of this method, there are no restrictions to whether a grid node is occupied by a dipole or not. This means that, in principle, the DDA method can be used for an approximate description of bodies of any shape and composition. It is assumed that the ith element is located at a point with a radius vector $\mathbf{r}_i$ and possesses a dipole polarizability α_i. Polarizabilities of higher orders (quadrupole, etc.) are taken as equal to zero. A dipole moment of every element $\mathbf{p}_i$ arises as a result of influence of a local field $\boldsymbol{E}_{\text{loc}}$, which is supposed to be monochromatic ($\sim e^{-i\omega t}$):

$$\mathbf{p}_i = \alpha_i \cdot \mathbf{E}_{\text{loc}}(\mathbf{r}_i). \tag{A2.1}$$

In its turn, $\mathbf{E}_{\text{loc}}$ is the sum of incidents and caused by all of the rest of the dipole fields:

$$\mathbf{E}_{\text{loc}}(\mathbf{r}_i) = \mathbf{E}_{\text{loc},i} = \mathbf{E}_{\text{in},i} + \mathbf{E}_{\text{dip},i} = \mathbf{E}_0 \exp(i\mathbf{k} \cdot \mathbf{r}_i) - \sum_{j \neq i} \mathbf{A}_{ij} \cdot \mathbf{p}_j, \tag{A2.2}$$

where $\mathbf{E}_0$ and $\mathbf{k}$ are the amplitude and wavevector of an incident wave (which is supposed to be a plane one but can be arbitrary in a general case), and the interaction matrix **A** has the following from (for $j \neq i$):

$$\mathbf{A}_{ij} \cdot \mathbf{p}_j = \frac{\exp(ikr_{ij})}{r_{ij}^3} \left\{ k^2 \mathbf{r}_{ij} \times (\mathbf{r}_{ij} \times \mathbf{p}_j) + \frac{1 - ikr_{ij}}{r_{ij}^2} \left[r_{ij}^2 \mathbf{p}_j - 3\mathbf{r}_{ij}(\mathbf{r}_{ij} \cdot \mathbf{p}_j) \right] \right\}, \qquad \text{(A2.3)}$$

where $k = \omega/c$. In fact, the matrix $\mathbf{A}'$ is a Green's function of Maxwell's equations with a dipole source in free space. It is important that electric permittivity of a metal particle and environment enter as the ratio $\varepsilon_i/\varepsilon_H$, which is contained in the expression for polarization (A2.1). Besides, the wavevector is to be multiplied by $\sqrt{\varepsilon_H}$ when the particle is located not in vacuum. Explicit expressions for polarizability α_i were developed by Purcell and Pennypacker (1973), Draine and Flatau (1994), and Draine and Goodman (1993):

$$\alpha_i^{CM} = \frac{3}{n} \frac{\varepsilon(\mathbf{r}_i) - 1}{\varepsilon(\mathbf{r}_i) + 2}, \qquad \text{(A2.4)}$$

where $n = 1/d^3$ is the dipole concentration and $\varepsilon(\mathbf{r}_i)$ is the value of electric permittivity in the location of the ith dipole. This expression allows one to describe precisely dielectric properties of a solid body approximated by infinite numbers of dipoles. Better results are given by the expression that considers radiative corrections (Collinge and Draine, 2003):

$$\alpha_i = \frac{\alpha_i^{CM}}{1 + \alpha_i^{CM}/d^3 \left[(b_1 + b_2\varepsilon(\mathbf{r}_i) + b_3 S\varepsilon(\mathbf{r}_i))(kd)^2 - \frac{2}{3} i (kd)^3 \right]}. \qquad \text{(A2.5)}$$

In (A2.5) $b_1 = -1.8915316$, $b_2 = 0.1648469$, and $b_3 = -1.7700004$ and S is the function of an incident plane wave propagation direction **a** and its polarization **e**:

$$S = \sum_{j=1}^{3} (e_j a_j)^2. \qquad \text{(A2.6)}$$

Application of this procedure for nanoparticles is approximate, and practically, extinction spectra of metal nanoparticles differ by 3%–5% from the precise result independently of the shape, composition, and size of particles under investigation.

Substituting Eqs. (A2.3) and (A2.2) into Eq. (A2.1), we obtain the matrix equation

$$\sum_{j\beta}[\mathbf{A}']_{i\alpha,j\beta}\mathbf{p}_{i\alpha} = \mathbf{E}_{\text{in},i\alpha}, \tag{A2.7}$$

where the matrix $\mathbf{A}'$ is obtained from the matrix $\mathbf{A}$ in Eq. (A2.3) and Greek letters denote the Cartesian coordinates. For a system of N dipoles, $\mathbf{E}$ and $\mathbf{P}$ in (A2.7) are $3N$ vectors and $\mathbf{A}'$ is a $3N \times 3N$ matrix. By solving a system of $3N$ complex linear equations, one can find the vector of dipole moments, the total dipole moment of the system, and other optical properties. In particular, extinction and absorption cross sections are expressed through a polarization vector and have the form

$$\begin{aligned} C_{\text{ext}} &= \frac{4\pi k}{|\mathbf{E}_0|^2}\sum_{j=1}^{N}\text{Im}(\mathbf{E}_{\text{in},j}\cdot\mathbf{p}_j) \\ C_{\text{abs}} &= \frac{4\pi k}{|\mathbf{E}_0|^2}\sum_{j=1}^{N}\left\{\text{Im}\left(\mathbf{p}_j\cdot\left(\overleftrightarrow{\alpha}_j^{-1}\right)\cdot\mathbf{p}_j\right) - \frac{2}{3}k^3|\mathbf{p}_j|^2\right\} \end{aligned}. \tag{A2.8}$$

Practically, the calculations are simplified substantially when fast-Fourier transform is applied for summing in Eq. (A2.2) and Eq. (A2.7) is solved by means of complex-conjugate gradients. All of these simplifications are implemented by Purcell and Pennypacker (1973), Draine and Flatau (1994), and Draine and Goodman (1993). Further details of the calculations and examples can be found, for example, in (Yang et al., 1995; Jensen et al., 1999) and in Chapter 9. The texts of all of the DDA method programs are at http://www.astro.princeton.edu/~draine/DDSCAT.6.0.html.

When calculating extinction and scattering spectra of nanoparticles of different shapes, it is convenient to normalize them to the effective geometrical cross section $A = \pi a_{\text{eff}}^2$:

$$Q_{\text{ext}} = \frac{C_{\text{ext}}}{A},\ Q_{\text{sca}} = \frac{C_{\text{sca}}}{A},\ Q_{\text{abs}} = \frac{C_{\text{abs}}}{A}, \tag{A2.9}$$

where the effective radius a_{eff} is determined from the equality of a particle volume to a volume of a sphere of the radius a_{eff}. This normalization was used in Chapter 9 in the course of the discussion of optical properties of polyhedral nanoparticles.

A2.2 The T-Matrix Method

Often, the T-matrix method is applied for a description of spherical nanoparticle clusters. In principle, this is a precise method and based on the Mie theory for each particle and on the vector spherical harmonics addition theorem (Quinten and Kreibig, 1993; Borghese et al., 1979; Bruning and Lo, 1971; Fuller and Kattawar, 1988; Hamid, 1996; Mackowski, 1994).

Initially, this method was suggested by Waterman (1971) and then improved substantially by Mishchenko and Travis (1994), Mishchenko et al. (1995), Mishchenko et al. (1996), Mackowski and Mishchenko (1996), Mackowski (2001). The procedure of the T-matrix analytical calculation for a cluster of spheres was described in detail by Mackowski (1994). The detailed bibliography of this method is presented in (Mishchenko et al., 2004).

We shall present the summary of the T-matrix method on the basis of the paper by Zhao et al. (2003). So, let us consider a cluster of an arbitrary shape of N_s spherical particles that is illuminated by a plane electromagnetic wave. A field scattered by the cluster $\mathbf{E}_\mathrm{s}$ is a superposition of fields scattered by separate cluster particles $\mathbf{E}_{\mathrm{s},i}$:

$$\mathbf{E}_\mathrm{s} = \sum_{i=1}^{N_\mathrm{S}} \mathbf{E}_{\mathrm{s},i}, \tag{A2.10}$$

where the field of every particle $\mathbf{E}_{s,i}$, in turn, can be expanded in vector spherical harmonics pinned to centers of corresponding particles:

$$\mathbf{E}_{\mathrm{s},i} = \sum_{n=1}^{\infty} \sum_{m=-n}^{n} \sum_{p=1}^{2} a^{i}_{mnp} \mathbf{h}_{mnp}(\mathbf{r}_i). \tag{A2.11}$$

Here, $\mathbf{h}$ denotes a divergent spherical wave of the degree n and order m, a^i is the expansion coefficient, p denotes the field polarization, and $p = 1$ and 2 denote transverse magnetic (TM) and transverse electric (TE) polarizations, correspondingly.

Since all spheres interact, a field scattered by the ith sphere is caused by both incident field and fields scattered by other spheres. Using the vector spherical harmonics addition theorem, all these fields can be rewritten in a series over vector spherical harmonics with a center in the considered ith sphere. Breaking summation over

n in Eq. (A2.11) at $n = N_{O,i}$, a system of linear equations of the following type is obtained:

$$a^i_{mnp} + \bar{a}^i_{np} \sum_{j=1,\, j\neq i}^{N_S} \sum_{l=1}^{N_{O,j}} \sum_{k=-l}^{l} \sum_{q=1}^{2} H^{ij}_{mnpklq} a^j_{klq} = \bar{a}^i_{np} p^i_{mnp}, \qquad \text{(A2.12)}$$

where the H_{ij} matrix is formed from coefficients entering the spherical harmonics addition theorem (on the basis of the spherical Hankel functions) and depends only on the distance and direction of the vector connecting the centers of the jth and ith spheres, $\bar{a}^i$ is the scattering coefficient of spherical waves of different multipolarity that is given by Mie theory, and, finally, p^i_{mnp} is the coefficient of an incident plane wave expansion in spherical harmonics related to the ith sphere center. The formal inversion of this system of equations gives the T-matrix of the cluster

$$a^i_{mnp} = \sum_{j=1}^{N_S} \sum_{l=1}^{N_{O,j}} \sum_{k=-1}^{l} \sum_{q=1}^{2} T^{ij}_{mnpklq} p^j_{klq}, \qquad \text{(A2.13)}$$

where the T^{ij} matrix is related to the center of the ith sphere. It can be transformed to the T-matrix related with the cluster center

$$T_{nl} = \sum_{i=1}^{N_S} \sum_{j=1}^{N_S} \sum_{n'=1}^{N_{O,j}} \sum_{l'=1}^{N_{O,i}} J^{0i}_{nn'} T^{ij}_{n'l'} J^{j0}_{l'l}, \qquad \text{(A2.14)}$$

where J^{0i} and J^{j0} are matrices formed from coefficients of the spherical Bessel functions addition theorem and the indices n, n' and l, l' denote the triple index characterizing the spherical wave. Knowing the T-matrix, one can find all desired optical properties of the nanospheres cluster. For example, the expression for the extinction efficiency takes the form

$$C_{\text{ext}} = \frac{2\pi}{k^2} \text{Re}\left(\sum_{nmp} T_{mnpmnp} \right). \qquad \text{(A2.15)}$$

The method of the T-matrix can be also applied to nonspherical particles. In this case, it becomes approximate, but very effective, for not very elongated particles.

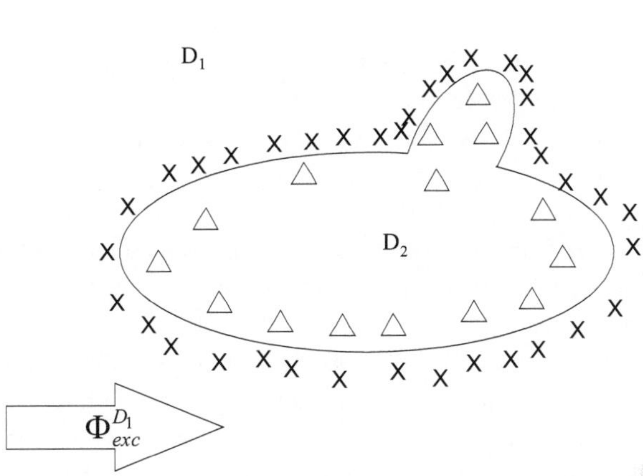

Figure A2.1 Geometry of the MMP method application. The multipole expansion centers for description of fields inside the external domain D_1 are shown by triangles and by crosses for description of fields inside the internal domain D_2.

A2.3 The Multiple Multipole Method

The MMP method (Hafner, 1990; Hafner, 1999; Wriedt, 1999; Moreno et al., 2002) makes use of the expansion of electromagnetic fields over basis functions, which in turn are based on spherical harmonics. An MMP method peculiarity is a great freedom of choice of these basis functions, which allows one to achieve a high accuracy of calculations at the cost of increasing the complexity of an algorithm.

The MMP method is applied for modeling linear systems with piecewise homogenous and isotropic optical properties. The whole domain where fields are to be calculated is divided into subdomains D_i where the constitutive constants ε_i and μ_i are invariable (the geometry of the MMP method is shown in Fig. A2.1). In each subdomain D_i, electromagnetic field potentials Φ_i^D are approximated by a linear superposition of N known *analytical* solutions of Maxwell's equations in a corresponding domain:

$$\Phi^{D_i}_{\text{approx}} = \Phi^{D_i}_{\text{exc}} + \sum_{k=1}^{N} a_{\text{k}}^{D_i} \phi_{\text{k}}^{D_i}, \qquad \text{(A2.16)}$$

where $\Phi^{D_i}_{\text{approx}}$ is an approximated value of the field potential and $\Phi^{D_i}_{\text{exc}}$ stands for the external excitation field. To determine coefficients $a_k^{D_i}$ of the expansion in basis functions $\varphi_k^{D_i}$, one should use boundary conditions at interfaces ∂D_{ij} separating the subdomains i and j. Thus, the MMP method is semianalytical since $\Phi_{\text{approx}}^{D_i}$ are analytical solutions of Maxwell's equations in whole space except the boundaries of particles. At the interfaces ∂D_{ij}, the boundary conditions are fulfilled only approximately. The MMP method can be also called a boundary method since only the boundary is to be discretized. The last circumstance results in reduction of computational costs. An important feature of the MMP method is that in the context of this method calculation accuracy can be estimated. Indeed, in the context of this method expansion coefficients are found as a result of minimization of boundary condition residuals. The minimal achievable residual gives the estimation of accuracy of the field calculation.

The only condition that is imposed on the basis functions is that they are to be solutions of Maxwell's equations. As for the rest they can be absolutely arbitrary. This arbitrariness is the most important peculiarity of this method since it determines flexibility of its application. For example, if an analytical solution of some problem is known, then in the course of solving a perturbed problem the known analytical solutions can be used as basis functions. In this respect, all of the known prior information can be included in a set of the basis functions, and as a result, a very precise solution can be obtained. As the name of the method implies, vector spherical harmonics are the main basis functions of it. The choice of these basis functions is related to their useful physical and mathematical properties. The vector spherical harmonics can be constructed through application of a corresponding differential operator (rot) to a solution of a scalar Helmholtz equation in spherical coordinates (Jackson, 1975). Thus constructed spherical harmonics describe the field of monochromatic dipole, quadrupole, etc., current and charge sources and have a singularity in the source location. Modeling of electrodynamics problems with the help of these sources is intuitive: light incident on the boundary ∂D_{ij} between different media is scattered into two media. The field inside the domain D_i is approximated by the multipole sources located outside this domain, that is, in the domain D_j. Often, centers of these multipoles are situated near the surface (see Fig. A2.1). This idea resembles the method of images, which is widely used in electrostatics. Asymptotic properties of the spherical harmonics are their other useful property: since they present emitting sources, they satisfy the Sommerfeld radiation condition at infinity automatically, and no special efforts are required to provide a correct behavior of the solution at infinity.

The choice of a number, disposition, and order of the multipole sources depends on the excitation field, boundary geometry, and constitutive constants. In the previous paragraph, we have already mentioned some intuitive physical reasons, which can be reformulated regarding the mathematical point of view: the source choice is to provide the boundary condition fulfillment with sufficient accuracy. This means that radial and angular characteristics of the vector spherical harmonics are to be adequate to complexity of modeled fields.

In the context of the MMP method, this is achieved by two steps. First, each basis function $\varphi_k^{D_i}$ can be constructed as a set of multipoles located at the same point. Such a set of multipoles is analogous to the set of multipoles used in Mie theory to describe scattering by single sphere (see Chapter 6). This is the first multipole expansion. Second, to avoid problems of bad convergence in nonspherical geometries, basis functions $\varphi_k^{D_i}$ with several (multiple) different centers are to be used. This explains the second part of the method name.

Usage of many multipoles with different positions provides great flexibility of the method. On the other hand, the flexibility should be used cautiously. Fields of multipolar sources of different orders with a center in one point are linearly independent. However, two multipole expansions with centers in different points are not independent. When the boundary condition is imposed on these systems of multipoles, badly conditioned matrices arise, leading to senseless results if special measures are not taken to achieve their regularity. Nevertheless, mathematical properties of the vector spherical harmonics allow moderating of these problems. These functions decrease with a distance, at that the decrease rate increases as the degree of a harmonic does. Thus, if centers of the multipole (MMP) expansions are distanced from each other, we obtain a well-conditioned problem. Thus, the right choice of the distances between multipoles is an important requirement to obtain correct results.

After the problem has been solved for several frequency values, positions of resonance peaks and, therefore, plasmon oscillations spectra can be found.

A2.4 The Finite-Difference Time Domain Method

In several recent years, this simulation method was successfully applied to solution of different problems of nano-optics and nanoplasmonics (Christensen, 1995; Kann et al., 1995; Furukawa and Kawata, 1996, 2001; Hirota et al., 2000; Nakano et al., 2001; Tanaka et al., 2001a, 2001b; Rockstuhl et al., 2010). In the context of this approach, Maxwell's equations in the time domain are solved by the method of finite differences. More precisely, application of this method begins with considering Amper's and Faraday's laws. For example, if a nanosystem with electric permittivity $\varepsilon(x, y, z)$ is considered, these two laws have the form

$$\mathrm{rot}\mathbf{E}(x, y, z, t) = -\frac{1}{c}\frac{\partial \mathbf{B}(x, y, z, t)}{\partial t} \tag{A2.17}$$

and

$$\mathrm{rot}\mathbf{H}(x, y, z, t) = -\frac{\varepsilon(x, y, z)}{c}\frac{\partial \mathbf{E}(x, y, z, t)}{\partial t}, \tag{A2.18}$$

correspondingly.

If these two vector equations are approximated with the help of finite differences in time and space domains, then we obtain six scalar finite-difference equations:

$$\begin{aligned}&\frac{E_z(x, y+\Delta y, z, t) - E_z(x, y-\Delta y, z, t)}{2\Delta y}\\&-\frac{E_y(x, y, z+\Delta z, t) - E_y(x, y, z-\Delta z, t)}{2\Delta z},\\&= -\frac{B_x(x, y, z, t+\Delta t) - B_x(x, y, z, t)}{c\Delta t}\end{aligned} \tag{A2.19}$$

$$\frac{E_x(x,y,z+\Delta z,t)-E_x(x,y,z-\Delta z,t)}{2\Delta z}-\frac{E_z(x+\Delta x,y,z,t)-E_z(x-\Delta x,y,z,t)}{2\Delta x}=-\frac{B_y(x,y,z,t+\Delta t)-B_y(x,y,z,t)}{c\Delta t}, \tag{A2.20}$$

$$\frac{E_y(x+\Delta x,y,z,t)-E_y(x-\Delta x,y,z,t)}{2\Delta x}-\frac{E_x(x,y+\Delta y,z,t)-E_x(x,y-\Delta y,z,t)}{2\Delta y}=-\frac{B_z(x,y,z,t+\Delta t)-B_z(x,y,z,t)}{c\Delta t}, \tag{A2.21}$$

$$\frac{H_z(x,y+\Delta y,z,t)-H_z(x,y-\Delta y,z,t)}{2\Delta y}-\frac{H_y(x,y,z+\Delta z,t)-H_y(x,y,z-\Delta z,t)}{2\Delta z}=\varepsilon(x,y,z)\left(\frac{E_x(x,y,z,t+\Delta t)-E_x(x,y,z,t)}{c\Delta t}\right), \tag{A2.22}$$

$$\frac{H_y(x+\Delta x,y,z,t)-H_y(x-\Delta x,y,z,t)}{2\Delta x}-\frac{H_x(x,y+\Delta y,z,t)-H_x(x,y-\Delta y,z,t)}{2\Delta y}=\varepsilon(x,y,z)\left(\frac{E_z(x,y,z,t+\Delta t)-E_z(x,y,z,t)}{c\Delta t}\right), \tag{A2.23}$$

and

$$\frac{H_x(x,y,z+\Delta z,t)-H_x(x,y,z-\Delta z,t)}{2\Delta z}-\frac{H_z(x+\Delta x,y,z,t)-H_z(x-\Delta x,y,z,t)}{2\Delta x}=\varepsilon(x,y,z)\left(\frac{H_y(x,y,z,t+\Delta t)-H_y(x,y,z,t)}{c\Delta t}\right). \tag{A2.24}$$

Now, knowing values of the fields in discrete points of a space grid, with the help of Eqs. (A2.19)–(A2.24) one can find values of the fields in the same space points at the moment $t+\Delta t$.

A numerical scheme represented above has only the first order of accuracy over Δt and Δx, Δy, and Δz, and to achieve the required accuracy, it is necessary to take very fine grids both for time and for space. In turn, it results in unfeasibly big matrix dimensions.

In modern realizations of the FDTD method, an explicit scheme of the second order of accuracy in Δt and Δx, Δy, and Δz, is used (Yee,

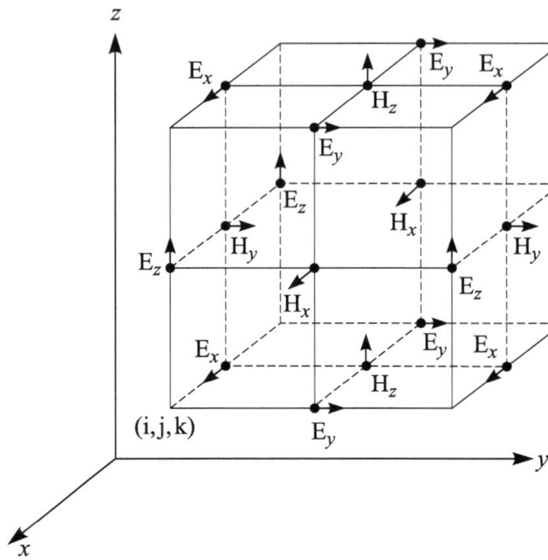

Figure A2.2 Positions of field components in a unit cell of the Yee grid.

1966). In the context of this approach, a special spatial grid (the Yee grid, see Fig. A2.2) is used.

In this grid, every node for the electric field is surrounded by four nodes for the magnetic field and vice versa (see Fig. A2.2). The second order of accuracy of time derivatives is achieved through a simple symmetric difference. As a result of this choice of a space-time grid, a discrete version of an x component of the equation (A2.17) will take the form

$$H_x|_{i,j,k}^{n+1/2} = H_x|_{i,j,k}^{n-1/2} + c\Delta t \times$$

$$\left[\left(E_y|_{i,j,k+1/2}^{n} - E_y|_{i,j,k-1/2}^{n}\right)/\Delta z - \left(E_z|_{i,j+1/2,k}^{n} - E_z|_{i,j-1/2,k}^{n}\right)/\Delta y\right], \quad \text{(A2.24)}$$

where the upper index corresponds to a time grid and three lower indices correspond to the space grid along the x, y, and z directions, correspondingly. From this equation it is seen how time and space derivatives are approximated by central derivatives; therefore the approximation has the second order of accuracy. Analogous equations are true also for other components of Eqs. (A2.17) and (A2.18). Thus, the whole FDTD method in the formulation of Yee (1966) is a method of the second order in domains with homogeneous values of electric permittivity and magnetic permeability.

For specific calculations, it is necessary to take into account that, to provide stability between the step in the time domain and the step of the space grid, the so-called Courant–Friedrichs–Lewy condition (Taflove and Hagness, 2005) is to be fulfilled:

$$\Delta t \leq \frac{1}{v\sqrt{\frac{1}{\Delta x^2} + \frac{1}{\Delta y^2} + \frac{1}{\Delta z^2}}}, \tag{A2.25}$$

where $v = c/\sqrt{\varepsilon\mu}$ is the speed of light in the medium.

By fulfillment of this condition, the FDTD method becomes a very reliable computational scheme. An important peculiarity of the FDTD method application is the presence of information on a field state for any time of simulation. This allows one to visualize easily the process of pulse propagation in nanoparticles and nanostructures before a stationary monochromatic regime is achieved.

An important feature of the FDTD method is the necessity to know field values, not only in a domain of nanoparticles and nanostructures, but also in the whole space. This is impossible due to purely technical reasons. To restrict the calculations by finite volumes and finite grids, in the context of the FDTD method the so-called absorbing boundary conditions (ABCs, Fig. A2.3) are used. Making use of these conditions allows one to avoid reflection from the walls of the simulation volume, which contains the considered nanoparticles and nanostructures. As a result, the nanosystems under consideration can be considered situated in an unbound space.

The more detailed explanation of why the ABCs are required is represented in Sadiku (2003). The detailed classification of different types of the boundary conditions is analyzed by Taflove and Hagness (2005).

The majority of the ABCs can be divided into two categories. One category of ABC is derived from differential equations. The other ABC category is related to the introduction of hypothetical material absorbers to the system.

At the present time, in course of simulation with the help of the FDTD method, the ABC based on making use of a PML is most

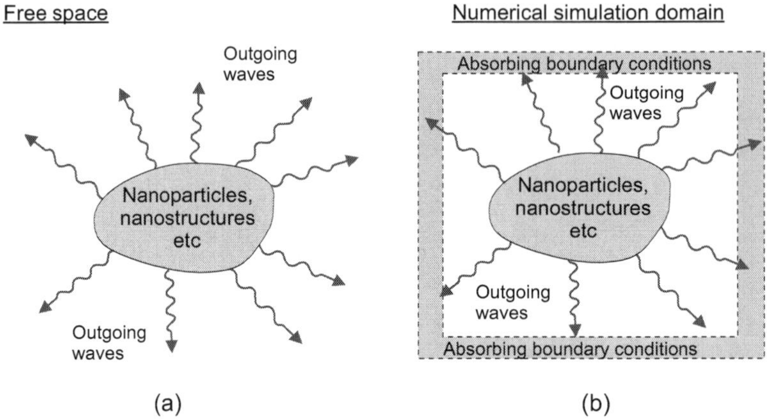

Figure A2.3 Illustration of the ABC operation by an example of a PML. (a) Nanoparticles and nanostructures in free space and (b) nanoparticles and nanostructures in the bounded simulation area. *Abbreviation*: PML, perfectly matched layer.

commonly used (Berenger, 1994, 1996, 1998). The PML ABC is realized by surrounding a calculation area by a layer of absorbing material, which weakens outgoing waves. At that, for properly chosen parameters of the PML, waves incident on the layer do not experience reflection for any angles of incidence and decrease rapidly, propagating inside the absorbing layer.

The reflection coefficient is equal to zero when absorbing layer parameters are related by the expression

$$\frac{\sigma_{\mathrm{e}}}{\varepsilon} = \frac{\sigma_{\mathrm{m}}}{\mu}, \tag{A2.26}$$

where σ_{e} and σ_{m} are electric and magnetic conductivities and ε and μ are electric and magnetic permittivities of the absorbing layer, which are, generally speaking, anisotropic tensors. The external boundary of the absorbing layer is considered perfectly conducting, that is, it reflects all waves into the absorbing layer. Let us note that the condition (A2.27) is schematic and applied only for determining certain components of corresponding tensors that are different for different sides of the absorbing layer (see Fig. A2.3).

One of the most important moments arising in the course of usage of the FDTD method for description of plasmon nanoparticles and nanostructures is a correct description of dispersion laws in metals. Contrary to frequency domain methods, introduction of dispersion laws into the FDTD method is possible only by making use of one or other analytical model of the dispersion law in metal, for example, by usage of the Drude–Lorentz model (see Chapter 3). To simulate nanoparticles with dispersion with the help of the FDTD method, the auxiliary differential equation (ADE) method is used frequently (Young and Nelson, 2001). In the frequency domain, a relation between electric induction and strength of the electric field has the form

$$\mathbf{D} = \varepsilon(\omega)\mathbf{E}. \tag{A2.27}$$

The ADE method is based on the conversion of Eq. (A2.28) to the time domain and its usage in combination with Maxwell's differential equation in the time domain. If the dispersion law is taken in the simplest Drude–Lorentz form

$$\varepsilon(\omega) = 1 - \frac{\omega_{pl}^2}{(\omega^2 + i\omega\gamma_0)} = \frac{\omega^2 + i\omega\gamma_0 - \omega_{pl}^2}{\omega^2 + i\omega\gamma_0}, \tag{A2.28}$$

then substituting it in Eq. (A2.28) and performing inverse Fourier transform, the differential equation that relates **D** and**E** can be obtained:

$$\gamma_0 \frac{\partial \mathbf{D}}{\partial t} + \frac{\partial^2 \mathbf{D}}{\partial t^2} = \frac{\partial^2 \mathbf{E}}{\partial t^2} + \gamma_0 \frac{\partial \mathbf{E}}{\partial t} + \omega_{\mathrm{pl}}^2 \mathbf{E}. \tag{A2.29}$$

Then, this equation is discretized using the central differences in a way that allows obtaining of an approximation of the second order of accuracy.

Making use of the ADE method, an electric field is calculated in two stages. At the first stage, the induction vector is obtained from a discretized version of the equation

$$\mathrm{rot}\mathbf{H} = \frac{1}{c}\frac{\partial \mathbf{D}}{\partial t}, \tag{A2.30}$$

which, for example, for the x component has the form

$$D_x|_{i,j,k}^{n+1} = D_x|_{i,j,k}^{n} + c\Delta t \times$$

$$\left[\left(H_y|_{i,j,k+1/2}^{n+1/2} - H_y|_{i,j,k-1/2}^{n+1/2}\right)/\Delta z - \left(H_z|_{i,j+1/2,k}^{n+1/2} - H_z|_{i,j-1/2,k}^{n+1/2}\right)/\Delta y\right]. \tag{A2.31}$$

For the rest of the components we have analogous equations.

At the second stage, the electric field is obtained by solving the discretized form of Eq. (A2.30). At that, to find a field at an instant $t = (n+1)\Delta t$, ($\mathbf{E}^{n+1}$), it is necessary to know both the induction at an instant $t = (n+1)\Delta t$, ($\mathbf{D}^{n+1}$) and at two preceding instants $t = (n-1)\Delta t, t = (n-2)\Delta t$, ($\mathbf{D}^{n-1}, \mathbf{D}^{n-2}$).

A2.5 Numerical Methods Based on the Integral Form of Maxwell's Equations

The methods considered above were more or less based on the differential form of Maxwell's equations. However, it is well known that Maxwell's equations can be written also in the integral form (Chew, 1999). With the help of an integral equation of this kind, a problem of atom spontaneous emission near a three-axial ellipsoid is considered solvable analytically (see Chapter 8). The discretized form of the integral Maxwell's equations is a rather effective instrument for description of nanoparticles' optical properties and their influence on emission of atoms and molecules (see the recent review (Girard, 2005)). By the way, the DDA method can be also regarded as one of approximated versions of this method.

Let us consider briefly main points of this approach in the context of one of its modifications (Martin et al., 1995), in which a tensor Green's function of an electric field $\overleftrightarrow{G}(\mathbf{r}, \mathbf{r}', \omega)$ is calculated. This Green's function satisfies the Helmholtz vector equation:

$$\nabla\times(\nabla\times\overleftrightarrow{G}(\mathbf{r}, \mathbf{r}', \omega)) - \left(\frac{\omega}{c}\right)^2 \varepsilon(\mathbf{r}, \omega)\overleftrightarrow{G}(\mathbf{r}, \mathbf{r}', \omega) = 4\pi\left(\frac{\omega}{c}\right)^2 \overleftrightarrow{1}\delta(\mathbf{r}-\mathbf{r}'). \tag{A2.32}$$

The tensor Green's function arises naturally in problems of atoms and molecules' emission in the presence of plasmonic nanoantennas. It can be shown (see, e.g., Chew, 1999) that this Green's function also satisfies the integral Lippmann–Schwinger–Dyson equation:

$$\overleftrightarrow{G}(\mathbf{r}, \mathbf{r}', \omega) = \overleftrightarrow{G}_0(\mathbf{r}, \mathbf{r}', \omega) - \int d\mathbf{r}''\overleftrightarrow{G}_0(\mathbf{r}, \mathbf{r}'', \omega) \times \left(\varepsilon(\mathbf{r}'', \omega) - 1\right)\overleftrightarrow{G}(\mathbf{r}'', \mathbf{r}', \omega). \tag{A2.33}$$

Here, integration is over the whole space, and $\overleftrightarrow{G}_0(\mathbf{r}, \mathbf{r}', \omega)$ is the Green's function of free space (Jackson, 1975):

$$\mathbf{G}_{0,ij}(\mathbf{r}, \mathbf{r}', \omega) = \left[(\delta_{ij} - \mathbf{n}_i\mathbf{n}_j) + \frac{(3\mathbf{n}_i\mathbf{n}_j - \delta_{ij})}{k^2} \times \left(\frac{1}{|\mathbf{r}-\mathbf{r}'|^2} - \frac{ik}{|\mathbf{r}-\mathbf{r}'|}\right)\right] \frac{e^{ik|\mathbf{r}-\mathbf{r}'|}}{|\mathbf{r}-\mathbf{r}'|}, \quad \text{(A2.34)}$$

where $\mathbf{n} = \frac{\mathbf{r}-\mathbf{r}'}{|\mathbf{r}-\mathbf{r}'|}$ is a unit vector directed from the point $\mathbf{r}$ to $\mathbf{r}'$. The dielectric function $\varepsilon(\mathbf{r}, \omega)$, which depends on spatial coordinates, describes dielectric properties of the system under consideration.

The discretized version of the Dyson equations has the form (Martin et al., 1995)

$$\overleftrightarrow{\mathbf{G}}_{ij} = \overleftrightarrow{\mathbf{G}}_{0,ij} - k^2 \sum_{p=1}^{N} \overleftrightarrow{\mathbf{G}}_{0,ip} \left(\varepsilon_p - 1\right) \overleftrightarrow{\mathbf{G}}_{pj} \Delta_p, \quad \text{(A2.35)}$$

where a scattering system is divided into N cells of volume Δ_p with the centers in the points $\mathbf{r}_p$, $p = 1 \ldots N$. The discrete variables $\overleftrightarrow{\mathbf{G}}_{ij}$, $\overleftrightarrow{\mathbf{G}}_{0,ij}$, ε_p are introduced to denote $\overleftrightarrow{\mathbf{G}}(\mathbf{r}_i, \mathbf{r}_j, \omega)$, $\overleftrightarrow{\mathbf{G}}_0(\mathbf{r}_i, \mathbf{r}_j, \omega)$, $\varepsilon(\mathbf{r}_p, \omega)$, correspondingly. The original effective method of solving (A2.36) was proposed by Martin et al. (1995). According to this method, the solutions of Eq. (A2.36) should be sought by the iterative method, considering each cell sequentially:

$$\overleftrightarrow{\mathbf{G}}_{ij}^{m} = \overleftrightarrow{\mathbf{G}}_{ij}^{m-1} - k^2 \overleftrightarrow{\mathbf{G}}_{im}^{m-1} (\varepsilon_m - 1) \overleftrightarrow{\mathbf{G}}_{mj}^{m} \Delta_m. \quad \text{(A2.36)}$$

$\overleftrightarrow{\mathbf{G}}_{ij}^{m-1}$ is used in Eq. (A2.37) instead of $\mathbf{G}_{0,ij}$ to take into account $m-1$ preceding cells of the scattering system. The solution of (A2.37) can be found easily. Assuming $I = m$, one can find $\overleftrightarrow{\mathbf{G}}_{mj}^{m}$ by solving a system of 3×3 equations. Values of the rest of the indices can be found immediately from (A2.37). Being performed for all of the cells, this procedure gives a value of the Green's function $\overleftrightarrow{\mathbf{G}}_{ij}^{N} = \overleftrightarrow{\mathbf{G}}_{ij}$ for a whole scattering system. This numerical scheme is quite stable in comparison with methods based on fulfillment of boundary conditions. It allows obtaining of $\overleftrightarrow{\mathbf{G}}_{ij}$ for very big physical systems.

By realization of this method, special attention should be paid to the divergence of $\overleftrightarrow{\mathbf{G}}_{0,ij}$ for $i = j$. The divergence requires a corresponding renormalization as it is described in details by Yaghjian (1980). The diagonal elements $\overleftrightarrow{\mathbf{G}}_{0,ij}$ play a very important role since they explain a substance depolarization, which is provoked by an external field. Indeed, a response of a dielectric material results in origination of a depolarization field in such a way that a complete field will satisfy the boundary conditions (Kittel, 1976).

The renormalization procedure depends on a discretization scheme and is reduced to modification of the Green's function $\mathbf{G}_0$ (Yaghjian, 1980)

$$\overleftrightarrow{\mathbf{G}}_0(\mathbf{r}_i, \mathbf{r}_i, \omega) \to \frac{\overleftrightarrow{L}}{\Delta_i k_0^2}, \tag{A2.37}$$

where $\overleftrightarrow{L}$ is a 3×3 matrix. In the case of a cubic grid

$$\overleftrightarrow{L} = \frac{1}{3}\begin{bmatrix} 1 & 0 & 0 \\ 0 & 1 & 0 \\ 0 & 0 & 1 \end{bmatrix}. \tag{A2.38}$$

By usage of Eq. (A2.38), the boundary conditions will be fulfilled automatically for any material and body shape.

A2.6 Other Numerical Methods

Above, only main (from our point of view) methods of calculating fields near nanoparticles and nano-objects of a complex shape are described. There are many other important methods, with the help of which optical properties of nanoparticles and their influence on emission of atoms and molecules can be investigated. Let us mention only the method of surface integral equations (Stratton, 1948; Chew, 1975; Poggio and Miller, 1973; Kern and Martin 2009) and the methods based on the differential theory of gratings (Petit, 1980; Maystre, 1984; Maystre and Neviere, 1978; Popov and Neviere, 2000). More detailed reviews of different methods and additional literature can be found in monographs (Mishchenko et al., 2004; Mishchenko et al., 2000; Barber and Hill, 1990). Let us note that the method of surface integral equations in its quasi-static version is described in more detail in Chapter 5 and is used (as "boundary elements method" [BEM]) widely for a description of plasmon oscillations in nanoparticles.

A2.7 Commercial Simulators for Nano-Optics and Nanoplasmonics

Above, we have shortly considered some popular algorithms that can be more or less easily implemented, even in small laboratories. However, there are plenty of commercially available tools that can be used for effective simulation and visualization of plasmonic nanostructures.

Among them, there are two main groups.

First are commercial simulators based on finite-element methods (FEMs):

1. Comsol Multiphysics by Comsol (http://www.comsol.com/)
2. HFSS by Ansoft (http://www.ansoft.com/products/hf/hfss/)
3. Microwave Studio by CST (http://www.cst.com/Content/Products/MWS/Overview.aspx)

Another group is based on the FDTD method:

1. SEMCAD-X by Speag (http://www.speag.com/products/semcad/solutions/)
2. Microwave Studio by CST (http://www.cst.com/Content/Products/MWS/Overview.aspx)

All these simulators are very powerful and can be used to solve very complicated three-dimensional problems of nanoplasmonics, but often it is difficult to understand what is running inside a solver and what is the precision of final results.

A2.8 Conclusion

Thus, in the present chapter main numerical methods for investigation of optical fields in the presence of complex nanoparticles and nanostructures were described briefly. Complicated physics of nanoplasmonics phenomena results in the fact that it is difficult to determine the accuracy of numerical methods in advance and calculations are to be necessarily performed in a context of several approaches. Nevertheless, in many cases the simulation results agree both with themselves and with experimental data, which is not a lot yet.

References

Barber, P.W., Hill, S.C. (1990) *Light Scattering by Particles: Computational Methods* (World Scientific, Singapore).

Berenger, J.P. (1994) A perfectly matched layer for the absorption of electromagnetic waves, *J. Comput. Phys.*, **114**, 185–200.

Berenger, J.P. (1996) Three-dimensional perfectly matched layer for the absorption of electromagnetic waves, *J. Comput. Phys.*, **127**, 363–379.

Berenger, J.P. (1998) An effective PML for the absorption of evanescent waves in waveguides, *IEEE Microwave Guided Wave Lett.*, **8**, 188–190.

Borghese, F., Denti, P., Toscano, G., Sindoni, O.I. (1979) Electromagnetic scattering by a cluster of spheres, *Appl. Opt.*, **18**, 116.

Bruning, J.H., Lo, Y.T. (1971) Multiple scattering of EM waves by spheres. I. Multipole expansion and ray-optical solutions, *IEEE Trans. Antennas Propag.*, **19**, 378–390.

Chew, W.C. (1999) *Waves and Fields in Inhomogenous Media* (IEEE Press Series on Electromagnetic Wave Theory), ISBN: 978-0-7803-4749-6 (Wiley-IEEE Press).

Christensen, D.A. (1995) Analysis of near field tip patterns including object interaction using finite-difference time-domain calculations, *Ultramicroscopy*, **57,** 189–195.

Collinge, M.J., Draine, B.T. (2003) Discrete-dipole approximation with polarizabilities that account for both finite wavelength and target geometry, arXiv:astro-ph/0311304 v1.

Draine, B.T., Flatau, P.J. (2003) *User Guide for the Discrete Dipole Approximation Code DDSCAT6.0.*

Draine, B.T., Goodman, J. (1993) Beyond Clausius–Mossotti-wave-propagation on a polarizable point lattice and the discrete dipole approximation, *Astrophys. J.*, **405**, 685–697.

Draine, B.T., Flatau, P.J. (1994) Discrete-dipole approximation for scattering calculations, *J. Opt. Soc. Am. A*, **11**, 1491–1499.

Bérenger, J.P. (1996) Three-dimensional perfectly matched layer for the absorption of electromagnetic waves, *J. Comput. Phys.*, **127**, 363–379.

Fuller, K.A., Kattawar, G.W. (1988) Consummate solution to the problem of classical electromagnetic scattering by an ensemble of spheres. I: linear chains, *Opt. Lett.*, **13**, 90–92.

Furukawa, H., Kawata, S. (1996) Analysis of image formation in a near-field scanning optical microscope: effects of multiple scattering, *Opt. Commun.*, **132** 170–178.

Furukawa, H., Kawata, S. (2001) Near-field optical microscope images of a dielectric flat substrate with subwavelength strips, *Opt. Commun.*, **196**, 93–102.

Girard, C. (2005) Near fields in nanostructures, *Rep. Prog. Phys.*, **68**, 1883–1933.

Hafner, Ch. (1990) *The Generalized Multipole Technique for Computational Electromagnetics* (Artech House, Boston).

Hafner, Ch. (1999) *Post-Modern Electromagnetics. Using Intelligent MaXwell Solvers* (Wiley, Chichester, UK).

Hamid, A.-K. (1996) Modeling the scattering from a dielectric spheroid by a system of dielectric spheres, *J. Electromagn. Waves Appl.*, **10**, 723–729.

Hirota, K., Milster, T.D., Zhang, Y., Erwin, J.K. (2000) Design of a near-field probe for optical recording using a 3-dimensional finite difference time domain method, Japan, *J. Appl. Phys.*, **39**, 973–975.

Jackson, J.D. (1975) *Classical Electrodynamics*, 2nd ed. (Wiley, New York), Chaps. 16.1–16.2, pp. 739–747.

Jensen, T., Kelly, L., Lazarides, A., Schatz, G.C. (1999) Electrodynamics of noble metal nanoparticles and nanoparticle clusters, *J. Cluster Sci.*, **10**, 295–317.

Kann, J.L., Milster, T.D., Froehlich, F.F., Ziolkowski, R.W., Judkins, J.B. (1995) Linear behavior of a near-field optical scanning system, *J. Opt. Soc. Am. A*, **12**, 1677–1682.

Kern, A.M., Martin, O.J.F. (2009) Surface integral formulation for 3D simulations of plasmonic and high permittivity nanostructures, *J. Opt. Soc. Am. A*, **26**, 732–740.

Kittel, C. (1976) *Introduction to Solid State Physics*, 5th ed. (Wiley, New York), Chap. 13.

Mackowski, D.W. (1994) Calculation of total cross sections of multiple-sphere clusters, *J. Opt. Soc. Am. A*, **11**, 2851–2861.

Mackowski, D.W. (2001) An effective medium method for calculation of the T matrix of aggregated spheres, *J. Quant. Spectrosc. Radiat. Transfer*, **70**, 441–464.

Mackowski, D.W., Mishchenko, M.I. (1996) Calculation of the T matrix and the scattering matrix for ensembles of spheres, *J. Opt. Soc. Am. A*, **13**, 2266–2278.

Martin, O.J.F., Girard, C., Dereux, A. (1995) Generalized field propagator for electromagnetic scattering and light confinement, *Phys. Rev. Lett.*, **74**, 526.

Maystre, D. (1984) *Progress in Optics, Rigorous Vector Theories of Diffraction Gratings*, XXI (Elsevier, Amsterdam).

Maystre, D., Neviere, M. (1978) Electromagnetic theory of crossed gratings, *J. Opt. (Paris)*, **9**, 301–309.

Mishchenko, M.I., Hovenier, J.W., Travis, L.D., eds. (2000a). *Light Scattering by Nonspherical Particles: Theory, Measurements, and Applications* (Academic Press, San Diego).

Mishchenko, M.I., Mackowski, D.W., Travis, L.D. (1995) Scattering of light by bispheres with touching and separated components, *Appl. Opt.*, **34**, 4589–4599.

Mishchenko, M.I., Travis, L.D., Lacis, A.A. (2004) *Scattering, Absorption, and Emission of Light by Small Particles* (Goddard Institute for Space Studies, New York).

Mishchenko, M.I., Travis, L.D. (1994) T-matrix computations of light scattering by large spheroidal particles, *Opt. Commun.*, **109**, 16–21.

Mishchenko, M.I., Travis, L.D., Mackowski, D.W. (1996) T-matrix computations of light scattering by nonspherical particles: a review, *J. Quant. Spectrosc. Radiat. Transfer*, **55**, 535–575.

Mishchenko, M.I., Videen, G., Babenko, V.A., Khlebtsov, N.G., Wriedt, T. (2004) T-matrix theory of electromagnetic scattering by particles and its applications: a comprehensive reference database, *J. Quant. Spectrosc. Radiat. Transfer*, **88**, 357–406.

Moreno, E., Erni, D., Hafner, C., Vahldieck, R. (2002) Multiple multipole method with automatic multipole setting applied to the simulation of surface plasmons in metallic nanostructures, *J. Opt. Soc. A*, **19**, 101.

Nakano, T., Yamakawa, Y., Tominaga, J., Atoda, N. (2001) Near-field optical simulation of super-resolution near-field structure disks, Japan, *J. Appl. Phys.*, **40**, 1531–1535.

Petit, R., ed. (1980) Electromagnetic theory of gratings, in *Topics in Current Physics*, Vol. 22 (Springer, Heidelberg).

Poggio, A.J., Miller, E.K. (1973) Integral equation solution of three dimensional scattering problems, in *Computer Techniques for Electromagnetics*, ed. R. Mittra (Pergamon Press, New York); Poggio, A.J., Miller, E.K. (1987) Integral equation solution of three dimensional scattering problems, in *Computer Techniques for Electromagnetics*, ed. R. Mittra (Hemisphere Publishing, New York).

Popov, E., Neviere, M. (2000) Differential theory for diffraction gratings: a new formulation for TM polarization with rapid convergence, *Opt. Lett.*, **25**, 598–600.

Purcell, E.M., Pennypacker, C.R. (1973) Scattering and absorption of light by non-spherical dielectric grains, *Astrophys. J.*, **186**, 705–714.

Quinten, M., Kreibig, U. (1993) Absorption and elastic scattering of light by particle aggregates, *Appl. Opt.*, **32**, 6173–6182.

Rockstuhl, C., Etrich, C., Helgert, C., Menzela, C., Paula, T., Fahra, S., Pertschb, T., Dorfmüllerc, J., Estebanc, R., Khunsinc, W., Vogelgesangc, R., Kernc, K., Dmitrievd, A., Bittkaue, K., Beckerse, T., Cariuse, R., Lederer, F. (2010) Large scale simulations in the realm of nanooptics, *Proc. of SPIE*, **7604**, 76040D.

Sadiku, M.N.O. (2003) *Numerical Techniques in Electromagnetics, Second Edition* (CRC Press, London).

Stratton, J.A. (1941) *Electromagnetic Theory* (McGraw-Hill, New York).

Taflove, A., Hagness, S.C. (2005) *Computational Electrodynamics: The Finite-Difference Time-Domain Method*, 3rd ed. (Artech House).

Tanaka, K., Ohkubo, T., Oumi, M., Nakajima, K., Hosaka, H., Itao, K. (2001) Numerical simulation on read-out characteristics of the planar aperture-mounted head with a minute scatterer, Japan, *J. Appl. Phys.*, **40**, 1542.

Waterman, P.C. (1971) Symmetry, unitarity, and geometry in electromagnetic scattering, *Phys. Rev. D*, **3**, 825.

Wriedt, T., ed. (1999) *Generalized Multipole Techniques for Electromagnetic and Light Scattering* (Elsevier, Amsterdam).

Yaghjian, A. (1980) Electric dyadic Green's functions in the source region, *Proc. IEEE*, **68,** 248–263.

Yang, W.-H., Schatz, G.C., Van Duyne, R.P. (1995) Discrete dipole approximation for calculating absorption and Raman intensities for small particles with arbitrary shapes, *J. Chem. Phys.*, **103**, 869–875.

Yee, K.S. (1966) Numerical solution of initial boundary value problems involving Maxwell's equations in isotropic media. *IEEE Trans. Antenna Propag.*, **ap-14**, 302–307.

Young, J.L., Nelson, R.O. (2001) A summary and systematic analysis of FDTD algorithms for linearly dispersive media, *IEEE Antennas Propag. Mag.*, **43**(1), 61–77.

Zhao, L.L., Kelly, K.L., Schatz, G.C. (2003) The extinction spectra of silver nanoparticle arrays: influence of array structure on plasmon resonance wavelength and width, *J. Phys. Chem. B*, **107**, 7343–7350.

Appendix A3

Acronyms and Terms Used Frequently in Nanoplasmonics, Nano-Optics, and Related Sciences

Acronyms

ABCs (absorbing boundary conditions) are applied in the course of usage of the FDTD (see Appendix A2) and other numerical methods to restrict solution of the space domain.

AFM (atomic force microscopy). The operation principle of an atomic force microscope is based on the usage of forces acting between substance atoms. Repulsive forces act at short distances ($\sim 10^{-8}$ cm) between two atoms, and attractive forces act at long distances. Analogous forces act between any approaching bodies. In a scanning atomic force microscope, these bodies are an examined surface and a needle moving close to it. Usually, a diamond needle is used in the device. The needle slides smoothly over the sample surface (it "scans" the surface). A change of the force that acts between the surface and the needle is registered and contains information of surface relief height (topography).

Nanoplasmonics
Vasily Klimov

ISBN 978-981-4267-16-8 (Hardcover), 978-981-4267-42-7 (eBook)
www.panstanford.com

ATR (attenuated total internal reflection) is used for surface plasmon excitation (see Chapter 4).

BEM (boundary elements method) makes use of surface integral equations. The method is applied to solve Maxwell's equations and to find plasmon oscillations spectra. The quasistatic BEM version is represented in Chapter 5.

CCD (charge-coupled device). The device is used for image fixation, both in household cameras and in scientific investigations.

DDA (discrete dipole approximation) is a numerical method of solving Maxwell's equations (see Appendix A2).

DNG (double negative). DNG metamaterials are the same as NIM (see also Chapter 11), that is, a metamaterial in which both electric permittivity and magnetic permeability are negative.

ENG (ε negative). An ENG metamaterial has negative electric permittivity (see NIM, DNG, MNG, and ENZ).

ENZ (ε near zero). An ENZ metamaterial has electric permittivity close to zero (see NIM, DNG, and ENG).

FCS (fluorescence correlation spectroscopy) has become widespread in biological investigations due to high sensibility of the method, possibility of working with nano- and subnanomolar concentrations, possibility of fluorescence registration from a very small volume ($\sim 10^{-15}$ L), possibility of fluorescent particle detection in a wide time range from microseconds (corresponding to structural transitions) to seconds (corresponding to bacterial cells motility in viscous solutions), and short measurement time (in tens of seconds one can judge fluorescence changes in an examined volume and their nature). The FCS registers the correlation function (CF) of fluorescence fluctuations in a confocal volume. CF analysis allows one to determine the absolute number of fluorescent particles and their motility in the solution, that is, to characterize particle distribution over their sizes and rates of intramolecular conformational transitions.

FDTD (finite-difference time domain). This method is an effective numerical one to solve Maxwell's equations in the time domain. The method is used commonly in nano-optics and nanoplasmonics (see Appendix A2).

FIB (focused ion beam) nanolithography allows creating of almost arbitrary nanostructures with a resolution of about 5 nm (see Chapter 2).

HMM (hyperbolic metamaterials) are metamaterials where eigenvalues of the electric permittivity tensor have different signs. It is shown that HMM can help to overcome the diffraction limit (see Chapter 13, Fig. 13.41) and to provide effective control over radiation of atoms and molecules.

ISC (intersystem crossing) is electronic excitation energy transfer between singlet and triplet molecular states.

IC (internal conversion) is electronic excitation energy transfer between molecular states with equal spins.

ITO (indium tin oxide). Thin films of the substance are transparent and colorless. Thicker samples have a color from yellow to grey. In the infrared region, it is a mirror. The main ITO advantage is a combination of electrical conductivity and optical transparency. It is used widely in different optical devices (displays, photoelectric cells, plasmon nanostructures).

LHM (left-handed material) is the same as NIM or DNG metamaterials (see Chapter 11).

LRSPPs (long-range surface plasmon polaritons) are surface plasmons propagating to great distances. They are necessary elements of plasmon wave-guiding devices.

LPRs (localized plasmon resonances) are plasmon oscillations in nanoparticles of finite volume. Contrary to SPRs (see Chapter 4), they have a discrete spectrum that is analogous to the spectrum of atoms and quantum dots (see Chapter 5 and the following chapters).

MDRs (morphology-dependent resonances) are high-quality oscillations in microspheres and other dielectric nanoparticles. Another name is whispering gallery modes (see WGM).

MEF (metal-enhanced fluorescence) is an effect of fluorescence quantum yield increasing and of reduction in fluorescence lifetime of weakly emitting fluorophores in the presence of metal (plasmonic) nanoparticles (see Appendix A1).

MMP (multiple multipole method) is a numerical method of solving of Maxwell's equations (see Appendix A2).

MNG (μ negative) metamaterial has negative magnetic permeability (see also LHM, ENG, NIM, and DNG).

NIR (near infrared) denotes electromagnetic waves with a wavelength range of 0.76–1.5 μcm. This band is important for living matter investigation since it has relatively small absorption, especially at wavelengths of 0.7–0.9 μm.

NIM (negative-index medium) has a negative refractive index. The other name is materials with simultaneously negative electric and magnetic permittivities (DNG), or "left-handed" materials. They form a basis for the creation of perfect lenses and related devices (see LHM, Chapters 11 and 13).

PMs (plasmonic molecules) are a type of plasmon oscillations in nanoparticle clusters that are similar to usual molecules (see Chapter 10).

PML (perfectly matched layer) is a very effective realization of ABS. It is applied to restrict calculation of the spatial domain in different numerical methods (see Appendix A2).

PMMA (polymethyl methacrylate) is a transparent plastic that is used frequently in the form of films in nano-optic experiments.

QD (quantum dot) is an isolated (zero-dimensional) nano-object possessing properties that differ substantially from bulk material properties of the same composition due to a so-called size effect. QD optical properties have a discrete character, and in this respect "quantum dots are artificial atoms, properties of which can be controlled" (Z. I. Alferov).

SPASER (surface plasmon amplification by stimulated emission of radiation) is a laser analog where plasmons are used instead of photons.

SEM (scanning electron microscopy) is a type of microscope that uses electrons instead of light to form an image.

SERS (surface-enhanced Raman scattering) is an effect of giant enhancement of the Raman scattering of light by molecules in the presence of nanostructured metal surfaces or discontinuous films.

SNOM (scanning near-field optical microscope) allows considerable increasing of microscope resolution since it is limited not by the wavelength of light but by the size of nanoholes ($\sim$50–100 nm) used to illuminate the sample.

SPP (surface plasmon polariton or surface phonon polariton) is a localized wave on a surface of materials where plasmon resonances or phonon resonances (resonance of crystal lattice oscillations) are present (see Chapter 4).

SPR (surface plasmon resonance) is a localized wave on a surface of some metals (see Chapter 4).

SRR (split-ring resonator) is used frequently for creating metamaterials with negative magnetic permeability (see Fig. 11.4).

STM (scanning transmission microscopy). In contrast to a usual scanning electron microscope, in a scanning transmission microscope an electron beam goes through a sample.

TDDFT (time-dependent density functional theory) is a method of calculation of solid bodies' electronic properties.

WGM (whispering gallery mode) is a high-quality resonance in dielectric microparticles. The modes are based on the effect of total internal reflection from the internal surface of particles. Sometimes, they are called morphology-depending resonances (see MDR).

Some Terms

Analyte is a component (a substance) that is sought or detected in a considered test. The analyte can be a chemical element, a chemical compound, an isotope, a substance, etc.

Antibodies are proteins that are synthesized in an organism after the penetration of **Antigens**. One of well-known methods of clinical research in modern medicine is antibody analysis, for which SPR can be effectively used (see Chapter 13).

Antigens are high-molecular colloidal substances that provoke synthesis of specific **Antibodies**, which react with them after penetration in an animal or a human. An obligatory antigenicity condition is their distinction from substances that are normally present in a recipient. Foreign proteins, first of all, some polysaccharides (mainly of a bacterial origin), and protein complexes with different chemical compounds can be classified as antigens.

Bi-Isotropic Media have constitutive equations where electric (**D**) and magnetic (**B**) inductions depend on gradients of both electric (**E**) and magnetic (**H**) fields: $\mathbf{D} = \varepsilon(\mathbf{E} + \zeta \text{rot}\mathbf{E})$; $\mathbf{B} = \mu(\mathbf{H} + \xi \text{rot}\mathbf{H})$. A particular case of bi-isotropic media is **Chiral Media**.

Chiral Media are media consisting of chiral objects. They are a particular case of bi-isotropic media. They are promising for creation of NIMs.

Chirality is a property of a system to not coincide with its mirror reflection after any movements or rotations. Simplest examples are a human hand or a spiral. The majority of biological molecules are chiral, and this property results in effects of rotation of the light polarization plane (optical activity) (see Chapter 11).

Fluorophore is a molecule part that allows this molecule to be fluorescent. This is a molecule functional group that absorbs energy of a certain wavelength and releases it at another wavelength after that. A quantity of energy and the wavelength depend on the fluorophore and its chemical environment. Fluorophores are used widely in biochemical investigations.

Hyperlens is an optical device that allows obtaining of an image with a resolution less than a wavelength at any distance from the device. The main element is anisotropic metamaterials having an electric permittivity tensor that has components of different signs for different directions (see HMM and Chapter 13).

Ligand (Latin *ligo*, or bind) is an atom, ion, or molecule bound immediately with one or several central (complexing) atoms in a complex compound. Ligands are used in sensors based on surface plasmons for analyte binding with the sensor surface (see Chapter 13).

Metactronics is a new trend where main elements of electron circuits are realized in the optical frequency region with the help of nanostructured ENZ metamaterials.

Metamaterials are artificial materials where nanoparticles of one or other shape are used instead of atoms. Nanoparticles can possess both electric and magnetic properties (artificial atoms or molecules). The possibility to control shapes of nanoparticles, which form metamaterials, and their spatial arrangement allows obtaining of substances possessing properties radically different from properties of natural materials (see Chapters 11 and 13). One should distinguish between **Metamaterials** and **Photonic Crystals**.

Nanoantenna is a nanodevice for efficiency increasing of interaction between near and far optical fields. It is applied both for efficiency improvement of nanoscale light sources (molecules, QDs) and for light focusing into nanoscale regions (super- and hyperlenses). The main element of nanoantennas is plasmon nanoparticles.

Nanolaser, also referred to as a plasmonic laser or SPASER-based laser, is a laser that has nanoscale dimensions. A nanolaser can be modulated quickly, and it has small sizes—these two properties make it an ideal candidate for on-chip optical computing (see Chapter 13).

Nucleotides are basic components of nucleic acids. Compounds consisting of two nucleotide molecules are called dinucleotides; of three, trinucleotides; of several, oligonucleotides; and of many, polynucleotides or nucleic acids.

Oligonucleotides—see **Nucleotides**.

Peptides are organic substances consisting of residues of the same or different amino acids linked by a peptide bound. On the basis of a number of residues, di-, tri-, tetra-, polypeptides, and proteins are recognized. Advances of peptide chemistry assist in solving a range of complex problems of modern biochemistry and molecular biology (e.g., decoding a genetic code).

Perfect Lens is a hypothetical optical device where a NIM slab is used for imaging without diffraction limitations (see Chapter 11).

Photonic Crystal is a crystal made of identical particles having sizes comparable with the light wavelength. Optical crystals can be found in nature among gemstones (opals). Optical properties of photonic crystals are similar to electronic properties of solids.

Polynucleotides—see **Nucleotides**.

Protease is a ferment that splits a protein into polypeptides.

Proteins are high-molecular natural organic substances that are constructed of amino acids and play a fundamental role in the structure and vital activities of organisms.

Quantum Bit (Q-bit) is a unit of quantum information. As a usual bit of information, the q-bit is based on two eigenstates denoted by $|0\rangle$ and $|1\rangle$, but at that, it can be in a state of their superposition, that is, in the state $\alpha|0\rangle + \beta|1\rangle$, where α are β are arbitrary complex numbers satisfying the condition $|\alpha|^2 + |\beta|^2 = 1$.

Quantum Computer is a calculating device that uses substantially quantum mechanical effects, such as a superposition principle or entanglement.

Streptavidin is a protein that is produced by *Streptomyces avidinii*. It possesses a very high affinity to biotin, and owing to this it is used in a complex with a plasmon nanoparticle for binding with a deoxyribonucleic acid (DNA) molecule that is treated by biotin (biotinylated DNA).

Superlens is an optical nanodevice that allows obtaining of an image with transverse resolution less than the wavelength but only at small distances from the device. The main element of the superlens is a metal nanofilm where plasmon resonances are possible (SPP or SPR) (see Chapters 11 and 13).

Terrylene Dye is a water-soluble dye that absorbs at the wavelength of 600 nm and therefore is almost ideal for experiments on detection of separate molecules and visualization of living cells.

Index